Springer Series in Materials Science

Volume 191

For further volumes:
http://www.springer.com/series/856

The Springer Series in Materials Science covers the complete spectrum of materials physics, including fundamental principles, physical properties, materials theory and design. Recognizing the increasing importance of materials science in future device technologies, the book titles in this series reflect the state-of-the-art in understanding and controlling the structure and properties of all important classes of materials.

Marta Castillejo · Paolo M. Ossi
Leonid Zhigilei
Editors

Lasers in Materials Science

Springer

Editors
Marta Castillejo
Instituto de Química Física Rocasolano, CSIC
Madrid
Spain

Leonid Zhigilei
Materials Science and Engineering
University of Virginia
Charlottesville, VA
USA

Paolo M. Ossi
Departimento di Energia
Politecnico di Milano
Milan
Italy

ISSN 0933-033X ISSN 2196-2812 (electronic)
ISBN 978-3-319-02897-2 ISBN 978-3-319-02898-9 (eBook)
DOI 10.1007/978-3-319-02898-9
Springer Cham Heidelberg New York Dordrecht London

Library of Congress Control Number: 2013955058

Printed on acid-free paper

Springer is part of Springer Science+Business Media (www.springer.com)

Preface

Lasers in Materials Science is the title of both this book and the Third International School, SLIMS-2012, held on S. Servolo Island, Venice (Italy) from July 8 to 15, 2012.

The selection of topics covered in the book and the combination of didactic introduction to the fundamentals of laser-materials interactions with up-to-date presentation of state-of-the-art techniques and emerging applications of laser processing reflect the content and spirit of the lectures and discussions at the School. One of the goals of this biennial school is to provide Ph.D. students and young research scientists working in the field of laser-materials interactions with robust fundamental knowledge that is often lacking in their training, so that they may profitably interact with colleagues working in areas neighboring their own research fields. The general area of *Lasers in Materials Science* spans fields where the interaction between laser radiation and matter plays a basic role to engineer new materials, or to enhance specific properties, mainly surface related, of irradiated matter. The laser community offers several established International Conferences where young researchers can meet with their peers, exchange experiences, establish collaborations, or display their own results to a qualified audience. However, a structured training opportunity, specifically geared toward young researchers, was lacking before the SLIMS series was established.

Focusing on the strong interplay between experimental and theoretical investigations of laser-induced phenomena, the program of the one-week residential School included 17 lectures on the fundamentals and principles of laser-materials interactions and laser materials processing. The syllabus covered the mechanisms, relevant experimental and computational techniques, as well as current and emerging applications in nanoscience, biomedicine, photovoltaics, analysis, and industry. The topics ranged from laser-surface and -bulk interactions, to the role of defects, nonlinear absorption phenomena, surface melting, vaporization, superheating, homogeneous and heterogeneous nucleation, phase explosion and plasma formation, nanosecond, femtosecond and attosecond laser pulses, film synthesis by pulsed laser deposition, nanoparticle nucleation, growth and assembling, laser nanostructuring of soft matter, development of new light and X-ray sources, free electron lasers, and laser interactions with biological tissues.

One of the distinctive features of SLIMS-2012 was the active participation of students in the activities of the School. This was facilitated by structured classroom discussions and ample opportunity for students to discuss their ongoing projects or research plans with the School lecturers in informal settings. The students presented posters that were displayed over the School duration in the lecture hall. All posters were discussed during three extensive poster sessions and at coffee breaks. The students also gave brief oral presentations highlighting key points of their research in dedicated sessions and participated in a competition for the Best Student Presentation Award (dedicated to the memory of Prof. Roger Kelly). The School was attended by 36 students from 13 countries, with 25 students coming from EU countries, six from the USA, and four from the Mediterranean Sea area.

All lecturers, coming from both leading research centers and academic institutions, are actively involved in research topics covered by their lectures. The School Directors are grateful to School lecturers for the attention they put in the preparation of truly didactic, though high level, presentations and for the relevant work they did to convert the didactic material into self-contained book chapters that offer excellent reviews of the different topics.

Venice International University (VIU) quarters at S. Servolo Island provided superior lecturing and logistic structures in a pleasant working ambience, immersed in a quiet, beautiful garden, a few minutes from the heart of the city. This confirmed to be strategic for the success of the School.

The positive evaluation of SLIMS-2012 by the participants stimulated the planning of the forthcoming Fourth International School on *Lasers in Materials Science*, SLIMS-2014 that will be held on S. Servolo Island from July 13 to 20, 2014 under the direction of N. M. Bulgakova, Y. Lu, P. Schaaf, and P. M. Ossi.

Madrid, Spain — M. Castillejo
Milan, Italy — P. M. Ossi
Charlottesville, USA — L. V. Zhigilei

Contents

Contributors

Nadezhda M. Bulgakova Institute of Thermophysics SB RAS, Prosp. Lavrentyev, 1, Novosibirsk 630090, Russia, e-mail: nbul@itp.nsc.ru

Anna Paola Caricato Dipartimento di Matematica e Fisica "Ennio De Giorgi", Università del Salento, Via Arnesano, 73100 Lecce, Italy, e-mail: Annapaola.Caricato@le.infn.it

Marta Castillejo Instituto de Química Física Rocasolano, CSIC, Madrid, Spain, e-mail: marta.castillejo@iqfr.csic.es

Maria Dinescu National Institute for Lasers, Plasma and Radiation Physics, Bucharest, Romania, e-mail: dinescum@ifin.nipne.ro

David B. Geohegan Center for Nanophase Materials Sciences and SHaRE Facility and Materials Sciences and Technology Divisions, Oak Ridge National Laboratory, Oak Ridge, TN 37831-6488, USA, e-mail: geohegandb@ornl.gov

Richard F. Haglund Department of Physics and Astronomy, Vanderbilt University, Nashville, TN 37235-1807, USA, e-mail: richard.haglund@ Vanderbilt.Edu

Wolfgang Kautek Department of Physical Chemistry, University of Vienna, Vienna, Austria, e-mail: wolfgang.kautek@univie.ac.at

Ion N. Mihailescu Lasers Department, National Institute for Lasers, Plasma and Radiations Physics, Bucharest, Romania, e-mail: ion.mihailescu@inflpr.ro

Antonio Miotello Dipartimento di Fisica, Università di Trento, 38100 Povo, Trento, Italy, e-mail: miotello@science.unitn.it

Mauro Nisoli Dipartimento di Fisica, Politecnico di Milano, Milan, Italy, e-mail: mauro.nisoli@fisi.polimi.it

Paolo M. Ossi Dipartimento di Energia, Politecnico di Milano, 20133 Milan, Italy, e-mail: paolo.ossi@polimi.it

Juergen Reif Brandenburgische Technische Universität, BTU Cottbus, Universitaetstrasse 1, 03046 Cottbus, Germany, e-mail: reif@tu-cottbus.de

Peter Schaaf Institute of Materials Engineering and Institute of Micro- and Nanotechnologies, Ilmenau University of Technology, Gustav-Kirchhoff-Strasse 5, 98693 Ilmenau, Germany, e-mail: peter.schaaf@tu-ilmenau.de

Koji Sugioka Laser Technology Laboratory, RIKEN, Wako, Saitama 351-0198, Japan, e-mail: ksugioka@riken.jp

Leo V. Zhigilei Department of Materials Science and Engineering, University of Virginia, Charlottesville, VA, USA, e-mail: lz2n@virginia.edu

Chapter 1
Laser Physics for Materials Scientists: A Primer

Richard F. Haglund

Abstract Laser processing of materials has achieved significant successes in pulsed laser deposition, micro- and nanostructuring and surface modification and analysis. However, materials scientists often do not think about the physics of those lasers, which determines their properties and therefore also the way in which these lasers can be employed in laser processing. This chapter discusses the essential theory of laser gain, oscillation and amplification, and provides examples drawn from lasers now frequently used in materials processing. The implications for the design of new lasers and new materials-processing strategies are considered, using the example of a picosecond laser system for polymer processing.

1.1 Introduction

To optimize laser processing for specific applications and materials, it is useful to understand how lasers function, not only to select the correct lasers for particular processes, but also to provide input to laser designers and builders for new developments. In the future, an increased understanding of laser-materials interactions as these relate to laser properties will feed back into the laser-building community, to enable materials scientists to play a role in:

- Designing and constructing broadly tunable laser systems, at reasonable cost, capable of being tuned to the parameters that optimize process throughput based on fundamental laser-matter interactions;
- Understanding how the choice of amplified versus oscillator-only laser systems enables or limits the application of lasers in new processing applications as well as efficiency and overall throughput; and

R. F. Haglund (✉)
Department of Physics and Astronomy, Vanderbilt University,
Nashville, TN 37235-1807, USA
e-mail: richard.haglund@Vanderbilt.Edu

M. Castillejo et al. (eds.), *Lasers in Materials Science*,
Springer Series in Materials Science 191, DOI: 10.1007/978-3-319-02898-9_1,

Table 1.1 Characteristics of lasers used in materials processing

Laser type	Pulse duration	Repetition frequency	Wavelength range (μm)	Fluence	Intensity
Fiber laser oscillators	CW-1 ps	CW-MHz	1.07–2.1	Modest	High
Amplified fiber lasers	Ns-fw	kHz-MHz	1.07–2.1	High	High
Diode lasers	CW-μs	CW-kHz	0.8–1.2	Modest	Low
Excimer oscillator	10–20 ns	200 Hz	0.308, 0.248, 0.193	High	Modest
Excimer amplifier	1–10 ps	10 Hz	0.248	High	High
Nd:YAG oscillators	5–20 ns	20 Hz	1.06, 0.532, 0.355, 0.266	High	Modest
Nd:YVO_4 oscillators	5–10 ns	20 kHz	1.06, 0.532, 0.355, 0.266	Modest	Low
Yb:YAG amplifiers	0.5–20 ps	20 MHz	1.06, 0.532, 0.355, 0.266	Modest	High
Ti:sapphire oscillator	10–100 fs	100 MHz	0.7–1.0	Low	High
Ti:sapphire amplifier	100–150 fs	5 kHz	0.7–1.0	High	High
Yb:YAG + OPA	10–100 ps	1–5 kHz	0.2–20	Modest	High

- Developing figures of merit that will enable comparisons among laser systems in selecting scalable tools for applications, such as thin-film deposition, where current laser processes are not competitive with conventional processing tools.

Meeting these challenges would not only enlarge the palette of materials that can be built, assembled and fabricated using the unique electronic and vibrational interactions of light with matter, but also expand the emphasis from bulk and thin-film processing to micro- and nanoscale materials modification where the incomparable precision of laser processing holds sway.

Lasers commonly used for processing advanced materials can be categorized according to their (1) temporal pulse structure; (2) laser frequency range; and (3) tradeoffs between intensity and fluence based on pulse energy and duration. The throughput or processing rate will, as will be shown later, depends on average power. Especially during the last decade, the number of different laser types used in materials processing has expanded significantly; commercially available laser systems cover a range of all these properties, as shown in Table 1.1.

Figure 1.1 illustrates the parameter space relevant to laser processing materials, spanned by laser pulse duration and laser intensity. Solid and dashed diagonal lines show contours of constant fluence. For example, the line 1 J/cm^2 is the line representing the threshold for many laser processes initiated by pulsed nanosecond lasers, as in pulsed laser deposition. Below the line representing a fluence of 10 mJ/cm^2, mesoscale atomic motion within a material is relatively unlikely; in this region, local phase changes and other subtle modifications to local materials properties can be achieved. Above the fluence of 100 J/cm^2, on the other hand,

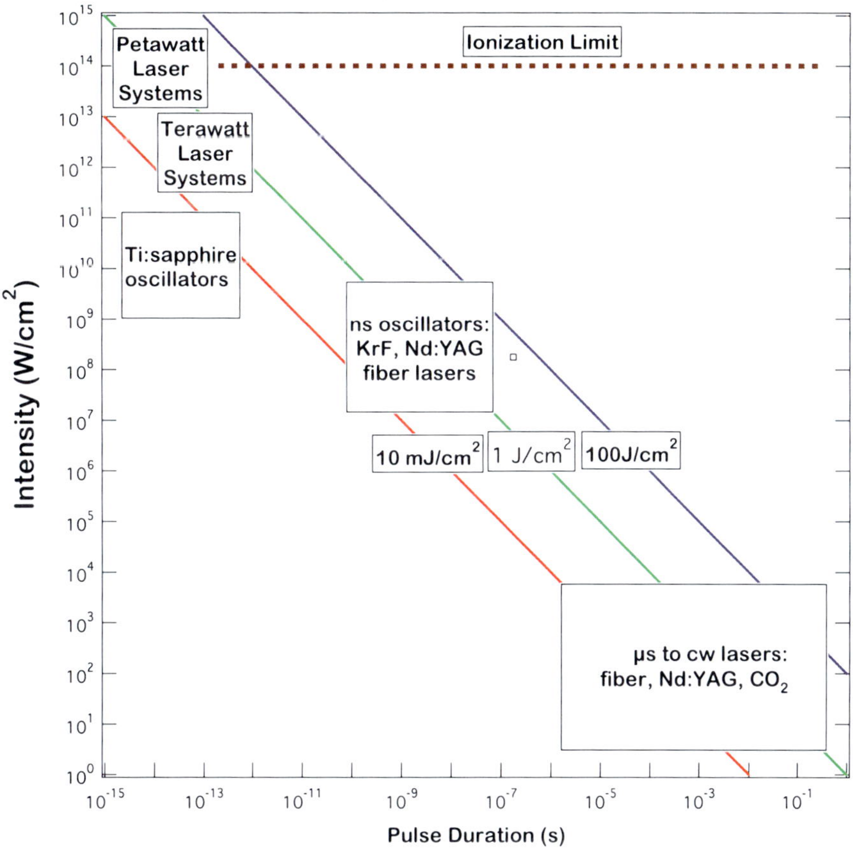

Fig. 1.1 A map of the parameter space occupied by current lasers used in materials processing. *Lines of constant fluence* are shown covering the range where most materials processing can occur

most materials modifications that are induced by lasers tend to be destructive for short- and ultrashort-pulse lasers, while at the limit of very long pulses, the changes tend to be thermal. As shown in Fig. 1.1, the laser systems listed in Table 1.1 nevertheless cover a wide range of the parameter space available for useful materials processing.

In this chapter, we first consider the fundamentals of laser interactions with materials as these relate to the choice of laser systems, and develop a simple figure-of-merit that makes it possible to see how various properties of a laser influence overall process throughput—and hence the economic merit of laser processing that can be compared to other processes. Next, we consider the fundamental theory of laser oscillators and amplifiers that are germanely to laser processing of materials, including ways of controlling laser pulse duration and repetition frequency. Finally, we present an example of a tunable, picosecond mid-infrared laser system based on currently available oscillators, amplifiers and parametric generators, illustrating how considering laser physics and laser properties can drive new modes of materials processing.

1.2 Fundamentals of Laser-Materials Interactions

Materials modification by lasers requires the motion of atoms, ions or molecules—which of necessity requires complex interactions in the material since photons carry very little momentum. To achieve atomic motion, three conditions must, in general, be met: First, a threshold intensity is needed to initiate the process, generally a few $MW \cdot cm^{-2}$. Second, vibrational energy—whether generated directly by infrared photon absorption or by multi-phonon cascades following electronic excitation—must be localized on a small group of atoms or a molecular-size cluster in the laser-irradiated solid for longer than a few vibrational periods. Third, the energy absorbed must be sufficient to initiate and sustain the breaking of bonds and the mesoscale motion of atoms, ions, molecules and clusters. In laser ablation, for example, the ejection of mesoscale volumes of material also creates a dense plume in which the interactions of atoms, ions, molecules and clusters with each other and with the laser light play a significant role.

Efficiency in materials modification and processing is enhanced by attending to the *hierarchical* character of laser-materials interactions. The crucial roles played by localized temporal and spatial excitation density in materials processing were first adumbrated by Stoneham and Itoh [1] and Itoh and Stoneham [2]. Their fundamental concept of localized excitation density is central to the idea of selective or non-thermal materials processing ranging from the visible-ultraviolet to the "molecular fingerprint" regions of the electromagnetic spectrum. This perspective is based on an *atomic-* or *molecular-scale* photon-matter interactions that generate *mesoscale* effects on time scales that are short compared to thermal equilibration times. Ultimately, the development of resulting in *macroscale* materials modifications on micro- to millimeter length scales.

Figure 1.2 shows schematically the sequence of processes through which the initial absorption of laser photons ultimately leads to macroscopic effects, such as those found in laser ablation, laser cutting and multiphoton structuring. By choosing the laser pulse duration or scanning speed to illuminate the target volume for a duration shorter than the thermal confinement time [3]

$$\tau_{thermal} = L_p^2 \Big/ D_{thermal} \tag{1.1}$$

(where L_{opt} is the optical penetration depth and $D_{thermal}$ is the thermal diffusivity), which in many technologically important materials is typically 0.1–10 μs, energy deposition into thermal modes of the target can be confined and the creation of a heat-affected damage zone by diffusion largely avoided, provided of course that the desired processing effect also occurs on a shorter time scale. Analogously,

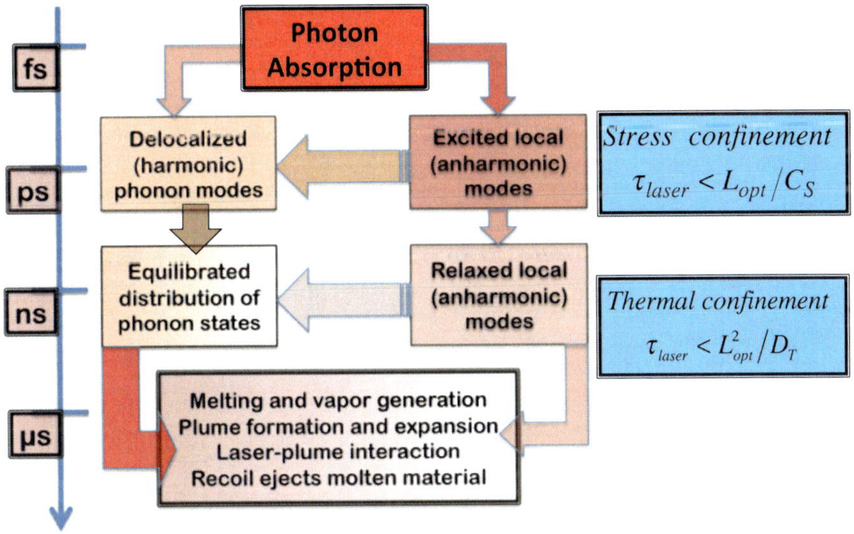

Fig. 1.2 Dynamics of photon absorption, electronic or vibrational excitation and relaxation processes leading to materials modifications such as ablation, melting and the formation of recoil-induced ejecta. The time scales relevant to stress confinement and thermal confinement are also shown

when the micropulse duration is shorter than the stress confinement time defined by [4–6]

$$\tau_{stress} = L_p / C_{sound} \tag{1.2}$$

(typically 001–0.1 μs), materials modification due to the propagation of photo-mechanically generated shock waves, spallation and exfoliation will likewise be limited to the volume in which the laser energy is absorbed.

When laser pulses excite a resonant electronic or vibrational mode of a solid, the persistence of the deposited energy in that mode and the specific relaxation mechanisms that relax or release the deposited energy determine whether or not a non-thermal process results from the laser excitation [7]. The time evolution of the laser-excited process follows from the basic quantum–mechanical result for the time rate of change of the yield of a particular process:

$$\frac{dN_0}{dt} = \eta N_0 \sigma_{(k)} \left(\frac{I}{\hbar\omega}\right)^k \tag{1.3}$$

where N_0 is the number of atoms or molecules per unit volume in the process volume, η is the quantum efficiency of the process, including loss channels; $\sigma_{(k)}$ the kth order cross section for the laser interaction with the material; and $\Phi = I/\hbar\omega$ is the photon flux, the number of photons per unit time per unit area.

From (1.3) it is evident that the rate at which a process occurs—critically important to the economics of materials processing—must then be proportional to average power. Since the probability for initiating nonlinear processes scales with powers of the intensity, rather than fluence, pulse durations of order a few picoseconds appear to have significant advantages for materials modification and processing. In particular, when the material is excited by ultrashort pulses, the initially spatially localized, anharmonic electronic or vibrational excitations are diluted only after enough time has elapsed for them to begin mixing with the delocalized, harmonic modes of the phonon bath *following* the laser pulse, whereas with nanosecond laser pulses, this relaxation begins already *during* the exciting laser pulse.

In choosing lasers for application to materials properties, it is useful to remember that the macroscopic observables—such as total yield—necessarily scale with energy deposited per unit volume (E/V):

$$Yield \propto \left(\frac{E}{V}\right) = F_L\alpha(\omega, I) \cong I_0\tau_L[\alpha_0(\omega) + \beta \cdot I(z,t)] \tag{1.4}$$

where F_L is the laser fluence, α and β are the linear and nonlinear absorption coefficients, ω and I are the laser frequency and intensity, respectively, and z and t are the relevant space (penetration depth) and time coordinates.

The interplay of laser energy and intensity can be understood in a qualitative way by calculating the effective temperature reached in a given target volume, where the effective temperature is simply the proxy for the energy required to effect a particular materials modification. The temperature reached by absorption of a single laser pulse, the number of pulses and the rate at which they are delivered by the laser turn out to be key parameters for calculating the effect of the laser-materials interaction, assuming that the laser energy is ultimately converted into processes that eventually reach thermal equilibrium. From the analytical solution to the one-dimensional heat-conduction equation, the target temperature during a single laser pulse reaches an average value [8]

$$\langle T \rangle = \sqrt{\frac{2}{\pi}}\frac{I_{abs}\sqrt{a \cdot \tau_L}}{\kappa}, \quad a \equiv \frac{\kappa}{C_v\rho_0} \tag{1.5}$$

If all the absorbed energy is converted into the desired materials modification (e.g., vaporization, melting, annealing), then the specific energy input per unit volume is given by

$$\frac{E_{abs}}{V} = C_p\rho_0\langle T \rangle = \rho_0\Omega \tag{1.6}$$

where Ω is the binding energy per atom, another material-dependent parameter. Once thermal equilibrium is reached—typically in a few picoseconds—the temperature rise in the laser-irradiated volume as a function of laser and materials parameters can be computed as follows:

$$\Delta T = \frac{F_{abs}(\omega) \cdot \alpha(\omega, I)}{C_v \rho_0} = \frac{F_{abs}(\omega) \cdot [\alpha(\omega_0) + \beta I_{abs}]}{C_v \rho_0}, I_{abs} \tau^{1/2} = \sqrt{\frac{\pi a}{2}} \Omega \equiv f(a, \Omega) \tag{1.7}$$

The processing *rate* Y—for example, the rate of material removal in a laser ablation process leading to the deposition of a thin film—for a laser producing pulses at a rate N_{pps} with energy E_L per pulse is given by

$$Y = \eta(\omega, I) \left(\frac{E}{V} \right) N_{pps} \cong \eta(\omega, I) \frac{E_L f(a, \Omega)}{L_{opt} A(F_0, \omega)} N_{pps} \tag{1.8}$$

Here L_{opt} is the optical absorption length, $A(F_0, \omega)$ is the laser spot size at the material modification threshold for the given laser frequency, and F_0 is the threshold fluence.

From (1.8), the yield (in units of processing events per unit time) depends critically both on the materials parameters $f(a, \Omega)$ and L_{opt}, and on the pulse repetition frequency, whereas the specific energy deposition benefits from non-linear effects in the target material. Hence, it becomes virtually axiomatic that the most efficient lasers for many materials processing protocols will be high-intensity (hence ultrashort pulse), high pulse-repetition frequency devices. Optimal laser processing conditions can be achieved by choosing short optical depth, small focal spots, and high pulse-repetition frequency to achieve efficient materials modification, as reflected in (1.8) that describes the overall processing rate.

1.3 Fundamentals of Laser Physics

The characteristics of the photon beam that emerges from the laser are determined by three essential components: the gain medium, the pump or excitation source, and the optical cavity. The abbreviated treatment here illustrates how the characteristics of all three contribute to the properties of the beam, and follows the general lines found in [9]; additional details can be found in [10–12].

1.3.1 Electromagnetic Waves in a Medium with Gain and Absorption

Consider an atomic, ionic or molecular system with two energy levels, E_1 and E_2, in thermal equilibrium. The relative population densities of the two levels N_1 and N_2 are given by Boltzmann's equation:

$$\begin{aligned} N_1 &= N_0 \exp[-(E_1 - E_0)/k_B T], N_2 = N_0 \exp[-(E_2 - E_0)/k_B T] \\ &\Rightarrow \frac{N_2}{N_1} = \exp[-(E_2 - E_1)/k_B T] \end{aligned} \tag{1.9}$$

where k_B is the Boltzmann constant, T is the absolute temperature, and N_0 is the ground-state population density. When an electric field interacts with such a two-level atomic system, its evolution in space and time is expressed by

$$E(\omega, t) = E_0 \exp\{i[\omega t - k'(\omega)z]\} \tag{1.10}$$

The propagation constant is related to the susceptibility of the two-level medium by

$$k'(\omega) = k + k\frac{\chi'(\omega) + i\chi''(\omega)}{2n^2} - \frac{i\alpha}{2} = k + k\frac{\chi'(\omega)}{2n^2} + \frac{i\gamma(\omega)}{2} - \frac{i\alpha}{2} \tag{1.11}$$

where the vacuum wave number is k and α is the distributed loss. The optical gain or loss at the frequency ω is thus proportional to the imaginary part of the propagation constant; the sign of $\gamma(\omega)$ also depends only on the relative size of the populations in states N_1 and N_2.

$$\gamma(\omega) \equiv k\frac{i\chi''(\omega)}{n^2} = \left[N_2 - \left(\frac{g_2}{g_1}\right)N_1\right]\frac{\pi c^2 A_{21}}{2\omega^2} g(\omega_0, \omega) \tag{1.12}$$

As an electromagnetic wave propagates through a medium with a complex dielectric function, the change in intensity is related to the average power absorbed per unit volume by

$$\frac{dI}{dz} = -\frac{\overline{Power}}{Volume} = -\frac{\omega\varepsilon_0}{2}\chi''|E(\omega, z)|^2 = I_0 \exp\{[\gamma(\omega) - \alpha]z\} \tag{1.13}$$

Thus if the gain at any given frequency exceeds the distributed loss, the incoming wave with intensity I_0 will be amplified; if the loss dominates, an incoming electromagnetic wave will be attenuated. The key to making a successful laser oscillator or amplifier is therefore to create a population inversion that makes the factor in braces in (1.13) positive; the second is to reduce the distributed losses α due to scattering, reflection and absorption in the laser medium and the optical resonator to a minimum.

1.3.2 Creating Gain in a Laser Medium

In any system of atoms, ions or molecules in thermal equilibrium, the population in state 2 is always exponentially less than that in state 1. Hence, creating gain, rather than loss, in a laser medium requires the creation of a highly non-equilibrium distribution of atoms. This is the function of the laser "pump," which may be

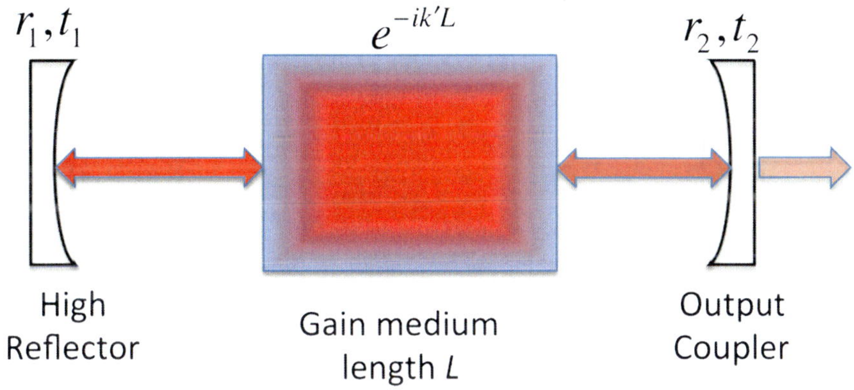

Fig. 1.3 Schematic of a Fabry–Perot resonator, showing how phase and amplitude accumulate with increasing numbers of round trips through the cavity

a flashlamp, an electron beam, an electric discharge, an electric current or even another laser. While flashlamps and electric discharges are still used in many nanosecond laser oscillators, the development of high-efficiency laser diodes operating in the near infrared has made optical pumping an increasingly popular choice, especially for mode-locked and fiber lasers, because of the uniformity and stability of their output and the efficiency with which the diode lasers can be coupled directly into the desired pump transition.

1.3.3 Laser Oscillators: Theory

The possibility of light amplification by stimulated emission leads naturally to the question of how to build an oscillator—that is, an optical device in which a very small input signal can grow into a self-sustaining optical beam, either continuous or pulsed. Such an oscillator may be either pulsed or continuous-wave (CW); regardless of which it is, the basic principles remain the same. As a model system (Fig. 1.3), we consider a simple laser cavity with a gain medium of length L, which is essentially a pair of resonator mirrors that are perfectly aligned. For the sake of definiteness, we assume that one mirror is a high reflector, with reflection and transmission coefficients (r_1, t_1), while the other mirror, the output coupler, is partially transmitting with coefficients (r_2, t_2). If we imagine the initially infinitesimally small light signal E_0 entering through mirror 1, with transmission coefficient t_1, by the time it passes through mirror 2 with transmission coefficient t_2 it has acquired an additional phase $\exp(-ik'L)$ that has both real and complex components and also incorporates the effects of optical gain at frequency ω and the distributed loss α. Describing the electric field in complex exponential notation, we can write the output wave as an infinite series of waves reflected back and forth within the Fabry–Perot cavity while allowing for partial transmission of the wave at mirror 2, as shown in Fig. 1.3. Summing the series yields

$$E_t = E_0 t_2 t_1 e^{-ik'L}\left(1 + r_2 r_1 e^{-2ik'L} + r_2^2 r_1^2 e^{-4ik'L} + \cdots\right) = \frac{E_0 t_2 t_1 e^{-i(\tilde{k}+\Delta k)L} e^{(\gamma-\alpha)L/2}}{1 - r_2 r_1 e^{-2i(\tilde{k}+\Delta k)L} e^{(\gamma-\alpha)L}} \tag{1.14}$$

For laser oscillation to occur, it is necessary that the denominator of this equation approach zero, so that a finite output can be generated by amplification of an infinitesimally small input signal. This can happen when the denominator vanishes, leading to two conditions for oscillation, one on the threshold gain required to overcome the distributed losses in the cavity, and the other on the phase of the waves:

$$\begin{aligned} r_1 r_2 \exp\{[\gamma_{th}(\omega) - \alpha]L\} &= 1(\text{amplitude}) \text{ and} \\ 2[k + \Delta k(\omega)]L &= 2\pi m, \quad m \text{ an integer (phase)} \end{aligned} \tag{1.15}$$

The phase condition effectively guarantees that a standing electromagnetic wave will be generated in the resonant cavity, and the amplitude condition says that the wave will grow in amplitude as long as the gain exceeds the threshold value. The *m*th frequency at which the cavity will oscillate can be derived from the phase condition to be:

$$[k + \Delta k(\omega)]L = kL\left[1 + \frac{\chi'(\omega)}{2n^2}\right] = m\pi \Rightarrow \omega_m = \omega\left[1 - \frac{\omega - \omega_0}{\Delta\omega}\frac{\gamma(\omega)}{k}\right] \tag{1.16}$$

If the laser resonator has a photon lifetime τ_0, we can use this to calculate the threshold population inversion as well, assuming that the population in the level N_1 is initially negligible:

$$\tau_0 = \frac{2nL}{c(1 - R_1 R_2 e^{-2\alpha L})} \approx \frac{2nL}{c[2\alpha L - \ln(R_1 R_2)]} \Rightarrow N_{2threshold} = \frac{8\pi}{A_{21}\lambda^2 g(\nu) c\tau_0} \tag{1.17}$$

In laser oscillators typical of those used in most materials-processing application, the mechanism described here results in the emission of a laser pulse that will persist until the population inversion has been extinguished, typically on a time comparable to the photon lifetime in the cavity (1.17). Figure 1.4 shows how the longitudinal modes of a Fabry–Perot cavity are modulated by the gain profile of the laser medium. Those modes on which the *net* gain exceeds the lasing threshold as defined by (1.15) will oscillate randomly; in a kind of Darwinian competition, various modes will oscillate until all are driven down to the threshold gain level.

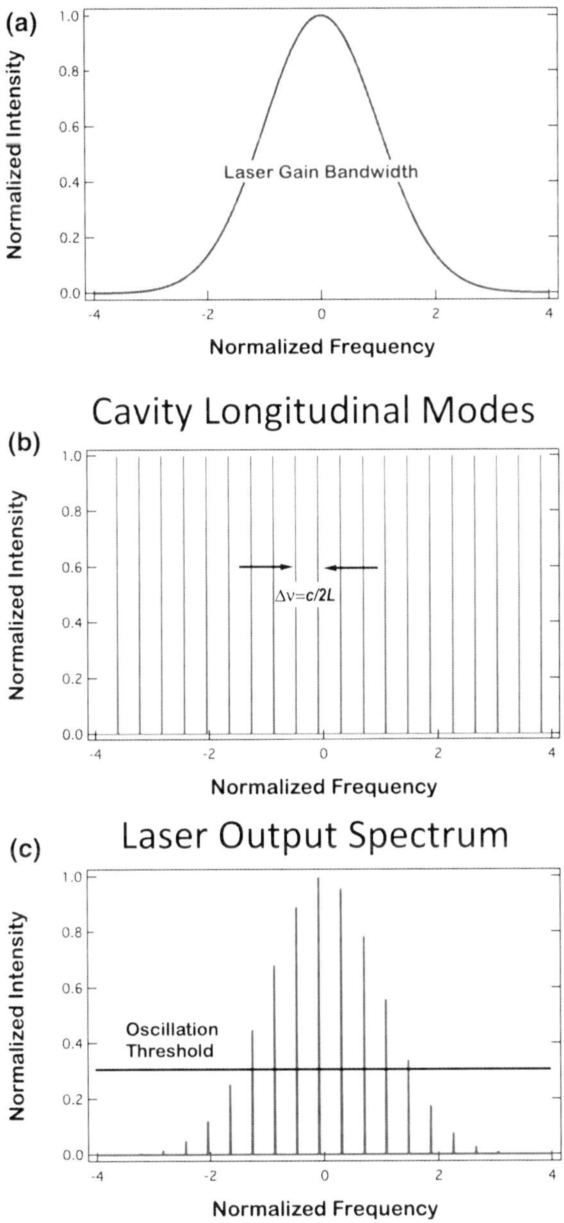

Fig. 1.4 **a** Schematic laser gain profile, derived from the atomic or molecular laser line. **b** Longitudinal cavity modes, spaced by the inverse of the cavity round-trip time. **c** Output spectrum convoluting the cavity modes with the laser gain profile; those modes that have intensity above threshold all cavity modes can lase in competition with one another

1.3.4 Mode-Locked Oscillators

Because the gain created in the excitation cycle is subject to statistical fluctuations, there may be substantial pulse-to-pulse variations in duration and energy. However, there is another approach that is designed to minimize these fluctuations, called *mode-locking*. The fundamental idea of mode-locking is the following: At a given frequency v, the gain $\gamma(v)$ satisfies

$$\gamma_{threshold}(v) = \alpha - \frac{1}{L} r_1 r_2 \Leftrightarrow \gamma(v) = (N_2 - N_1) \frac{c^2}{8\pi n^2 v^2 \tau_{spont}} g(v) \tag{1.18}$$

Moreover, we know that wherever the gain at a given frequency is sufficient to overcome the losses in the optical cavity, laser oscillations can occur over a range of frequencies v_q defined by

$$v_{q+1} - v_q = \frac{c}{2nL} \text{ or } \omega_q - \omega_{q-1} = \frac{\omega c}{L} \equiv \Omega \tag{1.19}$$

In essence, as long as there is sufficient gain, all the frequencies that differ from each other by the inverse of the round-trip propagation time in the cavity can oscillate. The total electric field at some arbitrary point in the cavity can be made periodic in the cavity round-trip time

$$E(t) = \sum_n E_n \exp[i(\omega_0 + n\Omega)t + \phi_n] = E(t+T), T = \frac{2\pi}{\Omega} = \frac{2L}{c} \tag{1.20}$$

provided that the phase ϕ_n, which normally fluctuate in a random fashion in the cavity, can be made equal to each other, that is *locked*. In this circumstance, the cavity modes E_n all have the same phase relationship with each other, and if the amplitude of all the electric field modes is constant, the total electric field is

$$E(t) = \sum_{-(N-1)/2}^{+-(N-1)/2} E_0 \exp[i(\omega_0 + n\Omega)]t = \exp\left[i\omega_0 t\right] \frac{\sin(N\omega t/2)}{\sin(\omega t/2)} \tag{1.21}$$

where ω_0 is the frequency at the center of the gain profile. The average oscillator output power, which is proportional to the square of the electric field, is then given by

$$P(t) \propto |E(t)|^2 = \frac{\sin^2(N\omega t/2)}{\sin^2(\omega t/2)} \tag{1.22}$$

What this means is that the power is emitted in a train of pulses separated from other by the cavity round-trip time, and where the individual pulse duration is $2L/cN$. Just as the intensity of light from N coherently interfering apertures is N^2 times the intensity from a single aperture, the power from N interfering modes is N^2 times the power from a single mode. Moreover, the pulse duration, approximated as the time from the peak to the first zero of the mode-locked train, is $1/N$ times the round-trip time T. Estimating the number of oscillating modes as the ratio of the transition linewidth $\Delta\omega$ to the intermode frequency spacing Ω, the pulse duration τ_0 is

$$\tau_0 \sim \frac{2\pi}{\Delta\omega} = \frac{1}{\Delta\nu} \tag{1.23}$$

Hence the larger the transition linewidth for the laser transition is, the shorter is the pulse that can be obtained by mode-locking. Importantly, in solid-state laser materials, this linewidth depends not only on the lasing atom or ion, but also on the host material in which the optically active ion is embedded. In addition, since the laser is being pumped continuously, the pulse-to-pulse variation in the mode-locked train tends to be substantially smaller than in laser oscillators in which the pumping or excitation cycle produces a single output pulse.

As long as the mode-locked laser cavity is continuously excited, pulses will be generated continuously at a frequency $c/2L$, which for typical cavity dimensions of tens of centimeters leads to pulse repetition frequencies from tens of MHz to a few GHz. The energies of these pulses are usually in the nJ range and must therefore be amplified to reach the typical J/cm^2 fluences required for laser processing. However, at the very highest pulse-repetition frequencies, even the thermal loading produced by the oscillator pulses is often unacceptably high, creating a requirement either to scan the beam at high speeds so that the requirements of thermal and stress confinement are satisfied, or to reduce the pulse-repetition frequency to an acceptable level. This is generally accomplished by time-dependent modification of the resonant cavity, as will be explained below.

1.3.5 Laser Amplifiers

The majority of lasers used in materials processing are oscillators, either continuous-wave (such as diode lasers) or pulsed. However, increasingly the advantages of mode-locked oscillators—pulse durations in the femtosecond and picosecond range, and excellent power stability—are generating interest in their use in materials processing. The chief difficulty with that is the output of mode-locked oscillators comes at very high pulse-repetition frequencies, leading to undesirable thermal loading for some applications, but very low pulse energies. This leads to the need for amplification.

We assume for purposes of discussion that we are dealing with a two-level system, and that for simplicity, the amplifier medium is homogeneously broadened. Then the processes we need to deal with are: the pumping rates per unit volume into level 2 and level 1, the radiative decay from those two levels, stimulated absorption or emission on the laser transition arising from interaction with a beam of radiation, and the population of the lower level by the process of spontaneous emission. With these assumptions, it is possible to write down a pair of rate equations that describe the time evolution of these two levels.

$$\begin{aligned} \frac{dN_2}{dt} &= R_2 - N^* \sigma_{21}(\omega_L - \omega_0)\frac{1}{\hbar\omega_L} - \frac{N_2}{\tau_2}, \quad \text{where } N^* \equiv N_2 - \frac{g_2}{g_1}N_1 \\ \frac{dN_1}{dt} &= R_1 - N^* \sigma_{21}(\omega_L - \omega_0)\frac{1}{\hbar\omega_L} - \frac{N_1}{\tau_1} + N_2 A_{21} \end{aligned} \tag{1.24}$$

Assuming that we are in a steady state, that is, that the lifetimes are long compared to whatever processes lead to loss of inversion, we have

$$\left.\begin{aligned} N_2 &= R_2\tau_2 - N^*\sigma_{21}\frac{I}{\hbar\omega_L}\tau_2 \\ N_1 &= R_1\tau_1 + N^*\sigma_{21}\frac{I}{\hbar\omega_L}\tau_2 + N_2A_{21}\tau_1 \end{aligned}\right\} \Rightarrow N^* = \frac{R_2\tau_2[1-(g_2/g_1)A_{21}\tau_1] - (g_2/g_1)R_1\tau_1}{1+\sigma_{21}\frac{I}{\hbar\omega_L}[\tau_2 + (g_2/g_1)\tau_1(1-A_{21}\tau_2)]} \tag{1.25}$$

Since this inversion density basically determines the magnitude of the gain and its duration, we can now notice that the numerator in the equation for N^* is the gain in the absence of the radiation field, which we denote $N^*(0)$, while the denominator contains a time-dependent term which is essentially the relaxation time for the population inversion:

$$\begin{aligned} N^*(I) &= \frac{N^*(0)}{1+\frac{I}{I_{sat}(\omega_L-\omega_0)}} \\ I_{sat}(\omega_L-\omega_0) &\equiv \frac{\hbar\omega_L}{\tau_R}\frac{1}{\sigma_{21}(\omega_L-\omega_0)}, \; \tau_R \equiv \tau_2 + \frac{g_2}{g_1}\tau_1[1-A_{21}\tau_2] \end{aligned} \tag{1.26}$$

The saturation intensity I_{sat} is a measure of how much the population inversion is affected by the presence of radiation in the cavity: If $I \ll I_{sat}$, the population inversion is changed very little by the intensity in the cavity; if $I \gg I_{sat}$, then the population inversion is strongly depleted by the presence of the radiation in the cavity. These (1.25) can be recast into a particularly useful form if we define a net pumping rate R^*, leading to an expression for the inversion rate per unit volume

$$\frac{dN^*}{dt} = R^* - \left(1+\frac{g_2}{g_1}\right)\frac{N^*}{\tau_R}\frac{I}{I_{sat}} + \text{spontaneous emission terms} \tag{1.27}$$

In any of the lasers commonly used in materials processing, the spontaneous emission terms are small compared to the other two terms, and can be safely neglected. Equation (1.26) essentially states that the inversion rate—which determines the gain—is proportional to the pumping rate diminished by a term that is inversely proportional to the relaxation time and directly proportional to the laser intensity in the cavity.

The practical effect of this situation for a laser amplifier is that as the intensity in the amplifier increases, the intensity gain for light traversing the amplifier in the z direction changes from exponential in the small-signal region to linear in the saturated region:

$$\frac{1}{I}\frac{dI}{dz} = \frac{\gamma_0(\omega - \omega_0)}{1 + (I/I_{sat})(\omega - \omega_0)} \Rightarrow \ln\left[\frac{I(z)}{I(0)}\right] = \gamma_0 z - \frac{I(z) - I(0)}{I_{sat}} \qquad (1.28)$$

There are two limiting cases to be considered: When the intensity in the amplifier is respectively small or large compared to I_{sat}. The solutions to (1.28) in the two cases are:

$$I(z) \approx I(0)\,\exp(\gamma_0 z),\, I \ll I_{sat} \quad \text{and} \quad I(z) \approx I(0) + (\gamma_0 z)I_{sat} \qquad (1.29)$$

This equation is strictly valid only for a continuous, rather than a pulsed, amplifier and for a homogeneously broadened gain medium. Nevertheless, rigorous analysis of the more complex cases of inhomogeneously broadened laser media and pulsed amplification comes to what is for all practical cases essentially the same conclusion: in the region of small signals, the gain increases exponentially with increasing distance z, while as the intensity in the amplifier rises to well above the saturation intensity, the gain is essentially linear. Plots of the two cases are shown in Fig. 1.5.

1.4 Laser Systems Used in Materials Processing

Historically, most lasers used in laser processing in industry have been oscillators; this is particularly true of lasers used in cutting, welding and similar applications, which are mostly of the continuous-wave (CW) type. However, nanosecond-pulsed, solid-state (e.g., Nd:YAG) and gas (e.g., KrF, ArF) lasers are also in widespread use. Increasingly, pulsed fiber lasers are also finding application both in research laboratories and on the factory floor. In this section, we briefly sketch the characteristics of four different types of laser oscillators to illustrate how different laser types embody the basic principles described in Sect. 1.2.

1.4.1 Laser Oscillators

Here we consider four examples of laser oscillators used in materials processing in order to illustrate the principles of laser physics from the preceding section. Further details about these specific lasers and other laser types may be found particularly in [10, 11].

Excimer lasers The rare-gas halide lasers have been part of the standard repertoire of lasers for thin-film processing since their first application in the deposition of thin films of high-temperature superconductors more than a quarter century ago. Rare-gas halide lasers—such as KrF, ArF and XeCl—are essentially two-level lasers in which the lower laser level is so weakly bound that it decays on a time scale short enough to prevent the accumulation of population in the lower

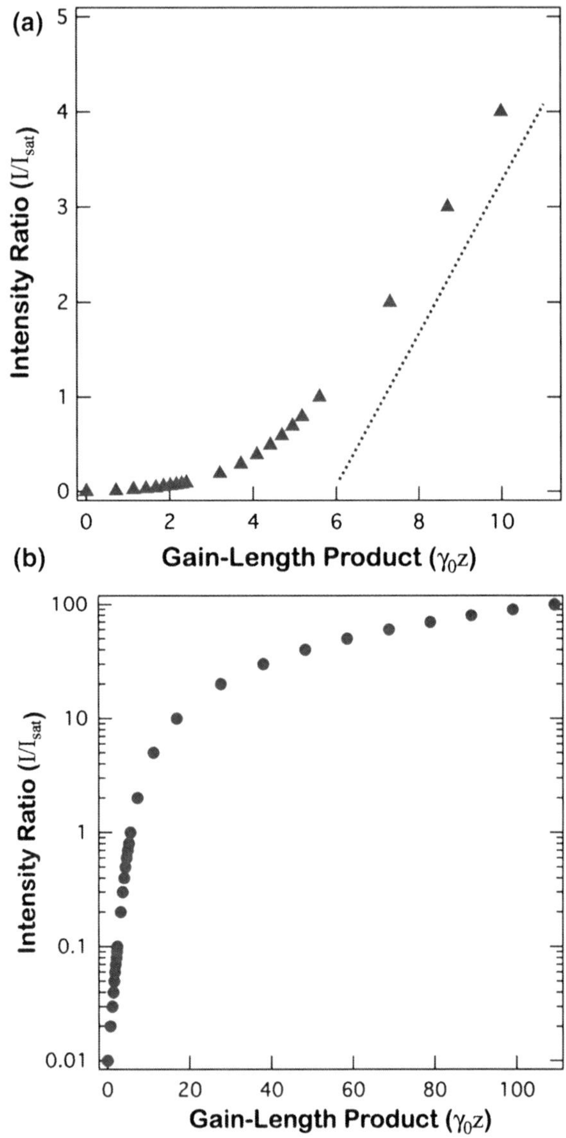

Fig. 1.5 Intensity normalized to the saturation intensity I_{sat} as a function of the gain-length product $\gamma_0 z$ in (**a**) the small-signal and (**b**) large-signal regimes

laser level [13, 14]. The energy-level diagram appropriate to this family of lasers is shown in Fig. 1.6a. The KrF laser, for example, is based on the three-body recombination reaction $Kr^+ + F^+ + Ar \rightarrow KrF^* + Ar$, where the ions may be created from the initial gas mixture by electron-beam excitation, or by preionization of the laser gas followed by an electric-field discharge. In current excimer laser designs, the preionization step seeds the discharge with 10^8–10^9 electrons—usually by a corona discharge from both sides of the laser cavity—so that the

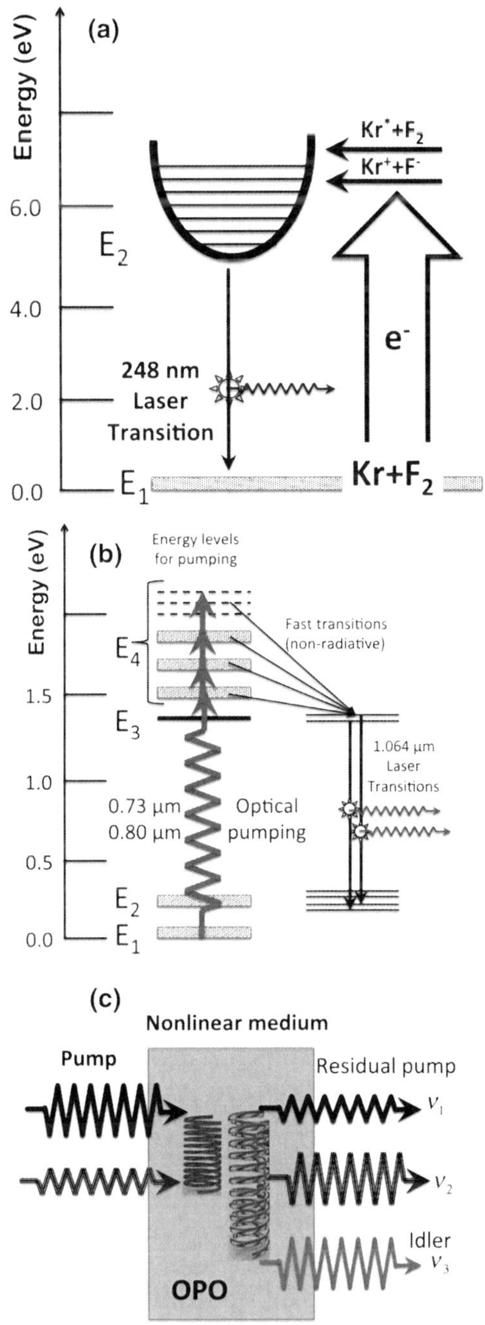

Fig. 1.6 Three different oscillator types. **a** Energy level diagram of the KrF rare-gas halide laser. **b** Energy level scheme for the Nd:YAG oscillator showing the highest gain 1.06 μm transition. **c** An optical parametric oscillator, showing the pump and seed beams, with outputs of the residual pump, signal and idler beams. The springs symbolize the wave-wave interactions in the nonlinear medium.

discharge can be homogeneous throughout the laser excitation volume; to facilitate the formation of a uniform discharge rather than plasma filaments, the gas mixture is typically 90 % He.

In a typical excimer laser, the laser cavity is bounded by plane mirrors. Since the round-trip gain is of order 100, the laser output is limited by the duration of the uniform discharge in the gain medium, typically 10–20 ns. This means, however, that the laser beam is of relatively poor quality compared to a continuous-wave laser oscillator, in which the very large number of round trips built up in the resonator tend to compress the radial mode volume and improve the output mode to the desirable and familiar TEM_{00} shape with a Gaussian radial distribution.

Nd:YAG lasers The Nd:YAG laser is one of a family of lasers that are based on laser transitions in rare-earth ions embedded in a variety of crystal hosts, including yttrium–aluminum-garnet (YAG), glass, and vanadate (YVO_4). These lasers are the classic examples of a four-level laser system as shown in the diagram in Fig. 1.6b. A recent development in multipass amplifiers has made possible a Yb:YAG laser producing over 800 W of output power in the infrared in 400 fs pulses in a compact, two-stage amplifier design [15]. One of the practical advantages of the rare-earth laser oscillators is that it is possible to generate harmonics that are often themselves intense enough for materials modification.

Optical parametric oscillators Optical parametric oscillators (OPOs) are increasingly popular as sources of moderately tunable radiation, especially for spectroscopy. However, with increasing crystal quality and output power, there is growing interest in using OPOs directly for materials research. An OPO is typically a singly- or doubly-resonant oscillator cavity in which a crystal with a large second-order susceptibility is driven by a pump beam, and possibly with a seed beam for the signal output. Well-known examples of such crystals are uniaxial, periodically poled $LiNbO_3$ or beta barium borate (BBO) pumped by a Nd:YAG laser at 1.064 or 0.532 μm wavelength. A parametric oscillator generates two output waves at the *signal* and *idler* wavelengths that simultaneously satisfy the requirements of collinear propagation and energy conservation according to:

$$\frac{n_{signal}}{\lambda_{signal}} = \frac{n_{pump}}{\lambda_{pump}} - \frac{n_{idler}}{\lambda_{idler}} \text{ and } \frac{1}{\lambda_{signal}} = \frac{1}{\lambda_{pump}} - \frac{1}{\lambda_{idler}} \tag{1.30}$$

Just as in an ordinary laser oscillator, the signal and idler wavelengths develop from random noise in the oscillator cavity that satisfies the condition for difference-frequency mixing. The input angle and cut of the nonlinear, birefringent crystal must be adjusted—typically by a combination of angle and/or temperature tuning—so that the requisite phase-matching condition between signal and idler are satisfied. While OPOs are still found primarily in analytical applications, their growing efficiency and output power/energy are leading to renewed interest in their use for laser processing [16].

Fiber lasers Fiber oscillators with output powers of a few Watts now dominate such laser-processing applications as marking. However, high-power fiber laser oscillators and large-mode-aperture systems (LMAs) are beginning to displace

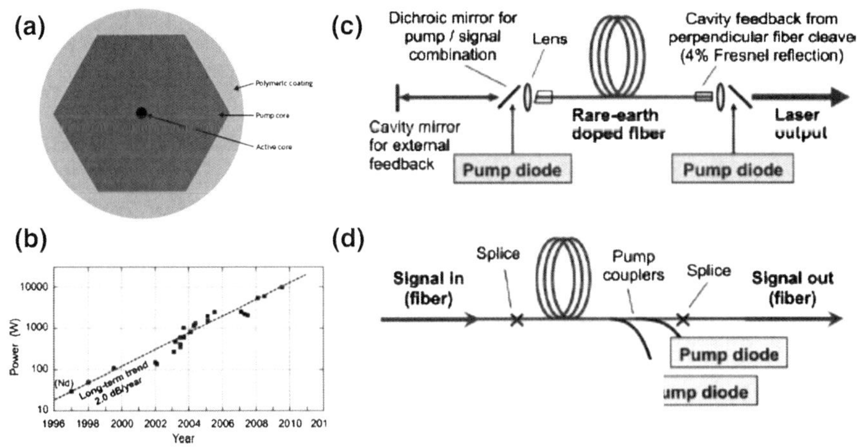

Fig. 1.7 **a** Cladding-pumped fiber geometry, from [19]. **b** Growth in fiber-laser output power as a function of year, from [17]. **c, d** Two different pumping schemes for large mode-area fiber amplifiers, also from [17]

Nd:YAG and CO_2 gas lasers that historically have been industry workhorses for high-power laser applications such as cutting and welding [17]. With powers in excess of 50 kW, these fiber lasers are much easier to incorporate into laser tools than lasers emitting free-space beams that may require complicated beam-steering optics in articulated mechanical arms to follow complex cutting or welding trajectories. Moreover, their compact construction, fundamental simplicity and durability are making it possible to use these lasers in applications for which older high-power laser types are too expensive or too complicated. While low-power fiber lasers have been under development since the 1960s and have been deployed in telecommunications applications since the 1970s, serious development of high-power fiber lasers became possible only with the discovery of the cladding-pumped fiber laser in 1988, shown schematically in Fig. 1.7a [18]. Current fiber lasers are based on a lasing transition in Yb ions in a variety of optical fibers, and emit at 1,070–1,080 nm; pumping in most cases is accomplished by 915 nm laser diodes that directly pump the ${}^2F_{7/2}$–${}^2F_{3/2}$ transition of the Yb^{3+} ions. Output powers are now well into the multi-kW range, and growing at nearly exponential rates, as suggested by Fig. 1.7b. The most powerful fiber lasers use an oscillator-amplifier configuration; two different designs for these large mode-aperture fiber amplifiers are shown in Fig. 1.7c. Moreover, rapid progress is being made in developing fibers with lasing wavelengths in the mid-infrared; these will be powerful pumps for tunable mid-IR parametric amplifiers [19].

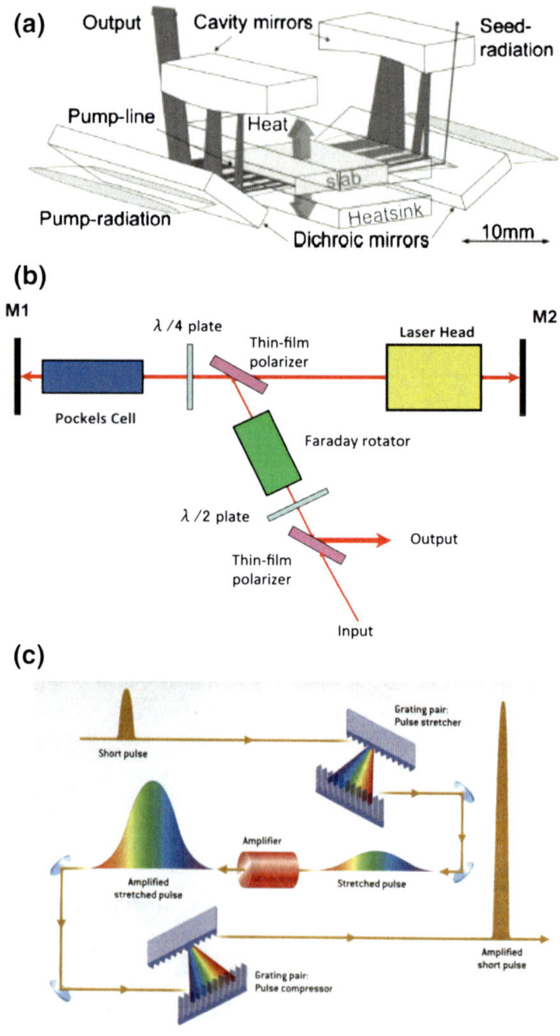

Fig. 1.8 Three different amplifier geometries: **a** the Innoslab® multipass geometry. **b** Regenerative amplifier, with a Faraday rotator to act as an optical diode. **c** the chirped-pulse amplifier, showing the stretcher grating pair, the amplifier medium and the grating-pair pulse compressor

1.4.2 Amplified Laser Systems

As pulse durations in the picosecond and femtosecond range have come into use, first as laboratory tools and more recently in a few industrial applications, master-oscillator, power-amplifier (MOPA) systems are becoming more common. The basic principles of laser gain and oscillation also apply here; however, most laser amplifier systems do not incorporate an optical cavity surrounding the gain medium. In the following paragraphs, we briefly note important characteristics of several of the most common types of laser amplifiers.

Multi-pass amplifiers One way of avoiding some of the problems of regenerative amplifiers is to use a multipass geometry like that shown in Fig. 1.8a [20]. This is in some ways a natural geometry for solid-state lasers that use cw laser diodes as a pump source, as the gain is being continually repumped, and thus is not depleted each time a pulse to be amplified passes through the gain medium. This generally requires a more complex optical arrangement and the overlap between the input or pump beam and the amplified pulses may not be as good as it is in a regenerative amplifier. While poor beam overlap may in principle lead to a loss of efficiency, in practice efficiencies of order 30 % may be achieved in multipass power amplifiers, comparable to the best regenerative devices. Moreover, the fact that amplification does not take place in an optical cavity means that problems of prepulse and amplified spontaneous emission are not a factor in multipass geometries.

Regenerative amplifiers are used to amplify laser pulses with durations ranging from femtoseconds to tens or even hundreds of picoseconds. As shown in Fig. 1.8b, amplification is carried out in an optical cavity into which the pulse to be amplified is injected, usually from a mode-locked oscillator. When the Pockels cell is off, incident laser pulses will exit the cavity via the polarizer after being reflected from the end mirror. If the Pockels cell is on because a quarter-wave voltage has been applied to it, the orthogonally polarized pulse returns to the intracavity polarizer and is trapped in the cavity, where it may make many passes through the laser gain medium. When a second quarter-wave voltage pulse is applied to the Pockels cell, the amplification cycle is ended and the amplified pulse exits the cavity. The Faraday rotator acts as an optical diode, ensuring that the amplified pulse travels only in the desired direction.

Chirped-pulse amplifiers To achieve the highest possible output intensities for femtosecond amplifiers, chirped-pulse amplifier (CPA) configurations are used, as shown schematically in Fig. 1.8c. The CPA solves the problem that amplification of a femtosecond laser pulse in a highly saturated gain medium is likely to exceed the damage threshold for the laser crystal, the laser optics, or both [21, 22]. For example, in a Ti:sapphire amplifier, the saturation fluence is of order 1 J/cm^2; hence the peak intensity of a 100 fs pulse at saturation fluence is of order 10^{13} Wcm^2—perilously close to the damage limits of most optical elements. This problem is solved in the CPA by stretching the oscillator pulse to the nanosecond regime in a dispersive medium (prism or grating pair), amplifying in one or more stages, and then temporally compressing the pulse using all-reflective dispersive media such as a diffraction grating pair. In this way, amplification takes place at moderate intensities, avoiding the optical damage issues, while high output intensity is achieved using all-reflective optics.

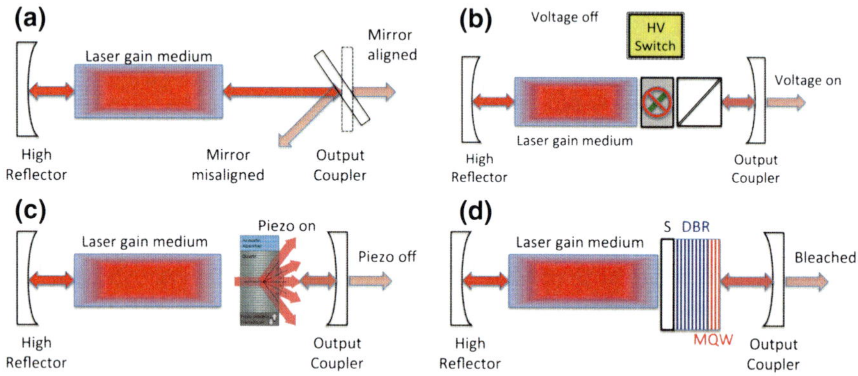

Fig. 1.9 Intracavity schemes for Q-switching. **a** Rotating mirror. **b** Electro-optic modulator. **c** Acousto-optic modulator. **d** Semiconductor saturable absorber mirror

1.4.3 Control of Laser Pulse Duration

In nanosecond and microsecond oscillators, control of the pulse generation is generally achieved by manipulating the time-dependent gain in the laser resonator, a practice often called "Q-switching" or "Q-spoiling." The idea is to allow the optical gain to be built up while not permitting oscillation, and then suddenly allowing the cavity to resonate. While the principle on which Q-switching is based is very different in concept from pulse-duration control by mode-locking, the electro-optic and acousto-optic modulators described here are also found as intracavity devices in mode-locked lasers.

The oldest of these techniques, illustrated in Fig. 1.9a, is the rotating mirror or prism. In the rotating mirror scheme, the rear high reflector in the laser cavity is rotated, so that positive feedback to the gain medium and light amplification can occur only during the brief interval of time when the output coupler and the high reflector are aligned. In between those brief intervals, the gain medium is pumped, allowing the buildup of gain far above threshold.

Electro-optic modulators make use of the Pockels effect, a second-order nonlinear effect in non-centrosymmetric crystals that allows for preferential selection of polarization in the same way that a quarter-wave plate can select for polarization. In a "Pockels cell," the polarization selectivity is created by applying a "quarter-wave voltage" that induces the necessary birefringence to block optical feedback between the high reflector and the output coupler. When the Pockels cell is turned on by applying the quarter-wave voltage, light passing through the Pockels cell converts vertically polarized light transmitted by the polarizer to circularly polarized light whose handedness is changed by reflection at the output coupler. When light can pass through the polarizer and Pockels cell with low loss, gain builds up in the cavity; horizontally polarized light is rejected by the polarizer and exits the cavity. The beam paths are shown schematically in Fig. 1.9b. The pulse durations achieved can be as short as 2–5 ns. However, the disadvantages of this modulation scheme is the insertion of the Pockels cell into the optical cavity with its high circulating power and attendant possibility of optical damage. Moreover, the Pockels effect is relatively weak, so that achieving the desired level of birefringence requires several kV to be applied.

The *acousto-optic* modulator is an intracavity device in which the lattice planes of a piezoelectric crystal are distorted by a radio-frequency drive signal applied at one end of the crystal. When the rf power is on, the crystal planes are slightly distorted and light incident on the crystal planes, typically at near grazing angles is diffracted out of the cavity. On the other hand, when the rf power is turned off, the cavity Q is high, and lasing can occur. As with the rotating mirror, while the Q of the laser resonator is low, gain builds up as long as the laser medium is being pumped; switching from low to high gain then allows all of the gain above threshold to be extracted for as long as the rf power remains off. Given the speed of sound of a few times 10^3 m/s, a 1 mm beam will be deflected out of the cavity for a few hundred nanoseconds. This switching time can be decreased by focusing the intracavity beam inside the acousto-optic crystal. A typical arrangement is seen in Fig. 1.9c.

Saturable absorbers are materials that bleach at high intensities when all the ground-state atoms (or molecules or quantum wells) have been pumped into an excited state, precluding further absorption. If inserted into the laser cavity, a saturable absorber blocks transmission as gain is building up—thus effectively lowering the Q of the cavity—until the gain is so far above threshold that the population in the upper level of the absorber begins to saturate. As positive feedback into the cavity continues, the absorber begins to bleach until it is essentially fully transparent, and a giant, Q-switched pulse develops. Solutions of dyes and doped crystals have been widely used. In recent years, the semiconductor saturable absorber mirror (SESAM) has become increasingly popular because it is readily adapted to short cavity lengths and thus to pulse durations as short as tens of picoseconds [23]. Such an absorber is illustrated schematically in Fig. 1.9d; the saturable element is the quantum well. The advantages of the SESAM are that it is easily fabricated, has no moving parts, requires no intracavity optics, and good mode discrimination because it has a turn-on time corresponding to many cavity round trips. Since the pulse-repetition frequency of the SESAM is determined by the pump beam for the laser, it can easily operate at repetition rates up to the MHz range.

For ultrashort laser pulses, it is also possible to optimize the temporal sequence of materials processing using techniques in which a single ultrafast laser pulse is converted by opto-electronic components into a train of coherent pulses that can interact sequentially with the material to be modified [24]. In a typical pulse shaper, an ultrashort laser pulse is spatially dispersed by a diffraction grating and then passed through a spatial light modulator (SLM) that imposes a phase delay on different portions of the wavefront [25]. This SLM is often an array of liquid

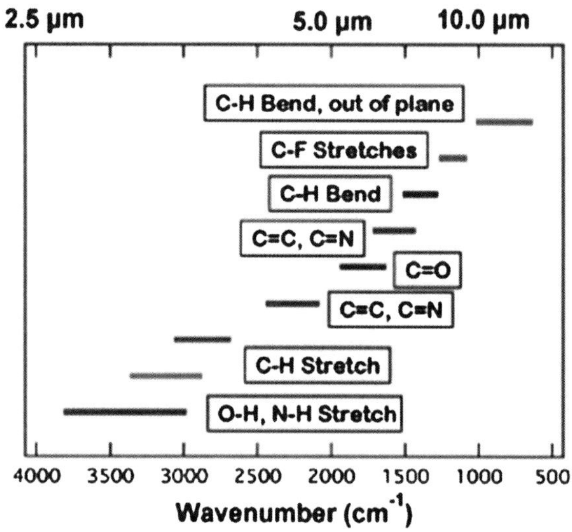

Fig. 1.10 Major vibrational bands in the "molecular fingerprint" region of the mid-infrared where polymer processing by resonant infrared excitation is possible

crystalline pixels that are individually controlled by a voltage specified via a computer. The temporal pulse sequence imposed on the input beam is then monitored by an autocorrelator or similar device. In coherent-control schemes, the SLM parameters are altered by computer control using a genetic or evolutionary algorithm based on a figure of merit that reflects the quality of the desired process outcome, such as an ion yield or a material removal rate.

1.5 A Tunable Picosecond Laser for Polymer Processing in the Mid-Infrared

To illustrate how the basic principles and systems concepts can be used to develop a novel table-top laser system for research in materials processing, we conclude by discussing a conceptual design for a table-top laser system that could replicate studies previously accomplished using an infrared free-electron laser for resonant, mid-infrared materials processing [26–28]. The system is designed to produce approximately 500 mJ/pulse at a pulse duration of 10 ps and a pulse repetition frequency of order 1 kHz; this would be adequate to reach measured thresholds for picosecond ablation of model polymers [29]. Such a system would significantly outperform the infrared free-electron laser at vastly reduced cost and complexity.

While laser processing of inorganic materials has historically been based on electronic transitions initiated by ultraviolet, visible or near-infrared lasers, processing of organic and polymer materials seems to be optimized in many cases by mid-infrared laser excitation. Most experiments in polymer or organic processing to this date have targeted O–H, C–H and N–H bands in the 3–4 μm region (see

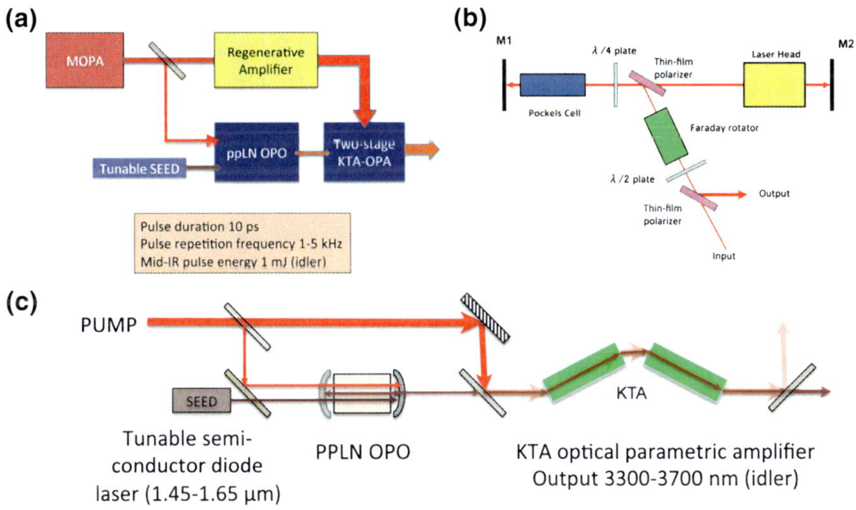

Fig. 1.11 Schematics for the ps-MIR laser system. **a** Block diagram of major system components. **b** Schematic of the regenerative amplifier. **c** Schematic of the optical parametric generator

Fig. 1.10); however, extension to other wavelength bands—such as the technologically important 5–6.5 and 8–9.5 μm stretching vibrations—with appropriate frequency-conversion elements will be highly desirable to meet laser-materials processing challenges in organic and polymer electronics and opto-electronics. In particular, developments in quasi-phase-matched frequency-conversion materials make it realistic to think about band-tunable picosecond lasers that can cover most of these absorption modes [30].

Picosecond, tunable, high-repetition lasers in the mid-infrared control the light-matter interactions selectively excite spectroscopically well-characterized vibrational states, in contrast to the multiplicity of often unknown states excited in multi- or multiple-photon *electronic* transitions. These localized, anharmonic vibrational states couple relatively slowly to the harmonic phonon bath. Mesoscale materials modifications can, in principle, be achieved by tuning the laser across the resonance to modify the local density of vibrational excitation.

It has now become possible to acquire from commercial sources all the components needed to build a tunable, mid-infrared (MIR) picosecond laser whose output is characterized by high intensity (up to 100 GW/cm^2), modest fluence (0.1–5 J/cm^2), and high pulse-repetition frequency (1–5 kHz). With these parameters and reasonable frequency-conversion efficiency, the average power in the targeted mid-infrared tuning range (2.7–3.7 μm) will need to be 10–20 W. These characteristics are sufficiently robust to enable not only proof-of-concept experiments in laser-materials interactions, but some modest scaling studies to test realistic laser-materials processing regimes at industrial scale.

Table 1.2 "Three-band" OPG output

Pump (nm)	Signal (nm)	Idler (nm)
1,064	1,670–1,690	2,872–2,932
1,064	1,520–1,540	3,442–3,560
1,064	1,450–1,470	3,852–3,996

Schematically, an example of a tunable, picosecond MIR laser is shown in Fig. 1.11a, and comprises three subsystems: the master-oscillator power-amplifier; a regenerative amplifier; and the optical parametric generator. The design is based on the successful operation of such a system used for proof-of-concept mid-infrared laser ablation and thin-film deposition experiments, first reported in 2010. We note in particular that the regenerative amplifier could be replaced by a multi-pass device; importantly, however, at pulse durations of a few picoseconds one avoids the need for chirped-pulse amplification with its added optical complexity.

The major subsystems of the ps-MIR laser are the oscillator-preamplifier (MOPA), the regenerative amplifier, and the optical parametric generator. The latter contains both the optical parametric oscillator (OPO), seeded by a tunable semiconductor diode laser, and the two-stage optical parametric amplifier for the mid-infrared generation. A collateral benefit of this design is that the pump laser can be used to provide harmonics of the Nd:YAG laser either for inorganic materials modification or for a narrow-band picosecond driver for time-resolved photoluminescence or other monitoring applications.

Based on currently available laser components, one can construct or acquire a MOPA with a linearly polarized, air-cooled fiber oscillator and high-energy fiber power amplifier. Several such MOPA systems are currently either available or under discussion for industrial applications, with typical maximum pulse energies of order 1 μJ, pulse duration of 1–10 ps and in some cases, a user-selectable pulse repetition frequency ranging from single-shot to 1 MHz. The regenerative amplifier is the most critical and most expensive component; however, the fact that the oscillator is based on a mode-locked pulse train should lead to very stable operation owing to favorable thermal (amplitude noise below ± 2 %) and optical ($M^2 < 1.3$) characteristics.

The regenerative amplifier shown schematically in Fig. 1.11b could comprise one or two side-pumped Nd:YAG laser amplifier modules, and would be entirely conventional in design, differing from the typical end-pumped configuration only in the fact that the pump diodes are integral to the side-pumped modules. The amplifier module selected would need to be designed for a few tens of Watts average power.

The optical parametric generator (OPG) shown schematically in Fig. 1.11c incorporates a tunable semiconductor diode laser seeding an optical parametric oscillator (OPO) based on periodically poled lithium niobate (ppLN), and a two-stage optical parametric amplifier (OPA) using potassium tantalate (KTA) as the frequency conversion material. These robust and widely used materials produce an idler beam in the 2.8–3.8 μm range and a signal beam in the 1,450–1,690 nm band with reasonable efficiencies. This gives adequate coverage of the O–H, N–H and C–H bands using the idler beam, and usable signal beam in the longer-wavelength telecommunications bands (Table 1.2). The OPG design is sufficiently flexible that other frequency-conversion elements could be incorporated to reach the C=C and N–N stretching vibrations (5–6.5 μm) and the fluorocarbon vibrational spectrum (8–9.5 μm). For example, quasi-phase matched GaAs could be substituted for the PPLN in the present design in order to reach even longer wavelengths [30].

1.6 Conclusion

An understanding of laser physics is the critical first step in using lasers with greatest efficiency to produced desired results in laser processing of materials. Basic considerations that relate the properties of lasers—attainable intensity and fluence, pulse duration and pulse-repetition frequency—to parameters of meso-scale and macroscale materials modifications suggest that requirements for stable, reproducible laser interactions with materials will probably lead away from many of the currently used lasers. In particular, the figure of merit suggested by (1.8) for laser processing rate clearly favors higher intensity (that is, shorter pulses) and the highest-possible pulse-repetition frequency. The versatility, growing tunability and increasing cost effectiveness of amplified, mode-locked laser systems will in all probability lead to their increasing adoption, at first in the research laboratory but inevitably in production environments.

Acknowledgments Research on infrared laser modification of materials at Vanderbilt is supported by AppliFlex LLC through a Phase II STTR grant from the National Science Foundation (IIP-0740683) and a Phase I SBIR grant from the Air Force Research Laboratories (FA9550-12-C-0006).

References

1. A.M. Stoneham, N. Itoh, Appl. Surf. Sci. **168**, 186–193 (2000)
2. N. Itoh, A.M. Stoneham, Radiat. Eff. Defects Solids **155**, 277–290 (2001)
3. L.V. Zhigilei, B.J. Garrison, J. Appl. Phys. **88**, 1281–1298 (2000)
4. L.V. Zhigilei, B.J. Garrison, Appl. Phys. Lett. **74**, 1341–1343 (1999)
5. L.V. Zhigilei, B.J. Garrison, Appl. Phy.-Mat. Sci. Proces. **69**, S75–S80 (1999)
6. L.V. Zhigilei, B.J. Garrison, Rapid Commun. Mass Spectrom. **12**, 1273–1277 (1998)
7. V.P. Carey, G. Chen, C. Grigoropoulos, M. Kaviany, A. Majumdar, Nanoscale Microscale Thermophys. Eng. **12**, 1–60 (2008)
8. E.G. Gamaly, A.V. Rode, B. Luther-Davies, J. Appl. Phys. **85**, 4213–4221 (1999)
9. A. Yariv, *Optical Electronics in Modern Communications* (Oxford University Press, New York, NY, 1997)
10. A.E. Siegman, *Lasers* (University Science Books, CA, 1986)
11. S. Hooker, C. Webb, *Laser Physics* (Oxford University Press, Oxford, 2010)
12. O. Svelto, S. Longhi, G. Della Valle, G. Huber, S. Kück, M. Pollnau, H. Hillmer, T. Kusserow, R. Engelbrecht, F. Rohlfing, J. Kaiser, R. Malz, M. Gerd, K. Mann, P. Simon,

C.K. Rhodes, F.J. Duarte, A. Borsutzky, J.A. L'huillier, M.W. Sigrist, H. Wächter, E. Saldin, E. Schneidmiller, M. Yurkov, R. Sauerbrey, J. Hein, M. Gianella, J. Helmcke, K. Midorikawa, F. Riehle, S. Steinberg, H. Brand, Lasers and Coherent Light Sources, in *Springer Handbook of Optics and Lasers*, ed. by F. Träger (Springer Verlag, Heidelberg, 2007)
13. C.A. Brau, J.J. Ewing, J. Chem. Phys. **63**, 4640–4647 (1975)
14. J.J. Ewing, C.A. Brau, Appl. Phys. Lett. **27**, 350–352 (1975)
15. P. Russbueldt, T. Mans, G. Rotarius, J. Weitenberg, H.D. Hoffmann, R. Poprawe, Opt. Express **17**, 12230–12245 (2009)
16. R.L. Byer, IEEE J. Sel. Top. Quantum Electron. **6**, 911–930 (2000)
17. D.J. Richardson, J. Nilsson, W.A. Clarkson, J. Opt. Soc. Am. B-Opt. Phy. **27**, B63–B92 (2010)
18. E. Snitzer, J. Less-Common Met. **148**, 45–58 (1989)
19. S.D. Jackson, Nat. Photonics **6**, 423–431 (2012)
20. P. Russbueldt, T. Mans, J. Weitenberg, H.D. Hoffmann, R. Poprawe, Opt. Lett. **35**, 4169–4171 (2010)
21. D. Strickland, G. Mourou, Opt. Commun. **56**, 219–221 (1985)
22. G. Mourou, D. Umstadter, Phys. Fluids B-Plasma Phys. **4**, 2315–2325 (1992)
23. U. Keller, K.J. Weingarten, F.X. Kartner, D. Kopf, B. Braun, I.D. Jung, R. Fluck, C. Honninger, N. Matuschek, J. Aus der Au, IEEE J. Sel. Top. Quantum Electron **2**, 435–453 (1996)
24. R. Stoian, A. Mermillod-Blondin, S.W. Winkler, A. Rosenfeld, I.V. Hertel, M. Spyridaki, E. Koudoumas, P. Tzanetakis, C. Fotakis, I.M. Burakov, N.M. Bulgakova, *Opt. Eng.* **44** (2005)
25. A.M. Weiner, Opt. Commun. **284**, 3669–3692 (2011)
26. M.R. Papantonakis, R.F. Haglund, Appl. Phys. A-Mater. Sci. Proces. **79**, 1687–1694 (2004)
27. S.L. Johnson, R.F. Haglund, Jr, Deposition of polymer and organic thin films using tunable, ultrafast mid-infrared lasers, in *Laser-Surface Interactions for New Materials Production: Tailoring Structure and Properties*, ed. by P. Ossi, A. Miotello (Springer Verlag, Berlin, 2009)
28. R.D. Torres, S.L. Johnson, R.F. Haglund, J. Hwang, P.L. Burn, P.H. Holloway, Crit. Rev. Solid State Mater. Sci. **36**, 16–45 (2011)
29. M. Duering, R. Haglund and B. Luther-Davies, Appl. Phys. A-Mater. Sci. Proces., doi:10.1007/s00339-013-7946-2 (2013).
30. K.L. Vodopyanov, O. Levi, P.S. Kuo, T.J. Pinguet, J.S. Harris, M.M. Fejer, B. Gerard, L. Becouarn, E. Lallier, Opt. Lett. **29**, 1912–1914 (2004)

Chapter 2
Material Response to Laser Energy Deposition (Thermal and Hyperthermal Processes)

Juergen Reif

Abstract We consider the laser absorption mechanisms, the energy dissipation processes, and the consequences for the target material after pulsed irradiation of—mainly—solid targets. The strong dependence on laser pulse duration is emphasized. Both, the classical thermodynamic equilibrium route (heating, melting, and evaporation) and hyperthermal processes beyond thermodynamic equilibrium (superheating, phase explosion, surface/bulk instability) are comprehensively discussed, under particular consideration of relevant time scales. The implication of different processes for laser ablation and material structuring is addressed.

2.1 Introduction

With the advent of the laser, a very powerful type of light source was created, with a power up to several kW or a pulse-energy in the kJ range. At the same time, it allowed for a tight focusing never encountered before, resulting in unmatched irradiance. Consequently, very soon strong interest was stimulated in using this powerful energy source in materials processing. Considered first almost as a type of concentrated, focused Bunsen burner, it became soon obvious, however, that for pulsed lasers, in particular, the effect can much more sophisticated. It is the topic of this chapter to shed some light on this complexity and to discuss the material response to intense, pulsed laser irradiation in some more detail. For this purpose, we will not deal with individual electronic bond-breaking or photo-chemical processes (those will be covered by the chapter by W. Kautek: Material Response to Laser Energy Deposition (Non-Thermal Processes) in this book) but concentrate

J. Reif (✉)
Brandenburgische Technische Universität (BTU), Cottbus and Cottbus JointLab,
Platz der Deutschen Einheit 1 03046 Cottbus, Germany
e-mail: reif@tu-cottbus.de
URL: https://www.physik.tu-cottbus.de/physik/xp2

M. Castillejo et al. (eds.), *Lasers in Materials Science*,
Springer Series in Materials Science 191, DOI: 10.1007/978-3-319-02898-9_2,

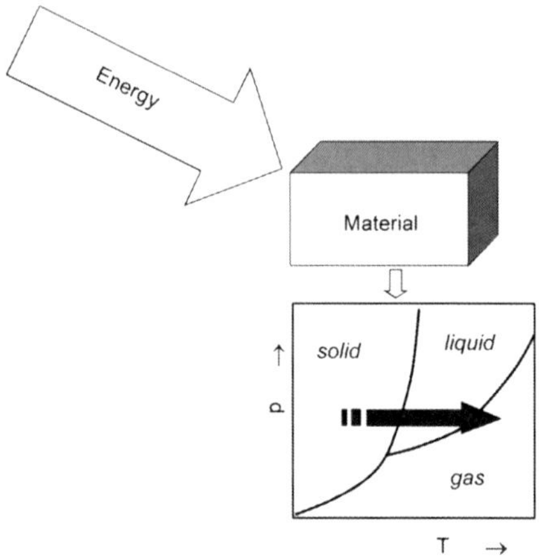

Fig. 2.1 Generalized phase diagram and influence of increasing temperature

on the response of the material as a continuum. Therefore, we will consider mainly thermal processes [1]. It will be shown, however, that the fast interaction of very short pulses can drive the irradiated matter into a state far from thermodynamic equilibrium. Then, we will define the material response as *hyperthermal*.

2.2 Basic Considerations

When laser light impinges on a material, at least part of its energy is absorbed, leading to an increase of internal energy of the target, classically denoted by the target temperature:

$$\Delta \mathrm{E} = \Delta U = cm\Delta T \tag{2.1}$$

($\Delta \mathrm{E}$ is the absorbed light energy, ΔU is the increase of internal energy, c, m are heat capacity and mass of the target, respectively, and ΔT is the increase in temperature).

In an atomistic view, temperature is a measure for the average kinetic energy of the target constituents. Consequently, an increase in temperature reduces the ratio of inter-atomic binding energy to kinetic energy, thus enabling phase transitions in the target, as indicated in Fig. 2.1 by the bold, horizontal arrow.

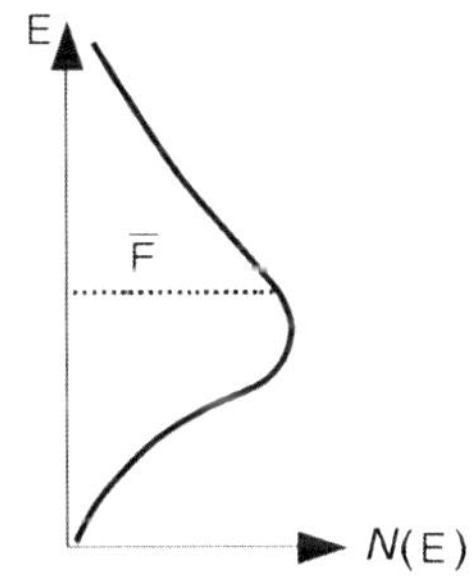

Fig. 2.2 Maxwell-Boltzmann distribution of kinetic energy E in an ensemble of N particles. The *dashed line* indicates the mean kinetic energy $\overline{\mathsf{E}}$

2.2.1 Thermodynamics

In classical thermodynamics, we always deal with systems in *thermal equilibrium*, i.e. all target constituent atoms behave like a statistical ensemble in equilibrium. If we consider a system of N particles, their velocities, and thus kinetic energies, follow a Maxwell-Boltzmann distribution (Fig. 2.2).

As indicated before, temperature is a measure of the mean kinetic energy of the target constituting particles:

$$T = \frac{2}{3k}\overline{\mathsf{E}} = \frac{2}{3k}\left(\frac{1}{2}m\overline{v^2}\right) \tag{2.2}$$

(k is Boltzmann's constant, $\overline{\mathsf{E}}$ is the mean kinetic energy, m, $\overline{v^2}$ are the particle mass and mean squared velocity, respectively).

In order to establish thermal equilibrium, i.e. a purely statistical distribution of kinetic energies, the particles need to interact and exchange their velocities. This occurs via collisions between particles or rather, in condensed matter, phonons (i.e. coherent particle oscillations).

Obviously, it takes several collisions before equilibrium is reached after a first excitation by absorption of a photon. This corresponds to typical time scales at the order of 10 … 100 ps.

The actual energy transfer from the incident irradiation to core vibrations is comparable to the optical excitation of molecular vibrations, as is indicated in Fig. 2.3a. Figure 2.3b indicates the transition from an individual vibration to a phonon.

In this picture, it appears straight forward to understand thermodynamic phase transitions: Once the vibrational energy is sufficiently large, i.e. the internal energy U about matches the maximum binding energy and total energy approaches 0, the particles can move around almost freely: this is the liquid state. When the total energy increases even more and becomes positive, the particles don't feel, ideally, any more binding forces and the inter-particle distance can increase almost arbitrarily: this is the gaseous state.

In a real system, the phase transitions are not really a smooth, steady increase of inter-particle distance. Instead, during melting resp. evaporation additional energy input is only used to overcome the residual binding energy, without significantly

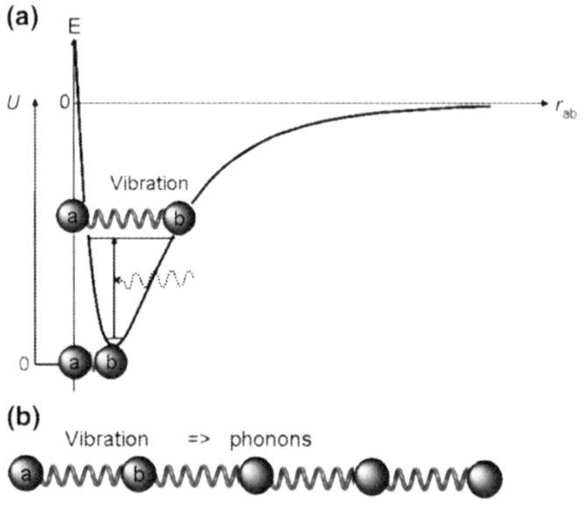

Fig. 2.3 **a** Schematic of the excitation of a molecular vibration by a photon (*dashed arrow*): the increased energy of an excited vibronic level corresponds to a varying distance between atoms *a* and *b*, limited by the binding potential (here a Lennard-Jones potential). **b** Continuation of a molecular vibration to a phonon (vibration of a molecular chain)

increasing the particle separation. This additional energy (melting/evaporation enthalpy) can be considered as an activation energy E_A beyond which the particle energy is increased, again, together with the number of widely moving particles. The situation is schematically shown in Fig. 2.4.

From the Boltzmann distributions shown schematically in Fig. 2.4 we see that the number of "liquid" resp. "gaseous" particles increases significantly, once the activation energy is overcome. This behavior can be analyzed by an Arrhenius equation, known from chemical kinetics, which describes the temperature-dependence of chemical reaction rates:

$$D = A\exp\{-\mathsf{E}_A/k_BT\} \tag{2.3}$$

(D is the reaction rate, A the so-called pre-exponential factor, accounting for the total collision rate, E_A the activation energy, k_B Boltzmann's constant and T the temperature).

The effect is demonstrated in Fig. 2.5, where the temperature dependence of melting and evaporation rates for silicon is shown in an Arrhenius plot, where the slopes indicate the activation energies, corresponding to the respective melting/ evaporation enthalpies.

2.2.2 Deposition of Laser Energy

The energy deposition from the incident laser radiation occurs via an increase of the target *electrons'* energy. Then, this energy has to be transferred to the target lattice. Generally, this transfer occurs on a time scale much slower than the electronic response. This is accounted for in the *two-temperature model*, introduced by Anisimov [2, 3], where the laser input [$S(t)$ in (2.4a, b)] heats the electrons, which immediately equilibrate by electron–electron collisions. Electron–phonon

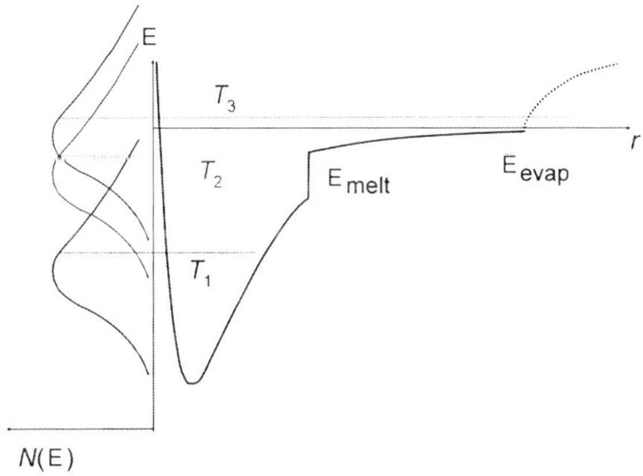

Fig. 2.4 Model potential describing phase transitions from a solid (strong binding, negative total energy) via a liquid (total energy ≈ 0) to a gas (positive energy). Indicated are the melting and evaporation barriers/activation energies. In the *left part*, BOLTZMANN distributions are shown for three temperatures. For the lowest temperature, T_1, practically all particles are in the bound state, for the highest temperature, T_3, most particles have only kinetic energy (gas)

Fig. 2.5 Arrhenius plot of melting (*open circles*) and evaporation (*full squares*) rates for silicon

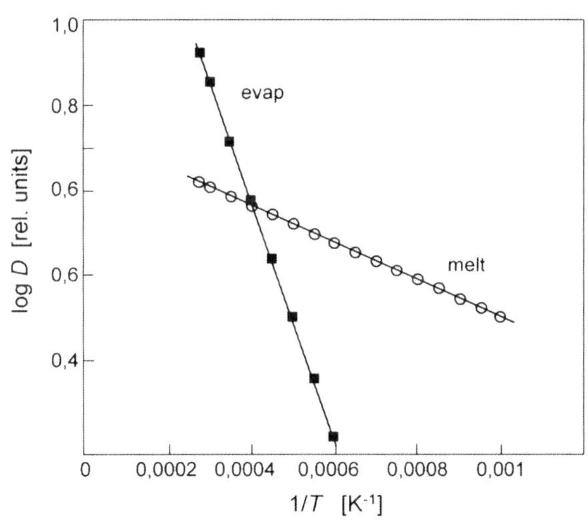

collisions then transfer the energy to the lattice which equilibrates by phonon–phonon collisions (Fig. 2.6). This is usually described by a pair of coupled equations (2.4a, b):

$$C_e(T_e)\frac{\partial T_e}{\partial t} = \nabla[K_e(T_e)\nabla T_e] - \gamma(T_e - T_l) + S(t) \tag{2.4a}$$

$$C_l(T_l)\frac{\partial T_l}{\partial t} = \nabla[K_l(T_l)\nabla T_l] - \gamma(T_e - T_l) \tag{2.4b}$$

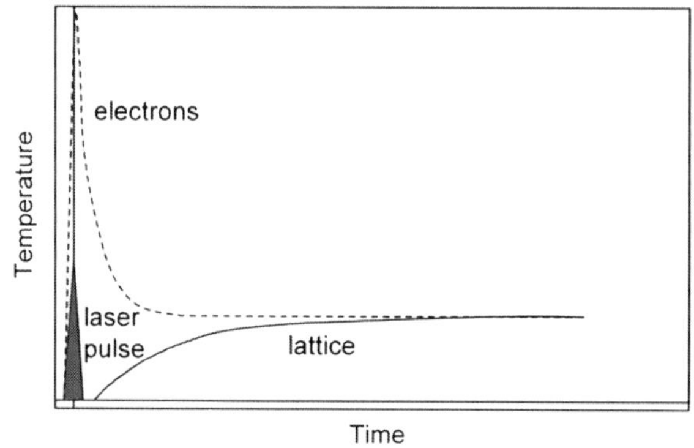

Fig. 2.6 Schematic representation of the Two-Temperature-Model. Only after a substantial time elapsed since the laser pulse (*grey shaded*) electrons (*dashed line*) and lattice (*solid line*) are in thermal equilibrium (for simplicity, we did not take into account, in this schematic, the effect of heat diffusion out of the considered volume)

C, K are the heat capacity and the thermal conductivity for electrons (index e) and lattice (index l), respectively. γ is the electron–phonon coupling constant. Note that, in contrast to classical heat diffusion [cf. (2.5)], the thermal conductivities are assumed to be not necessarily constant but may depend on electron and lattice temperature.

In the following of this chapter we will consider the situation when a steady state is reached after energy transfer to the lattice and phonon–phonon equilibration.

Classically, melting and evaporation occur when heat is transferred to the target through the surface and, then, is dissipated by heat flow into the target volume. This heat flow is associated with a corresponding temperature evolution. Consequently, melting and heating starts at the surface and penetrates into the target volume.

In contrast to the classical case, an excitation takes place in the whole absorption volume, given by the laser beam cross section and the penetration or absorption depth[1] δ, the reciprocal of absorption coefficient α. Typically, in silicon the penetration depth is some μm for visible light, in metals it corresponds to the skin depth, some percentage of the wavelength λ, and in transparent dielectrics it is the whole target. It is, however, not straight forward to assume that simply the absorption volume is molten and then evaporated during ablation. Already a simple estimate for metals or silicon indicates that this model is not really adequate, as shown in Table 2.1. Assuming that ALL absorbed energy is used to bring the material to the boil (i.e. corresponds to the sum of melting and evaporation enthalpies), we calculate the volume which should be ablated for an absorbed fluence of 1 J/cm^2.

[1] For simplicity, we assume a rectangular absorption volume, neglecting the details of the spatial beam profile and the exponential intensity decay according to Beer's law.

Table 2.1 Estimated ablation depth for several materials if absorbed fluence is fully converted in melting and evaporation

	Al	Au	Cu	Ni	Si
Melting enthalpy (kJ/kg)	397	65	205	303	1,792
Evaporation enthalpy (kJ/kg)	10,900	1,650	47,900	6,480	12,821
Density (kg/m^3)	2,700	19,300	8,920	8,900	2,336
Ablation depth @1 J/cm^2 (m)	3.0×10^{-7}	3.0×10^{-7}	2.3×10^{-7}	1.6×10^{-7}	2.9×10^{-7}

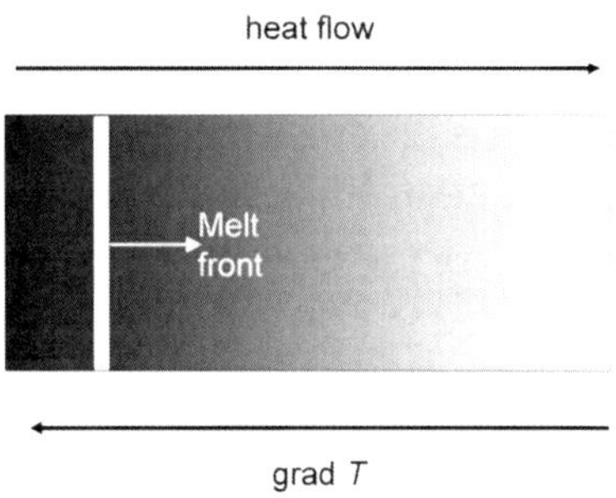

Fig. 2.7 Heat flow and melt front propagation in the presence of a temperature gradient

It appears that the ablation volume does not simply correspond to the absorption volume. So, laser ablation is *not simply* a "boiling-off" of the absorbing volume. Instead, we have to consider the energy transfer from the electrons to the lattice in more detail. First, the energy absorbed by the electrons is locally transferred to phonons in the absorption volume (electron–phonon collisions). The phonon-excitation is dissipated further into the target volume by phonon–phonon collisions, heating a larger and larger volume by heat diffusion, increasing the temperature in an increasing volume to overcome the evaporation enthalpy.

On a sufficiently slow time scale, i.e. when enough electron–phonon and phonon–phonon collisions establish thermal equilibrium, we can analyze the heat diffusion classically (Fig. 2.7).

The absorption–volume heating induces a temperature gradient ∇T. This induces the diffusion of heat according to

$$\partial T/\partial t = K\nabla^2 T \tag{2.5}$$

If, as usual, a pulsed laser of pulse duration τ is used, the total heat flow q can be determined:

$$q = q_0(T_1 - T_0)\exp\{-x^2/4K\tau\}^{1/2}, \tag{2.6}$$

with the diffusion length

$$\ell = 2(K\tau)^{1/2}, \tag{2.7}$$

which determines the effective heated volume. It is quite obvious, that melting and evaporation follows the heat diffusion, like in the classical case of external heating. This is indicated in Fig. 2.7 by the propagation of the melt front. In general, this is the case for a "modest" temperature gradient, and is the typical situation exploited for laser annealing of crystal defects. It is a typical case of equilibrium thermal processes, all occurring on a time scale which is slow compared to the time (resp. number of phonon–phonon collisions) required for the establishment of thermal equilibrium. Such classical melt-front propagation from the surface to the interior bulk, following heat diffusion, is usually termed "heterogeneous melting".

2.3 Beyond Thermal Equilibrium (*Hyperthermal* Processes)

All processes considered so far require that all dynamics is slow compared to the establishment of thermal equilibrium by a sufficient number of phonon–phonon collisions. The time scales are very sensitive to both the duration and the intensity of the laser pulses involved in the interaction.

2.3.1 Homogeneous Boiling

Let us, first, consider very strong excitation, i.e. very high laser intensity. Then, the absorption not only creates free electrons in the conduction band (for non-metals) but also can heat conduction electrons to rather high kinetic energy. Subsequent electron–electron and electron–phonon collisions cannot thermalize all that energy to the lattice. Instead, a non-negligible number of electrons remain with sufficient energy to penetrate well into the target bulk. We call those "ballistic electrons" because they behave just like high-kinetic energy particles obeying Newton's laws of motion. In contrast to the low energy case, they will not lose their excess kinetic energy close to the surface but will penetrate well inside the target bulk before the energy is transferred to the phonon bath and, correspondingly, converted to heat (Fig. 2.8).

As a consequence, we find "boiling" nucleation well distributed in the whole target bulk, even outside the original absorption region. Thus, boiling does not "propagate" from the surface deeper into the bulk but occurs simultaneously in a very large bulk region. Consequently, this is called "homogeneous boiling" [4]. The boiling centers are, however, surrounded by non-excited, cold matter, resulting in a strong confinement. Thus, the boiling centers experience a significant pressure increase from their environment, resulting in an increased boiling temperature, compared to that required for the lower-pressure environment. So, finally, we generate a "superheating" [5] of the excited volume which has to relax

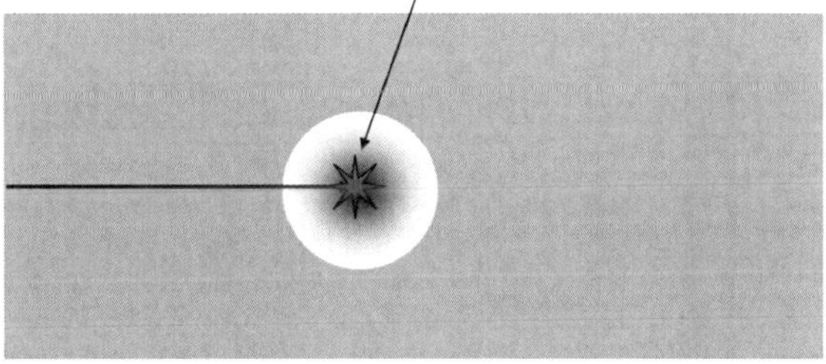

Fig. 2.8 Homogeneous nucleation

in an explosive energy release (phase explosion [6]). It should be noted here that, though this can be fully explained by thermodynamic principles, it is NOT really in thermodynamic equilibrium.

Nevertheless, the term "temperature" is often used to describe also processes which are *not in thermal equilibrium* (cf. superheating, phase explosion), which we will call "hyperthermal" in the further course of this chapter.

Hyperthermal: = *"Not in thermal equilibrium"*

- Hyperthermal processes are typical for situations occurring in a time shorter than equilibration time (usually a few generations of phonon–phonon collisions; i.e. several tens of picoseconds)
- Hyperthermal processes are, generally, observed for ultrashort (sub-picosecond) pulses or very high excitation.

Hyperthermal processes are characterized by a very rapid build-up of a highly energetic volume surrounded by a cold lattice. This corresponds to a steep gradient in atomic order, resulting in very fast relaxation; too fast for proceeding in thermal equilibrium.

2.3.2 Ultrashort Excitation: Self-organized Nano-structure Formation

The excitation by ultrashort laser pulses of duration of much less than one picosecond presents a very convenient way of studying hyperthermal processes. The pulse duration—and thus excitation—is much faster than any intrinsic relaxation processes (during the pulse duration, massive particles at thermal velocities will move less than a typical lattice spacing [a Si-atom at $T \approx 1{,}000$ K will move about one Bohr radius]). So, intrinsic dynamic processes will be almost completely decoupled from the mere excitation.

It has been shown both in theoretical simulations [7] as in ultrafast X-ray diffraction [8], that the high electronic excitation upon absorption of the laser pulse

Fig. 2.9 Aeolian dunes in a sandy basin at the Rocky Mountains (*Photo* by Gediminas Raçiukaitis)

immediately results in a significant softening of the crystalline binding. (Remember that "binding" is stable only for electrons in the ground state, cf. Figs. 2.3, 2.4). In fact, a soft state is created, being neither liquid nor solid. The individual atoms/cores can move almost freely over large distances (like in a liquid) but are still "attached" to their equilibrium position [8].

The rapid relaxation of the resulting instability has very much in common with—at first glance—very different other phenomena in nature: e.g. the formation of Aeolian dunes (Fig. 2.9), or the relaxation upon surface erosion by energetic ion beams [9]. Very complex, almost regular nano-structures are formed upon ablation (Laser-induced Periodic Surface Structures, LIPSS; Fig. 2.10).

It is commonly observed, that such LIPSS structures strongly depend on the polarization of the incident laser light [10], as is shown in Fig. 2.11.

In an attempt based on Sigmund's explanation of ion-beam induced regular surface structures [11], we have developed a model of self-organized structure formation. The surface instability between surface erosion (ablation) and surface smoothing (atomic diffusion) of the soft phase can be described by a system of semi-numerical of equations [12, 13]. The strong influence of the laser polarization can be introduced into this model by assuming an anisotropic energy coupling (light absorption or electron-lattice dissipation) [14] which can be provided, e.g., by near field effects [15] or field dependent electron mobility or stress.

Obviously, a simulation along this model can well reproduce the observed surface patterns, as is shown in Fig. 2.12.

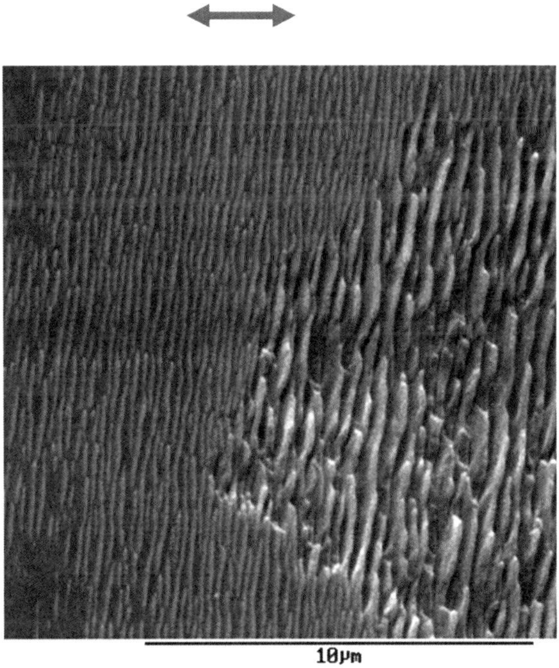

Fig. 2.10 Typical LIPSS structure after multi-pulse femtosecond-laser irradiation. The *double arrow* indicates the laser polarization. Note the different scales appear in ONE single laser spot

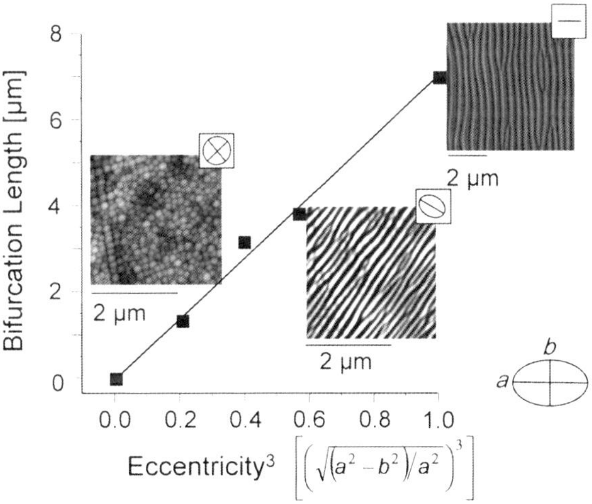

Fig. 2.11 Dependence of LIPSS on the polarization of the incident laser. For linear, elliptical, or circular polarization the length of the generated ripplets is proportional to the excentricity, given by the semi-axes of the polarization ellipse

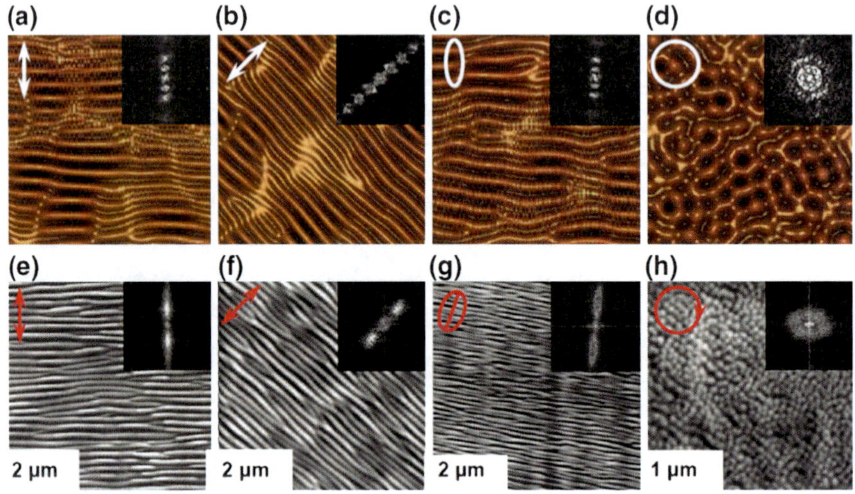

Fig. 2.12 LIPSS formation on silicon (*lower row*), using different polarization of the laser beam as indicated in the panels (*linear, tilted linear, elliptical, circular*), in comparison with simulated patterns [14] for similar conditions. The inserts represent Fourier spectra of the observed/generated patterns

2.4 Summary

We have reviewed in this chapter the relevant response phenomena in materials occurring during and upon laser irradiation. In contrast to merely electronic processes like, e.g., photochemical bond-breaking, here we review more or less a continuum response. As usual, our first approach is to consider the interaction in terms of thermodynamic, i.e. collective equilibrium effects. We show, that for very high excitation or very short pulses, the assumption of equilibrium is not really a good description of the material response. Instead, we have to consider non-equilibrium, but nevertheless continuum, processes, which we call "hyperthermal". The most prominent type is, certainly the phenomenon of phase explosion upon homogeneous bulk excitation/melting in contrast to the classical propagation of a melt front from the surface into the material.

More interesting, however, seems to be the generation of thermal non-equilibrium by the excitation with pulses shorter than typical equilibration times. Then, it appears, non-thermal relaxation results in self-organized formation of (regular) nano-structures on the irradiated resp. (slightly) ablated surface. The detailed mechanism of such nano-structure formation are, however still (2012) under discussion.

Acknowledgments The author gratefully acknowledges fruitful collaboration with Markus Ratzke, Tzanimir Arguirov, Michael Bestehorn, Sergej Varlamov, and, particularly, Olga Varlamova, which was essential for the preparation of this manuscript.

References

1. For this type of processes all textbooks on "heat" and thermodynamics will be an adequate reference
2. S.I. Anisimov, B.L. Kapeliovich, T.L. Perel'man, JETP Lett. **39**(2)
3. B. Rethfeld, K. Sokolowski-Tinten, D. von der Linde, S.I. Anisimov, Appl. Phys. A **79**, 767 (2004)
4. R. Kelly, A. Miotello, Appl. Phys. Lett. **67**, 3535 (1995)
5. R.W. Cahn, Nature **323**, 668 (1986)
6. A. Miotello, R. Kelly, Appl. Phys. A **69**, S67–S73 (1999)
7. H.O. Jeschke et al., Appl. Surf. Sci. **197–198**, 839 (2002)
8. A.M. Lindenberg et al., Science **308**, 392 (2005)
9. J. Erlebacher, M. Aziz, E. Chason, M.B. Sinclair, J.A. Floro, Phys. Rev. Lett. **82**, 2330 (1999)
10. J. Reif, O. Varlamova, F. Costache, Appl. Phys. A **92**, 1019 (2008)
11. P. Sigmund, J. Mater. Sci. **8**, 1545 (1973)
12. J. Reif, F. Costache, M. Bestehorn, in *Recent advances in laser processing of materials*, Chap. 9, ed. by J. Perriere, E. Millon, E. Fogarassy (Elsevier, Amsterdam, 2006), p. 275ff
13. O. Varlamova, F. Costache, J. Reif, M. Bestehorn, Appl. Surf. Sci. **252**, 4702 (2006)
14. J. Reif, O. Varlamova, S. Varlamov, M. Bestehorn, Appl. Phys. A **104**, 969 (2011)
15. J.Z.P. Skolski, G.R.B.E. Römer, J.V. Obona, V. Ocelik, A.J. Huis in't Veld, J.Th.M. de Hosson, Phys. Rev. B **85**, 075320 (2012)

Chapter 3
Non-Thermal Material Response to Laser Energy Deposition

Wolfgang Kautek and Oskar Armbruster

Abstract The absorption of sub-picosecond pulse laser radiation by the electronic system of dielectrics and metals leads to non-thermal processes such as ballistic transport, electron–electron collision, and electron emission across interfaces. Multi-photon excitation and impact ionization with subsequent avalanche ionization occur in dielectrics and single photon absorption in metals. Finally, electron–phonon-scattering sets in, electrons and lattice equilibrate, and thermal phenomena take over. The state of the current understanding of non-thermal phenomena is reviewed.

3.1 Introduction

The laser-matter interaction can be divided in a non-thermal (or pre-thermal) domain up to time periods of 1–10 ps and a thermal region at longer times [1–3]. Figure 3.1 shows the processes occurring during and after ultrafast laser excitation. The non-thermal processes start with the absorption of the radiation in the attosecond time domain. The absorption mechanism depends on the class of material irradiated as well as the intensity of irradiation. For ultrafast lasers, even moderate fluences allow nonlinear coupling of near-infrared radiation into wide-bandgap semiconductors and dielectrics. Multiphoton processes increase the effective absorption coefficient α_{eff} and thus decrease the threshold fluence F_{th} according to $F_{\text{th}} \propto 1/\alpha_{\text{eff}}$. This effect is stronger the shorter the pulse duration τ becomes. In a very simple approximation, α_{eff} for an n-photon process is related to the laser intensity I by $\alpha_{\text{eff}} \propto I^{n-1}$. Furthermore, I is related to the laser fluence F by $I \propto F\tau^{-1}$. We thus find $F_{\text{th}} \propto \tau^{(n-1)/n}$. Apparently, ultrashort pulses are necessary to study the non-thermal phenomenon of multi-photon excitation

W. Kautek (✉) · O. Armbruster
Department of Physical Chemistry, University of Vienna,
Währinger Straße 42 A-1090 Vienna, Austria
e-mail: wolfgang.kautek@univie.ac.at

M. Castillejo et al. (eds.), *Lasers in Materials Science*,
Springer Series in Materials Science 191, DOI: 10.1007/978-3-319-02898-9_3,

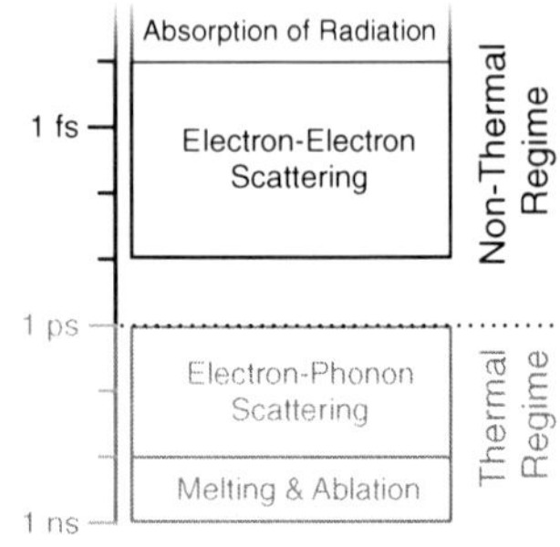

Fig. 3.1 Processes during and after ultrafast laser excitation

(Fig. 3.2b). A competing process to multi-photon excitation is impact ionization with subsequent avalanche ionization (Fig. 3.2c) [4–7]. The ultrafast excitation of metals creates hot electrons in a cold lattice, which can be described by the two-temperature model discussed later in this chapter. At high fluence ultrafast excitation, a solid-state plasma is formed in a thin surface layer of dielectrics and wide-bandgap semiconductors, resulting in Coulomb explosion, a phenomenon discussed later in this chapter. After these processes, electron–phonon-scattering sets in, i.e., electrons and lattice start equilibrating. All processes occurring during or after this are entitled thermal and are not the scope of this chapter.

The most commonly used ultrafast laser is the titanium-doped sapphire laser that radiates in the near infrared spectral region, most commonly around 800 nm. For low fluences, ablation is only expected for metals, narrow-bandgap semiconductors and materials with strongly nonlinear absorptivity. The advantage of ultrafast lasers in this regime is the prevention of plasma shielding which reduces the coupling efficiency of the radiation. Most dielectrics and wide-bandgap semiconductors, however, are transparent in this near-infrared spectral region.

3.2 Response of Metals

3.2.1 The Two-Temperature Model

The two-temperature model (TTM) [8] is successfully employed to describe the thermal balance and heat fluxes between laser-excited electrons and the lattice via electron–phonon coupling. The model assumes the validity of the heat equations on ultrashort time scales as well as thermalized electrons (i.e. the describability of the excitation states of the electrons by a temperature).

In metals and narrow-bandgap semiconductors, the photons are absorbed by electrons, which, in turn, transit to an excited state. The now excited electronic system thermalizes in a time span of typically 10 fs to 1 ps by electron–electron scattering. Electron–phonon scattering is slow compared to electron–electron scattering, resulting in a temporary disequilibrium between electron temperature

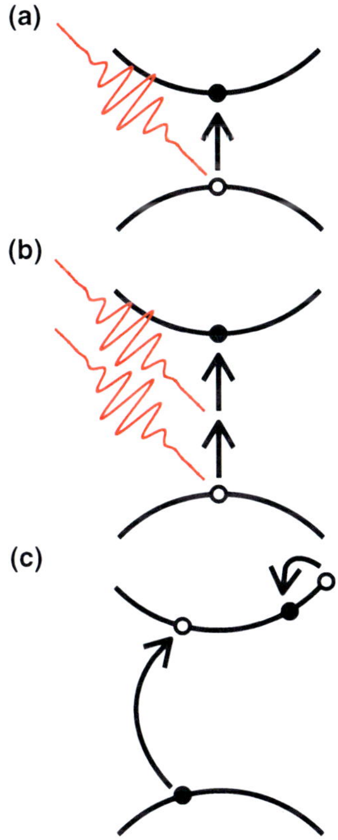

Fig. 3.2 Excitation mechanisms of semiconductors and dielectrics. **a** Single-photon absorption. **b** Multi-photon absorption (two-photon absorption as an example). **c** Impact and avalanche ionization

and lattice temperature. The coupled nonlinear equations for the electronic temperature T_e and the lattice temperature T_l can be written as

$$C_e(T_e) = \nabla[K_e(T_e, T_l)\nabla T_e] - G(T_e)[T_e - T_l] + S(\vec{r}, t) \tag{3.1}$$

and

$$C_l(T_l)\dot{T}_l = \nabla[K_l(T_l)\nabla T_l] + G(T_e)[T_e - T_l] \tag{3.2}$$

where C is the heat capacity and K is the thermal conductivity of the electrons (subscript e) and the lattice (subscript l), G is the electron–phonon coupling factor, and S is a source term, accounting for the energy deposition by the laser [3, 8].

Figure 3.3 shows a typical transient graph of T_e and T_l. The TTM allows a qualitative and in some cases even a quantitative description of the interaction of ultrafast laser radiation and matter. For example, the difference between the ablation threshold of noble metals and transition metals can be explained [9].

There are however experimental evidences that do not agree with calculations based on the TTM. For example, the TTM assumes thermalized electrons, which is not the case immediately after excitation. They rather need 10 fs to 1 ps to

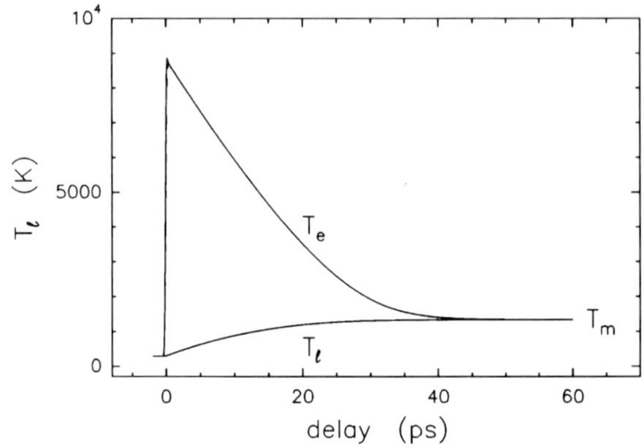

Fig. 3.3 Transient electron temperature T_e and lattice temperature T_l of a gold film with a thickness of 100 nm, excited by a laser pulse (200 fs, 400 nm, 23 mJ cm^{-2}) [9]

thermalize via electron–electron scattering [10–12]. The TTM has thus been phenomenologically extended to cover for example energy transport by ballistic electrons and scattering on grain boundaries [13, 14].

An often applied simplification of the TTM is the assumption of temperature-independent heat capacities and a constant electron–phonon coupling factor. It has been shown that this is not justified in many cases, as e.g., with aluminium, copper, gold, silver, nickel, platinum, tungsten and titanium [15]. The electronic heat capacity C_e was calculated according to

$$C_e(T_e) = \int_{-\infty}^{+\infty} d\varepsilon \frac{\partial f(\varepsilon, \mu, T_e)}{\partial Te} g(\varepsilon) \quad (3.3)$$

where $g(\varepsilon)$ is the electron density of states at the energy level ε, μ is the chemical potential at T_e and $f(\varepsilon, \mu, T_e) = (\exp[(\varepsilon - \mu)/(k_B T_e)] + 1)^{-1}$ is the Fermi distribution function. Due to the conservation of the number of electrons N_e, the chemical potential μ can be calculated according to

$$N_e = \int_{-\infty}^{+\infty} d\varepsilon f(\varepsilon, \mu, T_e) g(\varepsilon) \quad (3.4)$$

The electron density of states can be determined *ab initio* within the density functional theory employed e.g. the Vienna ab initio simulation package (VASP). In addition to the electronic heat capacity C_e, the electron–phonon coupling factor G is needed to calculate the electronic temperature T_e within the TTM:

$$G(T_e) = \frac{\pi \hbar k_B \xi \langle \omega^2 \rangle}{g(\varepsilon_F)} \int_{-\infty}^{+\infty} d\varepsilon - g^2(\varepsilon) \frac{\partial f}{\partial \varepsilon} \quad (3.5)$$

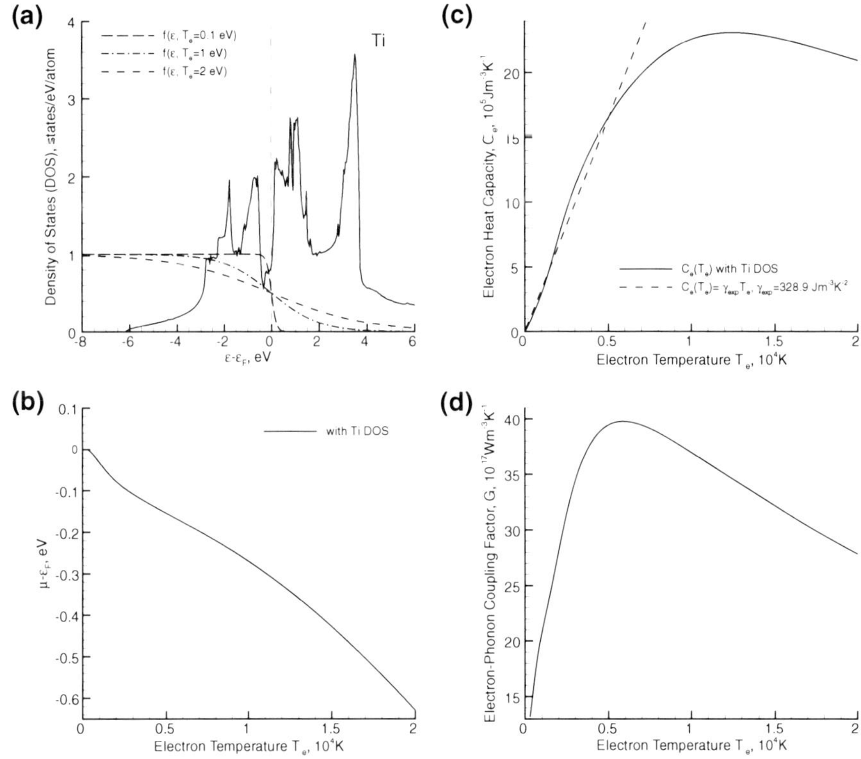

Fig. 3.4 Calculated electronic density of states g (**a**), chemical potential of electrons μ (**b**), electronic heat capacity G_e (**c**), electron phonon coupling factor G (**d**) of titanium [15]

where $\langle \omega^2 \rangle$ is McMillan's second moment of the photon spectrum and ξ is the electron–phonon mass enhancement parameter [15]. An example for titanium is represented in Fig. 3.4. It is clearly visible that neither the electronic heat capacity C_e nor the electron–phonon coupling factor G are well approximated by a constant.

3.2.2 Hot Electron Transport

Metals can be described by the free electron gas model. The non-equilibrium dynamics of hot electrons is determined by the competition between ballistic transport and hot-electron diffusion driven by the electron temperature gradient and electron–phonon coupling [13, 16]. The electrons show ballistic motion immediately after excitation [3]. The excited electrons thermalize by colliding with electrons close to the Fermi energy. This theory could be supported by an experiment where the change in reflectivity (which is proportional to the density of

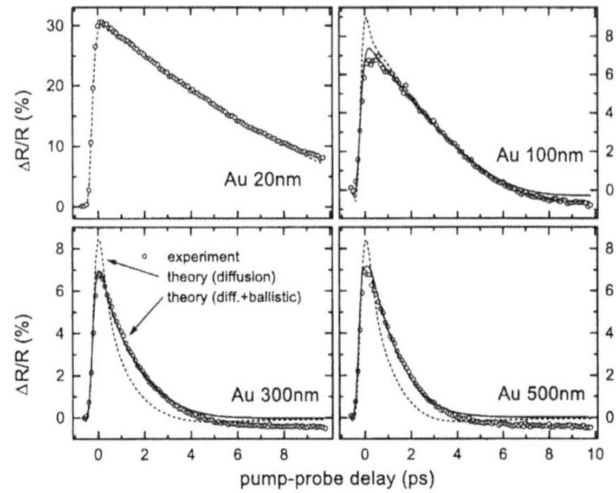

Fig. 3.5 Transient reflectivity of gold films with various thicknesses. *Dashed curves* TTM. *Solid curves* TTM with ballistic range of 105 nm. Pump pulse 400 nm, 200 fs. Probe pulse 500 nm, 100 fs, *p*-polarized [13]

free electrons at the surface) of thin gold films was measured for different pump-probe delays (Fig. 3.5) [13]. The depth of ballistic penetration into the bulk can be up to 100 nm. This means that ballistic hot electron transport has to be considered for gold film thicknesses of the order of 100 nm and becomes more pronounced with decreasing film thickness. A 20 nm thick film shows an almost linear time evolution in reflectivity change. This can be understood in terms of the TTM (see 3.1). The electronic temperature T_e is constant in the entire film and ∇T_e becomes zero when it is much thinner than the depth of ballistic motion. Thus, the first term on the right side of (3.1) can be neglected. Furthermore, the source term S vanishes for times greater than the pulse duration. The heat capacity of the electrons can be approximated with $C_e(T_e) = C_0\, T_e$ where $C_0 = \frac{(\pi^2 N_e k_B)}{(2T_f)}$ if $T_e \ll T_f = \frac{E_F}{k_B}$ holds. In this expression, N_e is the density of electrons, T_F is the Fermi temperature, E_F is the Fermi energy and k_B is the Boltzmann constant. The lattice temperature T_l is much smaller than the electronic temperature T_e and can thus be neglected in the second term on the right hand side of (3.1). The electron–phonon coupling factor G can be assumed as constant in a first order approximation. Taking this into account results in $C_0 \frac{\mathrm{d}T_e}{\mathrm{d}_t} = -G$ Solving this differential equation yields

$$T_e(t) = T_e^{(max)} - \frac{G}{C_0} t \tag{3.6}$$

This predicts a linear decay of the electronic temperature T_e with time after the laser excitation as has been observed in Fig. 3.5 (20 nm case). The slope yields the electron–phonon coupling factor G. Such experimental observations demonstrated on gold have also been reported for other noble metals like silver and copper [9, 13, 16, 17]. Transition metals such as nickel, chromium and molybdenum do not show ballistic transport of electrons. This explains why structuring of thin films of transition metals with ultrafast lasers is more precise than for noble metals. Noble metals, unlike transition metals, spread their energy by ballistic transport and electron diffusion. In metal-like thin films such as laser-deposited tetrahedral amorphous carbon and amorphous carbon nitride ballistic and diffusional electron transport lengths can be neglected in comparison to the penetration depth of light controlled by two-photon absorption [18].

As soon as the electron gas is thermalized, it can be described by the TTM and thus by an electronic temperature T_e. The dashed curves in Fig. 3.5 have been calculated this way. The gradient in electronic temperature ∇T_e is the driving force for the diffusion of electrons into deeper regions of the material. The diffusion length is determined by the strength of electron–phonon coupling. Experiments show that the TTM overestimates the change in reflectivity and thus the density of free electrons at the surface. This is an indication for an additional transport mechanism like ballistic motion [3].

3.2.3 Hot Electron Pressure

The thermal expansion of metals has two fundamental causes: the inharmonicity of the ion–ion interaction and the thermal pressure of the free electron gas. The electron pressure gives a dominant contribution to the thermal expansion at temperatures on the order of a few Kelvin. At elevated temperatures, however, it is negligible compared to the ionic contribution [19]. The main cause of thermal expansion (at thermal equilibrium between electrons and lattice) at temperatures close to 0 K is mainly governed by the pressure of the free electron gas. At higher temperatures than a few Kelvin, the electron pressure does not play a role and thermal expansion is mainly caused by the inharmonicity of the ion–ion interaction.

If however, the electrons and the lattice are not in thermal equilibrium, the electrons can have a very high temperature, causing a significant contribution of the electron gas pressure to thermal expansion, even at elevated lattice temperatures. The transient change in transmissivity of spheroid silver nanoparticles upon ultrafast laser excitation was studied. It is related to the elongation and thus to the two constitutes (i.e. inharmonicity of the ion–ion interaction and pressure of the free electron gas) of thermal expansion. The nanoparticles had a short axis of about 40 nm and a long axis of about 100 nm. They show surface plasmon resonances at 400 nm for the short axis and at 660 nm for the long axis.

The frequency-doubled pump pulse (400 nm) of an ultrafast laser system with a pulse duration of 150 fs was focussed on the nanoparticle array on a glass substrate. The probe pulse was a spectral part of an ultrashort white-light continuum pulse, yielding the change in transmission. Figure 3.6 shows the transmission transient for the long axis probed with a wavelength of 700 nm and the short axis probed with a wavelength of 550 nm.

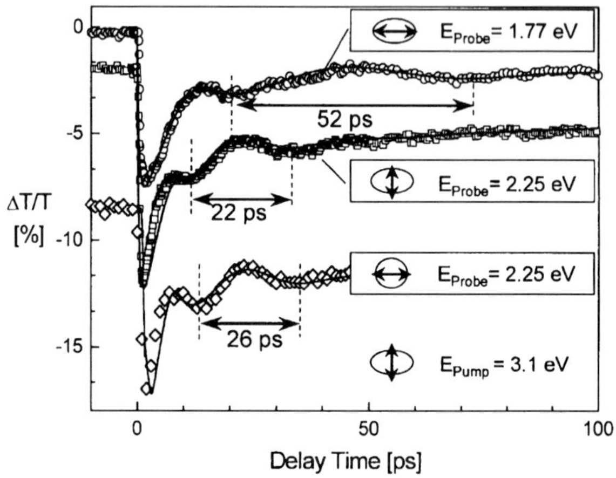

Fig. 3.6 Change of optical transmittivity measured at the given wavelength for the long axis (*top*), short axis (*middle*) of spheroid silver nanoparticles and of spherical silver nanoparticles (*bottom*). The results were obtained by ultrafast pump-probe spectroscopy [19]

Remarkably, the transient change in transmissivity shows distinct oscillations after an initial phase of relaxation. According to an elastic continuum model, the time dependence of the particle elongation Δx is given by a damped harmonic oscillator equation

$$\Delta\ddot{x} + 2\rho\Delta\dot{x} + \omega_0^2\Delta x = A\frac{\sigma}{m} \tag{3.7}$$

where ρ is the damping coefficient due to the substrate, ω_0 is the vibrational eigenfrequency, A is the surface area of the particle, σ is the stress, and m is the mass of the nanoparticle. Due to the linearity of the harmonic oscillator equation, the stress σ can be decomposed into stress from the lattice σ_1 and from the electrons σ_e and treated separately. The stress can be calculated according to

$$\sigma_{l,e} = -\gamma_{l,e}\int_{T_0}^{T_{l,e}} dTC_{l,e}(T) \tag{3.8}$$

where T_0 is the ambient temperature, $T_{l,e}$ are the temperature of lattice and electrons, $C_{l,e}$ are the heat capacity of lattice and electrons and $\gamma_{l,e}$ are the Grüneisen constants ($\gamma_l \cong 2.4$, $\gamma_e \cong 1.2$). The electronic and lattice temperature can be calculated according to the TTM (3.1 and 3.2). Figure 3.7 shows the calculated electronic and lattice temperature (top), the resulting elongation (middle) and the change in transmittivity (bottom). Apparently, the pressure of the hot electron gas, which reaches a peak temperature of more than 3000 K, has a significant influence on the experimentally observed change in transmittivity.

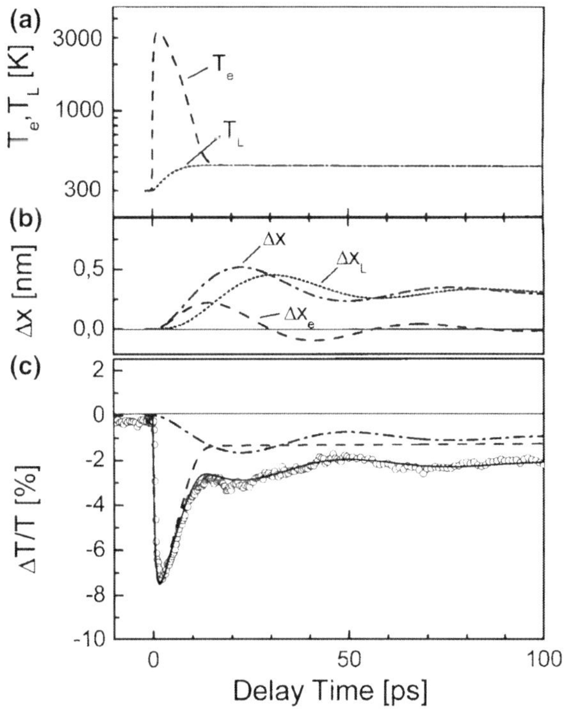

Fig. 3.7 Transient electronic (T_e) and lattice (T_l) temperature calculated for spheroidal silver nanoparticles upon ultrafast excitation based on the TTM (**a**). Resulting oscillations of the elongation according to (3.7) (**b**). Change of transmittivity resulting from elongation (**c**) [19]

3.2.4 Hot Electron Emission

Ultrafast excitation of metals can cause emission of hot electrons into the surrounding environment, either vacuum, gas or condensed phase (Fig. 3.8). These hot electrons exhibit the potential to initiate various physical and chemical processes depending on the environment they are emitted to. This process was investigated, e.g., for gadolinium films in vacuum. A pump-probe setup with ultrashort *s*-polarized pump laser pulses (800 nm, 55 fs) and *p*-polarized probe pulses (200 nm, 90 fs) was employed [20]. Figure 3.9 shows the time-of-flight photoelectron spectra for various pump-probe delays.

Highly efficient super-linear electron emission from metal electrodes into aqueous electrolyte solutions induced by visible sub-picosecond laser pulses (300 fs) with a photon energy ($h\nu \approx 2\,\text{eV}; 615\,\text{nm}$) less than the threshold of the linear photo-effect has been reported [21–24]. The emitted charge density was monitored by the charging of the electrochemical Helmholtz layer. It increased, e.g., on mercury from 10^{-9} to 10^{-5} C cm^{-2} in the intensity range $I = 4$–30 GW cm^{-2} (Fig. 3.10). At silver, 10^{-9}–10^{-7} C cm^{-2} were generated in the range of $I = 7$–70 GW cm^{-2}. The charge density grew practically linearly with the concentration of electron acceptors such as H_3O^+ and $CHCl_3$ at $I < 20$ GW cm^{-2}, because the capture of emitted electrons prevented them from

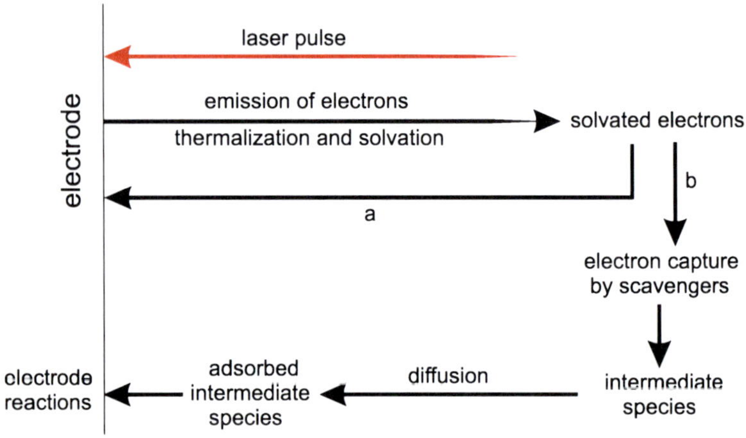

Fig. 3.8 Sequence of processes initiated by photo-thermoemission of electrons from metal into an electrolyte solution

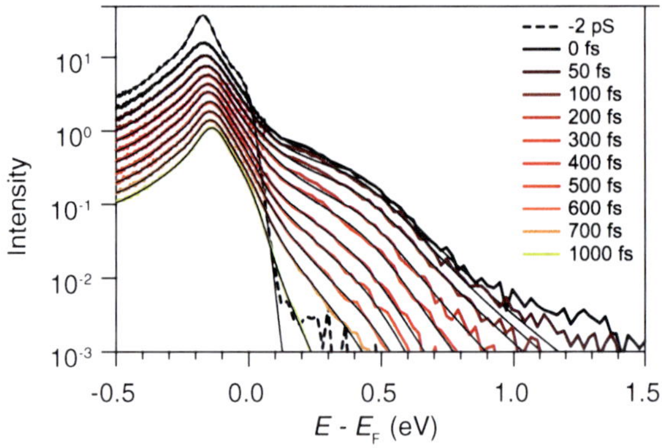

Fig. 3.9 Time-of-flight photoelectron spectra of gadolinium recorded for different pump-probe delay values [20]

returning to the electrode surface. One can assume an approximate equality of the hot electron current pulse duration with the electron–phonon relaxation time of the respective metal, 4 ps for Hg and 14 ps for Ag. With this and the measured emission charge densities one can derive maximum current density pulses of more than 10^6 A cm^{-2}. These results conform with theoretical predictions of charge-laser intensity, pulse duration, and electrode potential characteristics dependencies for thermoemission of electrons from a non-equilibrium electron gas from metals into electrolyte solutions [24]:

$$j = B\exp\left[-\frac{E_a - \alpha\left(E^0_{H_2/H^+} - E\right)}{k_B T}\right] \tag{3.9}$$

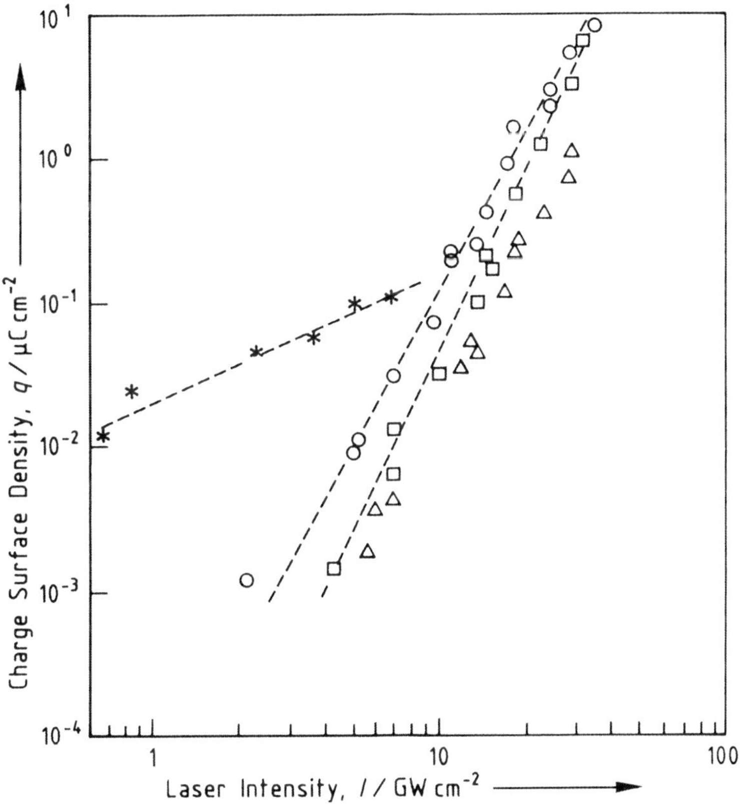

Fig. 3.10 Electron charge density emitted from mercury into aqueous 0.2 M K_2SO_4 by 300 fs laser pulses with intensity *I*. *Asterisk* ~4 eV, 308 nm. *Triangle, square, circle* ~2 eV, 615 nm. *Triangle* −0.7 V_{SCE}. *Square* −1.2 V_{SCE}. *Circle* −1.7 V_{SCE}. Electrode potentials versus the saturated calomel electrode (*SCE*) [21]

According to this model, the hot electron emission current density *j* exhibits a Tafel-like semi-logarithmic dependence on the electrode potential *E*.

3.3 Response of Dielectrics and Semiconductors

3.3.1 Impact/Avalanche and Multiphoton Ionization

Dielectrics and wide-bandgap semiconductors are transparent in and around the visible wavelength range. They become absorbing when high laser intensities are employed so that non-linear optical coupling occurs. This can be realized when ultrashort laser pulses are used.

The generation rate of the free electron density N_e due to photon absorption can be described by [25, 26]:

$$dN_e(t)/dt = \zeta N_e(t)I(t) + \sum_n \Xi(^n\sigma)I^n(t) - k_{ee}N_e^2(t) - k_{ep}N_e(t) \qquad (3.10)$$

The positive signs on the right hand side denote the generation terms and the negative the loss terms. The first term represents the impact and avalanche ionization with the avalanche coefficient ζ. It depends not only on the intensity I but also on the instantaneous free electron density $N_e(t)$. In contrast, the second term of (3.10) describes the n-photon absorption which is entirely independent of $N_e(t)$. Ξ is proportional to the n-photon absorption cross section $^n\sigma$. Only the lowest n-photon excitation bridging the band gap E_g, i.e. the lowest n satisfying $nhc/\lambda > E_g$, is practically significant. In the loss terms, k_{ee} is the coefficient of inelastic electron–electron scattering, and k_{ep} the coefficient of electron–phonon scattering.

The initial generation of free electrons can occur via thermal excitation, primarily from defect states within the gap [3], by resonant optical excitation from defect states, or by strong-field ionization [2, 27, 28]. When the laser electric field strength becomes sufficiently strong, it can perturb the Coulomb potential that is binding the electron to the nucleus. In this case the electron can tunnel through the suppressed barrier and ionization can occur [29]. When the field becomes even stronger the electron can directly escape into the continuum by an over-the-barrier ionization.

Impact ionization represented by the first term on the right hand side of (3.10) is a de-excitation mechanism for electrons which is practically the reverse Auger process (Fig. 3.2c) [1, 27, 28, 30]. The excited electron releases its excess energy by the generation of another free carrier in the conduction band. In the presence of a strong electric field of an ultrashort laser pulse, the free electrons are subject to a strong acceleration, collide with bound electrons in the valence band, and generate free electron–hole pairs. If the electric field is high enough, a chain (avalanche) reaction sets in and more and more free electrons are formed.

The second term in (3.10) accounts for a competitive generation mechanism, multiphoton (or n-photon) absorption (Fig. 3.2b). Two-photon absorption e.g. is a third-order process several orders of magnitude weaker than linear absorption. The strength of n-photon absorption depends on the nth power of the light intensity, thus it is a nonlinear optical process. Multiphoton ionization is, in contrast to impact and avalanche ionization, a deterministic process [4].

This contrasting behaviour could be demonstrated by the determination of ablation precision. The ablation depth per pulse (ablation rate) is of the order of the penetration depth of light, which, in turn, scales with the density of electrons in the conduction band. These electrons are either created by multiphoton or by impact ionization. Impact ionization implies an electron density that is dependent on the fluence and not on intensity. The ablation rate of barium borosilicate glass exposed to pulses with durations below 100 fs showed a strong dependence on the pulse duration (Fig. 3.11) indicating multiphoton absorption [4, 31]. A lateral extension

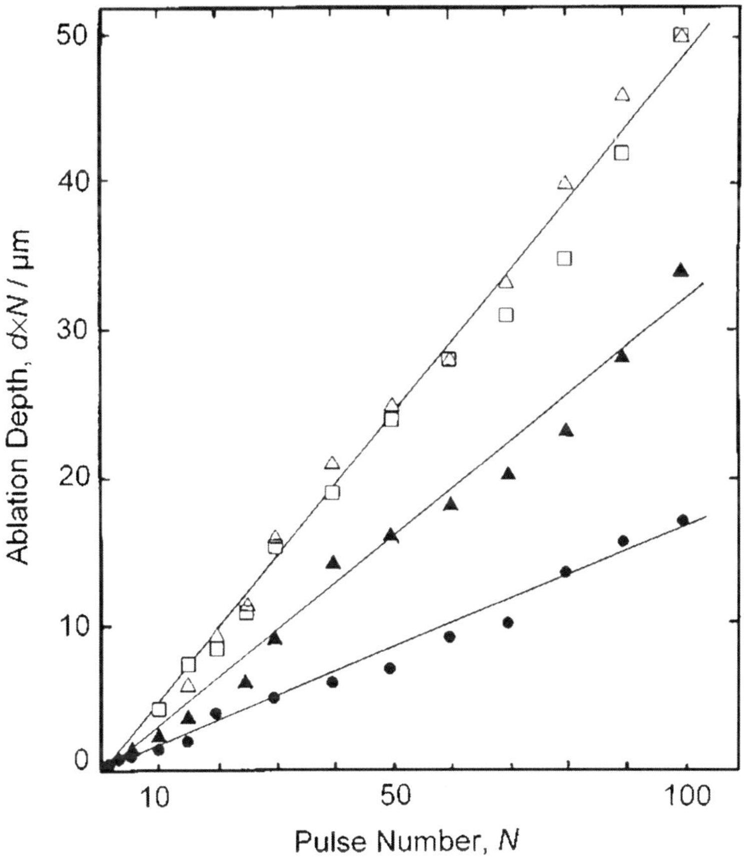

Fig. 3.11 Pulse duration dependence of the ablation depth per pulse of barium borosilicate glass illuminated with ultrashort pulses (780 nm, 4 J cm^{-2}). *Filled circle* 20 fs, *filled triangle* 50 fs, *triangle* 120 fs, *circle* 300 fs [4]

parameter q of the ablation craters served as a measure for the deterministic nature of the multiphoton coupling process [7]:

$$q = \frac{1}{2}\ln\left[\frac{F_0}{F_{th}}\right] \tag{3.11}$$

It is the ratio between the observed cross-section area of the ablated crater and the illuminated area of a Gaussian beam. The solid lines in Fig. 3.12 depict the predictions of (3.11). For pulse durations >100 fs the observed $q(F)$ dependence strongly deviates from the theoretically predicted straight line indicating a stochastic process. q-results obtained at shorter pulse durations show agreement with the prediction of (3.10) suggesting a deterministic process due to multiphoton absorption [5, 7].

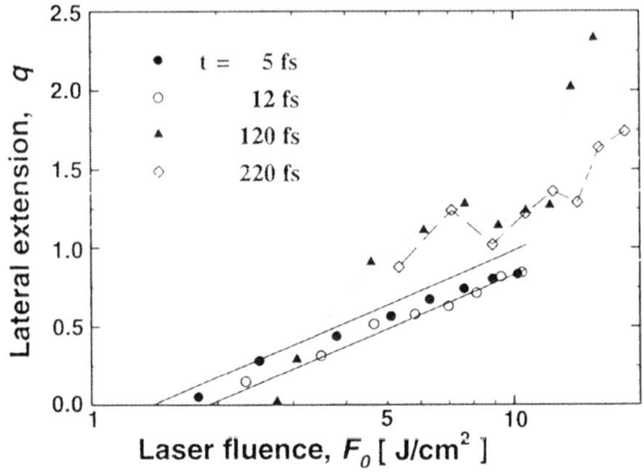

Fig. 3.12 Lateral extension parameter versus laser fluence for the ablation of fused silica at air (780 nm, $N = 50$), *filled circle* 5 fs, *circle* 12 fs, *filled triangle* 120 fs, *diamond* 220 fs. The straight lines are calculated from (3.11) with the values $F_{\text{th}}(5\ \text{fs}) = 1.4\ \text{J cm}^{-2}$ and $F_{\text{th}}(12\ \text{fs}) = 1.9\ \text{J cm}^{-2}$ [7]

3.3.2 Non-Thermal Melting

Thermal melting results in the phase transition of a substance from a solid to a liquid. In this classic melting process, lattice and electrons are in thermal equilibrium. Upon ultrafast illumination, strongly excited electrons can be created. This strongly influences the properties of the material, which is electronically softened until electron–phonon relaxation sets in. The material can rearrange and change its structure.

X-ray diffraction is a convenient technique to measure the crystallinity of a surface layer. The application of ultrashort X-ray pulses allows to probe the transient change in crystallinity upon ultrafast laser irradiation [32–34]. An ultrashort X-ray source was realized by the interaction of an ultrafast laser pulse with a silicon target. This generated the K_α-line of silicon (0.713 nm) with a duration similar to that of the exciting ultrafast laser pulse. The laser and the X-ray pulse were focussed with a variable delay on the same spot of an indium antimonide crystal. The first-order Bragg reflection in the (111) orientation was detected (Fig. 3.13). Apparently, the crystalline structure starts to decompose immediately after laser excitation with a time constant of the order of 1 ps, which is less than electron–phonon relaxation. This is a clear indication of non-thermal melting.

Non-thermal melting processes are also supported by a simulation of ablation by sub-picosecond laser pulses [35]. Rapid excitation of electrons can cause massive instability in the crystal lattice within a few 10 fs due to a perturbation of the interatomic bonds. Below ablation and above modification threshold fluences, this instability can relax on a several 100 fs time scale by surface reorganization in laser-induced periodic surface structures [36–38]. Molecular dynamics (MD) simulations on the basis of an electronic tight-binding Hamiltonian in real-space took into account all atomic degrees of freedom. Non-equilibrium occupation numbers for the energy levels of the system led to lattice dynamics on time-dependent potential energy surfaces. Thus, a theoretical framework was provided

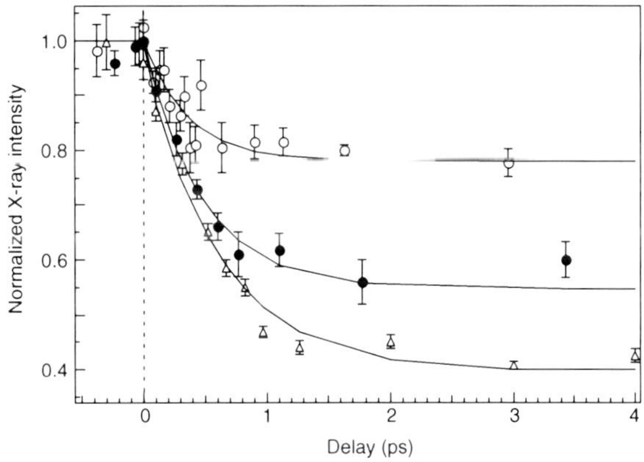

Fig. 3.13 Transient X-ray diffraction intensity for different depths: *circle* 120 nm, *filled circle* 170 nm, *circle* 350 nm [32]

for the treatment of strong non-equilibrium situations in materials where atomic and electronic degrees of freedom play an equally important role.

3.3.2.1 Formation of Laser-Induced Periodic Surface Structures

The formation of self-organized, laser-induced periodic surface structures (LIPSS) on various semiconductors was reported soon after the construction of the first operative laser [39]. The periodicity of LIPSS was approximately equal to the wavelength of the incident light. Such LIPSS can be called low spatial frequency LIPSS or LSFL. Investigations with different types of lasers (different gain media/ different wavelength and different operation modes from cw to ns pulses) of such structures followed on a variety of different material classes (semiconductors, metals, glasses, polymers) [40–49]. The LSFL were explained by the interference of the incident beam with surface-scattered waves originating from surface inhomogeneities [48, 50]. Scattering of the incident beam may be supported by microscopic rough surface features, by defects, spatial variations in the refractive index, etc. The interference between the various waves leads to an inhomogeneous energy input, which, together with a positive feedback mechanism, may generate surface instabilities [3]. For normal incidence irradiation, ripples with a periodicity Λ are formed for p-polarized light according to: $\Lambda = \lambda/n$.

An alternative theory describing the mechanisms of LSFL formation was developed [41–43] which criticises the above interference model. It is based on the electrodynamics of randomly rough surfaces, introducing a variation principle concerned with longitudinal fields responsible for local field corrections. When absorption of energy within the material occurs, it was found not clear why the surface-scattered wave should propagate with a wavelength of λ as opposed to λ/n (n is the refractive index). Further, a normal-incidence excitation would lead to a longitudinally polarized scattered wave, which does not satisfy the Maxwell equations. Actually, plane waves of any polarization, propagating parallel to an interface, do not satisfy the Maxwell equations. It is worth mentioning that this

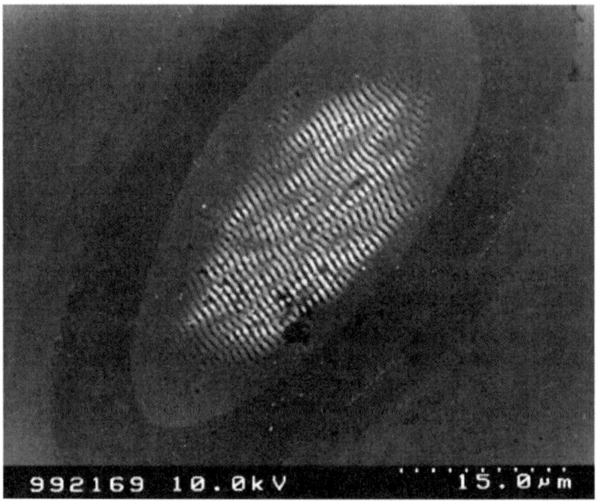

Fig. 3.14 Silicon surface irradiated with ultrashort laser pulses exhibiting laser-induced periodic surface structures (5 fs, 780 nm, 2.5 J cm^{-2}, $N = 5$) [36]

work is regularly cited erroneously in order to support LSFL formation by the interference model.

Besides LSFL formation, ultrafast lasers also trigger the formation of high spatial frequency LIPSS (HSFL), which have a periodicity far below the wavelength of the incident light. Along a Gaussian beam profile, intensity-dependent periodicities can be observed, i.e. sub-wavelength HSFL ($\Lambda \ll \lambda$) at the low intensity irradiation edges and the coarser LSFL ($\Lambda \approx \lambda$) at the high intensity irradiation centre with an abrupt transition in periodicities [51–53]. Figure 3.14 shows a scanning electron micrograph of a modified silicon surface that has been irradiated with 5 ultrafast pulses (5 fs, 780 nm, 2.5 J cm^{-2}). The sub-wavelength structuring of the material is clearly visible.

Systematic studies of the influence of various parameters influencing the formation of HSFL have been conducted [2, 37, 53–62]. HSFL are only observed on laser-treated surfaces where the pulse duration is less than the electron–phonon relaxation time [63]. In case of linearly polarized light, the direction of the HSFL follows the direction of polarization and is either normal or perpendicular, depending on the type of material. They can be either normal or perpendicular to LSFL [2]. Circularly polarized light produces nanodots rather than HSFL, elliptically polarized light induces nanorod-like structures [55, 61, 62]. The periodicity of HSFL increases with the number of laser pulses applied to a surface given that the fluence per pulse is kept constant [59, 60].

Although extensive studies on HSFL formation have been performed, no conclusive and general theory was formulated that describes their formation. Most studies solely concentrate on the influence of excitation parameters (such as polarization and double-pulse properties) on LIPSS formation.

The HSFL structures ($\Lambda \ll \lambda$) are still controversially discussed in literature, such as by interference effects along with transient changes in the optical properties during laser irradiation [64], second harmonic generation at compound semiconductors [65, 66], generation of surface excited waves [67, 68], transient nanoplasmonic model involving near-field enhancements generating plasma gratings [68], or self-organisation [37, 38, 53].

Investigations with high-performance ceramics like silicon carbide, aluminium nitride, and a composite compound SiC–TiC–TiB_2, revealed that a direct correlation between chemical composition and HSFL character exists [52, 69]. Moreover, femtosecond laser interaction with the silicon-water interface showed that also an adjacent condensed phase may have a physico-chemical influence on these phenomena [37, 70]. Sequential ultrafast irradiation of silicon surfaces immersed in water can lead to high-density regular arrays of 50 nm-scale rods [71]. It was proposed that the surface energy relaxation would be the driving force for self organisation on a non-thermally melted surface film [38]. This picture is becoming more detailed by the scanning probe observation of surface potential changes on Si of about 100 mV localised in the nm-scale. This suggests a mechanism related to a segregation-like accumulation of dopant concentration during the self-organised nanostructure formation and a crystal phase transformation of the silicon upon ablation, completely changing the material properties [72, 73]. Nanoscale ablation is obviously preferentially induced on the crest of the ripples where the local field intensity would be enhanced by the high surface curvature along the laser electric field direction, implying that localized nanofields (nanoscale potential changes) play a key role in nanostructuring [67, 68]. The bottom–up nanostructuring by ultrafast lasers may open up a wide field of applications in surface technology, such as the preparation of super water repellent surfaces [74].

3.3.3 Coulomb Explosion

Dielectrics and (wide-bandgap) semiconductors can undergo a transition to near-surface metal-like properties upon ultrafast illumination with high peak intensities. The coupling of the light to the solid-state plasma layer can generate strong photoemission of electrons [3, 75–77]. This charge separation results in a high surface electric field of several 10^8 V cm^{-1} due to the remnant excess of positive charge in the surface layer of the solid. If the electrostatic repulsion is greater than the binding energy of the lattice, the emission of a thin layer of surface ions occurs. This process is called Coulomb explosion. The electron sea of metals, on the other hand, can compensate the loss of emitted electrons and thus neutralize the surface electric field inhibiting Coulomb explosion.

The transient time-of-flight mass spectrometry yield for thermal and fast Al^+ ions emitted from sapphire is shown in Fig. 3.15. Fast electrons are emitted immediately after excitation, whereas thermal ions exhibit their peak emittance approximately 10 ps after the excitation. This time span between excitation and peak emission of thermal ions is governed by the electron–phonon relaxation time. The fact that the emission of fast electrons sets in immediately after excitation is a clear indication for a non-thermal process. Similar results where obtained for silicon dioxide. Metals and narrow-bandgap semiconductors show ion emission only after the electron–

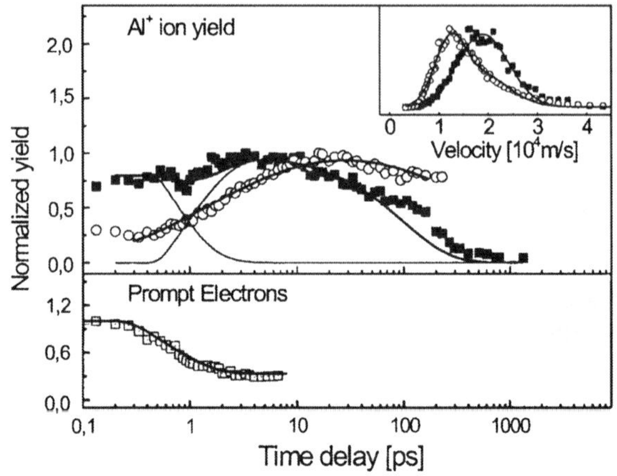

Fig. 3.15 Transient time-of-flight mass spectrometry Al^+ ion yield from sapphire target. *Circle* "thermal" ions, *filled square* Coulomb explosion [77]

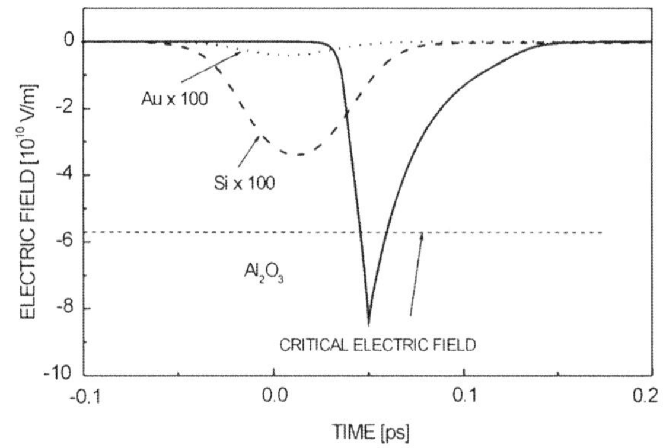

Fig. 3.16 The transient surface electric field of sapphire, silicon and gold upon ultrafast laser excitation (100 fs, 800 nm). The fluences where chosen to be above the ion emission threshold for the respective material (4 J cm^{-2} for sapphire, 0.8 J cm^{-2} for silicon and 1.2 J cm^{-2} for gold) [78]

phonon relaxation time suggesting a thermal mechanism of ablation. According to a universal model of charge dynamics the surface electric fields of sapphire, silicon and gold upon ultrafast excitations with fluences just above the ion emission thresholds could be calculated (Fig. 3.16) [78]. In this simulation, the surface electric field of silicon and gold is much less than the threshold of Coulomb explosion. It however remains not completely resolved whether Coulomb explosion is possible in metals and narrow-bandgap semiconductors.

3.3.4 Photochemical Ablation

The threshold of sub-picosecond ablation of transparent dielectrics coincides with the onset of avalanche or multiphoton ionization (see above). In contrast, there still exists some ambiguity for (bio)polymeric dielectrics [31, 79–86]. Ablation due to heating can occur at fluences less than that needed for avalanche or multiphoton

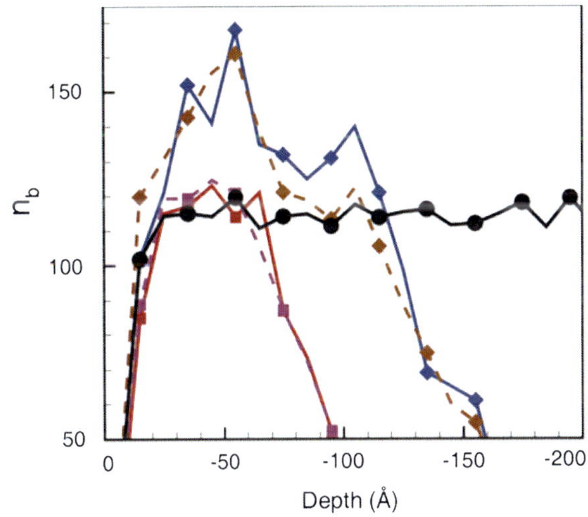

Fig. 3.17 Number of broken bonds per 1 nm thick layer as a function of depth. *Solid lines* photochemical. *Dashed lines* photothermal. *Filled square* 15 mJ cm^{-2}. *Filled diamond* 20 mJ cm^{-2}. *Filled circle* 10 % of bonds present in the 1 nm thick layer. This coincides with the number of bonds per layer that need to be broken for ablation [89]

ionization. Two-pulse ablation experiments on sub-picosecond ablation of poly-tetrafluoroethylene (PTFE), poly-methylmethacrylate (PMMA) and polyimide (PI) by 500 fs laser pulses at 248 nm resulted in a thermal relaxation time of about 30–50 ps [87]. It was mentioned, however, that within time interval of 200 ps after a single pulse there was no evidence of material removal from the surface. It should be noted, that additional experiments are necessary to identify photo-chemical mechanism [88].

The photochemical ablation behaviour of PMMA has been studied by combined molecular dynamics/Monte Carlo simulations [89]. There, photolysis of C–CO and C–CH_2 bonds were assumed. Depending on the bond broken, different small molecules are formed and ejected from the surface. Figure 3.17 shows the number of broken bonds per 10 Å thick surface layer as a function of the distance from the surface into the bulk. The simulation was carried out for two different laser fluences. The squares connected with solid lines correspond to the results obtained by the combined molecular dynamics/Monte Carlo simulation for a laser fluence of 15 mJ cm^{-2}. The squares that are interconnected with a dashed line are calculations for the same laser parameters, assuming a thermal model. The diamonds describe the same situation as the squares however with an increased fluence of 20 mJ cm^{-2}. The line marked with circles represents a fraction of 10 % of the bonds present in the material and it is assumed that ablation sets in as soon as 10 % of the bonds are broken. This line thus marks the threshold for ablation [89].

3.3.5 3D Stereolithography

Stereolithography is a rapid prototyping approach where a three-dimensional work piece is built up layer by layer through the materialization of raster-points. A contact-free implementation of stereolithography is scanning a laser beam (usually

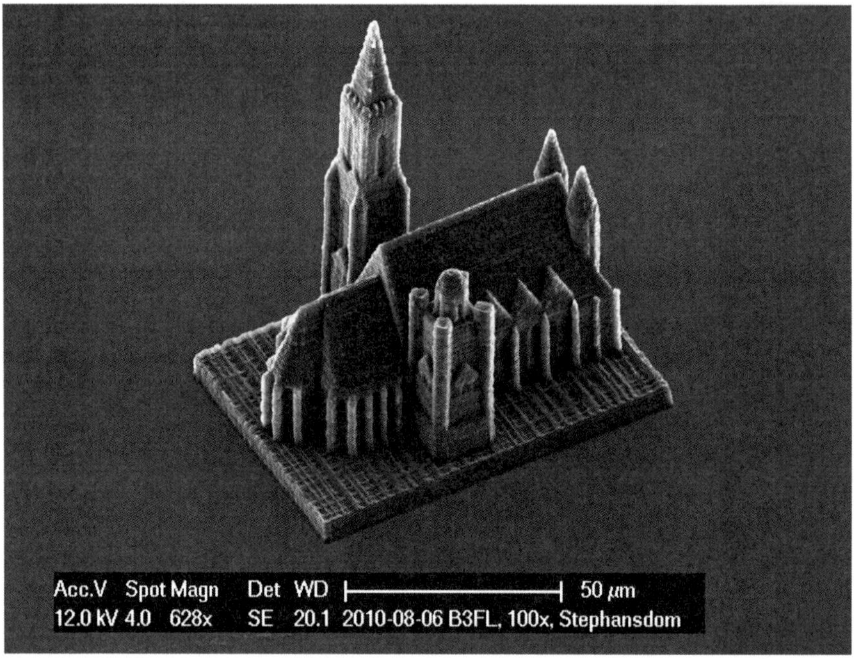

Fig. 3.18 Scanning electron micrograph of a micro-scale model of St. Stephen's Cathedral (Vienna, Austria) fabricated by two-photon-polymerization [107]

visible or ultraviolet radiation) over a resin, thus building up the structure layer by layer. The resin cross-links in a single-photon process and converts into the solid parts of the structure.

Multi-photon-polymerization allows to build three-dimensional structures by focussing an ultrafast laser beam into the bulk of a transparent (at the laser wavelength) photosensitive material [90–95]. Cross-linking takes place by multi-photon absorption and thus only in the focal volume of the laser. Different resins can be employed for multi-photon-polymerization. Normally they are viscous liquids like acrylates and epoxies. The minimum feature size that is achievable is well below 100 nm [96, 97]. The technique has been demonstrated by the fabrication of microscopic models (see Fig. 3.18) and optical components [98–106].

3.4 Conclusion

Non-thermal material response to laser energy deposition takes place in time periods less than 1–10 ps. It follows the absorption of laser radiation which occurs in the attosecond time domain. The absorption mechanism depends on the class of material irradiated as well as the intensity of irradiation. Ultrafast lasers allow nonlinear coupling of near-infrared radiation into wide-bandgap semiconductors and dielectrics by two competing processes, multi-photon excitation and impact ionization with subsequent avalanche ionization. The ultrafast excitation of metals creates hot electrons in a cold lattice, which can be described by the two-temperature model. After these processes, electron–phonon-scattering sets in, i.e., electrons and lattice start equilibrating, and thermal phenomena take over.

The non-equilibrium dynamics of hot electrons in metals is determined by the competition between ballistic and hot-electron diffusion driven by the electron temperature gradient and electron–phonon coupling. Transition metals do not show ballistic transport of electrons. Therefore, structuring of these with ultrafast lasers is more precise than of noble metals.

Ultrafast excitation can cause emission of hot electrons into the surrounding environment, either vacuum, gas or condensed phase. These hot electrons exhibit the potential to initiate various physical and chemical processes depending on the environment they are emitted to. Highly efficient thermo-electron emission from metal electrodes into aqueous electrolyte solutions induced by sub-picosecond laser pulses with a photon energy less than the threshold of the linear photo-effect can be monitored by the charging of the electrochemical Helmholtz layer. One can derive maximum current density pulses of more than a million A cm^{-2}.

Dielectrics can undergo a transition to near-surface metal-like properties upon ultrafast illumination with high peak intensities with subsequent photoemission of electrons. This charge separation results in a high surface electric field. The remnant excess of positive charge in the surface layer of the solid causes electrostatic repulsion and emission of a thin layer of surface ions (Coulomb explosion).

Rapid excitation of electrons can cause massive instability in the crystal lattice due to a perturbation of the interatomic bonds due to time-dependent potential energy surfaces. This instability can relax by surface reorganization in laser-induced periodic surface structures (LIPSS). When the periodicity is approximately equal to the wavelength of the incident light they are called low spatial frequency LIPSS or LSFL. They were explained by the interference of the incident beam with surface-scattered waves originating from surface inhomogeneities or by the electrodynamics of randomly rough surfaces, introducing a variation principle concerned with longitudinal fields responsible for local field corrections. Also so-called high spatial frequency LIPSS (HSFL) occur with a periodicity far below the wavelength of the incident light. Their origin is still under vivid discussion.

Ablation of polymers which represent transparent dielectrics can occur also by heating at fluences less than that needed for avalanche or multiphoton ionization.

Multi-photon-polymerization allows building three-dimensional structures by focussing an ultrafast laser beam into the bulk of a transparent photosensitive material which cross-links by multi-photon-polymerization.

References

1. S.K. Sundaram, E. Mazur, Nat. Mater. **1**, 217 (2002)
2. W. Kautek, M. Forster, Springer Ser. Mater. Sci. **135**, 189 (2010)
3. D. Bäuerle, *Laser Processing and Chemistry*, 4th edn. (Springer, Berlin Heidelberg, 2011)
4. W. Kautek, J. Krüger, M. Lenzner, S. Sartania, C. Spielmann, F. Krausz, Appl. Phys. Lett. **69**, 3146 (1996)
5. M. Lenzner, J. Krüger, S. Sartania, Z. Cheng, C. Spielmann, G. Mourou, W. Kautek, F. Krausz, Phys. Rev. Lett. **80**, 4076 (1998)
6. M. Lenzner, J. Krüger, W. Kautek, F. Krausz, Appl. Phys. A **69**, 465 (1999)
7. M. Lenzner, F. Krausz, J. Krüger, W. Kautek, Appl. Surf. Sci. **154**, 11 (2000)
8. S.I. Anisimov, B.L. Kapeliovich, T.L. Perelman, Sov. Phys., JETP **39**, 375 (1974)
9. S.-S. Wellershoff, J. Hohlfeld, J. Güdde, E. Matthias, Appl. Phys. A **69**, 99 (1999)
10. R.H.M. Groeneveld, R. Sprik, A. Lagendijk, Phys. Rev. B **51**, 11433 (1995)
11. B. Rethfeld, A. Kaiser, M. Vicanek, G. Simon, Phys. Rev. B **65**, 214303 (2002)
12. E. Carpene, Phys. Rev. B **74**, 027301 (2006)
13. J. Hohlfeld, S.-S. Wellershoff, J. Güdde, U. Conrad, V. Jähnke, E. Matthias, Chem. Phys. **251**, 237 (2000)
14. P.E. Hopkins, P.M. Norris, Appl. Surf. Sci. **253**, 6289 (2007)
15. Z. Lin, L.V. Zhigilei, V. Celli, Phys. Rev. B **77**, 075133 (2008)
16. M. Bonn, D.N. Denzler, S. Funk, M. Wolf, S.-S. Wellershoff, J. Hohlfeld, Phys. Rev. B **61**, 1101 (2000)
17. J. Krüger, D. Dufft, R. Koter, A. Hertwig, Appl. Surf. Sci. **253**, 7815 (2007)
18. M. Forster, L. Egerhazi, C. Haselberger, C. Huber, W. Kautek, Appl. Phys. A **102**, 27 (2011)
19. M. Perner, S. Gresillon, J. Marz, G.V. Plessen, J. Feldmann, J. Porstendorfer, K.-J. Berg, G. Berg, Phys. Rev. Lett. **85**, 792 (2000)
20. M. Lisowski, P. Loukakos, A. Melnikov, I. Radu, L. Ungureanu, M. Wolf, U. Bovensiepen, Phys. Rev. Lett. **95**, 137402 (2005)
21. A.G. Krivenko, J. Kruger, W. Kautek, V.A. Benderskii, Berichte der Bunsengesellschaft für physikalische Chemie **99**, 1489 (1995)
22. V.A. Benderskii, A.V. Benderskii, *Laser Electrochemistry of Intermediates* (CRC Press Inc., Boca Raton, 1995)
23. A.G. Krivenko, W. Kautek, J. Krüger, V.A. Benderskii, Russ. J. Electrochem. **33**, 394 (1997)
24. A.G. Krivenko, V.A. Benderskii, J. Krüger, W. Kautek, Russ. J. Electrochem. **34**, 1068 (1998)
25. M. Li, S. Menon, J.P. Nibarger, G.N. Gibson, Phys. Rev. Lett. **82**, 2394 (1999)
26. R. Stoian, Doctoral Dissertation
27. L.V. Keldysh, Sov. Phys. JETP **20**, 1307 (1965)
28. P.B. Corkum, Phys. Rev. Lett. **71**, 1994 (1993)
29. S. Augst, D. Strickland, D.D. Meyerhofer, S.L. Chin, J.H. Eberly, Phys. Rev. Lett. **63**, 2212 (1989)
30. L.V. Keldysh, Sov. Phys. JETP **7**, 788 (1958)
31. J. Krüger, W. Kautek, Adv. Polym. Sci. **168**, 247 (2004)
32. A. Rousse, C. Rischel, S. Fourmaux, I. Uschmann, S. Sebban, G. Grillon, P. Balcou, E. Förster, J.P. Geindre, P. Audebert, J.C. Gauthier, D. Hulin, Nature **410**, 65 (2001)
33. K. Sokolowski-Tinten, C. Blome, C. Dietrich, A. Tarasevitch, M. Horn Von Hoegen, D. Von Der Linde, A. Cavalleri, J. Squier, M. Kammler, Phys. Rev. Lett. **87**, 225701 (2001)
34. A.M. Lindenberg, Science **308**, 392 (2005)
35. H.O. Jeschke, M.E. Garcia, M. Lenzner, J. Bonse, J. Krüger, W. Kautek, Appl. Surf. Sci. **197–198**, 839 (2002)
36. J. Bonse, S. Baudach, J. Krüger, W. Kautek, M. Lenzner, Appl. Phys. A **74**, 19 (2002)

37. W. Kautek, P. Rudolph, G. Daminelli, J. Krüger, Appl. Phys. A **81**, 65 (2005)
38. W. Kautek, Springer Ser. Opt. Sci. **129**, 215 (2007)
39. M. Birnbaum, J. Appl. Phys. **36**, 3688 (1965)
40. M. Csete, Z. Bor, Appl. Surf. Sci. **133**, 5 (1998)
41. J.F. Young, J.E. Sipe, H.M. van Driel, Phys. Rev. B **30**, 2001 (1984)
42. J.E. Sipe, J.F. Young, J.S. Preston, H.M. van Driel, Phys. Rev. B **27**, 1141 (1983)
43. J.F. Young, J.S. Preston, H.M. van Driel, J.E. Sipe, Phys. Rev. B **27**, 1155 (1983)
44. P.M. Fauchet, A.E. Siegman, Appl. Phys. A **32**, 135 (1983)
45. H.M. van Driel, J.E. Sipe, J.F. Young, Phys. Rev. Lett. **49**, 1955 (1982)
46. J.F. Young, J.E. Sipe, J.S. Preston, H.M. van Driel, Appl. Phys. Lett. **41**, 261 (1982)
47. N.R. Isenor, Appl. Phys. Lett. **31**, 148 (1977)
48. D.C. Emmony, R.P. Howson, L.J. Willis, Appl. Phys. Lett. **23**, 598 (1973)
49. M. Siegrist, G. Kaech, F.K. Kneubühl, Appl. Phys. **2**, 45 (1973)
50. M. Oron, G. Sorensen, Appl. Phys. Lett. **35**, 782 (1979)
51. C.R. Phipps, B.S. Lukyan'chuk, *Laser Ablation and its Applications* (Springer, New York, 2007)
52. P. Rudolph, W. Kautek, Thin Solid Films **453–454**, 537 (2004)
53. J. Reif, F. Costache, Adv. Atom. Mol. Opt. Phys. **53**, 227 (2006)
54. F. Costache, M. Henyk, J. Reif, Appl. Surf. Sci. **186**, 352 (2002)
55. O. Varlamova, F. Costache, J. Reif, M. Bestehorn, Appl. Surf. Sci. **252**, 4702 (2006)
56. J. Bonse, A. Rosenfeld, C. Grebing, G. Steinmeyer, N.A. Mailman, G.A. Botton, H.K. Haugen, J. Appl. Phys. **106**, 074907 (2009)
57. J. Bonse, A. Rosenfeld, J. Krüger, J. Appl. Phys. **106**, 104910 (2009)
58. J. Bonse, J.M. Wrobel, K.-W. Brzezinka, N. Esser, W. Kautek, Appl. Surf. Sci. **202**, 272 (2002)
59. J. Reif, O. Varlamova, M. Bounhalli, T. Arguirov, AIP Conf. Proc. **1278**, 446 (2010)
60. J. Reif, O. Varlamova, M. Ratzke, M. Schade, H.S. Leipner, T. Arguirov, Appl. Phys. A **101**, 361 (2010)
61. O. Varlamova, J. Reif, S. Varlamov, M. Bestehorn, Appl. Surf. Sci. **257**, 5465 (2011)
62. J. Reif, O. Varlamova, S. Varlamov, M. Bestehorn, Appl. Phys. A **104**, 969 (2011)
63. M. Forster, N. Faure, E. Audouard, R. Stoian, W. Kautek, in publication (2013)
64. Q. Wu, Y. Ma, R. Fang, Y. Liao, Q. Yu, X. Chen, K. Wang, Appl. Phys. Lett. **82**, 1703 (2003)
65. A. Borowiec, H.K. Haugen, Appl. Phys. Lett. **82**, 4462 (2003)
66. X. Jia, T.Q. Jia, Y. Zhang, P.X. Xiong, D.H. Feng, Z.R. Sun, J.R. Qiu, Z.Z. Xu, Opt. Lett. **35**, 1248 (2010)
67. G. Miyaji, W. Kobayashi, K. Miyazaki, Electrochim. Acta **53**, 167 (2007)
68. G. Miyaji, K. Miyazaki, Appl. Phys. Lett. **91**, 123102 (2007)
69. P. Rudolph, K.-W. Brzezinka, R. Wäsche, W. Kautek, Appl. Surf. Sci. **208–209**, 285 (2003)
70. G. Daminelli, J. Krüger, W. Kautek, Thin Solid Films **467**, 334 (2004)
71. M. Shen, J.E. Carey, C.H. Crouch, M. Kandyla, H.A. Stone, E. Mazur, Nano Lett. **8**, 2087 (2008)
72. J. Reif, M. Ratzke, O. Varlamova, F. Costache, Mater. Sci. Eng. **134**, 114 (2006)
73. O. Varlamova, F. Costache, M. Ratzke, J. Reif, Appl. Surf. Sci. **253**, 7932 (2007)
74. M. Groenendijk, Laser Tech. J. **5**, 44 (2008)
75. J. Reif, Opt. Eng. **28**, 1122 (1989)
76. J.C. Zink, J. Reif, E. Matthias, Phys. Rev. Lett. **68**, 3595 (1992)
77. R. Stoian, A. Rosenfeld, D. Ashkenasi, I.V. Hertel, N.M. Bulgakova, E.E.B. Campbell, Phys. Rev. Lett. **88**, 097603 (2002)
78. N.M. Bulgakova, R. Stoian, A. Rosenfeld, I.V. Hertel, W. Marine, E.E.B. Campbell, Appl. Phys. A **81**, 345 (2005)
79. S. Küper, M. Stuke, Appl. Phys. B **44**, 199 (1987)
80. S. Küper, M. Stuke, Microelectron. Eng. **9**, 475 (1989)
81. R. Sauerbrey, G.H. Pettit, Appl. Phys. Lett. **55**, 421 (1989)

82. W. Kautek, S. Mitterer, J. Krüger, W. Husinsky, G. Grabner, Appl. Phys. A **58**, 513 (1994)
83. S. Baudach, J. Bonse, W. Kautek, Appl. Phys. A **69**, S395 (1999)
84. S. Baudach, J. Bonse, J.r. Krüger, W. Kautek, Appl. Surf. Sci. **154–155**, 555 (2000)
85. S. Baudach, J. Krüger, W. Kautek, Rev. Laser Eng. **29**, 705 (2001)
86. S. Gaspard, M. Forster, C. Huber, C. Zafiu, G. Trettenhahn, W. Kautek, M. Castillejo, Phys. Chem. Chem. Phys. **10**, 6174 (2008)
87. S. Preuss, M. Spath, Y. Zhang, M. Stuke, Appl. Phys. Lett. **62**, 3049 (1993)
88. N. Bityurin, B.S. Luk'yanchuk, M.H. Hong, T.C. Chong, Chem. Rev. **103**, 519 (2003)
89. M. Prasad, P.F. Conforti, B.J. Garrison, Appl. Phys. A **92**, 877 (2008)
90. S. Hong-Bo, M. Shigeki, M. Hiroaki, Appl. Phys. Lett. **74**, 786 (1999)
91. H.-B. Sun, S. Kawata, Adv. Polym. Sci. **170**, 169 (2004)
92. J. Koch, E. Fadeeva, M. Engelbrecht, C. Ruffert, H. Gatzen, A. Ostendorf, B.N. Chichkov, Appl. Phys. A **82**, 23 (2006)
93. R. Inführ, N. Pucher, C. Heller, H. Lichtenegger, R. Liska, V. Schmidt, L. Kuna, A. Haase, J. Stampfl, Appl. Surf. Sci. **254**, 836 (2007)
94. R. Liska, M. Schuster, R. Inführ, C. Turecek, C. Fritscher, B. Seidl, V. Schmidt, L. Kuna, A. Haase, F. Varga, H. Lichtenegger, J. Stampfl, J Coat. Technol. Res. **4**, 505 (2007)
95. P.J. Bártolo (ed), *Stereolithography* . (Springer, 2011)
96. B. Tan, K. Venkatakrishnan, J. Micromech. Microeng. **16**, 1080 (2006)
97. W. Haske, V.W. Chen, J.M. Hales, W. Dong, S. Barlow, S.R. Marder, J.W. Perry, Opt. Express **15**, 3426 (2007)
98. M. Deubel, G. Von Freymann, M. Wegener, S. Pereira, K. Busch, C.M. Soukoulis, Nat. Mater. **3**, 444 (2004)
99. M. Straub, L.H. Nguyen, A. Fazlic, M. Gu, Opt. Mater. **27**, 359 (2004)
100. A. Ostendorf, B.N. Chichkov, Photonics. Spectra. **40**, 72 (2006)
101. R. Kiyan, C. Reinhardt, S. Passinger, A.L. Stepanov, A. Hohenau, J.R. Krenn, B.N. Chichkov, Opt. Express **15**, 4205 (2007)
102. C. Reinhardt, R. Kiyan, S. Passinger, A.L. Stepanov, A. Ostendorf, B.N. Chichkov, Appl. Phys. A **89**, 321 (2007)
103. C.-F. Li, X.-Z. Dong, F. Jin, W. Jin, W.-Q. Chen, Z.-S. Zhao, X.-M. Duan, Appl. Phys. A **89**, 145 (2007)
104. M. Farsari, A. Ovsianikov, M. Vamvakaki, I. Sakellari, D. Gray, B.N. Chichkov, C. Fotakis, Appl. Phys. A **93**, 11 (2008)
105. H. Luo, Y. Li, H.-B. Cui, H. Yang, Q.-H. Gong, Appl. Phys. A **97**, 709 (2009)
106. A. Mautner, X. Qin, H. Wutzel, S.C. Ligon, B. Kapeller, D. Moser, G. Russmueller, J. Stampfl, R. Liska, J. Polym. Sci. Part A: Polym. Chem. **51**, 203 (2013)
107. J. Stampfl, Vienna University of Technology, Austria

Chapter 4
Atomic Movies of Laser-Induced Structural and Phase Transformations from Molecular Dynamics Simulations

Chengping Wu, Eaman T. Karim, Alexey N. Volkov and Leonid V. Zhigilei

Abstract Molecular dynamics (MD) simulations of laser-materials interactions are playing an important role in investigation of complex and highly non-equilibrium processes involved in short pulse laser processing and surface modification. This role is defined by the ability of MD simulations to reveal in-depth information on the structural and phase transformations induced by the laser excitation and, at the same time, to provide clear visual representations, or "atomic movies," of laser-induced dynamic processes. This chapter provides a brief overview of recent progress in the description of laser coupling and relaxation of photo-excited states in metals, semiconductors, insulators and molecular systems within the general framework of the classical MD technique and presents several examples of MD simulations of laser melting, generation of crystal defects, photomechanical spallation, explosive boiling and molecular entrainment in laser ablation. Possible directions of future progress in atomistic modeling of laser-materials interactions and the potential role of MD simulations in the design of an integrated multiscale computational model capable of accounting for interrelations between processes occurring at different time- and length-scales are discussed.

4.1 Introduction

Computer modeling is playing an increasingly important role in the development of the theoretical understanding of laser-materials interactions and the advancement of laser applications. The need for computer modeling is amplified by the complexity of the material response to the rapid laser energy deposition. Short pulse laser irradiation brings the target material to a state of strong electronic, thermal, and

C. Wu · E. T. Karim · A. N. Volkov · L. V. Zhigilei (✉)
Department of Materials Science and Engineering, University of Virginia, 395 McCormick Road, Charlottesville, VA 22904-4745, USA
e-mail: lz2n@virginia.edu

M. Castillejo et al. (eds.), *Lasers in Materials Science*,
Springer Series in Materials Science 191, DOI: 10.1007/978-3-319-02898-9_4,

mechanical non-equilibrium and triggers a cascade of interrelated processes that may involve structural and phase transformations, such as melting and resolidification, generation of crystal defects, fracture and disintegration of the region of the laser energy deposition due to photomechanical stresses (front surface spallation), vaporization and explosive boiling of strongly overheated surface region, ionization and plasma formation. Computer modeling of this diverse range of processes is challenging and requires a combination of different computational techniques.

Computational models used in the simulations of laser-materials interactions can be separated into three distinct groups: (1) quantum mechanics based (ab initio) electronic structure calculation methods, (2) classical atomistic and coarse-grained MD techniques, and (3) continuum-level kinetic and hydrodynamic models.

The electronic structure calculations are capable of reproducing the initial material response to the laser excitation, including transient changes in interatomic bonding and corresponding modifications of material properties [1–3], non-thermal structural transformations [3–8], bond-breaking/rearrangement and atomic ejection under conditions of strong electronic excitation [9–11], as well as relaxation of excited states and electron–phonon equilibration [12, 13]. The high computational cost of electronic structure calculations, however, limits the size of the computational systems treated in ab initio simulations to several hundreds of atoms, thus preventing realistic modeling of laser-induced structural and phase transformations.

Continuum models of laser-materials interactions, on the other hand, are capable of computationally efficient treatment of laser-induced processes at experimental time and length-scales. These models are typically based on the solution of a set of partial differential equations describing the laser energy deposition, heat transfer, thermoelastic and plastic deformation, photolytic and pyrolytic chemical reactions, phase transformations (melting, resolidification, vaporization and volume ablation processes), as well as ablation plume expansion, e.g., [14–24]. Hydrodynamic models that adopt multi-phase equations of state to describe the evolution of thermodynamic parameters of the material have also been used for simulation of laser-induced phase transformations, including photomechanical spallation, vaporization, and ablation [25–33]. The predictive power of the continuum-level models, however, is limited by the need for a priory knowledge of all the processes that may take place during the simulations. The fundamental understanding of the mechanisms and kinetics of complex and highly non-equilibrium structural and phase transformations occurring in the irradiated target is often lacking, making it difficult to design a reliable continuum description of laser-materials interactions.

Given the limitations of the continuum and ab initio approaches to modeling of laser-materials interactions, classical MD simulation technique [34–36] has emerged as a promising alternative approach capable of providing insights into collective atomic dynamics responsible for laser-induced structural and phase transformations. The MD technique is based on the numerical integration of the classical equations of motion for all atoms in the system. The interatomic interaction is described by a potential energy function that defines the equilibrium structure and thermodynamic properties of the material. The interatomic potentials are often designed with the help of ab initio simulations and are fitted to reproduce some of the basic properties of the materials of interest (e.g., elastic moduli, cohesive energy, energies of formation/migration of point defects, coefficient of thermal expansion, melting temperature). Once the interatomic potential is chosen and the initial conditions (positions and velocities of all particles in the system) are defined, the equations of motion for all atoms are solved numerically and atomic trajectories (positions and velocities as a function of time) are obtained. The main strength of the MD method is that it does not require any assumptions about the processes taking place in the systems that are investigated. This characteristic of the MD technique presents a significant advantage over the continuum-level methods where all relevant processes have to be known (and described mathematically) before the simulations can be performed.

The atomic trajectories generated in MD simulations provide a clear visual picture, or "atomic movies," of the laser-induced processes and, at the same time, can be used to study the evolution of temperature, pressure, structure and phase state of the irradiated material, thus revealing complete mechanistic information on the complex phenomenon of laser-materials interactions. The MD technique has been actively used in investigations of laser melting and resolidification [37–48], generation of crystal defects [49–51], photomechanical spallation [40, 45, 48, 52–58], and ablation [39, 43, 45, 48, 52, 59–91] of various material systems.

A comprehensive review of the applications of the MD technique to simulation of laser-materials interactions has been provided in a book published as a follow up on the first Venice School on Lasers in Materials Science [36]. In this chapter, therefore, the discussion of the computational models in Sect. 4.2 is limited to a brief overview of methods developed for the description of laser coupling to different materials and relaxation of photo-excited states within the classical atomistic or coarse-grained MD technique. Recent results from simulations of laser melting, generation of crystal defects (dislocations, vacancies and interstitials), photomechanical spallation, explosive boiling and molecular entrainment in laser ablation are discussed in Sect. 4.3. The ability of MD simulations to provide visual representations of laser-induced dynamic processes is highlighted by providing series of snapshots taken at different moments during the MD simulations. Finally, in Sect. 4.4, the potential role of the classical MD simulations in the general approach of multiscale modeling of laser-materials interactions is discussed.

4.2 Representation of Laser Excitation in Classical Molecular Dynamics

The interatomic interactions, described in classical MD by pair or many-body potential energy functions [92–94], are ultimately defined by the electronic structures of the materials. The electrons and quantum effects, however, are not explicitly represented in the classical MD method. Therefore, the optical properties of an irradiated material, as well as the mechanisms and kinetics of the relaxation of optically excited states, can not be obtained in the course of an MD simulation, but have to be assumed prior to the simulation and incorporated into the model. The diversity of physical mechanisms of laser coupling and relaxation/thermalization of the absorbed laser energy necessitates the design of material-specific computational approaches for the description of the laser excitation within the general framework of the classical MD method. A brief overview of computational models developed for MD simulations of laser interactions with molecular systems, metals, semiconductors and insulators is provided below.

In molecular systems, photoexcitation of a particular vibrational mode or a group of modes can occur through direct resonant absorption of infrared (IR) irradiation [95] or internal conversion of molecular chromophores electronically excited by ultraviolet (UV) irradiation. The vibrational excitation is followed by a rapid, within picoseconds, intra-molecular redistribution of the deposited energy within the excited molecule [96, 97] and a slower, within tens to hundreds of picoseconds, inter-molecular energy transfer leading to the vibrational cooling of the excited molecule [98–100].

The relatively fast, compared to the timescale of the collective molecular ejection in laser ablation, rate of the vibrational thermalization of the deposited laser energy may serve as a justification for simulation of the laser excitation by fast temperature increase in the surface region of the irradiated target, an approach adopted in a number of atomistic MD simulations of short pulse laser ablation of molecular targets [65, 71, 101]. An internal vibrational excitation followed by the dynamic energy transfer to the translational degrees of freedom has been used in simulations of laser-induced molecular desorption from an oxygen crystal [102].

The finite rate of the vibrational relaxation of photo-excited molecules is accounted for in a coarse-grained "breathing sphere" model developed for simulation of laser interactions with molecular targets [60, 75]. In this model, each molecule is represented by a spherical particle, with parameters of inter-particle interaction potential chosen to approximately reproduce the physical properties of a molecular target. In order to simulate molecular excitation by photon absorption and vibrational relaxation of the excited molecules, an additional internal degree of freedom is added to each molecule by allowing the particles to change their radii, or to "breathe." The laser irradiation is simulated by vibrational excitation of molecules that are randomly chosen during the laser pulse duration within the penetration depth appropriate for a given wavelength. The vibrational excitation is modeled by depositing a quantum of energy equal to the photon energy into the kinetic energy of the internal breathing mode of a given molecule. In the case of UV laser irradiation, the breathing mode can be considered as the recipient of the energy released by internal conversion from electronically excited states. The parameters of a potential function attributed to the internal motion control the characteristic frequency of the breathing mode and, as a result, define the rate of the conversion of the internal energy of the molecules excited by the laser to the translational and internal motions of the surrounding molecules. The breathing sphere model has been actively used in investigations of laser desorption, ablation, and spallation of molecular targets [52, 53, 55, 60–62, 66–69, 74, 75] and polymer solutions [72, 86, 87, 103, 104]. A description of ionization mechanisms has also been included into the breathing sphere model and used for investigation of the processes that control the yield of ions in matrix-assisted laser desorption/ionization (MALDI) mass spectrometry technique [105–107]. Several examples of the applications of the coarse-grained MD to investigation of laser ablation of molecular targets, including polymer and carbon nanotube solutions, are provided in Sects. 4.3.4 and 4.3.5.

An alternative result of the photon absorption, photodissociation of the excited molecule into fragments, has also been considered in coarse-grained MD simulations. In early simulations of laser ablation of polymer targets, the pressure generated by products of photolytic reactions is reproduced through activation of repulsive interactions between the molecules in the absorption region of the target [108, 109]. More recently, a methodology for semi-quantitative representation of photochemical reactions (bond scission followed by abstraction and recombination reactions) within the coarse-grained MD model has been developed and applied for investigation of UV laser ablation of a chlorobenzene [70, 75, 110] and poly(methyl methacrylate) [81, 88].

Substrate-assisted laser-driven ejection of a layer of volatile molecular or atomic material has been investigated in MD simulations performed for various systems, including a water ice film deposited on a Au substrate [111–113], a generic fluid-substrate two-dimensional system [114], a liquid Ar film deposited on a Pt substrate [115, 116], and a solid Xe overlayer deposited on a Si substrate [117]. The laser energy deposition is simulated by instantaneous [111, 112, 114–116] or gradual [113] temperature increase in the substrate. A realistic temperature profile defined by the laser heating and cooling due to the thermal conduction to the bulk of the substrate has also been simulated by solving the heat conduction equation with a source term describing irradiation of a Si substrate by a 5 ns laser pulse [117]. The simulations suggest that energy transfer from an absorbing substrate to a transparent layer can result in superheating and fast vaporization (or explosive boiling) of a part of the layer adjacent to the hot substrate, leading to the separation and ejection of the layer from the substrate.

In metals, the absorption of laser light by the conduction band electrons is followed by quick, within femtoseconds, equilibration among the electrons (establishment of Fermi-Dirac distribution) and slower, from fractions of a picosecond to several tens of picoseconds (depending on the strength of electron–phonon coupling [118]), equilibration between the electrons and atomic vibrations (phonons). Several computational approaches have been developed for the description of the electronic excitation, thermalization of the absorbed laser energy, and electron contribution to the heat conduction in metals within the framework of the classical MD method [38, 40, 58, 73, 79, 119–121]. In particular, the model suggested in [40, 73], is based on the well-known two-temperature model (TTM) [122] that describes the time evolution of the lattice and electron temperatures in the irradiated target by two coupled nonlinear differential equations. In the combined TTM–MD model, MD substitutes the TTM equation for the lattice temperature. The TTM equation for the electron temperature is solved by a finite difference method simultaneously with MD integration of the equations of motion of atoms. The electron temperature enters a coupling term that is added to the MD equations of motion to account for the energy exchange between the electrons and the lattice. The cells in the finite difference discretization are related to the corresponding volumes of the MD system, and the local lattice temperature is defined for each cell from the average kinetic energy of thermal motion of atoms. In simulations of laser intertactions with bulk targets, e.g., [45, 46, 48, 49, 51, 55], the atomistic representation is used only in the surface region of the target, where active processes of laser melting, resolidification and/or ablation take place, whereas the heat transfer in the deeper region of the target is represented by the conventional TTM or one-temperature thermal diffusion equation. The TTM–MD model has been successfully applied for investigation of laser melting [38, 40–42, 44–48], generation of crystal defects [49–51], photomechanical spallation and ablation [40, 45, 48, 55, 58, 73, 78, 79, 90, 91] of metal targets. Several examples of the applications of the TTM–MD model are provided in Sects. 4.3.1–4.3.3.

The adaptation of classical MD for simulation of laser interactions with semiconductors and insulators is more challenging compared to metals, as the models should account for the formation and diffusion of electron-hole pairs, localized weakening of interatomic bonding, ionization and generation of free electrons, absorption of the laser energy by the free electrons, as well as the energy coupling between the excited electrons and ions. First models developed for Si are based on a local treatment of individual excitation events and include a description of bond weakening, ionization and electron–ion recombination [64, 123], as well as transition of Si to the metallic state upon melting and a stochastic (Monte Carlo) treatment of carrier diffusion and scattering [76, 82]. The development of atomistic models for laser interaction with insulators has been limited to exploratory work for fused silica [124] and LiF [125] films, where a continuum description of the laser coupling and generation of free electrons is combined with MD treatment of atoms that accounts for the energy transfer from the excited electrons to ions [124, 125] and local changes in interatomic interactions due to the ionization [124].

4.3 Atomic Movies from MD Simulations of Laser-Material Interactions

One of the main advantages of the MD technique is in its ability to provide complete atomic-level information on the material response to short pulse laser irradiation. A mere visual analysis of snapshots obtained in MD simulations is often sufficient for drawing preliminary ideas on the physical processes responsible for laser material modification or removal. These ideas can then be used to design more advanced data analysis methods capable of quantitative thermodynamic or structural characterization of the laser-induced processes. In this section, we exploit the ability of the MD method to provide the atomic-level "movies" of the material behavior under conditions of fast laser energy deposition to develop a better intuitive understanding of the mechanisms of laser melting, generation of crystal defects, photomechanical spallation and ablation of irradiated targets.

4.3.1 Laser Melting

Although melting is a common and well-studied phenomenon that plays a prominent role in many laser processing applications, the microscopic mechanisms and kinetics of laser-induced melting has been and still remain subjects of active discussions in literature, e.g., [41, 126–136]. Under conditions of slow energy deposition by long (nanoseconds and longer) laser pulses, the melting process starts from surfaces and internal crystal defects and proceeds by propagation of melting fronts at low levels of superheating above the equilibrium melting temperature, T_m. The kinetic description of the melting process in this case reduces to a linear relationship between the velocity of the melting front and the superheating [137, 138], or the melting process is considered to be heat flow limited and the superheating at the melting front is neglected altogether. Irradiation by short (pico- and femtosecond) laser pulses, however, can result in extremely high heating rates exceeding 10^{14} K/s, making the time of the temperature increase shorter than the time needed for any significant advancement of the melting front. The homogeneous nucleation of liquid regions inside the superheated crystal can be expected to make the dominant contribution to the melting process induced by the short laser pulses.

Experimental probing of the short pulse laser melting with optical, X-ray and electron diffraction time-resolved techniques have provided valuable information on the characteristic time-scales and mechanisms of melting occurring under conditions of strong superheating [126–135]. The important role of MD simulations has been to provide detailed atomic-level information on the rapid laser-induced phase transformations in the irradiated targets and to assist in interpretation of the results of the experimental probing. Indeed, MD simulations of laser melting and resolidification of bulk metal targets and thin films of various thickness [37–51] has revealed a wealth of information on the microscopic mechanisms and kinetics of laser melting. In particular, the relative contributions of the homogeneous and heterogeneous melting mechanisms have been analyzed and related to the irradiation conditions in simulations performed for different metals (Ni, Au, Al, Cu and Cr), target geometries (thin films [40–42, 44, 47, 50, 56], bulk targets [45, 48, 49] and metal bilayer systems [46, 51]), and laser pulse durations (from 200 fs to 150 ps). The lattice distortions associated with relaxation of laser-induced stresses have been found to significantly reduce the crystal stability against melting, resulting in homogeneous nucleation of the liquid phase at temperatures close to the equilibrium melting temperature [41]. The calculations of the changes in the diffraction profiles and density correlation functions during the melting process [44] have been used to establish direct connections between the results of MD simulations and time-resolved electron diffraction experiments [129, 131, 132].

The atomic-level view of the two distinct melting mechanisms identified in MD simulations, heterogeneous melting by propagation of melting fronts from external surfaces of the target and homogeneous melting by nucleation of liquid regions inside a strongly superheated crystalline material, is provided in Fig. 4.1a, b, where snapshots from two simulations performed for a 20 nm single crystal Au film irradiated with a 200 fs laser pulse are shown for absorbed fluences of 45 and 180 J/m^2 [47]. In both simulations, the ballistic energy transfer by nonthermal electrons and relatively weak electron–phonon coupling in Au result in the establishment of a uniform distribution of the electron and lattice temperatures throughout the 20 nm film during the time of the electron–phonon equilibration. At a fluence of 45 J/m^2 (Fig. 4.1a), the lattice temperature increases due to the electron–phonon energy exchange and reaches the maximum value of 1.16 T_m by the time of ~20 ps. Two melting fronts emerge from the free surfaces of the film and propagate toward the center of the film. The velocities of the melting fronts gradually decrease from their maximum values of ~100 m/s down to zero as the temperature decreases (due to transformation of the thermal energy to the heat of melting) and eventually saturates at T_m by the time of 500 ps. The snapshot shown in Fig. 4.1a for 500 ps corresponds to the final state of the partially melted film with a crystalline layer located in the middle. At a higher fluence of 180 J/m^2 (Fig. 4.1b), the high level of the electronic excitation leads to a transient increase of the strength of the electron–phonon coupling [118] and to a much faster lattice heating. By the time of 2 ps the lattice temperature exceeds the level of ~1.25 T_m that corresponds to the limit of thermal stability of the crystal lattice [41, 44, 136, 139, 140], resulting in a rapid collapse of the crystal structure within the next several picoseconds. Rapid nucleation and growth of liquid regions throughout the film are preceded by thermally activated generation of a large number of vacancy-interstitial pairs that can be identified in the snapshots shown for the time of 4 ps in Fig. 4.1b. The high density of point defects introduces lattice distortions and reduces stability of the crystal structure against melting [47].

The effect of high density of grain boundaries in nanocrystalline metals on the characteristics of the laser melting process is illustrated in Fig. 4.1c, d, where snapshots from MD simulations performed for Au films with nanocrystalline structure are shown for the same irradiation conditions as the ones discussed above for single crystal films. Although at all laser fluences the melting process starts from the grain boundary regions as soon as the lattice temperature approaches and exceeds the equilibrium melting temperature, the contribution of the grain boundary melting to the overall melting process is very different in the low fluence regime (below or close to the threshold for the complete melting of the film) and in the high fluence regime (significantly above the melting threshold).

At low laser fluences, when the maximum lattice temperature does not exceed the limit of crystal stability against homogeneous melting, the microstructure of the film is found to have a much stronger effect on the characteristics of the melting process. In particular, in contrast with the partial melting of about a half of

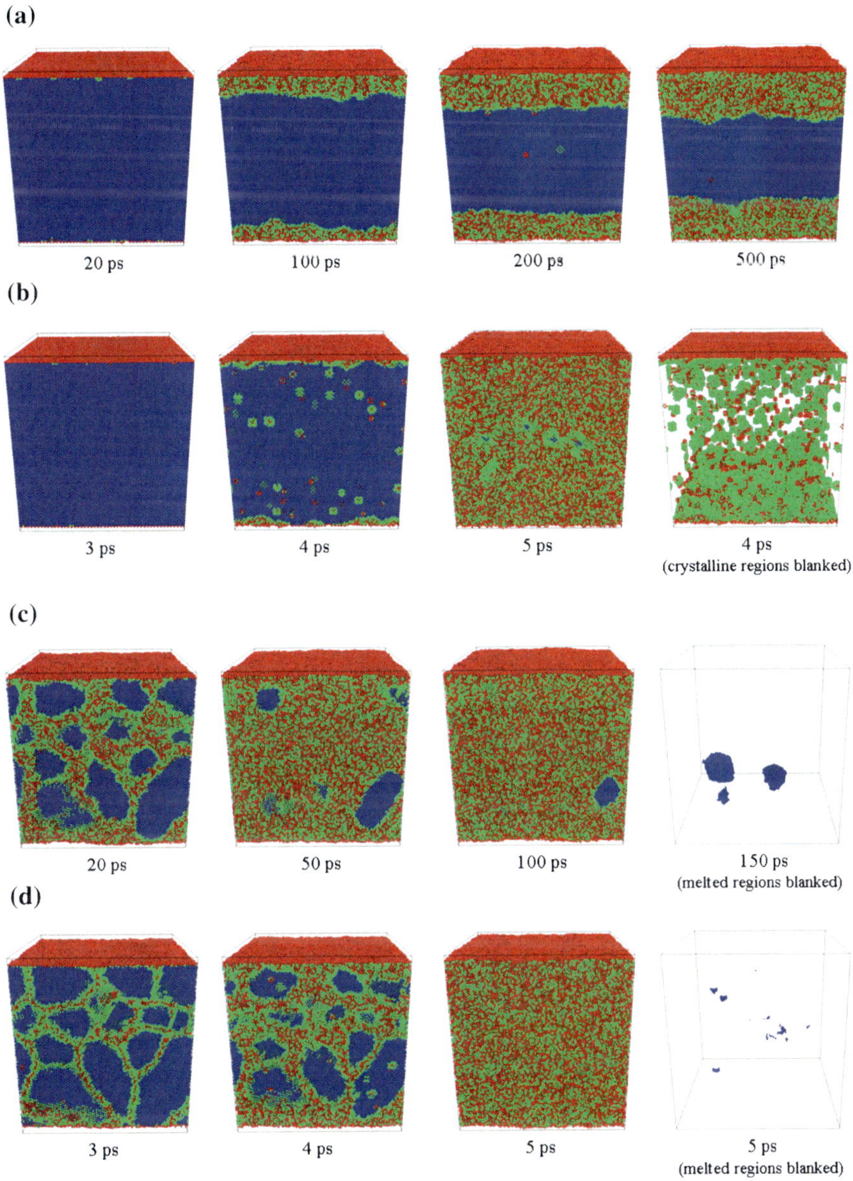

Fig. 4.1 Snapshots of atomic configurations generated in TTM–MD simulations of the melting process in a 20 nm single crystal (**a**, **b**) and nanocrystalline (**c**, **d**) Au films irradiated by 200 fs laser pulses at absorbed fluences of 45 J/m^2 (**a**, **c**) and 180 J/m^2 (**b**, **d**). Atoms are colored according to the local structure: *blue* atoms have local crystalline surroundings, *red* and *green* atoms are at the free surfaces, near crystal defects (grain boundaries, vacancies and interstitials) or belong to the liquid phase. In the *right panel* of (**b**), the atoms in the crystalline parts of the film are blanked to provide a clear view of the emerging point defects and disordered regions. In the *right panels* of (**c**) and (**d**), the atoms that belong to the liquid phase are blanked to expose the remaining crystalline regions. The interstitials and vacancies show up in the snapshots as *large green clusters* and *smaller red clusters*, respectively. The simulations of laser melting of Au films are discussed in [47]

a 20 nm single crystal Au film illustrated in Fig. 4.1a, the nanocrystalline film undergoes complete melting at the same absorbed fluence of 45 J/m^2, Fig. 4.1c. Surprisingly, the melting of the nanocrystalline film is found to continue even after the temperature of the film drops below the equilibrium melting temperature at $\sim$30 ps and the last crystalline regions (e.g., the ones shown in the right panel in Fig. 4.1c) continue to shrink under conditions of about 6 % undercooling below the melting temperature and disappear by a time of 250 ps. This unusual melting behavior of the nanocrystalline films is explained in [47] based on thermodynamic analysis of the stability of small crystalline clusters surrounded by undercooled liquid.

At high laser fluences, e.g., Fig. 4.1d, the kinetics and mechanisms of melting are only weakly affected by the nanocrystalline structure of the film. Although the grain boundary melting in nanocrystalline films results in a moderate decrease of the size of the crystalline grains at the initial stage of the laser heating, the overall timescale of the melting process is largely defined by the fast temperature increase and rapid (within several picoseconds) collapse of the crystal structure as soon as the lattice temperature exceeds the limit of superheating [41, 44, 47, 136, 139, 140].

4.3.2 Generation of Crystal Defects

The fast rate of energy deposition in short pulse laser processing applications may result not only in the rapid melting and resolidification of a surface region of an irradiated target but also lead to the generation of strong compressive stresses. The laser-induced stresses are particularly high in the regime of stress confinement [45, 52, 55, 75, 141–145], when the time of the laser heating (defined by the laser pulse duration, τ_p, or the time of the electron–phonon equilibration, τ_{e-ph}, whichever is longer) is shorter than the time required for the mechanical relaxation (expansion) of the heated volume, $\tau_s \sim L_p/C_s$, where C_s is the speed of sound in the target material and L_p is the effective depth of the laser energy deposition (optical penetration depth or, in the case of metals, the depth of diffusive/ballistic energy transport during the time of the electron–phonon equilibration [42, 45, 55]). The relaxation of the laser induced stresses, taking place simultaneously with melting and resolidification of the surface region of the irradiated target, can result in the generation of a high density of crystal defects (vacancies, interstitials, dislocations, grain boundaries) and corresponding modification of physical and mechanical properties of the surface layer.

Small size of the laser-modified region (on the order 100s of nm) in femtosecond and picosecond laser processing of metals enables, with appropriate design of boundary conditions, the direct MD modeling of the laser-induced generation of crystal defects. Indeed, the emission of partial dislocations from the melting front [36, 146], plastic deformation of films heated by a laser pulse [147], generation of nanocrystalline structure [50] and dislocations [43, 63] around regions of localized laser energy deposition, strong supersaturation of a surface layer with vacancies

[49], and transformation of a lattice-mismatched interface into a three-dimensional periodic array of stacking fault pyramids outlined by stair-rod partial dislocations [51] have been observed in MD simulations. In two examples considered below, the generation of a region of high vacancy concentration and the emission of partial dislocations from a melting front are illustrated by the results of TTM MD simulations performed for bulk Cr and Ag targets.

The generation of vacancies in a surface region of a Cr target experiencing transient melting and rapid resolidification in response to the irradiation by a 200 fs laser pulse at an absorbed fluence of 850 J/m^2 is illustrated in snapshots shown in Fig. 4.2. The atoms that retain the original body-centered cubic (bcc) local structure are blanked in the snapshots to provide a clear view of the crystal defects and the melted part of the target. The first snapshot is taken at 40 ps after the laser pulse, the time when the depth of the melted layer reaches its maximum. The red spot in the lower part of the melted layer corresponds to a void generated during the passage of the tensile component of the laser-induced stress wave through the liquid–crystal interface. The void collapses by about 90 ps after the laser pulse and is not present in the snapshot shown for 120 ps. As discussed in Sect. 4.3.3, at higher laser fluences the growth and coalescence of voids can result in the separation (spallation) and ejection of a liquid layer from the target. The fluence of 850 J/m^2, however, is just below the threshold for laser spallation of the Cr target [48]. A strong temperature gradient created in the surface region of the target by the short pulse laser irradiation results in a fast cooling and epitaxial resolidification of the melted region. The solidification front advances with velocity that increases with increasing degree of undercooling below the equilibrium melting temperature and reaches its maximum value of about 80 m/s by the end of the resolidification process at about 330 ps, when the temperature at the solidification fronts drops down to about 0.8 T_m.

Two types of crystal defects can be identified in the snapshots shown in Fig. 4.2. The first type is exhibited by atomic planes with elevated energy appearing below the melted region in the snapshot shown for 40 ps. Detailed analysis of the atomic configurations [49] reveals that these planes correspond to the intrinsic stacking faults generated as a result of multiple internal slips on {110} crystallographic planes in response to the rapid uniaxial expansion of the crystal during the dynamic relaxation of the laser-induced stresses. The intrinsic stacking faults are unstable in an unstrained bcc crystal [49, 148] and disappear shortly after the tensile component of the laser-induced stress wave leaves the surface region of the target. Only isolated islands of the stacking faults can be observed in the snapshot shown for 120 ps and no stacking faults are present after 160 ps. The second type of crystal defects is vacancies that appear as clusters of atoms with elevated energy in the snapshots shown in Fig. 4.2. The vacancies are largely concentrated in the region of the target that experienced transient melting and resolidification (above the dashed line in the snapshots). The distribution of vacancies in the top 40 nm surface region of the target at the end of the simulation is shown in Fig. 4.3. The distribution confirms that the overwhelming majority of vacancies are located in the resolidified part of the target. Analysis of the

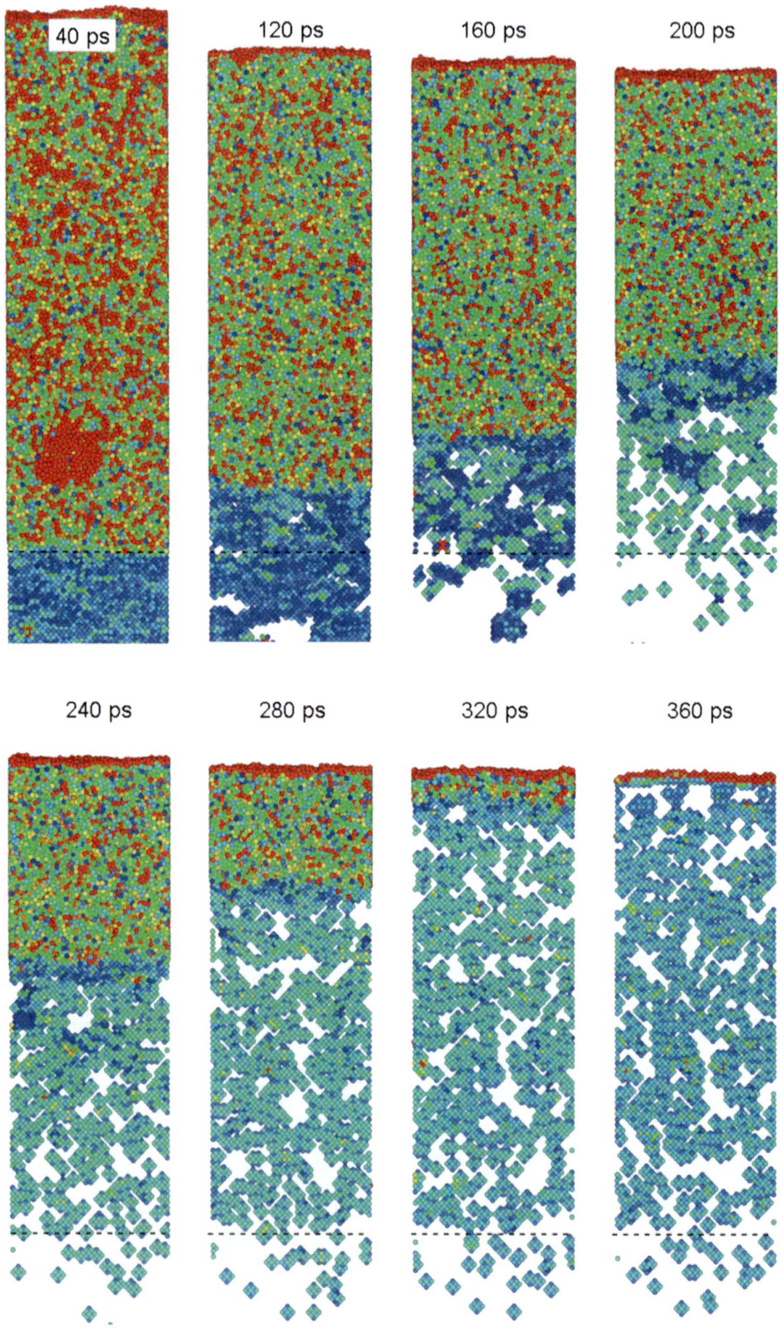
40 ps
120 ps
160 ps
200 ps
240 ps
280 ps
320 ps
360 ps

◀ **Fig. 4.2** Snapshots of atomic configurations generated in TTM–MD simulations of melting and resolidification of a bulk Cr target irradiated by a 200 fs laser pulse at an absorbed fluence of 850 J/m^2. The atomic configurations are quenched for 1 ps to reduce thermal noise in atomic positions and energies. Atoms are colored according to their potential energies and the atoms that have low potential energy and belong to local configurations with the original bcc structure are blanked to expose crystal defects. The *dashed lines* mark the maximum depth of melting reached in the simulation at about 40 ps. Each *blue ball* (a cluster of atoms with elevated energies) in the snapshots corresponds to a vacancy

mechanisms of the vacancy formation indicates that the majority of vacancies are generated at the rapidly advancing solidification front and are stabilized by the fast cooling of the surface region.

The observation of very high vacancy concentrations, up to 2–5 vacancies per 1,000 atoms, in the surface regions of the irradiated targets may have important practical implications, including the formation of nanovoids and degradation of the mechanical properties of the surface region of the target in the multi-pulse irradiation regime. The generation of vacancies, therefore, may be related to experimental observations of the incubation effect, when the laser fluence threshold for ablation/damage decreases with increasing number of laser pulses applied to the

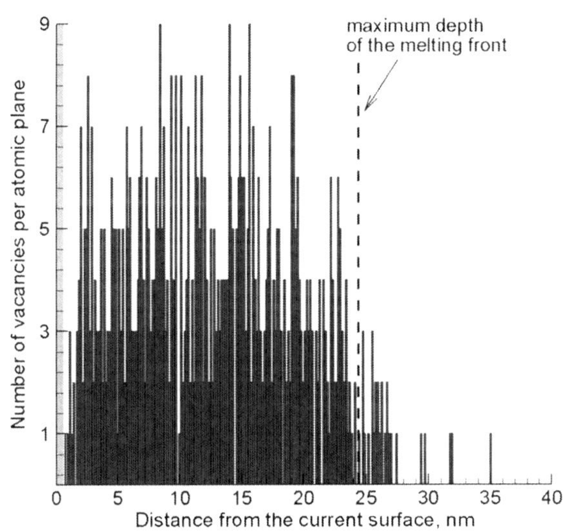

Fig. 4.3 Distribution of vacancies in the surface region of a bulk Cr target irradiated by a 200 fs laser pulse at an absorbed fluence of 850 J/m^2 at 400 ps after the laser pulse. The bars in the histogram show the number of vacancies in individual (001) atomic planes oriented parallel to the irradiated surface. The number of atoms in a defect-free (001) plane is 900. The *gray area* in the *left* part of the figure corresponds to the surface layer of the target where identification of vacancies is not possible. The *dashed line* marks the depth of the region that experiences a transient melting and resolidification during the first 330 ps of the simulation. The snapshots of atomic configurations for this simulation are shown in Fig. 4.2

same area, e.g., [149–153]. The high density of vacancies generated in the surface region may also play an important role in the redistribution of impurities or mixing/alloying in multi-component or composite targets.

To illustrate the sensitivity of the generation of crystal defects in short pulse laser processing to the crystal structure of the target material, snapshots from a simulation performed for a face-centered cubic (fcc) Ag target are shown in Fig. 4.4. In contrast to the bcc Cr target, where only a transient appearance of intrinsic stacking faults has been observed, the snapshots shown in Fig. 4.4 provide a clear view of a massive emission of partial dislocations from the melting front formed during the first tens of picoseconds after irradiation with a 100 fs laser pulse. Similarly to Fig. 4.2, the atoms that retain the original fcc local structure are blanked in Fig. 4.4 and the green planes extending down from the melted surface region correspond to low-energy stacking faults left behind by the partial dislocations (yellow lines) propagating from the melting front.

The physical conditions leading to the emission of the dislocations can be understood by considering the evolution of temperature and pressure in the irradiated target shown in Fig. 4.5a, b. In accord with the discussion of the stress confinement regime provided above, the fast energy transfer from the excited electrons to the lattice vibrations and the corresponding temperature increase in the surface region of the target (Fig. 4.5a) lead to the buildup of high compressive stresses (Fig. 4.5b). These stresses then relax by driving a compressive stress wave deeper into the bulk of the target and inducing an unloading tensile wave that forms due to the interaction of the compressive stresses with free surface of the target and follows the compressive component. The compressive and tensile components of the pressure wave leave the surface region by about 100 ps and pass through the boundary separating the MD and continuum parts of the model [154, 155] without any noticeable reflection. The onset of the emission of dislocations in the first snapshots of Fig. 4.4 coincides with the time when the tensile component of the stress wave passes through the melting front at 50–90 ps. In the single crystal fcc target oriented along $\langle 100 \rangle$ direction, the partial dislocations are activated in four different active $\{111\}$ slip planes. Interactions between the dislocations propagating along the different slip planes result in the formation of immobile dislocation segments (so-called stair-rod dislocations) that ensure stability of the dislocation configurations generated during the initial spike of temperature and thermoelastic stresses. These stable dislocation configurations remain in the target after it cools to the ambient temperature, resulting in the hardening of the laser-treated surface.

The contrast between the observation of the active emission of dislocations in the fcc Ag target and the absence of dislocations in the bcc Cr target is related to the dependence of the stress required to overcome the resistance of the crystal to the movement of dislocations (the so called Peierls stress) on the crystal structure. It is generally low in fcc crystals, where 12 close-packed $\{111\}\langle 1\bar{1}0 \rangle$ slip systems are present, but is higher in bcc crystals. Note that the generation of vacancies in the process of rapid solidification of the transiently melted surface layer, discussed above for the Cr target, is also observed in simulations performed for fcc metals.

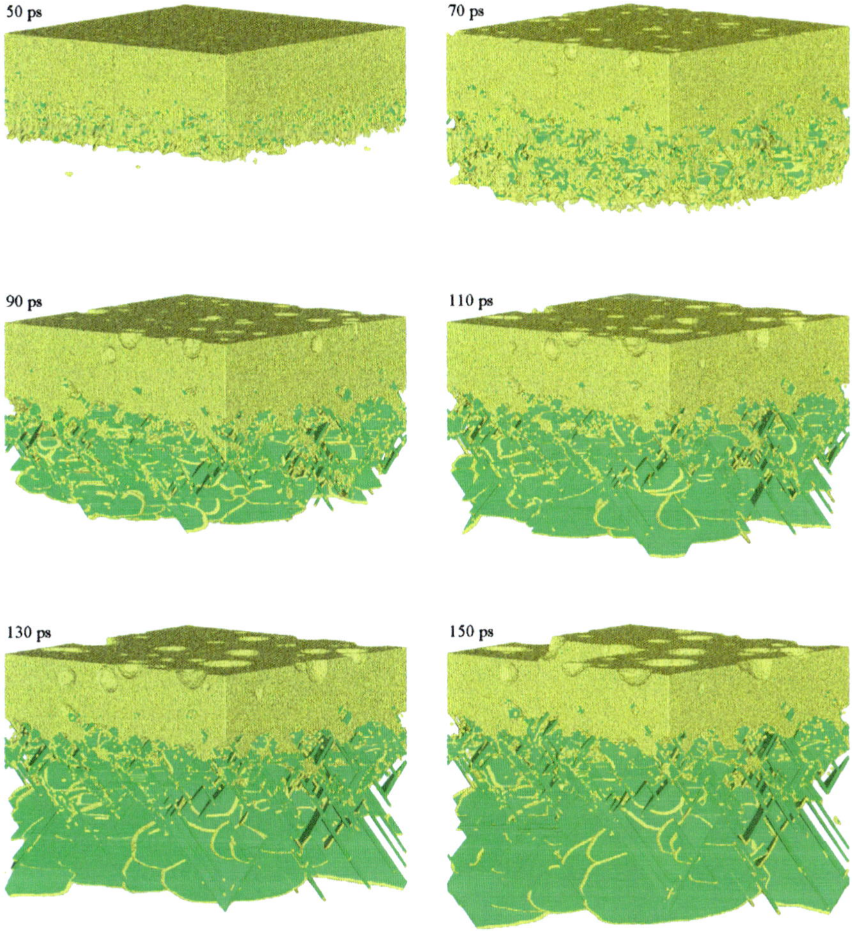

Fig. 4.4 Snapshots of atomic configurations generated in a TTM–MD simulation of a bulk Ag target irradiated by a 100 fs laser pulse at an absorbed fluence of 900 J/m^2. Only a part of the computational cell in the vicinity of the melting front (below 40 nm from the initial surface) is shown in the snapshots. The atoms are colored according to the local structure: *yellow atoms* belong to the melted part of the target or defect configurations (mostly cores of partial dislocations), *green atoms* have local hcp structure and belong to the stacking faults left behind by partial dislocations. The atoms that have local fcc structure are blanked to provide a clear view of the emission of the dislocations from the melting front. The contour plots of temperature, pressure and density evolution in this simulation are shown in Fig. 4.5

4.3.3 Photomechanical Spallation

The laser-induced stresses generated under conditions of stress confinement (see Sect. 4.3.2, Fig. 4.5b, and [45, 55, 141, 142]) can be sufficiently strong to cause mechanical fracture of a solid material or disruption and sputtering of a melted

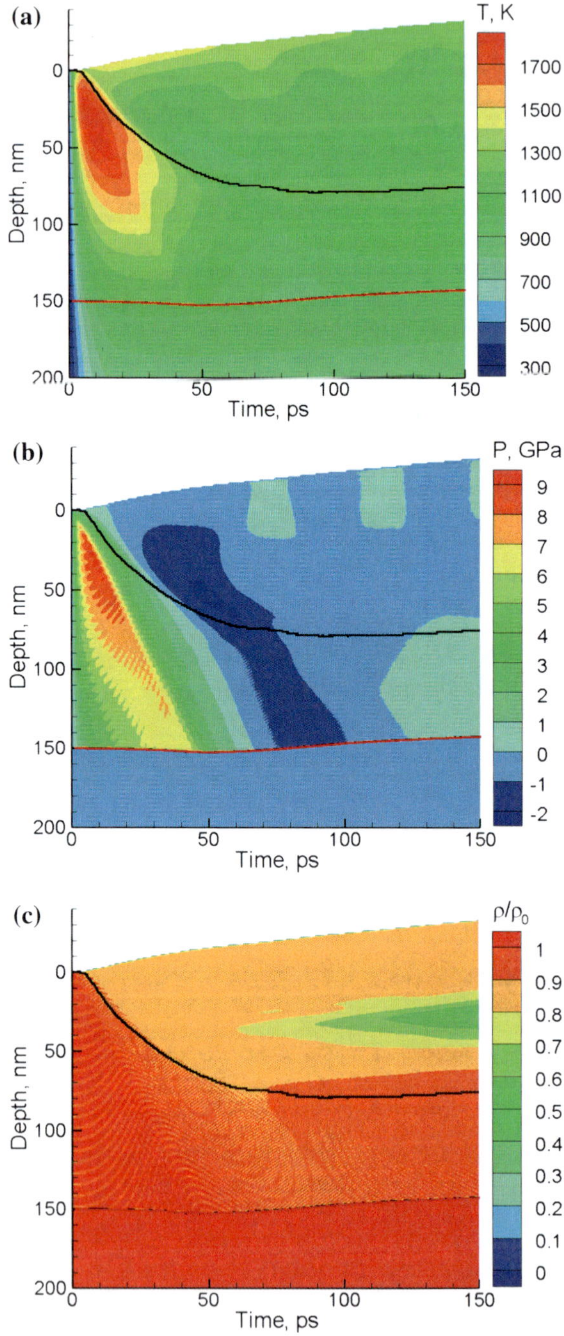

Fig. 4.5 Temperature, pressure and density contour plots in a TTM–MD simulation of a bulk Ag target irradiated by a 100 fs laser pulse at an absorbed fluence of 900 J/m^2. The density scale is normalized to the initial density before the irradiation, ρ_0. The laser pulse is directed along the Y axes from the *top* of the contour plots. The *red solid curves* separate the continuum (TTM) and atomistic (TTM–MD) parts of the computational system. The *black solid curves* separate the melted surface region from the crystalline bulk of the target. Snapshots of atomic configurations for this simulation are shown in Figs. 4.4 and 4.6b

layer. The ejection of large liquid droplets or solid particulates caused by the relaxation of the laser-induced stresses is often called photomechanical ablation or spallation [45, 52, 55, 141–145]. First MD simulations of laser spallation performed for molecular targets [52, 53, 69] revealed the dynamic process of nucleation, growth and coalescence of multiple voids in a subsurface region of the target leading to the separation and ejection of a melted layer. The results of subsequent simulations performed for metal films [40, 55, 56, 58, 90, 156] and bulk targets [45, 48, 55, 73], silicon [82], and systems with interatomic interactions described by the Lennard-Jones potential [54, 85, 157] suggest that spallation is a general process that can occur in a wide class of materials. Moreover, the void size distributions in targets as dissimilar as molecular solids [55] and metals [156] are found to follow almost identical evolution in the course of spallation, suggesting that the process of void nucleation, growth and coalescence may be similar in different materials.

The processes responsible for laser spallation are illustrated by three series of snapshots shown in Fig. 4.6 for TTM–MD simulations performed for Ag and Al targets irradiated by 100 fs laser pulses. To capture the evolution of multiple voids in the sub-surface region of the target, the simulations are performed for relatively large computational systems, where the atomistic parts of the TTM–MD model consist of 84.2 and 106.7 million atoms and have dimensions of $98.7 \times 98.7 \times 150$ nm and $93.9 \times 93.9 \times 200$ nm for the Ag and Al targets, respectively. The continuum (TTM) part of the model extends down to 4 μm in all simulations. The two simulations for the Ag target are performed at absorbed fluences of 850 and 900 J/m^2, just below and above the spallation threshold. The sharp increase of the electron–phonon coupling in Ag in the electronically excited state [118] creates the conditions for the confinement of the deposited laser energy in a relatively shallow surface region of the target, as can be seen in the temperature contour plot shown in Fig. 4.5a. The rapid heating of the lattice due to the energy transfer from the excited electrons results in the build up of high compressive stresses in the surface region of the target, as can be seen in Fig. 4.5b. The relaxation of the compressive stresses in the presence of free surface leads to the generation of an unloading tensile wave that propagates from the surface of the target and increases its strength with depth. At a certain depth under the surface the tensile stresses exceed the dynamic strength of the melted metal, leading to the generation of multiple voids. The voids can be observed at a depth of ~40–60 nm in the first snapshots shown in Figs. 4.6a and 4.6b, as well as in the top parts of the snapshots shown in Fig. 4.4. The appearance of the voids is also apparent in the density contour plot shown in Fig. 4.5c, where a low density region can be seen to appear at ~60 ps and expand as voids grow and coalesce. The depth of the laser-induced void nucleation and growth does not coincide with the depth where the maximum tensile stresses are generated, Fig. 4.5b. Rather, the voids are generated closer to the surface, where the ability of the melt to withstand the dynamic loading is diminished due to higher temperatures [40, 45, 48, 55].

In the simulation performed at 850 J/m^2, below the threshold for laser spallation of Ag, the growth of the voids slows down with time and turns into recession at about 800 ps. At the same time, the fast cooling of the surface region creates conditions for fast epitaxial solidification of the melted region (see snapshot for 900 ps in Fig. 4.6a), with the velocity of the solidification front increasing with increasing undercooling. The solidification front reaches the deepest voids at about 500 ps and crosses the region where the voids are evolving during the following 1,000 ps. As a result, by the time of 1,500 ps (see the corresponding snapshot in Fig. 4.6a) all the remaining voids are trapped by the solidification front and are completely surrounded by the crystalline material. By this time, the temperature of

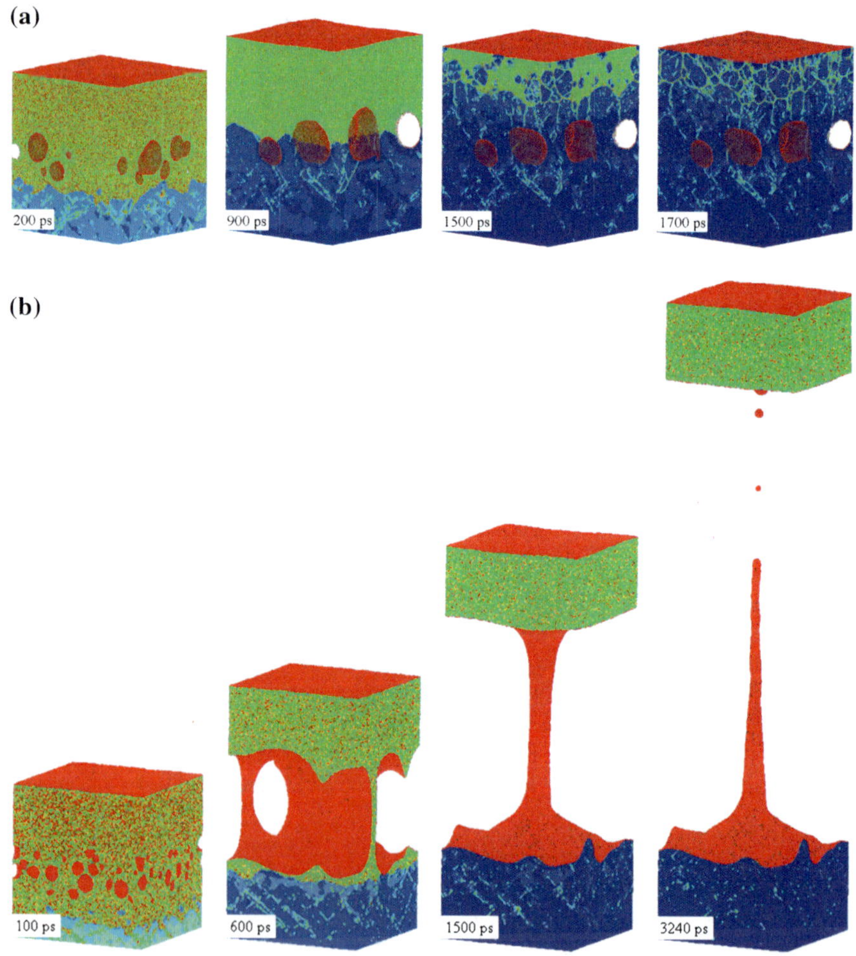

Fig. 4.6 Snapshots of atomic configurations generated in TTM–MD simulations of bulk Ag (**a**, **b**) and Al (**c**) targets irradiated by 100 fs laser pulses at absorbed fluences of 850 J/m^2 (**a**) and 900 J/m^2 (**b**, **c**). Only *top* parts of the computational cells are shown in the snapshots. The atoms are colored according to their potential energy in (**a**, **b**) and by local density of their immediate surroundings in (**c**)

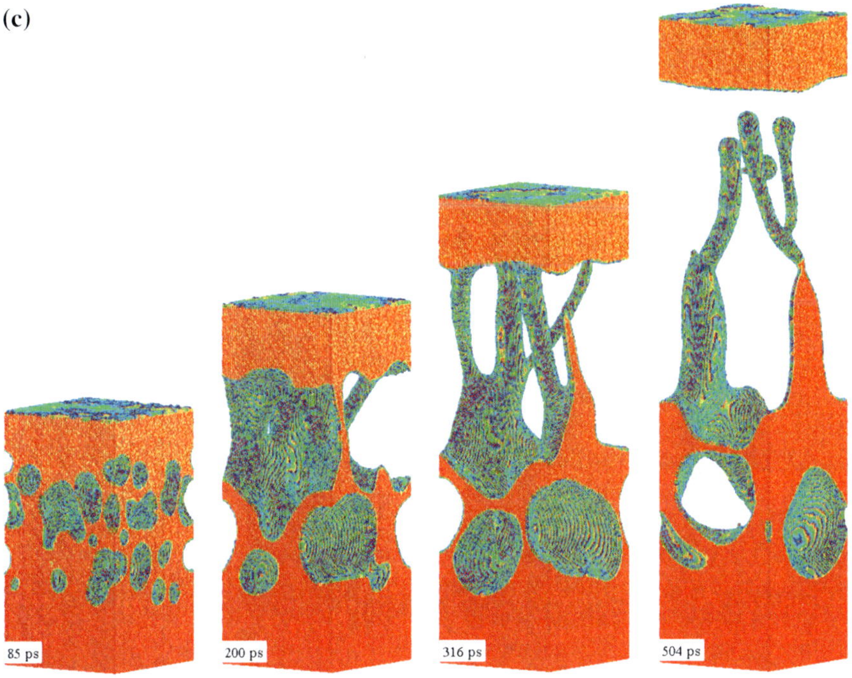

Fig. 4.6 continued

the remaining liquid layer drops down to 0.69 T_m, triggering massive homogeneous nucleation of randomly oriented crystalline regions. The homogeneous nucleation and growth of multiple crystallites result in a rapid solidification of the remaining liquid and generation of nanocrystalline structure of the top layer of the target (see snapshot for 1,700 ps in Fig. 4.6a). The voids captured by the solidification front increase the volume of the surface region, leading to an effective "swelling" of the irradiated target by about 17 nm. The computational prediction of the generation of porous structure of a surface region of the irradiated target provides a hint for explaining the recently reported experimental observation of surface swelling, or "frustrated ablation," in Al and Ag targets irradiated by 100 fs laser pulses [158]. The larger thermal conductivity and smaller melting depth near the spallation threshold in Ag, as compared to Al, make it possible to explore the swelling behavior with smaller computational systems and shorter simulation times.

In experiments, a non-uniform laser energy deposition within the laser spot and restraining effect of the cold periphery of the laser spot on the separation and ejection of the liquid layer accelerated during the initial relaxation of the laser induced stresses are likely to suppress the onset of spallation and to extend the range of laser fluences where the effect of surface swelling is observed. In the simulations, on the contrary, the use of periodic boundary conditions facilitates percolation of voids in the lateral directions and separation of the liquid layer from the target. As a result, a relatively small increase of the laser fluence, from 850 to 900 J/m^2, leads to the ejection of a 55 nm-thick liquid layer from the Ag target, Fig. 4.6b. In this simulation, the growth and coalescence of the voids lead to the formation and eventual break down of an elongated bridge between the substrate and the top liquid layer. The expansion of the bridge hardly decelerates the layer that moves away from the target with a velocity of about 100 m/s. Although the ejected liquid layer is stabilized in the simulation by the use of the periodic boundary conditions, one can expect a rapid decomposition of the layer into liquid droplets under experimental conditions, where the variation of local fluence within the laser spot and the dynamics of the layer separation from the target are likely to cause disruption in the thin liquid shell expanding from the irradiated area of the target.

As the laser fluence increases above the spallation threshold, the size of the region affected by the void formation is increasing and the thickness of the ejected liquid layer is decreasing. These trends are illustrated in Fig. 4.6c, where snapshots are shown for a simulation performed for an Al target at a laser fluence that is about 25 % above the spallation threshold. The appearance, growth, and coalescence of voids in this case proceed at a higher temperature and lead to the formation of a complex structure of interconnected liquid regions. The top liquid layer separates from the target by 500 ps and moves away with a velocity of about 650 m/s. Although this simulation has not been continued until the complete resolidification, we can speculate that the competition between the solidification of the surface region and a relatively slow collapse/flattening of the foamy liquid structure generated in the course of spallation may be responsible for the formation of complex nanoscale surface morphology commonly observed in femtosecond laser processing experiments, e.g., [159–162].

4.3.4 Phase Explosion and Cluster Ejection

The photomechanical spallation discussed above is responsible for the onset of the collective material removal (or ablation) from the target in the regime of stress confinement. Further increase of the laser fluence above the spallation threshold results in the separation and ejection of thinner layers and/or droplets from the target and, at a sufficiently high laser fluence, induces transition to a different ablation regime commonly called the regime of “phase explosion” or “explosive boiling.” In this regime, the surface region of the irradiated target is overheated above the limit of its thermodynamic stability, leading to an explosive decomposition of the overheated melted material into a mixture of vapor and liquid droplets. The transition to the regime of phase explosion is signified by an abrupt change in the composition of the ejected plume, from liquid layers and large droplets to a mixture of individual atoms, small clusters and droplets [45, 48, 52]. The sharp (threshold-like) increase in the amount of the vapor-phase atoms at the transition from photomechanical spallation to the phase explosion regime reflects the difference in the physical mechanisms responsible for the material ejection in these two regimes. An explosive release of vapor, rather than the relaxation of photomechanical stresses, provides the main driving force for the collective ejection of the overheated surface region of the target in the regime of phase explosion.

In the ablation by longer laser pulses, in the absence of stress confinement, the spallation is not activated and the explosive boiling is the mechanism responsible for the direct transition from the regime of surface evaporation (desorption) to the collective ejection of a mixture of vapor and liquid droplets (ablation) [45, 52, 69, 74, 75, 86, 89]. Experimental observations of the existence of a threshold fluence for the onset of droplet ejection, as well as a steep increase of the rate of the material removal at the threshold, have been interpreted as evidence of the transition from normal vaporization to the phase explosion in nanosecond laser ablation [163–166].

The thermodynamic conditions leading to the onset of the phase explosion and the dependence of the ablation process on irradiation conditions and properties of the target material have been extensively investigated in MD simulations [39, 43, 45, 48, 52, 59–91]. One of the findings of the simulations is the existence of a well-defined threshold fluence for the transition from surface evaporation to the ablation regime [45, 52, 62, 69]. This threshold behavior is related to the sharp increase in the rate of homogeneous nucleation of the vapor phase in a narrow temperature range close to about 90 % of the critical temperature, which has been predicted based on the classical nucleation theory [166–169] and confirmed in simulations [45, 48, 170]. In contrast with the assumptions of the classical nucleation theory, however, the release of the vapor phase does not involve the formation of well-defined spherical bubbles but proceeds through the formation and decomposition of a transient foamy structure of interconnected liquid regions [74, 75, 86, 171]. Similarly to the fast melting at the limit of crystal stability discussed in Sect. 4.3.1, where the homogeneous melting proceeds through the collapse of the crystal structure on the picosecond timescale, there is no time for the vapor phase regions rapidly evolving in the course of the phase explosion to minimize the liquid–vapor interfacial energy and to form spherical bubbles.

The ablation dynamics in the phase explosion regime is illustrated in Fig. 4.7, where snapshots are shown for a MD simulation of laser ablation of a molecular target irradiated by a 50 ps laser pulse at an absorbed laser fluence of 80 J/m^2, which is about 2.3 times above the ablation threshold [74, 86]. Computational system used in the simulation consists of 22 million "breathing sphere" molecules (see Sect. 4.2) and has initial dimensions of 169 $\times$ 169 $\times$ 109 nm. With the optical penetration depth of 50 nm, the irradiation conditions in this simulation correspond to the absence of stress confinement and the material ejection is driven by the release of the vapor phase in the phase explosion. In the first snapshot, shown for 200 ps after the beginning of the laser pulse, one can see the expansion of the top layer of the target where the deposited energy density is sufficient to completely vaporize the material. Deeper into the target, the energy density is not high enough to vaporize the material, but is sufficient to induce an phase explosion

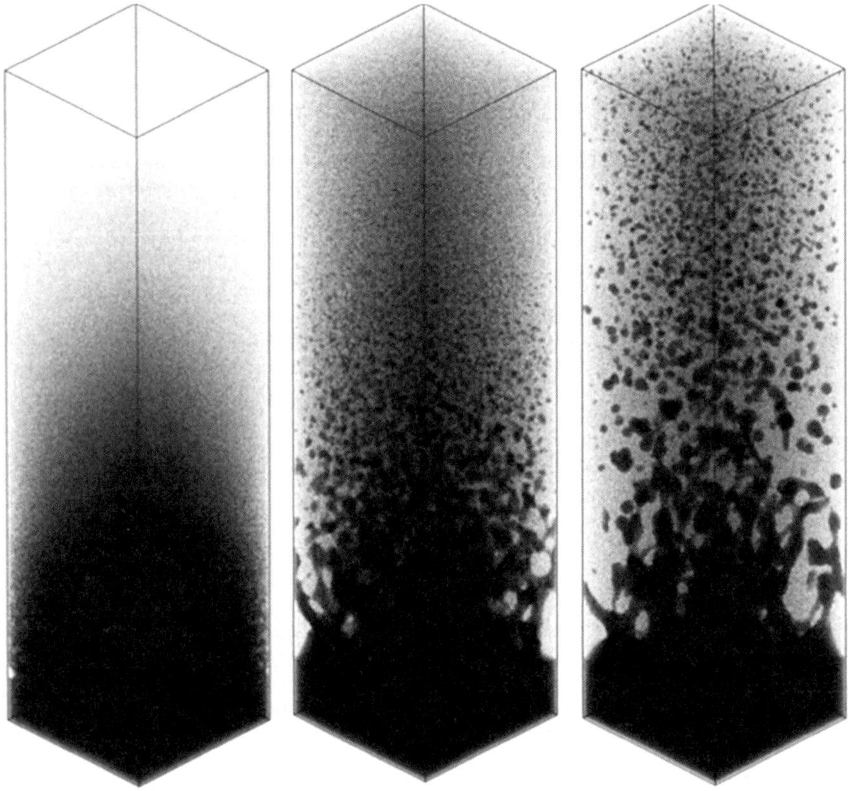

Fig. 4.7 Snapshots from a MD simulation of laser ablation of a molecular target irradiated by a 50 ps laser pulse at an absorbed fluence of 80 J/m^2. The optical penetration depth assumed in the simulation is 50 nm. Each molecule is shown as a *black dot*. Only a central part of the computational cell near the irradiated surface is shown in the snapshots

of the overheated material. The expansion of this part of the target is driven by the rapid release of vapor and proceeds through appearance of density fluctuations and formation of a transient foamy structure of interconnected liquid regions surrounded by vapor. The foamy structure eventually decomposes into individual liquid droplets, clusters and vapor-phase molecules that join the multi-component ablation plume moving away from the target (see snapshots for 600 and 1,000 ps in Fig. 4.7).

A prominent feature of the ablation process illustrated in Fig. 4.7 (and observed in MD simulations performed for different target materials, from metals to multi-component molecular systems [45, 48, 52, 74, 75, 86]) is the effect of spatial segregation of clusters/droplets of different sizes in the ablation plume. The front part of the expanding plume mostly consists of the vapor-phase molecules and small clusters, the medium size clusters are localized in the middle of the expanding plume, and the large liquid droplets formed at the final stage of the plume development tend to be slower and are closer to the original surface. The effect of the spatial segregation of clusters/droplets of different sizes in the ablation plume has been analyzed in detail for simulations performed for molecular targets with both long (no stress confinement) [74] and short (stress confinement) [172] laser pulses and has been shown to be defined by the variation of the mechanisms of the cluster formation with the depth of origin (and the energy density deposited by the laser pulse) of the material contributing to different parts of the ablation plume. The cluster segregation effect, predicted in the simulations, can be related to the results of plume imaging experiments [173–177], where splitting of the plume into a fast component with optical emission characteristic for neutral atoms and a slow component with blackbody-like emission attributed to the presence of hot clusters [178] is observed. Similarly, and consistently with the results of the simulations discussed above, a layered structure of the plume (vaporized layer followed by small particles and larger droplets) observed in nanosecond laser ablation of water and soft tissue [179] is attributed to the succession of phase transitions occurring at different depth in the irradiated target [179, 180].

An interesting observation in the simulations is that the larger droplets in the plume tend to have substantially higher internal temperatures as compared to the smaller droplets and clusters [74, 172]. Given that the larger droplets are ejected from deeper parts of the target, where the energy density deposited by the laser pulse is smaller, this observation is surprising from the first sight and calls for analogy with so-called Mpemba effect named after a secondary school student in Tanzania who observed that hot milk freezes faster than cold milk when placed into a refrigerator at the same time [181]. Similarly to the Mpemba effect, different pathways of the thermal energy dissipation are activated at different levels of the initial thermal energy, leading to the faster cooling of the parts of the plume where the initial overheating is higher. In particular, the stronger overheating of the upper part of the ablation plume leads to a more vigorous phase explosion with a larger fraction of the released vapor-phase molecules, a faster expansion, and a more efficient cooling, as compared to a slower evaporative cooling of the large droplets ejected in the tail of the plume.

4.3.5 Matrix-Assisted Pulsed Laser Evaporation

In the previous sections, the mechanisms of laser spallation and ablation are discussed for one-component metals and molecular targets. Many systems of practical importance, however, have a more complex composite structure and/or consist of multiple components. Molecular dynamics technique describes material with atomic- or molecular-level resolution and, therefore, the extension of this technique to investigation of laser interactions with multi-component targets of arbitrary structural complexity is relatively straightforward. Recent computational study of laser interactions with frozen dilute polymer solutions [72, 86, 87, 103,

104, 171] provides an example of adopting the MD model for multi-component molecular targets. This study was motivated by the need for understanding the mechanisms of molecular transport in matrix-assisted pulsed laser evaporation (MAPLE) technique for deposition of polymer and nanocomposite thin films [182, 183] and, in particular, the origin of unexpectedly large roughness of the films deposited by MAPLE [87, 103, 104, 184–193]. The simulations reveal that, contrary to the original picture of the ejection and transport of individual polymer molecules [182, 183], the deposition of polymer films is only possible in the ablation regime, when the polymer molecules are ejected as parts of polymer-matrix clusters/droplets generated in the process of the explosive disintegration of the overheated matrix [86]. The entanglement of the polymer molecules is found to facilitate the formation of intricate elongated viscous droplets that can be related to the "nanofiber" or "necklace" polymer features observed in SEM images of films deposited by MAPLE [103, 104, 186]. Moreover, in-flight molecular rearrangements in the ejected polymer-matrix droplets [87] and the dynamic processes occurring upon landing of the droplets on a substrate [104] have been connected to the formation of characteristic surface features observed in MAPLE experiments, such as wrinkled "deflated balloons," "collapsed pipes," and interconnected polymer filaments [87, 104, 186, 188–190].

The simulations of MAPLE has been recently extended to targets loaded with carbon nanotubes (CNTs) [171] and the ability of the MAPLE technique to transfer large structural elements that may be required for deposition of nano-structured films and coatings has been demonstrated. To enable the investigation of the mechanisms of the ejection and transfer of CNTs in MAPLE, the coarse-grained model for laser interactions with molecular systems has been integrated with a mesoscopic dynamic model recently developed for CNTs [194–196]. The model represents each individual CNT as a sequence of stretchable cylindrical segments. The dynamic behavior of CNTs is governed by a mesoscopic force field that accounts for the internal stretching, bending, and buckling of nanotubes as well as for the van der Waals interactions among the CNTs. The simulations performed for MAPLE targets loaded with CNTs of different length (from 16 to 150 nm) and at different CNT concentrations (from 1 to 17 wt %) demonstrate that, similarly to MAPLE of polymer molecules [86], the ejection CNTs is driven by the explosive boiling of the overheated matrix. The CNTs with length comparable to or even exceeding the laser penetration depth can be efficiently entrained in the expanding matrix plume, lifted off from the target, and transferred to the substrate. Moreover, the ejection of CNT bundles and large tangles has been observed in simulations performed for targets containing networks of interconnected CNT bundles.

The dynamics of disintegration of the network structures driven by the explosive boiling of the matrix material is illustrated in snapshots shown in Fig. 4.8. The MAPLE target used in this simulation is composed of a network of bundles consisting of more than a thousand of 150-nm long (10, 10) single-walled CNTs immersed into a frozen matrix represented by about 20 million molecules. The simulation is performed at the same laser fluence of 80 J/m^2 that is used in the

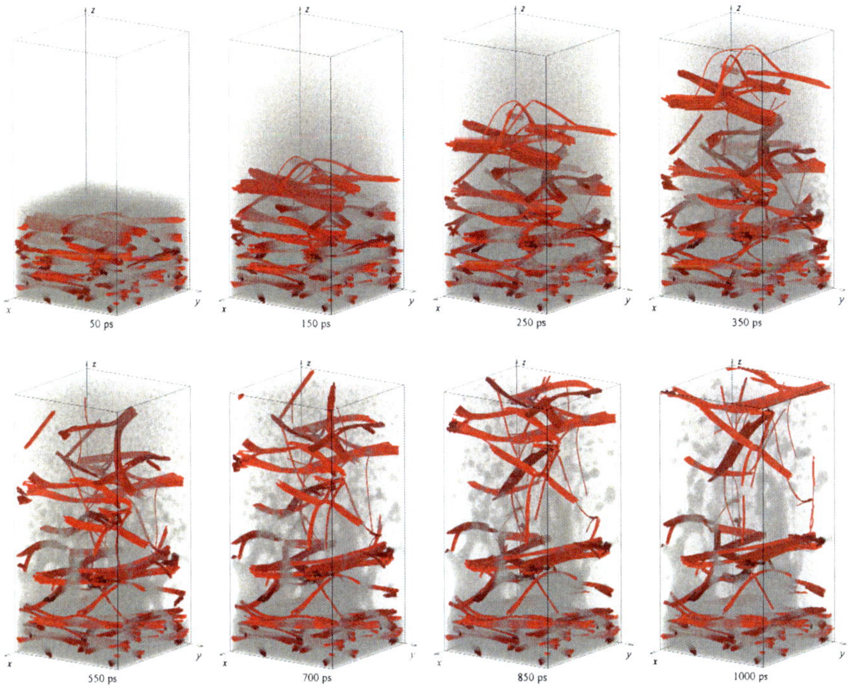

Fig. 4.8 Snapshots from a MD simulation of the ejection of CNTs from a MAPLE target loaded with 17 wt % of 150 nm long CNTs and irradiated by a 50 ps laser pulse at an absorbed fluence of 80 J/m^2. The CNTs in the target are arranged into a continuous network of bundles that is embedded into a frozen matrix. The nanotubes are shown as *red* cylinders and the matrix molecules are shown as small *gray dots*. This and other simulations of MAPLE of CNTs are discussed in [171]

simulation of laser ablation of one-component molecular matrix illustrated in Fig. 4.7. The visual picture of the explosive boiling of the overheated matrix, leading to the formation and disintegration of a transient foamy structure of interconnected liquid regions, is similar in the two simulations. The rapid expansion of the matrix plume in the simulation illustrated in Fig. 4.8 entrains the CNT bundles and tears several large CNT tangles out of the continuous network. The mass of the largest CNT tangle ejected in the simulation exceeds 50 MDa (not counting the mass of the attached matrix molecules), even though the simulation is performed for a relatively short laser penetration depth of 50 nm. No signs of any significant splitting or thinning of the thick bundles present in the initial network structure is observed in the simulation, suggesting that large fragments of the network ejected from the target can be expected to be transported to the substrate. This computational prediction is consistent with experimental observations of MAPLE deposition of CNT bundles and agglomerates with sizes comparable to or exceeding the laser penetration depth in the target [186, 197, 198]. The survival of large CNT structures held together by relatively weak van der Waals forces suggests that fragile structural elements or molecular agglomerates with complex secondary structures can be transferred and deposited to the substrate with the MAPLE technique.

4.4 Concluding Remarks and Future Directions

The development of several computational approaches for the description of laser coupling and relaxation of photo-excited states in metals, semiconductors, insulators and molecular systems within the general framework of the classical MD technique, combined with the fast growth of the available computing resources, puts MD modeling at the forefront of computational investigation of laser-materials interactions. The examples of recent applications of MD technique to investigation of laser melting, generation of crystal defects, photomechanical spallation, explosive boiling and molecular entrainment in laser ablation, briefly reviewed in this chapter, demonstrate the ability of atomic/molecular-level simulations to provide insights into the mechanisms of laser-materials interactions, to assist in interpretation of experimental observations, and to clarify some of the research questions of direct relevance to practical applications. Clear visual representations of laser-induced dynamic processes provided by series of snapshots taken at different moments during the simulations [199] is a bonus feature of MD simulations that offers a path towards the development of an intuitive picture of the complex phenomenon of laser-materials interactions.

Future progress in atomistic simulations of laser-materials interactions is likely to take advantage of the ability of massively parallel simulations to model increasingly large systems (hundreds of nanometers or 10^8–10^9 atoms) for longer simulation times (tens of nanoseconds), thus approaching the experimental time- and length-scales of processes induced by laser energy deposition. The increased accuracy of interatomic potentials in the description of phase transformations and material properties under conditions of high temperature and high pressure, characteristic of laser-materials interactions, is likely to enable material-specific computational predictions that can be quantitatively related to experimental data. The design of novel mesoscopic dynamic models may result in expansion of the domain of applicability of the MD-type coarse-grained dynamic simulations into the area of laser interactions with complex multi-phase and multi-component systems, such as nanocomposite materials and biological tissue.

Finally, MD simulations can play a key role in the design of an integrated multiscale computational model capable of accounting for interrelations between processes occurring at different time- and length-scales and providing a comprehensive picture of laser-materials interactions. In particular, classical MD simulations can serve as a bridge between the ab initio electronic structure calculations, which reveal the transient changes in the interatomic bonding and the ultrafast atomic dynamics in the electronically excited states, and continuum-level modeling of the effective (macroscopic) material response to the laser excitation.

To enable the incorporation of the predictions of ab initio calculations into the framework of MD simulations, new computational methods should be developed for the description of changes in the interatomic interactions due to the electronic excitations, as well as modification of the thermophysical and transport properties of materials in electronically excited states. Recent work in this direction includes incorporation of the contribution of so-called electron pressure [200] exerted on atoms due to the sharp electron temperature gradient generated by femtosecond laser irradiation of metals into the TTM–MD model [201–204] and the design of electron temperature dependent interatomic potentials for Si [205–207] and W [208] based on ab initio calculations performed at different levels of electronic excitation. The electron temperature dependences of the thermophysical material properties (electron–phonon coupling factor, electron heat capacity, and heat conductivity) [118], revealed in the computational analysis based on first-principles electronic structure calculations of the electron density of states, have been incorporated in the TTM–MD model and resulted in an improved agreement between the computational predictions and experimental observations [47, 132, 209, 210].

Further advancements of computational models are needed to achieve more accurate representation of the material behavior and properties in the electronically excited state in MD simulations. In particular, the uncertainty in the contribution of non-equilibrium electrons to the initial ultra-fast energy redistribution, inherent to the TTM–MD model (e.g., see discussion of the ballistic energy transport in [46, 51, 211, 212]), may be resolved by adding spatial dependence to the kinetic approaches based on the solution of the Boltzmann equation for electrons [213, 214] or Monte-Carlo treatment of individual photo-excitation and scattering events [215] and incorporating these approaches into MD models. The effect of the electronic excitation in metals cannot be entirely described by the introduction of electron pressure and additional modifications of the effective interatomic interactions should be introduced to match the predictions of ab initio calculations and experimental probing. Similarly, it may be difficult to design an adequate description of the effect of laser excitation on interatomic bonding in semiconductors and dielectrics based solely on the concept of electronic temperature, making it necessary to consider local changes in interatomic bonding in the vicinity of the optically excited states, particularly at low excitation densities.

The extension of the integrated multiscale computational approach to larger time- and length-scales that are beyond the reach of atomistic modeling can be achieved by using the detailed information on laser-induced structural and phase transformations, revealed in MD simulations, in the design of continuum-level models. The continuum modeling of laser-materials interactions have to include assumptions on the kinetics of phase transformations far from equilibrium, evolution of photomechanical damage under the action of laser-induced tensile stresses, characteristics of the ablation plume generated as a result of the explosive decomposition of the overheated surface region in laser ablation, etc. The results of MD simulations can be formulated in terms of the thermodynamic parameters such as pressure, temperature, and density distributions, as well as the structural/

phase transformations in the material, and can be used to provide the necessary information for a reliable description of fast nonequilibrium processes within a continuum model. The information on the kinetics of the homogeneous and heterogeneous melting processes [41, 44, 45, 47, 136], evolution of void size distribution in laser spallation [55, 156], and the cluster size distributions in the ablation plume [74, 86, 174] can provide ideas for the design of new constitutive relations for continuum modeling of the material behavior under conditions of rapid temperature variations and ultrafast mechanical loading realized in short-pulse laser processing.

The parameters of the ablation plume (velocity, spatial and size distributions of the ejected atoms/molecules, clusters, and droplets) predicted in MD simulations can also be used as initial conditions in the direct simulation Monte Carlo (DSMC) method [216] capable of following the evolution of the ablation plume on the time- and length-scales characteristic for experimental conditions, up to hundreds of microseconds and millimeters [217–225]. First applications of the combined MD-DSMC model for simulation of laser interactions with molecular systems [75, 174, 226–228] have demonstrated the ability of the model to reveal connections between the processes occurring at the initial stage of the plume formation and during the long-term plume expansion. In particular, the initial generation of clusters in the phase explosion, predicted in MD simulations, is found to provide cluster precursors for condensation during the long-term plume expansion, thus eliminating the three-body collision bottleneck in the cluster growth process. The presence of clusters facilitates the collisional condensation and evaporation processes, thus affecting the cluster composition of the plume as well as the overall dynamics of the plume expansion.

Acknowledgments The authors acknowledge financial support provided by the National Science Foundation (USA) through Grants DMR-0907247, CBET-1033919, and CMMI-1301298 and by the Air Force Office of Scientific Research through Grant FA9550-10-1-0541. Computational support was provided by the Oak Ridge Leadership Computing Facility (projects MAT009 and MAT048) and the National Science Foundation through the Extreme Science and Engineering Discovery Environment (projects TG-DMR110090 and TG-DMR130010).

References

1. V. Recoules, J. Clérouin, G. Zérah, P.M. Anglade, S. Mazevet, Phys. Rev. Lett. **96**, 055503 (2006)
2. F. Bottin, G. Zérah, Phys. Rev. B **75**, 174114 (2007)
3. P.B. Hillyard, D.A. Reis, K.J. Gaffney, Phys. Rev. B **77**, 195213 (2008)
4. P.L. Silvestrelli, A. Alavi, M. Parrinello, D. Frenkel, Phys. Rev. Lett. **77**, 3149–3152 (1996)
5. J.S. Graves, R.E. Allen, Phys. Rev. B **58**, 13627 (1998)
6. H.O. Jeschke, M.E. Garcia, in *Chapter 7 Nonlinear Optics, Quantum Optics, and Ultrafast Phenomena with X-Rays*, ed. by B.W. Adams (Springer, New York, 2003), pp. 175–214
7. T. Dumitrica, A. Burzo, Y. Dou, R.E. Allen, Phys. Status Solidi B **241**, 2331–2342 (2004)
8. H.O. Jeschke, M.S. Diakhate, M.E. Garcia, Appl. Phys. A **96**, 33–42 (2009)

9. B.R. Torralva, R.E. Allen, J. Mod. Opt. **49**, 593–625 (2002)
10. Y. Dou, B.R. Torralva, R.E. Allen, Chem. Phys. Lett. **392**, 352–357 (2004)
11. H.O. Jeschke, M.E. Garcia, M. Lenzner, J. Bonse, J. Krüger, W. Kautek, Appl. Surf. Sci. **197–198**, 839–844 (2002)
12. Z. Lin, R.E. Allen, J. Phys.: Condens. Matter **21**, 485503 (2009)
13. A.J. Neukirch, Z.Y. Guo, O.V. Prezhdo, J. Phys. Chem. C **116**, 15034–15040 (2012)
14. R.K. Singh, J. Narayan, Phys. Rev. B **41**, 8843–8858 (1990)
15. A. Peterlongo, A. Miotello, R. Kelly, Phys. Rev. E **50**, 4716–4727 (1994)
16. J.R. Ho, C.P. Grigoropoulos, J.A.C. Humphrey, J. Appl. Phys. **78**, 4696–4709 (1995)
17. X. Xu, G. Chen, K.H. Song, Int. J. Heat Mass Transfer **42**, 1371–1382 (1999)
18. V.A. Gnatyuk, T. Aoki, O.S. Gorodnychenko, Y. Hatanaka, Appl. Phys. Lett **83**, 3704–3706 (2003)
19. I.H. Chowdhury, X. Xu, Num. Heat Transfer A **44**, 219–232 (2003)
20. A. Bogaerts, Z. Chen, R. Gijbels, A. Vertes, Spectrochim. Acta B **58**, 1867–1893 (2003)
21. N. Bityurin, B.S. Luk'yanchuk, M.H. Hong, T.C. Chong, Chem. Rev. **103**, 519–552 (2003)
22. K.N. Vonatsos, D.I. Pantelis, Appl. Phys. A **80**, 885–889 (2005)
23. O.A. Bulgakova, N.M. Bulgakova, V.P. Zhukov, Appl. Phys. A **101**, 53–59 (2010)
24. N.M. Bulgakova, R. Stoian, A. Rosenfeld, I.V. Hertel, in *Chapter 4 in Laser-Surface Interactions for New Materials Production: Tailoring Structure and Properties*, vol. 130, Springer Series in Materials Science, ed. by A. Miotello, P.M. Ossi (Springer, New York, 2010), pp. 81–97
25. K. Eidmann, J. Meyer-ter-Vehn, T. Schlegel, S. Huller, Phys. Rev. E **62**, 1202–1214 (2000)
26. Y. Kondoh, T. Yabe, J. Maehara, T. Nakamura, Y. Ogata, Phys. Rev. E **68**, 066408 (2003)
27. F. Vidal, T.W. Johnston, J.-C. Kieffer, F. Martin, Phys. Rev. B **70**, 184125 (2004)
28. J.P. Colombier, P. Combis, F. Bonneau, R. Le Harzic, E. Audouard, Phys. Rev. B **71**, 165406 (2005)
29. M.E. Povarnitsyn, T.E. Itina, M. Sentis, K.V. Khishchenko, P.R. Levashov, Phys. Rev. B **75**, 235414 (2007)
30. A.N. Volkov, L.V. Zhigilei, J. Phys.: Conf. Ser. **59**, 640–645 (2007)
31. A.N. Volkov, C. Sevilla, L.V. Zhigilei, Appl. Surf. Sci. **253**, 6394–6399 (2007)
32. M.E. Povarnitsyn, T.E. Itina, K.V. Khishchenko, P.R. Levashov, Phys. Rev. Lett. **103**, 195002 (2009)
33. J.P. Colombier, P. Combis, E. Audouard, R. Stoian, New J. Phys. **14**, 013039 (2012)
34. M.P. Allen, D.J. Tildesley, *Computer Simulation of Liquids* (Clarendon Press, Oxford, 1987, 1990)
35. D. Frenkel, B. Smit, *Understanding molecular simulation: from algorithms to applications* (Academic Press, San Diego, 1996)
36. L.V. Zhigilei, Z. Lin, D.S. Ivanov, E. Leveugle, D.H. Duff, D. Thomas, C. Sevilla, S.J. Guy, in *Chapter 3 in Laser-Surface Interactions for New Materials Production: Tailoring Structure and Properties*, vol. 130, Springer Series in Materials Science, ed. by A. Miotello, P.M. Ossi (Springer, New York, 2010), pp. 43–79
37. C.F. Richardson, P. Clancy, Mol. Sim. **7**, 335–355 (1991)
38. H. Hakkinen, U. Landman, Phys. Rev. Lett. **71**, 1023–1026 (1993)
39. X. Wang, X. Xu, J. Heat Transfer **124**, 265–274 (2002)
40. D.S. Ivanov, L.V. Zhigilei, Phys. Rev. B **68**, 064114 (2003)
41. D.S. Ivanov, L.V. Zhigilei, Phys. Rev. Lett. **91**, 105701 (2003)
42. D.S. Ivanov, L.V. Zhigilei, Appl. Phys. A **79**, 977–981 (2004)
43. X. Wang, Y. Lu, J. Appl. Phys. **98**, 114304 (2005)
44. Z. Lin, L.V. Zhigilei, Phys. Rev. B **73**, 184113 (2006)
45. L.V. Zhigilei, Z. Lin, D.S. Ivanov, J. Phys. Chem. C **113**, 11892–11906 (2009)
46. D.A. Thomas, Z. Lin, L.V. Zhigilei, E.L. Gurevich, S. Kittel, R. Hergenröder, Appl. Surf. Sci. **255**, 9605–9612 (2009)
47. Z. Lin, E.M. Bringa, E. Leveugle, L.V. Zhigilei, J. Phys. Chem. C **114**, 5686–5699 (2010)
48. E.T. Karim, Z. Lin, L.V. Zhigilei, AIP Conf. Proc. **1464**, 280–293 (2012)

49. Z. Lin, R.A. Johnson, L.V. Zhigilei, Phys. Rev. B **77**, 214108 (2008)
50. D.S. Ivanov, Z. Lin, B. Rethfeld, G.M. O'Connor, T.J. Glynn, L.V. Zhigilei, J. Appl. Phys. **107**, 013519 (2010)
51. C. Wu, D.A. Thomas, Z. Lin, L.V. Zhigilei, Appl. Phys. A **104**, 781–792 (2011)
52. L.V. Zhigilei, B.J. Garrison, J. Appl. Phys. **88**, 1281–1298 (2000)
53. A.G. Zhidkov, L.V. Zhigilei, A. Sasaki, T. Tajima, Appl. Phys. A **73**, 741–747 (2001)
54. S.I. Anisimov, V.V. Zhakhovskii, N.A. Inogamov, K. Nishihara, A M. Oparin, Yu.V. Petrov, Pis'ma Zh. Eksp. Teor. Fiz. **77**, 731 (2003) [JETP Lett. **77**, 606–610 (2003)]
55. E. Leveugle, D.S. Ivanov, L.V. Zhigilei, Appl. Phys. A **79**, 1643–1655 (2004)
56. A.K. Upadhyay, H.M. Urbassek, J. Phys. D **38**, 2933–2941 (2005)
57. H.Y. Lai, P.H. Huang, Appl. Surf. Sci. **254**, 3067–3073 (2008)
58. B.J. Demaske, V.V. Zhakhovsky, N.A. Inogamov, I.I. Oleynik, Phys. Rev. B **82**, 064113 (2010)
59. E. Ohmura, I. Fukumoto, Int. J. Japan Soc. Prec. Eng. **30**, 128–133 (1996)
60. L.V. Zhigilei, P.B.S. Kodali, B.J. Garrison, J. Phys. Chem. B **101**, 2028–2037 (1997)
61. L.V. Zhigilei, B.J. Garrison, Appl. Phys. Lett. **71**, 551–553 (1997)
62. L.V. Zhigilei, P.B.S. Kodali, B.J. Garrison, Chem. Phys. Lett. **276**, 269–273 (1997)
63. E. Ohmura, I. Fukumoto, I. Miyamoto, Int. J. Japan Soc. Prec. Eng. **32**, 248–253 (1998)
64. R.F.W. Herrmann, J. Gerlach, E.E.B. Campbell, Appl. Phys. A **66**, 35–42 (1998)
65. X. Wu, M. Sadeghi, A. Vertes, J. Phys. Chem. B **102**, 4770–4778 (1998)
66. L.V. Zhigilei, P.B.S. Kodali, B.J. Garrison, J. Phys. Chem. B **102**, 2845–2853 (1998)
67. L.V. Zhigilei, B.J. Garrison, Rapid Commun. Mass Spectrom. **12**, 1273–1277 (1998)
68. L.V. Zhigilei, B.J. Garrison, Appl. Phys. Lett. **74**, 1341–1343 (1999)
69. L.V. Zhigilei, B.J. Garrison, Appl. Phys. A **69**, S75–S80 (1999)
70. Y.G. Yingling, L.V. Zhigilei, B.J. Garrison, J. Photochem. Photobiol. A **145**, 173–181 (2001)
71. S. Kristyan, A. Bensura, A. Vertes, Theor. Chem. Acc. **107**, 319–325 (2002)
72. T.E. Itina, L.V. Zhigilei, B.J. Garrison, J. Phys. Chem. B **106**, 303–310 (2002)
73. C. Schäfer, H.M. Urbassek, L.V. Zhigilei, Phys. Rev. B **66**, 115404 (2002)
74. L.V. Zhigilei, Appl. Phys. A **76**, 339–350 (2003)
75. L.V. Zhigilei, E. Leveugle, B.J. Garrison, Y.G. Yingling, M.I. Zeifman, Chem. Rev. **103**, 321–347 (2003)
76. P. Lorazo, L.J. Lewis, M. Meunier, Phys. Rev. Lett. **91**, 225502 (2003)
77. N.N. Nedialkov, P.A. Atanasov, S.E. Imamova, A. Ruf, P. Berger, F. Dausinger, Appl. Phys. A **79**, 1121–1125 (2004)
78. X. Xu, C. Cheng, I.H. Chowdhury, ASME Trans. J. Heat Transfer **126**, 727–734 (2004)
79. C. Cheng, X. Xu, Phys. Rev. B **72**, 165415 (2005)
80. X.W. Wang, J. Phys. D **38**, 1805–1823 (2005)
81. Y.G. Yingling, B.J. Garrison, J. Phys. Chem. B **109**, 16482–16489 (2005)
82. P. Lorazo, L.J. Lewis, M. Meunier, Phys. Rev. B **73**, 134108 (2006)
83. N.N. Nedialkov, P.A. Atanasov, Appl. Surf. Sci. **252**, 4411–4415 (2006)
84. S.I. Anisimov, V.V. Zhakhovskii, N.A. Inogamov, K. Nishihara, YuV Petrov, V.A. Khokhlov, J. Exp. Theor. Phys. **103**, 183–197 (2006)
85. M.B. Agranat, S.I. Anisimov, S.I. Ashitkov, V.V. Zhakhovskii, N.A. Inogamov, K. Nishihara, YuV Petrov, V.E. Fortov, V.A. Khokhlov, Appl. Surf. Sci. **253**, 6276–6282 (2007)
86. E. Leveugle, L.V. Zhigilei, J. Appl. Phys. **102**, 074914 (2007)
87. E. Leveugle, A. Sellinger, J.M. Fitz-Gerald, L.V. Zhigilei, Phys. Rev. Lett. **98**, 216101 (2007)
88. M. Prasad, P. Conforti, B.J. Garrison, J. Appl. Phys. **101**, 103113 (2007)
89. L. Zhang, X. Wang, Appl. Surf. Sci. **255**, 3097–3103 (2008)
90. M. Gill-Comeau, L.J. Lewis, Phys. Rev. B **84**, 224110 (2011)
91. X. Li, L. Jiang, Appl. Phys. A **109**, 367–376 (2012)
92. B.J. Garrison, D. Srivastava, Annu. Rev. Phys. Chem. **46**, 373–394 (1995)

93. A.F. Voter, MRS Bull. **21**, 17–18 (1996)
94. S.B. Sinnott, D.W. Brenner, MRS Bull. **37**, 469–473 (2012)
95. S.L. Johnson, M.R. Papantonakis, R.F. Haglund, in *Chapter 8 in Laser-Surface Interactions for New Materials Production: Tailoring Structure and Properties*, vol. 130, Springer Series in Materials Science, ed. by A. Miotello, P.M. Ossi (Springer, New York, 2010), pp. 177–202
96. R.E. Wyatt, C. Iung, C. Leforestier, Acc. Chem. Res. **28**, 423–429 (1995)
97. V. Kozich, W. Werncke, J. Dreyer, K.-W. Brzezinka, M. Rini, A. Kummrow, T. Elsaesser, J. Chem. Phys. **117**, 719–726 (2002)
98. H. Kim, D.D. Dlott, Y. Won, J. Chem. Phys. **102**, 5480–5485 (1995)
99. J.C. Deàk, L.K. Iwaki, S.T. Rhea, D.D. Dlott, J. Raman Spectr. **31**, 263–274 (2000)
100. E.L. Sibert, R. Rey, J. Chem. Phys. **116**, 237–257 (2002)
101. M. Sadeghi, X. Wu, A. Vertes, J. Phys. Chem. B **105**, 2578–2587 (2001)
102. Ł. Dutkiewicz, R.E. Johnson, A. Vertes, R. Pędrys, J. Phys. Chem. A **103**, 2925–2933 (1999)
103. E. Leveugle, L.V. Zhigilei, A. Sellinger, J.M. Fitz-Gerald, Appl. Surf. Sci. **253**, 6456–6460 (2007)
104. A. Sellinger, E. Leveugle, J.M. Fitz-Gerald, L.V. Zhigilei, Appl. Phys. A **92**, 821–829 (2008)
105. R. Knochenmuss, L.V. Zhigilei, J. Phys. Chem. B **109**, 22947–22957 (2005)
106. R. Knochenmuss, L.V. Zhigilei, J. Mass Spectrom. **45**, 333–346 (2010)
107. R. Knochenmuss, L.V. Zhigilei, Anal. Bioanal. Chem. **402**, 2511–2519 (2012)
108. B.J. Garrison, R. Srinivasan, J. Appl. Phys. **57**, 2909–2914 (1985)
109. B.J. Garrison, R. Srinivasan, Appl. Phys. Lett. **44**, 849–851 (1984)
110. Y.G. Yingling, B.J. Garrison, J. Phys. Chem. B **108**, 1815–1821 (2004)
111. Y. Dou, L.V. Zhigilei, N. Winograd, B.J. Garrison, J. Phys. Chem. A **105**, 2748–2755 (2001)
112. Y. Dou, L.V. Zhigilei, Z. Postawa, N. Winograd, B.J. Garrison, Nucl. Instr. Meth. B **180**, 105–111 (2001)
113. Y. Dou, N. Winograd, B.J. Garrison, L.V. Zhigilei, J. Phys. Chem. B **107**, 2362–2365 (2003)
114. K.M. Smith, M.Y. Hussaini, L.D. Gelb, S.D. Allen, Appl. Phys. A **77**, 877–882 (2003)
115. X. Gu, H.M. Urbassek, Appl. Phys. B **81**, 675–679 (2005)
116. X. Gu, H.M. Urbassek, Appl. Surf. Sci. **253**, 4142–4149 (2007)
117. O. Stein, Z. Lin, L.V. Zhigilei, M. Asscher, J. Phys. Chem. A **115**, 6250–6259 (2011)
118. Z. Lin, L.V. Zhigilei, V. Celli, Phys. Rev. B **77**, 075133 (2008)
119. F. Gao, D.J. Bacon, P.E.J. Flewitt, T.A. Lewis, Modelling Simul. Mater. Sci. Eng. **6**, 543–556 (1998)
120. D.M. Duffy, A.M. Rutherford, T. Phys. Cond. Matter **19**, 016207 (2007)
121. J.D. Schall, C.W. Padgett, D.W. Brenner, Mol. Simulat. **31**, 283–288 (2005)
122. S.I. Anisimov, B.L. Kapeliovich, T.L. Perel'man, Sov. Phys. JETP **39**, 375–377 (1974)
123. R. Holenstein, S.E. Kirkwood, Y.Y. Tsui, R. Fedosejevs, Proc. SPIE **5579**, 688–695 (2004)
124. Y. Wang, X. Xu, L. Zheng, Appl. Phys. A **92**, 849–852 (2008)
125. Y. Cherednikov, N.A. Inogamov, H.M. Urbassek, J. Opt. Soc. Am. B **28**, 1817–1824 (2011)
126. S. Williamson, G. Mourou, J.C.M. Li, Phys. Rev. Lett. **52**, 2364–2367 (1984)
127. M.B. Agranat, S.I. Ashitkov, V.E. Fortov, A.V. Kirillin, A.V. Kostanovskii, S.I. Anisimov, P.S. Kondratenko, Appl. Phys. A Mater. Sci. Process. **69**, 637–640 (1999)
128. B. Lin, H.E. Elsayed-Ali, Surf. Sci. **498**, 275–284 (2002)
129. B.J. Siwick, J.R. Dwyer, R.E. Jordan, R.J.D. Miller, Science **302**, 1382–1385 (2003)
130. K. Sokolowski-Tinten, C. Blome, J. Blums, A. Cavalleri, C. Dietrich, A. Tarasevich, I. Uschmann, E. Förster, M. Kammler, M. Horn-von-Hoegen, D. von der Linde, Nature (London) **422**, 287–289 (2003)
131. J.R. Dwyer, C.T. Hebeisen, R. Ernstorfer, M. Harb, V. Deyirmenjian, R.E. Jordan, R.J.D. Miller, Phil. Trans. R. Soc. A **364**, 741–778 (2006)

132. J.R. Dwyer, R.E. Jordan, C.T. Hebeisen, M. Harb, R. Ernstorfer, T. Dartigalongue, R.J.D. Miller, J. Mod. Optics **54**, 905–922 (2007)
133. C.Y. Ruan, Y. Murooka, R.K. Raman, R.A. Murdick, Nano Lett. **7**, 1290–1296 (2007)
134. W.L. Chan, R.S. Averback, D.G. Cahill, A. Lagoutchev, Phys. Rev. B **78**, 214107 (2008)
135. R. Ernstorfer, M. Harb, C.T. Hebeisen, G. Sciaini, T. Dartigalongue, R.J.D. Miller, Science **323**, 1033–1037 (2009)
136. D.S. Ivanov, L.V. Zhigilei, Phys. Rev. Lett. **98**, 195701 (2007)
137. D.Y. Sun, M. Asta, J.J. Hoyt, Phys. Rev. B **69**, 024108 (2004)
138. F. Celestini, J.-M. Debierre, Phys. Rev. E **65**, 041605 (2002)
139. S.-N. Luo, T.J. Ahrens, T. Çağın, A. Strachan, W.A. Goddard III, D.C. Swift, Phys. Rev. B **68**, 134206 (2003)
140. A.B. Belonoshko, N.V. Skorodumova, A. Rosengren, B. Johansson, Phys. Rev. B **73**, 012201 (2006)
141. A. Vogel, V. Venugopalan, Chem. Rev. **103**, 577–644 (2003)
142. G. Paltauf, P.E. Dyer, Chem. Rev. **103**, 487–518 (2003)
143. I. Itzkan, D. Albagli, M.L. Dark, L.T. Perelman, C. von Rosenberg, M.S. Feld, Proc. Natl. Acad. Sci. U.S.A. **92**, 1960–1964 (1995)
144. A.A. Oraevsky, S.L. Jacques, F.K. Tittel, J. Appl. Phys. **78**, 1281–1290 (1995)
145. R. Cramer, R.F. Haglund, F. Hillenkamp, Int. J. Mass Spectrom. Ion Processes **169/170**, 51–67 (1997)
146. W.H. Duff, L.V. Zhigilei, J. Phys.: Conf. Ser. **59**, 413–417 (2007)
147. S.I. Anisimov, V.V. Zhakhovskii, N.A. Inogamov, K. Nishihara, Yu. V. Petrov, Appl. Surf. Sci. **253**, 6390–6393 (2007)
148. V. Vitek, Philos. Mag. **21**, 1275–1278 (1970)
149. D. Ashkenasi, M. Lorenz, R. Stoian, A. Rosenfeld, Appl. Surf. Sci. **150**, 101–106 (1999)
150. P.T. Mannion, J. Magee, E. Coyne, G.M. O'Connor, T.J. Glynn, Appl. Surf. Sci. **233**, 275–287 (2004)
151. S.E. Kirkwood, A.C. van Popta, Y.Y. Tsui, R. Fedosejevs, Appl. Phys. A **81**, 729–735 (2005)
152. J. Krüger, D. Dufft, R. Koter, A. Hertwig, Appl. Surf. Sci. **253**, 7815–7819 (2007)
153. G. Raciukaitis, M. Brikas, P. Gecys, M. Gedvilas, Proc. SPIE **7005**, 70052L (2008)
154. L.V. Zhigilei, B.J. Garrison, Mat. Res. Soc. Symp. Proc. **538**, 491–496 (1999)
155. C. Schafer, H.M. Urbassek, L.V. Zhigilei, B.J. Garrison, Comp. Mater. Sci. **24**, 421–429 (2002)
156. L.V. Zhigilei, D.S. Ivanov, E. Leveugle, B. Sadigh, E.M. Bringa, in *High-Power Laser AblationV*, ed. by C.R. Phipps. Proceedings of SPIE, vol. 5448, pp. 505–519 (2004)
157. D. Perez, L.J. Lewis, Phys. Rev. B **67**, 184102 (2003)
158. J.-M. Savolainen, M.S. Christensen, P. Balling, Phys. Rev. B **84**, 193410 (2011)
159. A.Y. Vorobyev, C. Guo, Phys. Rev. B **72**, 195422 (2005)
160. A.Y. Vorobyev, C. Guo, Appl. Phys. A **86**, 321–324 (2007)
161. Q.-Z. Zhao, S. Malzer, L.-J. Wang, Opt. Express **15**, 15741–15746 (2007)
162. Y. Dai, M. He, H. Bian, B. Lu, X. Yan, G. Ma, Appl. Phys. A **106**, 567–574 (2012)
163. K.H. Song, X. Xu, Appl. Surf. Sci. **127–129**, 111–116 (1998)
164. J.H. Yoo, S.H. Jeong, X.L. Mao, R. Greif, R.E. Russo, Appl. Phys. Lett. **76**, 783–785 (2000)
165. C. Porneala, D.A. Willis, Appl. Phys. Lett. **89**, 211121 (2006)
166. N.M. Bulgakova, A.V. Bulgakov, Appl. Phys. A **73**, 199–208 (2001)
167. M.M. Martynyuk, Sov. Phys. Tech. Phys. **21**, 430–433 (1976)
168. A. Miotello, R. Kelly, Appl. Phys. A **69**, S67–S73 (1999)
169. R. Kelly, A. Miotello, J. Appl. Phys. **87**, 3177–3179 (2000)
170. B.J. Garrison, T.E. Itina, L.V. Zhigilei, Phys. Rev. E **68**, 041501 (2003)
171. L.V. Zhigilei, A.N. Volkov, E. Leveugle, M. Tabetah, Appl. Phys. A **105**, 529–546 (2011)
172. L.V. Zhigilei, Mat. Res. Soc. Symp. Proc. **677**, AA2.1.1–AA2.1.11 (2001)
173. S. Noël, J. Hermann, T. Itina, Appl. Surf. Sci. **253**, 6310–6315 (2007)

174. T.E. Itina, K. Gouriet, L.V. Zhigilei, S. Noël, J. Hermann, M. Sentis, Appl. Surf. Sci. **253**, 7656–7661 (2007)
175. S. Amoruso, R. Bruzzese, C. Pagano, X. Wang, Appl. Phys. A **89**, 1017–1024 (2007)
176. O. Albert, S. Roger, Y. Glinec, J.C. Loulergue, J. Etchepare, C. Boulmer-Leborgne, J. Perriere, E. Millon, Appl. Phys. A **76**, 319–323 (2003)
177. N. Jegenyes, J. Etchepare, B. Reynier, D. Scuderi, A. Dos-Santos, Z. Tóth, Appl. Phys. A **91**, 385–392 (2008)
178. D. Scuderi, R. Benzerga, O. Albert, B. Reynier, J. Etchepare, Appl. Surf. Sci. **252**, 4360–4363 (2006)
179. I. Apitz, A. Vogel, Appl. Phys. A **81**, 329–338 (2005)
180. A. Vogel, I. Apitz, V. Venugopalan, in *Oscillations, Waves and Interactions*, ed. by T. Kurz, U. Parlitz, U. Kaatze (Universitätsverlag Göttingen, Göttingen, 2007), pp. 217–258
181. E.B. Mpemba, D.G. Osborne, Phys. Educ. **4**, 172–175 (1969)
182. A. Piqué, R.A. McGill, D.B. Chrisey, D. Leonhardt, T.E. Mslna, B.J. Spargo, J.H. Callahan, R.W. Vachet, R. Chung, M.A. Bucaro, Thin Solid Films **355/356**, 536–541 (1999)
183. D.B. Chrisey, A. Piqué, R.A. McGill, J.S. Horwitz, B.R. Ringeisen, D.M. Bubb, P.K. Wu, Chem. Rev. **103**, 553–576 (2003)
184. D.M. Bubb, P.K. Wu, J.S. Horwitz, J.H. Callahan, M. Galicia, A. Vertes, R.A. McGill, E.J. Houser, B.R. Ringeisen, D.B. Chrisey, J. Appl. Phys. **91**, 2055–2058 (2002)
185. K. Rodrigo, P. Czuba, B. Toftmann, J. Schou, R. Pedrys, Appl. Surf. Sci. **252**, 4824–4828 (2006)
186. A.T. Sellinger, E.M. Leveugle, K. Gogick, L.V. Zhigilei, J.M. Fitz-Gerald, J. Vac. Sci. Technol. A **24**, 1618–1622 (2006)
187. R. Pate, A.D. Stiff-Roberts, Chem. Phys. Lett. **477**, 406–410 (2009)
188. A.P. Caricato, M. Anni, M.G. Manera, M. Martino, R. Rella, F. Romano, T. Tunno, D. Valerini, Appl. Surf. Sci. **255**, 9659–9664 (2009)
189. A.P. Caricato, G. Leggieri, M. Martino, A. Vantaggiato, D. Valerini, A. Cretì, M. Lomascolo, M.G. Manera, R. Rella, M. Anni, Appl. Phys. A **101**, 759–764 (2010)
190. D.M. Bubb, J. Corgan, S.Y. Yi, M. Khan, L. Hughes, U. Gurudas, M. Papantonakis, R.A. McGill, Appl. Phys. A **100**, 523–531 (2010)
191. S. Guha, D. Adil, N.B. Ukah, R.K. Gupta, K. Ghosh, Appl. Phys. A **105**, 547–554 (2011)
192. Y. Guo, A. Morozov, D. Schneider, J.W. Chung, C. Zhang, M. Waldmann, N. Yao, G. Fytas, C.B. Arnold, R.D. Priestley, Nature Mater. **11**, 337–343 (2012)
193. K.B. Shepard, Y. Guo, C.B. Arnold, R.D. Priestley, Appl. Phys. A **110**, 771–777 (2013)
194. L.V. Zhigilei, C. Wei, D. Srivastava, Phys. Rev. B **71**, 165417 (2005)
195. A.N. Volkov, L.V. Zhigilei, J. Phys. Chem. C **114**, 5513–5531 (2010)
196. A.N. Volkov, L.V. Zhigilei, ACS Nano **4**, 6187–6195 (2010)
197. A.T. Sellinger, A.H. Martin, J.M. Fitz-Gerald, Thin Solid Films **516**, 6033–6040 (2008)
198. Á.P. del Pino, E. György, L. Cabana, B. Ballesteros, G. Tobias, Carbon **50**, 4450–4458 (2012)
199. Animated sequences of snapshots from MD simulations of laser melting, spallation, and ablation can be found at http://www.faculty.virginia.edu/CompMat/Resources.html
200. L.A. Falkovsky, E.G. Mishchenko, J. Exp. Theor. Phys. **88**, 84–88 (1999)
201. Y. Gan, J.K. Chen, Appl. Phys. Lett. **94**, 201116 (2009)
202. Y. Gan, J.K. Chen, J. Appl. Phys. **108**, 103102 (2010)
203. G.E. Norman, S.V. Starikov, V.V. Stegailov, J. Exp. Theor. Phys. **114**, 792–800 (2012)
204. G. Norman, S. Starikov, V. Stegailov, V. Fortov, I. Skobelev, T. Pikuz, A. Faenov, S. Tamotsu, Y. Kato, M. Ishino, M. Tanaka, N. Hasegawa, M. Nishikino, T. Ohba, T. Kaihori, Y. Ochi, T. Imazono, Y. Fukuda, M. Kando, T. Kawachi, J. Appl. Phys. **112**, 013104 (2012)
205. L. Shokeen, P.K. Schelling, Appl. Phys. Lett. **97**, 151907 (2010)
206. L. Shokeen, P.K. Schelling, J. Appl. Phys. **109**, 073503 (2011)
207. L. Shokeen, P.K. Schelling, Comput. Mater. Sci. **67**, 316–328 (2013)
208. S. Khakshouri, D. Alfè, D.M. Duffy, Phys. Rev. B **78**, 224304 (2008)
209. Z. Lin, L.V. Zhigilei, Proc. SPIE **6261**, 62610U (2006)

210. Z. Lin, L.V. Zhigilei, Appl. Surf. Sci. **253**, 6295–6300 (2007)
211. C. Suárez, W.E. Bron, T. Juhasz, Phys. Rev. Lett. **75**, 4536–4539 (1995)
212. Z. Chen, V. Sametoglu, Y.Y. Tsui, T. Ao, A. Ng, Phys. Rev. Lett. **108**, 165001 (2012)
213. B. Rethfeld, A. Kaiser, M. Vicanek, G. Simon, Phys. Rev. B **65**, 214303 (2002)
214. W. Wendelen, B.Y. Mueller, D. Autrique, B. Rethfeld, A. Bogaerts, J. Appl. Phys. **111**, 113110 (2012)
215. N. Medvedev, U. Zastrau, E. Förster, D.O. Gericke, B. Rethfeld, Phys. Rev. Lett. **107**, 165003 (2011)
216. G.A. Bird, *Molecular Gas Dynamics and the Direct Simulation of Gas Flows* (Clarendon Press, Oxford, 1994)
217. H.M. Urbassek, D. Sibold, Phys. Rev. Lett. **70**, 1886–1889 (1993)
218. O. Ellegaard, J. Schou, H.M. Urbassek, Appl. Phys. A **69**, S577–S582 (1999)
219. T.E. Itina, J. Appl. Phys. **89**, 740–746 (2001)
220. T.E. Itina, W. Marine, M. Autric, J. Appl. Phys. **82**, 3536–3542 (1997)
221. A.A. Morozov, Phys. Fluids **19**, 087101 (2007)
222. G.A. Lukyanov, Y. Khang, D.V. Leshchev, S.V. Kozyrev, A.N. Volkov, N.Y. Bykov, O.I. Vakulova, Fuller. Nanotub. Car. N **14**, 507–512 (2006)
223. T.E. Itina, M. Sentis, W. Marine, Appl. Surf. Sci. **252**, 4433–4438 (2006)
224. A.N. Volkov, G.M. O'Connor, T.J. Glynn, G.A. Lukyanov, Appl. Phys. A **92**, 927–932 (2008)
225. A.N. Volkov, L.V. Zhigilei, Appl. Phys. A **110**, 537–546 (2013)
226. M.I. Zeifman, B.J. Garrison, L.V. Zhigilei, J. Appl. Phys. **92**, 2181–2193 (2002)
227. M.I. Zeifman, B.J. Garrison, L.V. Zhigilei, Appl. Surf. Sci. **197/198**, 27–34 (2002)
228. T. E. Itina and L. V. Zhigilei, J. Phys.: Conf. Ser. **59**, 44–49 (2007)

Chapter 5
Continuum Models of Ultrashort Laser–Matter Interaction in Application to Wide-Bandgap Dielectrics

Nadezhda M. Bulgakova and Vladimir P. Zhukov

Abstract This chapter is aimed to provide a basic introduction into the principles of modeling approaches which have been developed for getting insight into various interconnected processes initiated inside transparent materials under the action of ultrashort laser pulses with consequences in volumetric modification of material structure. In view of extreme complexity of the problem, modification mechanisms and their driving processes are still far from complete understanding and require further considerable research efforts. Here we focus our consideration on established approaches that treat matter as a continuum medium. They include models describing laser beam propagation through a non-linear transparent glass or crystal with kinetics of electron plasma generation upon beam focusing and attempts to consider further material evolution with insights into thermodynamic state, stress dynamics, and plastic deformations. We underline that the quality of the final structures is determined by the synergetic action of laser excitation/relaxation kinetics, thermodynamics, and mechanics. The chapter does not pretend to completeness and aims to outline main ideas, achievements, and most intriguing findings which are still waiting for explanations and theoretical treatments.

N. M. Bulgakova (✉)
Optoelectronics Research Centre, University of Southampton,
Southampton SO17 1BJ, UK
e-mail: nbul@itp.nsc.ru

N. M. Bulgakova
Institute of Thermophysics SB RAS, 1 Lavrentyev Avenue,
Novosibirsk, Russia630090,

V. P. Zhukov
Institute of Computational Technologies SB RAS, 6 Lavrentyev Avenue,
Novosibirsk, Russia630090,

V. P. Zhukov
Novosibirsk State Technical University, 20 Karl Marx Avenue,
Novosibirsk, Russia630073,

M. Castillejo et al. (eds.), *Lasers in Materials Science*,
Springer Series in Materials Science 191, DOI: 10.1007/978-3-319-02898-9_5,

5.1 Introduction

Since 1996 when it was demonstrated that tightly focused femtosecond laser pulses could induce a local internal increase of the refractive index inside bulk transparent glasses [1], the interaction of ultrashort laser pulses with transparent optical materials has attracted a lot of attention as a powerful tool for modification of material properties [2] resulting in generation of surface [3–7] and volume [5, 8–13] periodic structures, densification and refractive index changes [1, 14–20], formation of micro- and nanovoids [9, 21–27], phase transitions (crystallization in amorphous materials and amorphization of crystalline ones) [23, 28]. This gives rise to numerous technological applications based on three-dimensional photonic structures in bulk optical materials, such as waveguides [1, 14–20], Bragg gratings [29, 30], Fresnel zone plates [31], waveplates based on volume nanogratings (VNG) [32], splitters [33], couplers [34, 35], amplifiers [36], rewritable optical memories [37, 38], and computer-generated holograms [39]. As seen from the above citations, the field of laser writing of optical structures in glasses has been rapidly developing during last 15 years and the laser-written structures become the key elements of integrated photonic devices.

While tremendous achievements have been demonstrated toward laser-writing techniques and assembling integrated photonic devices, the physical mechanisms underlying glass modifications have not been fully understood. Here we discuss general principles of applying continuum approaches for numerical modeling of the variety of interconnected processes induced in wide-bandgap dielectric materials by ultrashort pulse laser radiation, starting from the excitation stage and extending to microsecond timescales when a final structure is imprinted into the material matrix. Apart from the fact that the theory and modeling of the laser-induced processes can be the cost reducing tools which may allow choosing the optimal conditions and most appropriate materials for particular desired modifications, they provide a detailed physical understanding of the phenomenon and required material properties for technological applications. We underline, that in view of extreme complexity of the problem, modification mechanisms and their driving processes are still far from complete understanding and require further considerable research efforts as well as developing novel advanced models of both continuum and atomistic/molecular kinds. It should be noticed that this chapter can be considered as a continuation of [40] where the general principles of continuum modeling in application to ultrashort pulsed laser ablation of solids are discussed.

5.2 Ultrafast Laser Excitation of Wide-Bandgap Dielectrics

First we consider the fundamental aspects of ultrafast laser excitation specific for inorganic dielectric materials with an assessment of the state-of-art and further research directions. Successful development of applications based on laser-induced micro- and nanomodifications of transparent materials requires deep understanding of the whole chain of the intricate processes initiated in dielectrics by fs laser pulses and extending up to millisecond time scales with formation of permanent mechanically deformed and/or chemically modified states. The mentioned chain of the processes starts from material photoionization with creation of seed free electrons. The latters absorb laser energy and, at proper conditions, can produce secondary electrons in collisions with neutral atoms of dielectric matrix, thus generating electron avalanche. This results in considerable change of optical response of the laser-irradiated region towards its "metallization". Laser beams focused inside the bulk of a transparent material can experience self-focusing starting from a material-dependent threshold power. However, the beam self-focusing collapse is arrested by scattering from the laser produced plasma. In some dielectrics, active recombination of free electrons starts already at sub-ps time scales while in other dielectrics the free electron gas survives up to hundreds of ps [41]. Rapid recombination of free electrons leads to swift heating of the photoexcited region that occurs at ps timescale when heat conduction effects are negligible. As a result of rapid heat release into the atomic subsystem and corresponding pressure rise, thermoelastic stress waves are generated which, depending on the heating level and heat localization, can either completely dissipate, or lead to significant plastic deformations of the material, or even to mechanical damage in the form of micro- and nanovoids in the energy-release zone [42].

It should be underlined that the stress waves not only induce deformations in a hot laser-excited region but also create a hoop stress in an extended cold zone around it. The thermomechanical effects terminate at ~ 10 ns after the laser pulse action when the three-dimensional pressure waves propagate a distance of several micrometers and substantially dissipate. However, under some experimental conditions the 'mechanical scenario' can repeat at microsecond timescale after laser pulse termination when the locally released energy spreads due to heat conduction and reaches the regions of 'cold' deformation. Softening of the deformed regions can cause secondary redistribution of matter in the laser-affected zone [20] and even emission of secondary thermoelastic waves.

The main mechanisms involved in generation of free carriers in wide-bandgap dielectrics irradiated by visible and near-IR fs laser pulses are the multiphoton and avalanche ionizations. To describe temporal evolution of the free electron density $n_e(t)$, a simple but intuitive rate equation can be written in the following form [43–47]:

$$\frac{\partial n_e(t)}{\partial t} = \sigma_k I^k(t) + \delta I(t) n_e(t) . \tag{5.1}$$

Here $I(t)$ is laser intensity; σ_k and δ are the multiphoton ionization (MPI) cross section and the avalanche coefficient respectively: k is the number of photons required for an MPI event. This equation represents a simplest way for estimating the experimental conditions taking into account generating non-homogeneous profiles of the electron density in the surface layer or inside the bulk depth and changing the optical response of laser-excited matter [43–52]. The excitation process can conventionally be divided into two stages. At the initial stage, free electrons are generated via the multiphoton ionization while avalanche (collisional) ionization develops when a definite "seed" level of the free-electron density is reached which is sufficient to efficiently absorb laser light.

At relatively high radiation intensities, the tunneling ionization mechanism can dominate over multiphoton ionization as determined by the Keldysh parameter $\gamma = \omega(2m_{\mathrm{eff}}E_g)^{0.5}/(eE_{\mathrm{L}})$ [53–56] where m_{eff} is the effective electron mass; E_{L} is the electric field of the laser wave; E_{g} is the band gap width in dielectric materials or ionization potential of individual atoms or molecules; here and everywhere below the symbol e denotes the elementary (positive) charge. The Keldysh parameter can be presented as a ratio between the characteristic time that an electron takes to overcome the energy barrier (the ionization potential whose value can vary in the strong-wave field) and the electromagnetic wave field period. At $\gamma \gg 1$, the MPI mechanism prevails while at $\gamma \ll 1$ the tunneling mechanism becomes dominating. It is widely accepted that, for the tunneling ionization to produce a noticeable effect, the condition $\gamma < 0.5$ should be met. However, a number of experimental facts indicate that multiphoton ionization can dominate even at $\gamma \ll 1$ [55]. These facts gave rise to numerous generalizations of the Keldysh theory of photo-ionization of dielectric materials [56], as well as for atoms and molecules in the gas phase [55].

It must be admitted that the tunneling mechanism of ionization can play a significant role upon focusing on the surface layer of material in vacuum or a low-pressure gas environment. At focusing into material bulk or on its surface in the presence of a dense ambient gas or a liquid surrounding, ionization of medium starts before focus at reaching a definite level of laser intensity. As a result, a high-energy beam is attenuated in its way to the laser focus that leads to the intensity constraints at levels at which the MPI mechanism prevails unambiguously (so-called clamping effect, see [57–59] and Sect. 5.3). It should be noted that, for relatively long wavelengths of femtosecond laser pulses, toward the mid-IR range, the MPI rate strongly decreases and tunneling ionization will inevitably play the dominant role. Although the femtosecond lasers at mid-IR wavelengths are seldom used today, this trend should be mentioned for providing more complete understanding.

The role of avalanche ionization in the breakdown of dielectrics in ultrafast irradiation regimes is still debated [60–63]. Some authors completely deny its existence for sub-ps pulsed irradiation regimes [61] while the others assert that its effects can be even more pronounced with decreasing pulse duration due to a decrease of the potential barrier in a strong laser field (cold avalanche) [63]. Electron recombination in inorganic dielectrics proceeds in the form of trapping in localized states (the recombination rate may be expressed as $R_e = n_e/\tau_{\mathrm{tr}}$ where τ_{tr} is a characteristic recombination time), accompanied by creation of excitons, color centers, non-bridging oxygen hole centers, and other defects [64–67]. Defect generation leads to pronounced incubation effects manifested as a decrease of the damage threshold in multipulse irradiation regimes [66]. For some materials with very short trapping time (e.g., for fused silica where $\tau_{tr} \approx 150$ fs [67]), the incubation effect, i.e. accumulation of the defect states and their preferential ionization, can appear already during fs laser pulse action.

Based on the above consideration, the kinetic scenario of excitation of an inorganic dielectric material can be described as following. Multiphoton ionization, whose order is determined by the ratio between the bandgap width and photon energy $\hbar\omega$, results in excitation of electrons from the valance band to a low-energy state within the conduction band. At high laser intensities when $\gamma \leq 1$, the tunneling mechanism of ionization becomes dominating in photo-production of free carries. The free electrons may now efficiently absorb laser radiation due to inverse bremsstrahlung. When an electron has absorbed a sufficient amount of the laser energy ($>E_g$), it can collisionally ionize a neutral atom of the material matrix. The development of avalanche (collisional multiplication of free electrons) changes optical properties of the laser-excited region of the material. Spatiotemporal dynamics of optical parameters can be described within the Drude formalism via the complex dielectric function $\varepsilon^*(n_e)$ whose value can be seen as contributions from unexcited matter and generated dense plasma [47, 68]:

$$\varepsilon^*(n_e) \cong 1 + (\varepsilon_g - 1)\left(1 - \frac{n_e}{n_{val}}\right) - \frac{n_e}{n_{cr}}\left(1 + i\frac{1}{\omega\tau_c}\right)^{-1}. \tag{5.2}$$

Here $n_{cr} = \varepsilon_0 n_0 m_{eff}\omega^2/e^2$ is the critical electron density n_{val} is the total valence-band electron density in the unexcited state; n_0 is the refractive index; ε_g is the dielectric function of unexcited material; ε_0 is the vacuum permittivity; $\omega\tau_c$ is the damping factor [64] determined by the finite electron collision time τ_c (see comments in Sect. 5.3.2). We underline that the reflection coefficient from an inhomogeneous dense plasma at the sample surface layer should be calculated within a multilayer reflection model [42, 68, 69].

Several examples of successful models constructed to simulate laser-induced excitation of surfaces of dielectric and semiconducting materials and their heating dynamics can be found in [41–47, 49–52, 70–72]. They are usually based on the rate equations for charge-carrier generation and recombination and an analog of the two-temperature model either for the average charge-carrier energy or the temperature as a measure of the average energy under the conditions of incomplete thermalization within the electron subsystem. In this section, the main processes in wide-bandgap dielectrics have been outlined for the case of laser beam focusing on the sample surface. Actually the dynamics of laser-induced excitations of dielectric materials is much richer that motivates their utterly wide applications in various optical technologies. In the following sections we concentrate on the peculiarities of optical material modifications by ultrashort laser pulses upon focusing in the bulk depth.

5.3 Volume Modifications of Wide-Bandgap Dielectrics

The regimes of focusing of ultrashort laser pulses inside transparent crystals and glasses which result in local volume modifications of the properties of the irradiated sample are of paramount interests for applications in photonics and optoelectronics among which the main well-established application field is direct writing of waveguide structures, based on a controlled change in the refractive index in laser-modified zones [14–20, 34, 73–80]. Formation of a phase object embedded in the dielectric matrix is caused by rearrangement of bonds in the sample with the displacement of atoms and corresponding change in density, accumulation of stresses, and appearance of defect states. Improved control over modifications in laser-irradiated materials requires detailed studies of both individual laser-induced processes and their interrelations on the timescales from photo-excitation to imprinting a final 3D structure into the matrix of the original material. In the last years, a progress has been achieved in theoretical modeling of laser-excited processes, both during the laser pulse propagation through a transparent material sample and post-irradiation effects though understanding of mechanisms and dynamics of modifications are far from being compete and require further considerable efforts. Here we review the main approaches for investigations of in-volume laser-induced processes and outline future modeling directions.

5.3.1 Propagation of Focused Laser Beams Through Non-linear Absorbing Media

Several types of modeling approaches can be listed which have been developed for studies of propagation of an electromagnetic wave in transparent materials with accounting laser energy absorption. A simple quasi-analytical model [58] can be useful for an estimative analysis of the geometry of laser energy absorption regions and light transmission through the sample. The model relates $\partial I(z,r,t)/\partial t$ and $\partial n_e(z,r,t)/\partial t$ to I and n_e (r and z are the radius and propagation distance of the laser beam). For the regimes considered in [58], it was unambiguously shown that, in transparent solids, the laser intensity is strongly clamped to the maximal reached levels of order of 5×10^{13} W/cm^2 due to non-linear absorption. Another important observation is that the absorbed laser energy is proportional to the pulse duration. It must be underlined that a more rigorous approach based on solving the non-linear Schrödinger equation supports the latter conclusion, indicating however that at longer pulses the laser energy is deposited into a more localized region [20].

The models of laser light propagation in transparent media based on the non-linear Schrödinger equation (NLSE) are widely utilized for studying the processes of laser excitation of dielectrics in the regimes of modification. The NLSE is an asymptotic parabolic approximation of Maxwell's equations [81] applicable for describing unidirectional propagation of slowly varying envelopes of laser pulses. This equation describes the self-focusing effect which manifests itself as a laser beam collapse at beam energies beyond a critical value particular for a Kerr medium with the positive non-linear refractive index n_2. We note that for transparent crystals and glasses the n_2 values are typically in the range of 10^{-16}–10^{-14} cm^2/W. To account for additional physical effects such as a small non-paraxiality, plasma defocusing, multiphoton ionization, etc., the additional terms are introduced to the scalar models based on the NLSE [18, 19, 82–84]. An important detailed review of NLSE application for various laser beam propagation conditions is given in [85].

A generalized NLSE which takes into account radiation losses for generation of electron plasma on the beam way and plasma-induced changing of the permittivity of the medium can be written in the cylindrically symmetric form as [18, 19, 82–85]:

$$\begin{aligned} \frac{\partial -\bar{C}}{\partial z} &= \frac{i}{2k_0} T^{-1}\left(\frac{\partial^2}{\partial r^2} + \frac{1}{r}\frac{\partial}{\partial r}\right)\bar{\in} - \frac{ik''}{2}\frac{\partial^2 -\bar{C}}{\partial t^2} \\ &+ \frac{ik_0 n_2 T}{n_0}\left[(1-f_R)\left|-\bar{C}\right|^2 + f_R \int_{-\infty}^{t} R(t-\tau)\left|-\bar{C}\right|^2 d\tau\right] -\bar{C} \\ &- \frac{\sigma}{2}(1+i\omega\tau_c)T^{-1}(n_e -\bar{C}) - \frac{1}{2}\frac{W_{\mathrm{PI}}(|-\bar{C}|)E_g}{\left|-\bar{C}\right|^2} -\bar{C} \end{aligned} \tag{5.3}$$

where $-\bar{C}$ is the complex envelope of the electric field strength of the light wave which is assumed to be slowly varying in time. For a Gaussian beam with cylindrical symmetry one has

$$-\bar{C}(r,t,0) = -\bar{C}_0 \exp\left(-r^2/w^2 - t^2/\tau_{\mathrm{L}}^2 - ik_0 r^2/2f\right). \tag{5.4}$$

Here $-\bar{C}_0^2 = 2E_{\mathrm{L}}/(\pi w^2 \tau_{\mathrm{L}}\sqrt{\pi/2})$ is the input pulse intensity; E_{L} is the pulse energy; $w = w_b(1 + d^2/z_f^2)^{1/2}$ and w_b are the beam radius at the distance d from the geometric focus and the beam waist respectively; the curvature radius f and the focusing distance d are related as $f = (d + z_{\mathrm{f}}^2/d)$; z_{f} is the Rayleigh length; τ_{L} is the pulse duration (half-width determined by a decrease in the field envelope by 1/ e times compared to the maximum value); $k_0 = n_0\omega/c$ and ω are the wave number and the frequency of the carrier wave; n_0 is the refractive index of the medium; c is speed of light; the parameter k$''$ describes the second-order group velocity dispersion; $E_g = E_{g0} + e^2 -\bar{C}^2/(2cn_0\varepsilon_0 m_{\mathrm{r}}\omega^2)$ is the effective ionization potential in the electromagnetic wave field expressed here via the electric field envelope [19]; m_{r} is the reduced mass of the electron and hole. Equation (5.3) takes into account the beam diffraction in the transverse direction, group velocity dispersion, the optical Kerr effect with a term corresponding to the delayed (Raman) response of the non-linear material (characterized by the parameter f_{R}), plasma defocusing, energy absorption due to photoionization and inverse bremsstrahlung. The operator $T = 1 + (i/\omega) \times (\partial/\partial t)$ describes the self-steepening effects. The inverse bremsstrahlung process is described in the frames of the Drude model with the absorption cross section $\sigma = k_0 e^2 \omega\tau_c/[n_0^2\omega^2\varepsilon_0 m_e(1 + \omega^2\tau_c^2)]$. The characteristic collisional time of electrons τ_c is a variable value dependent on electron energy and density (see comments in Sect. 5.3.2).

It should be noted that the linear term in (5.3) gives only an approximate estimation of the absorption efficiency when the free electron concentration considerably increases as the influence of the electron concentration on the absorption cross section is not taken into account. Additionally, the possibility of multiphoton absorption by free electrons is neglected which can be important at relatively high radiation intensities [86]. However, at laser beam focusing into the sample volume, the clamping effect limits the attainable intensities [57–59]. The rate equation describing generation and recombination kinetics of free electrons can be written as:

$$\frac{\partial n_e}{\partial t} = \left[W_{\mathrm{PI}}(|\bar{C}|) + \frac{\sigma n_e}{(1 + m_r/m_e)E_g}|\bar{C}|^2\right]\frac{n_{\mathrm{at}} - n_e}{n_{\mathrm{at}}} - \frac{n_e}{\tau_{\mathrm{tr}}}. \tag{5.5}$$

Here n_{at} is the atomic density in the undisturbed material matrix. Equation (5.5) takes into account free electron production in the processes of photoionization and avalanche as well as electron recombination in a trapping-like process associated with local deformations of the atomic lattice (see Sect. 5.2). The rate of photoionization W_{PI} can be described by the Keldysh formalism [53, 54] or in a simplified form for purely multiphoton ionization regimes when the clamping effect limits laser intensity levels to $\gamma \gtrsim 1$ [58].

Numerical investigations based on the NLSE allow elucidating important features of laser pulse propagation through transparent solids such as filamentation [83, 85], clamping [42, 83, 85], strong dependence of the laser energy deposition geometry on pulse duration [19, 42] for different irradiation conditions. Remarkable is the temporal dynamics of laser energy deposition into bulk dielectrics in the modification regimes [19, 42]. On an example of fused silica, it has been demonstrated that only a small fraction of the pulse leading edge, containing 10–15 % of the pulse energy, is absorbed with a high efficiency near and in front of the geometric focus. Due to strong defocusing scattering of the electron plasma generated by the pulse leading edge, the rest laser beam does not fall into the region near the geometric focus. However, as a result of the self-focusing effect, the later parts of the beam are absorbed before the geometric focus and, integrally, they generate the second region of efficient absorption (compare Figs. 11 and 12 in [42]). An important consequence of the complex correlation between self-focusing and plasma defocusing effects is that the local intensity over the whole pulse does not exceed app. 5×10^{13} W/cm^2, pointing once more to unavoidable intensity clamping. In the context of the clamping effect, the problem of the efficient delivery of laser energy into a local region inside transparent samples remains open. In particular, at high numerical apertures (NA $\gtrsim 1$) the laser light may be concentrated to a small focal volume with consequences of strong material damage [23].

The validity of the NLSE for ultrashort laser beams focused inside transparent crystals and glasses can be broken down in many situations that is conditioned by neglecting some small terms upon its derivation from Maxwell's equations. The condition of a slowly varying envelope limits applications of the NLSE to relatively long laser pulses. For pulse durations of order of 10 fs and shorter, either the NLSW has to be generalized with additional terms to accounting features of such extremely short pulses or, more appropriate, the complete set of Maxwell's equations are to be used for describing light propagation through a non-linear medium. Another strong limitation imposed on using the NLSE is the requirement of unidirectionality of the light beam. This requirement makes impossible to apply the NLSE to describing tightly focused beams as well as to the cases when dense electron plasma is generated causing light scattering to large angles. Maxwell's equations are free of the above limitations.

To describe laser beam propagation through an absorbing ionizable medium, Maxwell's equations are appropriately supplemented to account for multiphoton ionization, multiphoton absorption (that is the depletion of the laser beam due to multiphoton ionization), the Kerr effect, and plasma dispersion while the optical response of the plasma is described in the frames of a plasma fluid model [87]. Maxwell's equations and the plasma fluid equations are coupled via the free electron current. However, comparing the codes with the NLSE and Maxwell's equations in application to the same irradiation conditions [88], it has been shown that the NLSE considerably overestimates the generated electron plasma density and, as a result, the locally absorbed laser energy that may lead to misinterpretation of the simulation results. The mentioned overestimation is caused by the fact that the NLSE does not take into account laser light scattering to large angles which becomes significant at relatively high electron densities.

The complete set of Maxwell's and electron plasma dynamics equations can be solved using a Finite-Difference Time-Domain (FDTD) algorithm. At present, such models represent the best choice for modeling tightly-focused laser beams which can potentially generate dense electron plasma inside transparent dielectric materials. In the three-dimensional (3D) geometry, the model allows to elucidate the effects of laser light polarization [87]. However, such 3D modeling is extremely time and labor consuming and requires unreasonable computer memory resources. In the next section we present a detailed description of a new 2D model based on Maxwell's equations which however accounts for laser light polarization, including the electron oscillatory motion [88, 89]. Compared to similar 3D codes, the 2D model is a much more time- and cost-efficient tool which allows elucidating many important features of laser light absorption inside transparent solids accompanied by dense electron plasma formation.

5.3.2 2D Model of Electron Plasma Generation upon Laser Beam Focusing Inside Transparent Solids

The basics of the model are essentially similar to those reported in [87] with further development by taking into account avalanche ionization and light dispersion. The model is two-dimensional (2D) that implies cylindrical symmetry of laser intensity distribution with, however, electron oscillations along laser polarization direction. Maxwell's equations for laser beam propagation through a nonlinear absorbing medium can be written in the following form [87–89]:

$$\frac{1}{c}\frac{\partial \vec{D}}{\partial t} - i\frac{\omega}{c}\vec{D} = -\frac{4\pi}{c}\vec{j} + \mathrm{rot}\vec{H} - \frac{8\pi}{c}W_{\mathrm{PI}}E_g\frac{\vec{E}}{|E^2|}, \tag{5.6}$$

$$\frac{1}{c}\frac{\partial \vec{H}}{\partial t} - i\frac{\omega}{c}\vec{H} = -\mathrm{rot}\vec{E}. \tag{5.7}$$

Here we use the standard denotations with E, D, H, and j to be the electric field, the electric displacement field, the magnetic field, and the electric current respectively; all other parameters as in Sect. 5.3.1. In (5.6, 5.7) we assume that the electric field of the beam wave can be presented in the form $\vec{E} = (\vec{E}_0 e^{-i\omega t} + \vec{E}_0^* e^{i\omega t})/2$ and for the plane wave the laser field intensity is expressed as

$$I = \frac{cn_0}{8\pi}\left|\vec{E}_0^2\right|.$$

Taking into account the energy clamping effect which saturates the laser intensity at levels with the Keldysh parameter $\gamma > 1$, the multiphoton mechanism of ionization is accepted with the rate $W_{\mathrm{PI}} = \sigma_k I^k(n_{\mathrm{at}} - n_e)/n_{\mathrm{at}}$ reduced by the available ionization centers. Note that we limit consideration to single ionization per atom as a higher order ionization implies a lower ionization cross section whose theory has not yet been developed. Below the simulation results are presented for fused silica for which $k = 6$, $\sigma_6 = 2 \times 10^{-47}$ cm^9/(s W^6) [83], and the atomic (not molecular) density of 6.6×10^{22} cm^{-3} are adopted. For convenience, the term W_{PI} is rewritten to the form $W_{\mathrm{PI}} = W_{\mathrm{PI0}}(|E^2|/E_*^2)^k(n_{\mathrm{at}} - n_e)/n_{\mathrm{at}}$. Here $W_{\mathrm{PI0}} = 3.7 \times 10^{34}$ cm^{-3} s^{-1} and $E_*^2 = 8\pi I_*/n_0 c$ with $I_* = 3.5 \times 10^{13}$ W/cm^2 to be the laser intensity at which the Keldysh parameter $\gamma = 1$. The intensity dependent band gap width $E_g = E_{g0}(1 + e^2|E^2|/(4m_r\omega^2))$ can be rewritten as $E_g = E_{g0}(1 + |E^2|/(4E_*^2))$; $E_{g0} = 9$ eV and $m_r = 0.64m_e$ [83].

The electric displacement field can be presented as

$$\vec{D} = \vec{E} + \sum_m \vec{P}_m + \vec{P}_{\mathrm{nl}}. \tag{5.8}$$

The linear part of the medium polarization is modeled as a set of oscillators

$$\frac{\partial \vec{P}_m}{\partial t} - i\omega\vec{P}_m = \vec{V}_{Pm}, \tag{5.9}$$

$$\frac{\partial \vec{V}_{Pm}}{\partial t} - i\omega\vec{V}_{Pm} = -\omega_m^2(\vec{P}_m - B_m\vec{E}), \tag{5.10}$$

where $\vec{P}_m$ and $\vec{V}_{Pm}$ are the local material response and its derivative. For fused silica $m = 1, 2, 3$; $\omega_1 = 27.539$ fs^{-1}; $\omega_2 = 16.21$ fs^{-1}; $\omega_3 = 0.19034$ fs^{-1};

$B_1 = 0.6962$; $B_2 = 0.4079$; and $B_3 = 0.8975$ [83]. In a particular case of $\vec{E} = \text{const}$, the linear refractive index is expressed as

$$n_0^2 = 1 + \sum_m \frac{B_m \omega_m^2}{\omega_m^2 - \omega^2} = 1 + \sum_m \frac{B_m \lambda_m^2}{\lambda^2 - \lambda_m^2}$$

and for fused silica at $\lambda = 800$ nm we have $n_0 = 1.45$.

In the general case, the non-linear polarization part of (5.8) takes into account the optical Kerr effect with a term corresponding to the delayed Raman-Kerr optical response (see [19, 85]). Here in order to decrease the computation costs, we restrict our consideration to relatively small numerical apertures when the contribution of the delayed Raman-Kerr response is small. Hence, in our case

$$\vec{P}_{\text{nl}} = \frac{c}{4\pi} n_0^2 n_2 |\vec{E}^2| \vec{E}. \tag{5.11}$$

The equations for the density and momentum of free electrons are written as

$$\frac{\partial n_e}{\partial t} = W_{\text{PI}} + W_{\text{av}} - \frac{n_e}{\tau_{\text{tr}}}, \tag{5.12}$$

$$\frac{\partial (n_e \vec{\mathrm{v}})}{\partial t} - i\omega n_e \vec{\mathrm{v}} = -n_e \frac{e}{m_e} \vec{E} - n_e \frac{\vec{\mathrm{v}}}{\tau_c}, \tag{5.13}$$

$$\vec{j} = -n_e e \vec{\mathrm{v}}. \tag{5.14}$$

For fused silica $\tau_{\text{tr}} = 150$ fs [67, 90] is used in modeling.

A special comment must be made on the electron collision time τ_c. In the general case, this value is electron density and energy dependent and may be treated similar to [62] with, however, a serious caution in respect of the electron temperature issues. In the fs-laser excited band-gap materials, the electron energy distribution may stay far from equilibrium during the whole pulse duration [54] that does not allow a simple Maxwellian approach for deriving analytical expressions for this characteristic time. The electron collision time is one of the core parameters of the Drude model which is used for evaluation of the electron density in pump-probe measurements of transient dynamics of phase shift and transmission signal intensity [61, 91–93]. For fused silica glass, the *averaged* values of τ_c reported in literature scatter in the range from 0.2 to 23.3 fs [18, 48, 61, 67, 83, 94–97] that is definitely determined by the excitation conditions. As a result, attempts to evaluate the levels of electron plasma densities may lead to an error more than an order of magnitude [88]. A more detailed analysis of Drude-based evaluations of laser-induced electron plasma density upon ultrafast laser excitation will be given in [88]. Here we accept the τ_c value to be equal to 1.27 fs, accounting that the damping factor $\omega\tau_c = 3$ [19]. It must be underlined that, according to simulations, the change in the τ_c value affects but insignificantly the maximum electron density and intensity distributions while influencing the balance between multiphoton and collisional ionization mechanisms. This is apparently due to a self-consistent nature of beam focusing, self-focusing, plasma absorption and defocusing mechanisms.

From the collisional ionization cross section (see Sect. 5.3.1), one may derive the following avalanche ionization rate

$$W_{av} = \frac{e^2\tau_c n_e|\vec{E}^2|}{n_0 c\varepsilon_0 m_e E_g(1+\omega^2\tau_c^2)(1+m_r/m_e)}\frac{n_{at}-n_e}{n_{at}}. \quad (5.15)$$

In (5.15) all the parameters are expressed in SI units. The system of (5.6)–(5.14) was solved for the cylindrically symmetric case for fused silica irradiated by femtosecond laser pulses with linear polarization of light at 800 nm wavelength along the x axis, assuming that $\vec{E}^2$ weakly depends on the azimuthal angle φ. The components of the electric field depend on φ as $E_r = \hat{E}_r(r,z,t)\cos\varphi$; $E_\varphi = \hat{E}_\varphi(r,z,t)\sin\varphi$; $E_z = \hat{E}_z(r,z,t)\cos\varphi$. The validity of cylindrical symmetry is controlled by ensuring the condition $||\hat{E}_r^2| + |\hat{E}_z^2| - |\hat{E}_\varphi^2|| \ll |\hat{E}_r^2| + |\hat{E}_z^2| + |\hat{E}_\varphi^2|$. The incoming laser beam is focused inside the bulk at the distance d from the surface and at the sample surface ($z = 0$) it corresponds to the linear polarized light:

$$\hat{E}_r = E_{in}\exp(-r^2/w^2 - t^2/\tau_L^2 - ik_0 r^2/(2f)), \quad \hat{E}_\varphi = -\hat{E}_r. \quad (5.16)$$

Accordingly, $E_y = 0$; $E_x = \hat{E}_r$; $E_{in} = [16E_L/(n_0 cw^2\tau_L\sqrt{\pi/2})]^{1/2}$; all other parameters are defined in Sect. 5.3.1. Note that the generated free electrons oscillate along the x axis while the problem symmetry implies cylindrically symmetric distribution of laser intensity. At the other boundaries, the zero boundary conditions lead to the reflected waves and the necessity of using a large computational region (r_0, z_0). To avoid this, the conditions of the zero Riemann invariants on the incoming characteristics are applied:

$$n_0E_r - B_\varphi = 0; \quad n_0E_\varphi + B_r = 0 \text{ at } z = z_0;$$
$$n_0E_\varphi - B_z = 0; \quad n_0E_z + B_\varphi = 0 \text{ at } r = r_0.$$

It must be noted that in (5.12) and (5.13) the terms of the $\mathrm{div}(\vec{v}n_e)$ and $\partial(n_e v_i v_j)\partial x_j$ kinds and the term involving the Lorentz force have been disregarded. The contributions of the convective terms for the typical irradiation regimes were analyzed and it was found that the plasma density change due to these terms was of order of 10^{-4} as compared to the results with disregarding such terms. As for the Lorentz force, its contribution to the electron current (5.13) is smaller by the factor of v/c as compared to the electric field force and thus it may be safely neglected. For simulations, an implicit numerical scheme was used in the frames of the FDTD method.

Below the results of simulations are presented which are characteristic for the two modification regimes of fused silica, of a relatively low energy pulse (LEP) when volume nanograting structures are obtained at multipulse irradiation ($E_L = 1$ μJ, 150 fs pulse duration [8]) and of a higher energy pulse (HEP) above which one may expect the transition to a strong modification [19] resulted from material melting that may involve creation of void-like structures by single laser pulses ($E_L = 2.5$ μJ, 80 fs). Note that the power of both pulses studied is well above the self-focusing threshold (2.8 MW for fused silica [85]).

A comment should be made on the effect of focusing conditions found in the present simulations. Here we limit simulations to the beam waist of 1 μm. Attempts of beam waist decreasing lead to violation of cylindrical symmetry caused by electron current and, additionally, to the energy nonconservation in the simulation process. For the regimes with NA $\gtrsim 0.35$ the problem becomes essentially three-dimensional. The realization of a 3D code for describing laser-induced plasma generation inside transparent solids which requires extremely large computational resources is now under development.

Figures 5.1 and 5.2 present snapshots from the calculations [89] which show respectively how the free electron plasma is developing during the laser pulse propagation through the focal zone and how plasma created by the beam front affects propagation of the rest beam. We compare the instantaneous distribution maps of the free electron density (Fig. 5.1) and the laser intensity (Fig. 5.2) for two pulses whose energy differs by 2.5 times. The more energetic pulse has shorter duration so its peak intensity in vacuum is app. 4.6 times higher as compared to the lower-energy pulse. The geometric focus is marked by the white dashed lines while the location of the laser pulse maximum is indicated by the black dot on the z axis. Several important aspects can be noticed. In the LEP regime (Fig. 5.1a, 410 fs), ionization is evidently induced by the very front of the laser pulse when it is approaching the geometrical focus. In the HEP case (Fig. 5.1b) at the same time moment the ionized region is located well before the geometrical focus. This difference is explained by the fact that in the HEP regime the laser intensity level capable to induce considerable ionization is reached earlier relative to the geometric focus as compared to the LEP case. As a result, the pulse front is depleted of energy earlier in space (see Fig. 5.2) and, thus, cannot induce noticeable ionization by this time moment closer to the focus. One can notice that, at the early stages of the LEP propagation, the electron plasma develops rather in the direction toward the laser whereas in the HEP regime the ionization front moves forward with the laser beam. Interesting is that the maximum levels of the plasma density are essentially the same for the two cases and remain considerably subcritical.

Another important difference which has allowed us to realize the features of VNG formation (see [89] for the details) is that in the LEP case the density distribution of the developed electron plasma (at $t \geq 500$ fs) is almost quasi-uniform within the excited region whereas in the HEP case ionization is more localized and one may recognize the signs for double focusing with arising the second maximum (see Fig. 5.1b). During further evolution, the more energetic pulse penetrates deeper to the focal zone with creation plasma behind the geometric focus [88]. Plasma decays quite slowly and, as the simulations show, in three hundred femtoseconds after reaching its maximum it still stays at a level of one-third of the maximum value.

Interesting is that in the LEP regime the snapshots resemble a known hydrodynamic picture of flow over the blunt body where a plasma "body" generated by the

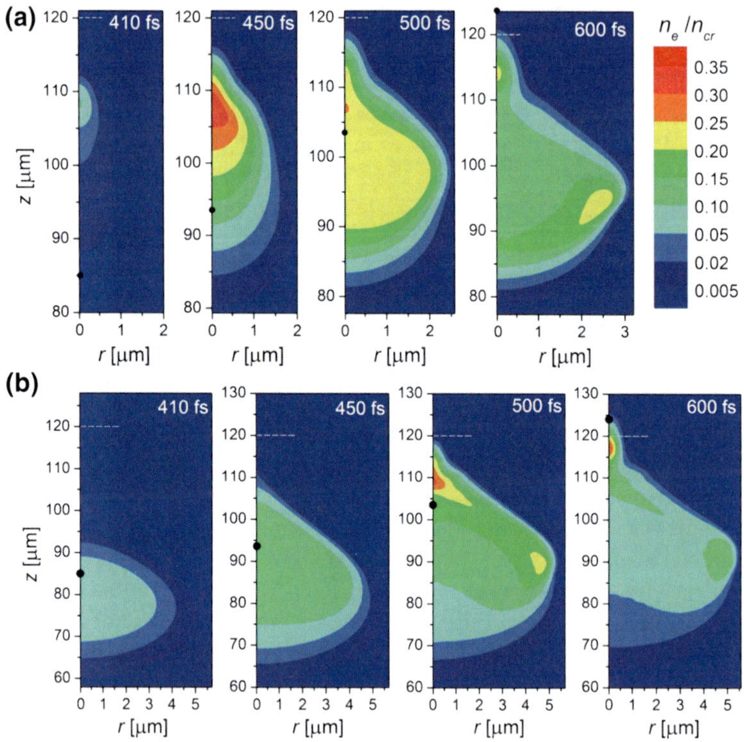

Fig. 5.1 Dynamics of free electron density in fused silica for $E_L = 1$ μJ, $\tau_L = 150$ fs (**a**) and $E_L = 2.5$ μJ, $\tau_L = 80$ fs (**b**) [89]. The laser beam propagates from the *bottom*. The electron density is normalized by the critical electron density, $n_{cr} = \varepsilon_0 m_e \omega^2/e^2$, whose value is 1.74×10^{21} cm^{-3} for the laser wavelength of 800 nm. Geometric focus of the beam (marked by *dashed lines*) is located at the distance $z = 120$ μm from the sample surface. At time moment $t = 0$, the maximum of the beam is at $z = 0$ (sample surface) and its location upon beam propagation is marked by the *black dots*. Calculations are started at a negative time moment to ensure that the intensity of the beam front cannot induce any excitation at the initial time moments

beam front completely displaces the rest beam from the plasma region (Fig. 5.2a). The situation is even more dramatic for the HEP regime where strong plasma scattering completely displaces the beam from the beam axis to its periphery from where it again experiences self-focusing behind the plasma region (Fig. 5.2b). Beam self-focusing after strong defocusing (scattering) by the electron plasma is conditioned by the fact that, according to the simulation results, though the laser energy absorbed in the sample during the beam propagation can integrally reach several dozens of percent depending on the beam energy, the beam power still stays well above the self-focusing regime. The simulations clearly demonstrate a strong intensity clamping effect: for the beams with considerably different power the maximum intensity levels differ insignificantly and safely fall to the regimes of *multiphoton ionization*, thus supporting the results of simplified modeling [58].

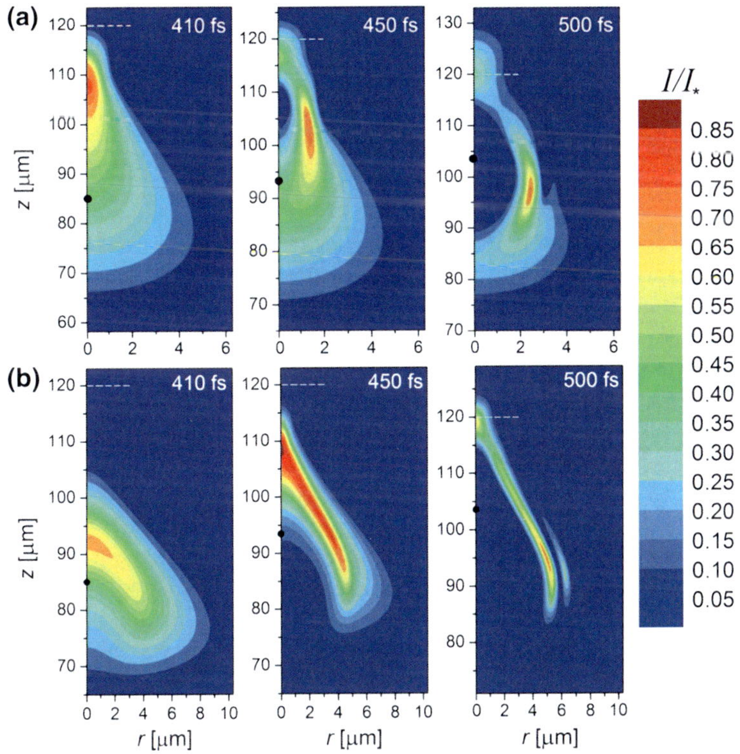

Fig. 5.2 Instantaneous maps of the laser beam intensity for the conditions of Fig. 5.1 [89]. **a** $E_L = 1$ μJ, $\tau_L = 150$ fs; **b** $E_L = 2.5$ μJ, $\tau_L = 80$ fs. The intensity is normalized by the value I_* at which the Keldysh parameter $\gamma = 1$ ($I_* = 3.5 \times 10^{13}$ W/cm^2)

The presented model as well as those reported in [18–20, 23, 41–47, 49–52, 54, 58, 62, 71, 72, 83–87, 94] both simplified and more comprehensive have a great potential for supplementing experimental studies and is a powerful tool to predict and foresee the underlying physics of the observed phenomena. Many detailed aspects which can be overlooked in experiments due to limited resolutions of up-to-date measurement techniques may be revealed with the help of numerical simulations.

5.3.3 Single-Pulse Material Heating and Laser-Induced Stresses

In the model presented above, the absorbed laser energy distribution integrated over the simulation time can be converted to the lattice temperature map, assuming that, at picosecond timescale, all the locally absorbed energy is spend solely to heat the lattice after electron-lattice thermalization and electron recombination. On the other hand, once the energy absorbed due to electronic excitation is stored in a localized volume of the sample, the level of heating of the material matrix can be evaluated, assuming the definite levels of the density of free electrons and their average energy (temperature). The energy balance may be written in the following form:

$$c_p\rho(T^* - T_0) = n_e e(E_e + E_{g0}). \tag{5.17}$$

Here c_p is the lattice heat capacity, the average energy of free electrons E_e and the band gap energy are expressed in electron-Volt. In (5.17) it is assumed that all the energy from the electronic subsystem including the bandgap energy is finally transmitted to the lattice. Although free electrons are first trapped to the self-trapped exciton (STE) states, the STE population decays at subnanosecond time after excitation [90] and only a small fraction of the STEs (app. 10^{-3}) actually turns to the defect states such as E′-centers [98]. Thus, in (5.17) the energy accumulated in the defect states after single pulse excitation is disregarded. Figure 5.3 [88] shows an example of such E_e–n_e diagram for fused silica which matches pairs of the free electron temperature and density securing material heating to a threshold temperature T^*: annealing (1,400 K), softening (1,858 K [99]), melting (2,006 K [100]), sublimation (2,523 K [100]).

According to a number of studies [42, 45, 54], the typical values of free electron energy in wide-bandgap dielectrics upon excitation with laser pulses of $\geq$100 fs duration are in the range 5–15 eV. This is explainable from the viewpoint of ionization kinetics. Free electrons absorbing photons may produce secondary electrons by collisional ionization if their energy exceeds the material band gap. This sets upper limits on the average electron energy to app. $1.5E_g$ at relatively long pulse durations (at least $\geq$100 fs) when electrons have time to gain enough energy from the beam for developing the avalanche process [62]. The electron number density is much more debated. The pump-probe experiments report the maximum electron densities of 5×10^{19} cm^{-3} or even lower upon beam focusing into the glass volume (see, e.g., [93]) while the simulation results obtained in the frames of different models give values in the range $\sim$(2–8) $\times$ 10^{20} cm^{-3} [19, 20, 83, 85, 87]. We notice that in the typical modification regimes already single laser pulses produce material expansion (see, e.g., Fig. 5 in [19]). This unambiguously indicates material irreversible expansion in the laser-affected zone which can be produced only if matter is heated at least above the annealing point or even close or above the softening point. Hence, one may roughly circumscribe the electron plasma parameters upon volumetric laser processing of fused silica by the shaded region in Fig. 5.3 where, the higher is the pulse energy, the closer the material state is to the softening point (both higher the electron density and energy). We note that the regimes with plasma parameters above the sublimation point and, partially, above the melting point will result in void (or bubble) generation [88]. Such diagrams can be drawn for any transparent material interesting from the viewpoint of laser processing in order to get a view of the post-irradiation material state.

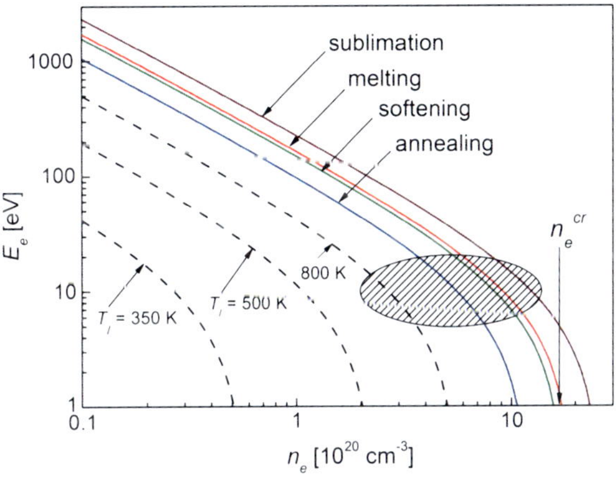

Fig. 5.3 The diagram matching the energy and density of free electrons excited in fused silica for reaching different levels of heating (annealing, softening, melting, and sublimation points). The *dashed lines* are added for heating silica glass from 300 K to several fixed temperature levels (350, 500, and 800 K)

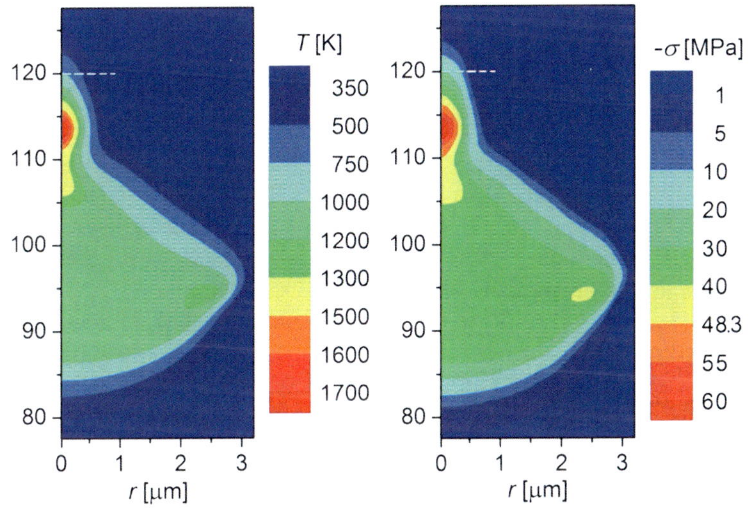

Fig. 5.4 The maps of the temperature (*left*) and thermal stress (*right*) after the laser beam propagation through fused silica sample for the LEP regime that corresponds to Figs. 5.1a and 5.2a. The stress level of 48.3 MPa is the static tensile strength value tabulated for fused silica

Using Expression (5.17), the spatial distributions of the lattice temperature may be mapped while, applying the thermoelastoplastic model [20, 101], we can obtain instantaneous distributions of the material stress. An example is given in Fig. 5.4 for the LEP case which corresponds to Figs. 5.1a and 5.2a. As can be seen, the matter stays below the melting point (Fig. 5.4, left). The compressive stress has a maximum of $\sim$65 MPa that is higher than the material tensile strength of 48.3 MPa tabulated for fused silica under the normal conditions (Fig. 5.4, right). The tensile stress which is generated upon material expansion from the compressed state may have the same or lower amplitude as compared to preceded compressed stress. Although the transient tensile stress may exceed the material static strength, materials can usually withstand the levels of the dynamic stresses much higher than the static ones. However, the induced high stress gradients of order of (60–70) MPa/μm (Fig. 5.4b) will result in formation of a strong compressive wave which will lead to rarefaction of the regions of the enhanced temperature and creation of a densified envelope surrounding the expanded core [20]. If the matter is heated above the melting point, its strength drops by several orders of magnitude. Hence, it may be expected that at higher laser energies bubble/void formation may be observed (see, e.g., [19, 27]). According to simulations, our HEP regime (see Figs. 5.1b, 5.2b) falls to a boundary above which bubble/void formation may be already expected [88]. We underline that, for bubble/void formation, reaching stresses of order of the Young's modulus is far not obligatory though such stress levels may be attainable by beam focusing with high numerical apertures (NA $\gtrsim$ 1), resulting in creation of warm dense matter conditions [23, 102].

5.3.4 Comments on Multipulse Irradiation Regimes

Laser writing of local structures of modified matter in transparent materials for different application purposes is performed mainly in a gentle manner by multipulse irradiation at relatively low beam energies ($\leq$1 μJ) in order to avoid material failure. Depending on a desired final structure, writing is performed either into a fixed volume inside the sample or by moving the sample relative to the laser beam focus with different scanning speeds. Under such conditions, accumulation of laser-induced modifications from pulse to pulse is a topic of prime importance for designing structures with desired properties.

To provide crack-free laser writing of permanent structures with positive refractive index changes for waveguiding applications, laser irradiation should be applied gently in an accumulative manner avoiding conditions of material failure. Several accumulation mechanisms responsible for gentle modification of transparent materials towards waveguiding properties are established among which heat accumulation is best demonstrated [15, 76, 103–109]. At relatively high repetition rates, the energy absorbed at the focal volume from each pulse has no time to diffuse out completely before the subsequent pulse, thus forming a point source of heat. The process of heat accumulation upon waveguide writing can be controlled by several means: variation of pulse energy, pulse repetition rate, scanning speed, and focusing conditions. A simple analysis of the heat accumulation effect can be performed on the basis of the heat flow equation. The characteristic time of heat propagation by the distance r can be estimated as $t_{heat} \sim r^2(\rho c/\lambda)$ where ρ, c, and λ are the material density, heat capacity, and thermal conductivity respectively. For fused silica, the heat wave propagates by $\sim$1 μm during first microsecond after the laser pulse termination while complete dissipation of heat from the focal volume of few micrometer size may take time up to dozens of microseconds (note that, according to the heat flow equation, dissipation is determined by both the thermal conductivity and the temperature gradient and is slowing down with decreasing the latter). For unmoved silicate glass samples at a pulse energy sufficient for noticeable heating of a localized focal zone, pulse repetition rate of order of 100 kHz and higher is required for pronounced accumulation of heat [106]. For fixed laser pulse energy, the higher the repetition rate is, the fewer pulses are required for considerable accumulative heating of the sample within the focal volume. The morphologies of laser-induced modifications with and without the heat-accumulation regimes may be completely different [108]. The structures produced at low repetitions rates are strongly localized. In fact, only a zone absorbing laser energy is modified while surrounding material looks essentially unchanged. At high repetition rates when the conditions for heat accumulation are realized, an extended volume around the light absorbing region shows signs of modification where one can recognize two distinct zones, the interior with indications of material melting and a halo whose outer boundary is presumably determined by reaching the softening temperature upon laser exposure (see Fig. 1 in [108]).

Translating the sample relative to the laser pulse imposes further restrictions on the minimal repetition rate. The requirement here is the number of pulses coupling with the same sample volume upon translation which should be sufficient for heat accumulation inducing local modification. Increased laser pulse energy or a higher repetition rate result in larger radius of heat modified zone and allow faster translating for producing a similar final structure that is beneficial for practical applications [75]. This process must be thoroughly controlled as excessive heat accumulation causes material failure with formation of random or organized voids [109]. On the other hand, such regimes are exploited for writing microfluidic channels for biosensing and biomicrochip applications [78, 110, 111]. However, in such regimes heat accumulation cannot be considered separately from the formation of highly stressed states of matter resulting in stress wave emission with possible consequences in the form of plastic deformations. Another parameter considerably influencing the modification structure is focusing. As an example, a laser pulse with the energy of only 20 nJ applied at tight focusing (0.65 NA) induces a similar refractive index change (though in a smaller volume) as the 1-μJ pulse with loose focusing ($\sim$0.1 NA) [103]. Hence, a tighter focusing, being energetically advantageous as compared to loose focusing conditions, results also in miniaturization of written structures. This has especially strikingly demonstrated with NA $>$ 1 [23, 102].

In spite of the evidence that accumulative processes play a key role in glass modification, the accumulative modification mechanisms have not completely been understood. Additionally, it must be remarked that in multipulse irradiation regime each subsequent pulse arrives to already modified matter as compared to the previous pulses. Thus, a gradual accumulation of the defect states with larger excitation cross sections makes excitation easier for subsequent pulses. On the other hand, the refractive index also varies due to defect generation [67], local densification/rarefaction, and stress accumulation, thus affecting laser pulse propagation through the modified volume. However, such aspects of laser micromachining still require further studies, both experimental and theoretical, while their modeling is still in a very primeval state. More detailed reviews of this and other issues on ultrafast laser writing in transparent materials can be found in [112, 113].

5.4 Concluding Remarks

In this chapter we have reviewed an intricate interconnection of the physical processes which can be initiated by ultrashort laser pulses inside transparent solids with consequences in the form of structural material modifications. Some modification kinds have already been utilized in various applications such as integrated optical devices while the others are still waiting for adoption to practice. In the majority of technologies based on laser directly-written material modifications, practice develops well ahead as compared to fundamental understanding of underlying physics and chemistry of the phenomena. On one hand, this is conditioned by possibilities of using new discovered structures and phenomena for growing practice demands but, on the other hand, the effectiveness of laser-based direct writing is of limited utility namely because of lacking knowledge-based optimization. One of the striking examples is volume nanograting formation inside fused silica glass whose nature is still highly debated. After discovery in 2003 [8], the VNGs, in particular in fused silica, have become an exciting object for research [see, e.g., 9, 10, 114–116] and an important element for a variety of existing and promising applications in optics and photonics [117]. The origin of the VNGs is attributed to several mechanisms such as interference of plasma waves with laser light [8, 9], formation of nanoplasmas followed by self-organization into nanoplanes through the memory effects [10, 38], and interference of the two modes of ultrashort-living excitons-polaritons [118]. All these explanations remain to be unconvincing.

We note that the VNG explanation by light interference with the main mode of plasma oscillations requires the near-critical electron density and the free electron energies of hundreds or even thousands eV [8]. Such electron energies lead to creation of multiply charged ions, complete braking of bonds inside the excited area, followed by hydrodynamics similar to that described in [23, 102]. The exciton–polariton interaction [118] may be important only at low densities of low-energetic free electrons which rapidly transform into self-trapped excitons after excitation to the conduction band [90, 119]. At $n_e \geq 10^{19}$ cm^{-3}, the exciton–polariton interaction is to be heavily screened by the electron plasma when it begins to efficiently absorb laser energy. The nanoplasma mechanism of nanograting formation may be initiated between the two extremes mentioned above with evolution toward the critical density at nanoplane sites [10, 38] though the evolution routes remain vague. Modeling presented in Sect. 5.3.2 has allowed us to get aware on a paramount role of collective behavior of free electron plasma excited on the surface and inside transparent solids, the topic which still remains unexplored though well-established in dense plasma physics. As a result, a new mechanism of VNG formation based on scattering of the laser light by developing electron plasma culminating with ionization scattering instability [120] has been proposed [89]. Additionally, writing anisotropy inside isotropic materials with tilted laser pulses observed in [11] has also been explained on the basis of plasma scattering effects within the model described in Sect. 5.3.2 [89]. We stress that the above explanations would not be possible without the comprehensive modeling efforts.

Returning to the importance of the different continuum approaches, we refer readers to [40] and again underline the incipient stage in modeling efforts to investigate the whole route of laser-induced evolution of transparent materials towards obtaining new unusual practicable properties. Finally, we repeat that the theory and modeling of the laser-induced processes can be the cost reducing tools which may allow choosing the optimal conditions and most appropriate materials for particular desired modifications and provide a detailed physical understanding of the phenomenon and required material properties for technological applications.

The authors acknowledge financial support of the European Commission funding under the 7th Framework Programme (Marie Curie International Incoming Fellowship grant of the principal author, No. 272919) and of the Russian Foundation for Basic Research (RFBR project No. 12-01-00510) which made possible to develop the above-described comprehensive modeling tool.

References

1. K.M. Davis, K. Miura, N. Sugimoto, K. Hirao, Opt. Lett. **21**, 1729 (1996)
2. R.R. Gattass, E. Mazur, Nat. Photonics **2**, 219 (2008)
3. P.A. Temple, M.J. Soileau, IEEE J. Quantum Electron. **17**, 2067 (1981)
4. N.C. Kerr, B.A. Omar, S.E. Clark, D.C. Emmony, J. Phys. D **23**, 884 (1990)
5. J. Gottmann, D. Wortmann, M. Hörstmann-Jungemann, Appl. Phys. A **255**, 5641 (2009)
6. B. Dusser, Z. Sagan, H. Soder, N. Faure, J.P. Colombier, M. Jourlin, E. Audouard, Opt. Express **18**, 2913 (2010)
7. S.K. Das, K. Dasari, A. Rosenfeld, R. Grunwald, Nanotechnology **21**, 155302 (2010)
8. Y. Shimotsuma, P.G. Kazansky, J.R. Qiu, K. Hirao, Phys. Rev. Lett. **91**, 247405 (2003)
9. Y. Shimotsuma, K. Hirao, J.R. Qiu, P.G. Kazansky, Mod. Phys. Lett. B **19**, 225 (2005)
10. V.R. Bhardwaj, E. Simova, P.P. Rajeev, C. Hnatovsky, R.S. Taylor, D.M. Rayner, P.B. Corkum, Phys. Rev. Lett. **96**, 057404 (2006)
11. P.G. Kazansky, W.J. Yang, E. Bricchi, J. Bovatsek, A. Arai, Y. Shimotsuma, K. Miura, K. Hirao, Appl. Phys. Lett. **90**, 151120 (2007)
12. R.S. Taylor, C. Hnatovsky, E. Simova, P.P. Rajeev, D.M. Rayner, P.B. Corkum, Opt. Lett. **32**, 2888 (2007)
13. G. Cheng, K. Mishchik, C. Mauclair, E. Audouard, R. Stoian, Opt. Exp. **17**, 9515 (2009)
14. K. Miura, J. Qiu, H. Inouye, T. Mitsuyu, K. Hirao, Appl. Phys. Lett. **71**, 3329 (1997)
15. C.B. Schaffer, A. Brodeur, J.F. García, E. Mazur, Opt. Lett. **26**, 93 (2001)
16. Z. Wang, K. Sugioka, Y. Hanada, K. Midorikawa, Appl. Phys. A **88**, 699 (2007)
17. W. Yang, C. Corbari, P.G. Kazansky, K. Sakaguchi, I.C.S. Carvalho, Opt. Exp. **16**, 16215 (2008)

18. S.W. Winkler, I.M. Burakov, R. Stoian, N.M. Bulgakova, A. Husakou, A. Mermillod-Blondin, A. Rosenfeld, D. Ashkenasi, I.V. Hertel, Appl. Phys. A **84**, 413 (2006)
19. I.M. Burakov, N.M. Bulgakova, R. Stoian, A. Mermillod-Blondin, E. Audouard, A. Rosenfeld, A. Husakou, I.V. Hertel, J. Appl. Phys. **101**, 043506 (2007)
20. A. Mermillod-Blondin, I.M. Burakov, Y.P. Meshcheryakov, N.M. Bulgakova, E. Audouard, A. Rosenfeld, A. Husakou, I.V. Hertel, R. Stoian, Phys. Rev. B **77**, 104205 (2008)
21. E.N. Glezer, E. Mazur, Appl. Phys. Lett. **71**, 882 (1997)
22. W. Watanabe, K. Itoh, Opt. Exp. **10**, 14 (2002)
23. S. Juodkazis, K. Nishimura, S. Tanaka, H. Misawa, E.G. Gamaly, B. Luther-Davies, L. Hallo, P. Nicolai, V.T. Tikhonchuk, Phys. Rev. Lett. **96**, 166101 (2006)
24. R. Graf, A. Fernandez, M. Dubov, H.J. Brueckner, B.N. Chichkov, A. Apolonski, Appl. Phys. A **87**, 21 (2007)
25. H.Y. Sun, J. Song, C.B. Li, J. Xu, X.S. Wang, Y. Cheng, Z.Z. Xu, J.R. Qiu, T.Q. Jia, Appl. Phys. A **88**, 285 (2007)
26. A. Mermillod-Blondin, J. Bonse, A. Rosenfeld, I.V. Hertel, Y.P. Meshcheryakov, N.M. Bulgakova, E. Audouard, R. Stoian, Appl. Phys. Lett. **94**, 041911 (2009)
27. M. Beresna, M. Gecevičius, N.M. Bulgakova, P.G. Kazansky, Opt. Express **19**, 18989 (2011)
28. V. Koubassov, J.F. Laprise, F. Théberge, E. Förster, R. Sauerbrey, B. Müller, U. Glatzel, S.L. Chin, Appl. Phys. A **79**, 499 (2004)
29. L. Sudrie, M. Franco, B. Prade, A. Mysyrowicz, Opt. Commun. **171**, 279 (1999)
30. K. Kintaka, J. Nishii, Y. Kawamoto, A. Sakamoto, P.G. Kazansky, Opt. Lett. **27**, 1394 (2002)
31. E. Bricchi, J.D. Mills, P.G. Kazansky, B.G. Klappauf, Opt. Lett. **27**, 2200 (2002)
32. M. Beresna, P.G. Kazansky, Opt. Lett. **35**, 1662 (2010)
33. J. Liu, Z. Zhang, S. Chang, C. Flueraru, C.P. Grover, Opt. Comm. **253**, 315 (2005)
34. A.M. Streltsov, N.F. Borrelli, Opt. Lett. **26**, 42 (2001)
35. K. Minoshima, A.M. Kowalevicz, E.P. Ippen, J.G. Fujimoto, Opt. Exp. **10**, 645 (2002)
36. R. Osellame, S. Taccheo, G. Cerullo, M. Marangoni, D. Polli, R. Ramponi, P. Laporta, S. De Silvestri, Electron. Lett. **38**, 964 (2002)
37. Y. Shimotsuma, M. Sakakura, K. Miura, J. Qiu, P.G. Kazansky, K. Fujita, K. Hirao, J. Nanosci. Nanotechnol. **7**, 94 (2007)
38. R. Taylor, C. Hnatovsky, E. Simova, Laser Photonics Rev. **2**, 26 (2008)
39. W.J. Cai, A.R. Libertun, R. Piestun, Opt. Express **14**, 3785 (2006)
40. N.M. Bulgakova, R. Stoian, A. Rosenfeld, I.V. Hertel, in *Laser-Surface Interactions for New Material Production*, vol. 130, ed. by A. Miotello, P.M. Ossi, Springer Series in Material Science (Springer, Berlin 2009), p. 81
41. I.M. Burakov, N.M. Bulgakova, R. Stoian, A. Rosenfeld, I.V. Hertel, Appl. Phys. A **81**, 1639 (2005)
42. N.M. Bulgakova, R. Stoian, A. Rosenfeld, Quantum Electron. **40**, 966 (2010)
43. A.S. Epifanov, Sov. Phys. JETP **35**, 897 (1975)
44. B.G. Gorshkov, A.S. Epifanov, A.A. Manenkov, Sov. Phys. JETP **49**, 309 (1979)
45. D. Arnold, E. Cartier, Phys. Rev. B **46**, 15102 (1992)
46. B.C. Stuart, M.D. Feit, A.M. Rubenchik, B.W. Shore, M.D. Perry, Phys. Rev. Lett. **74**, 2248 (1995)
47. B.C. Stuart, M.D. Feit, S. Herman, A.M. Rubenchik, B.W. Shore, M.D. Perry, Phys. Rev. B **53**, 1749 (1995)
48. S.S. Mao, F. Quéré, S. Guizard, X. Mao, R.E. Russo, G. Petite, P. Martin, Appl. Phys. A **79**, 1695 (2004)
49. N.M. Bulgakova, R. Stoian, A. Rosenfeld, I.V. Hertel, W. Marine, E.E.B. Campbell, Appl. Phys. A **81**, 345 (2005)
50. A. Vogel, J. Noack, G. Hüttman, G. Paltauf, Appl. Phys. B **81**, 1015 (2005)
51. M. Mero, W. Rudolph, D. Ristau, K. Starke, Phys. Rev. B **71**, 115109 (2005)
52. N.S. Shcheblanov, T.J.Y. Derrien, T.E. Itina, AIP Conf. Proc. **1464**, 79 (2012)

53. L.V. Keldysh, Sov. Phys. JETP **20**, 1307 (1965)
54. A. Kaiser, B. Rethfeld, M. Vicanek, G. Simon, Phys. Rev. B **61**, 11437 (2000)
55. K. Mishima, M. Hayashi, J. Yi, S.H. Lin, H.L. Selzle, E.W. Schlag: Phys. Rev. A **66**, Paper 053408 (2002)
56. V.E. Gruzdev, Phys. Rev. B **75**, 205106 (2007)
57. A. Becker, N. Akozbek, K. Vijayalakshmi, E. Oral, C.M. Bowden, S.L. Chin, Appl. Phys. B **73**, 287 (2001)
58. D.M. Rayner, A. Naumov, P.B. Corkum, Opt. Express **13**, 3208 (2005)
59. Q. Sun, F. Liang, R. Vallee, S.L. Chin, Opt. Lett. **33**, 2713 (2008)
60. M. Lenzner, J. Krüger, S. Sartania, Z. Cheng, C. Spielmann, G. Mourou, W. Kautek, F. Krausz, Phys. Rev. Lett. **80**, 4076 (1998)
61. F. Quéré, S. Guizard, P. Martin, Europhys. Lett. **56**, 138 (2001)
62. G.M. Petrov, J. Davis, J. Phys. B **41**, 025601 (2008)
63. P.P. Rajeev, M. Gertsvolf, P.B. Corkum, D.M. Rayner, Phys. Rev. Lett. **102**, 083001 (2009)
64. S.C. Jones, P. Braunlich, R.T. Casper, X.-A. Shen, P. Kelly, Opt. Engin. **28**, 1039 (1989)
65. K. Tanimura, H. Fujiwara, T. Suzuki, Nucl. Instrum. Meth. B **116**, 26 (1996)
66. J. Krüger, M. Lenzner, S. Martin, M. Lenner, C. Spielman, A. Fiedler, W. Kautek, Appl. Surf. Sci. **208**, 233 (2003)
67. P. Martin, S. Guizard, Ph. Daguzan, G. Petite, P. D'Oliveira, P. Maynadier, M. Pedrix, Phys. Rev. B **55**, 5799 (1997)
68. K. Sokolowski-Tinten, D. von der Linde, Phys. Rev. B **61**, 2643 (2000)
69. Sh. Furman, A.V. Tikhonravov, *Basics of Optics of Multilayer Systems* (Editions Frontières, Gif-sur-Yvette, 1992)
70. N.M. Bulgakova, R. Stoian, A. Rosenfeld, I.V. Hertel, E.E.B. Campbell, Phys. Rev. B **69**, Paper 054102 (2004)
71. C. Mézel, A. Bourgeade, L. Hallo, Phys. Plasmas **17**, 113504 (2010)
72. B. Chimier, O. Utéza, N. Sanner, M. Sentis, T. Itina, P. Lassonde, F. Légaré, F. Vidal, J.C. Kieffer, Phys. Rev. B **84**, 094104 (2011)
73. C. Florea, K.A. Winick, J. Lightwave Tech. **21**, 246 (2003)
74. W. Watanabe, T. Asano, K. Yamada, K. Itoh, J. Nishii, Opt. Lett. **28**, 2491 (2003)
75. L. Shah, A. Arai, S. Eaton, P. Herman, Opt. Express **13**, 1999 (2005)
76. S.M. Eaton, H.B. Zhang, P.R. Herman, F. Yoshino, L. Shah, J. Bovatsek, A.Y. Arai, Opt. Express **13**, 4708 (2005)
77. K. Suzuki, V. Sharma, J.G. Fujimoto, E.P. Ippen, Opt. Express **14**, 2335 (2006)
78. H.Y. Sun, F. He, Z.H. Zhou, Y. Cheng, Z.Z. Xu, K. Sugioka, K. Midorikawa, Opt. Lett. **32**, 1536 (2007)
79. C. Mauclair, G. Cheng, N. Huot, E. Audouard, A. Rosenfeld, I.V. Hertel, R. Stoian, Opt. Express **17**, 3531 (2009)
80. A. Ferrer, A.R. de la Cruz, D. Puerto, W. Gawelda, J.A. Valles, M.A. Rebolledo, V. Berdejo, J. Siegel, J. Solis, J. Opt. Soc. Am. B **27**, 1688 (2010)
81. R. Temam, A. Miranville, *Mathematical Modeling in Continuum Mechanics* (Cambridge University Press, UK, 2005)
82. A.L. Gaeta, Phys. Rev. Lett. **84**, 3582 (2000)
83. A. Couairon, L. Sudrie, M. Franco, B. Prade, A. Mysyrowicz, Phys. Rev. B **71**, 125435 (2005)
84. V.P. Kandidov, S.A. Shlenov, O.G. Kosareva, Quantum Electron. **39**, 205 (2009)
85. A. Couairon, A. Mysyrowicz, Phys. Rep. **441**, 47 (2007)
86. H. Bachau, A.N. Belsky, I.B. Bogatyrev, J. Gaudin, G. Geoffroy, S. Guizard, P. Martin, Y.V. Popov, A.N. Vasil'ev, B.N. Yatsenko, Appl. Phys. A **98**, 679 (2010)
87. K.I. Popov, C. McElcheran, K. Briggs, S. Mack, L. Ramunno, Opt. Express **19**, 271 (2010)
88. N.M. Bulgakova, V.P. Zhukov, S.V. Sonina, Yu.P. Meshcheryakov, to be published
89. N.M. Bulgakova, V.P. Zhukov, Yu.P. Meshcheryakov, Appl. Phys. B (2013). doi:10.1007/s00340-013-5488-0

90. D. Grojo, M. Gertsvolf, S. Lei, T. Barillot, D.M. Rayner, P.B. Corkum, Phys. Rev. B **81**, 212301 (2010)
91. V.V. Temnov, K. Sokolowski-Tinten, P. Zhou, B. Rethfeld, V.E. Gruzdev, A. El-Khamawy, D. von der Linde, Proc. SPIE **5448**, 1119 (2004)
92. V.V. Temnov, K. Sokolowski-Tinten, P. Zhou, A. El-Khamhawy, D. von der Linde, Phys. Rev. Lett. **97**, 237403 (2006)
93. D.G. Papazoglow, I. Zergioti, S. Tsortzakis, Opt. Lett. **32**, 2055 (2007)
94. A.Q. Wu, I.H. Chowdhury, X. Xu, Phys. Rev. B **72**, 085128 (2005)
95. Q. Sun, H. Jiang, Y. Liu, Z. Wu, H. Yang, Q. Gong, Opt. Lett. **30**, 320 (2005)
96. A. Mermillod-Blondin, C. Mauclair, J. Bonse, R. Stoian, E. Audouard, A. Rosenfeld, I.V. Hertel, Rev. Sci. Instrum. **82**, 033703 (2011)
97. L. Sudrie, A. Couaron, M. Franco, B. Lamouroux, B. Prade, S. Tzortzakis, A. Mysyrowicz, Phys. Rev. Lett. **89**, 186601 (2002)
98. G. Petite, S. Guizard, P. Martin, F. Quéré, Phys. Rev. Lett. **83**, 5182 (1999)
99. X.R. Zhang, X. Xu, A.M. Rubenchik, Appl. Phys. A **79**, 945 (2004)
100. J. Siegel, K. Ettrich, E. Welsch, E. Matthias, Appl. Phys. A **64**, 213 (1997)
101. Y.P. Meshcheryakov, N.M. Bulgakova, Appl. Phys. A **82**, 363 (2006)
102. E.G. Gamaly, A. Vailionis, V. Mizeikis, W. Yang, A.V. Rode, S. Juodkazis, High Energy Density Phys. **8**, 13 (2012)
103. C.B. Schaffer, J.F. García, E. Mazur, Appl. Phys. A **76**, 351 (2003)
104. R.R. Gattass, L.R. Cerami, E. Mazur, Opt. Express **14**, 5279 (2006)
105. I. Miyamoto, A. Horn, J. Gottmann, J. Laser Micro/Nanoeng. **2**, 7 (2007)
106. H. Zhang, A.M. Eaton, J. Li, P.R. Herman, J. Phys: Conf. Ser. **59**, 682 (2007)
107. M. Sakakura, M. Shimizu, Y. Shimotsuma, K. Miura, K. Hirao, Appl. Phys. Lett. **93**, 231112 (2008)
108. M. Shimizu, M. Sakakura, M. Ohnishi, Y. Shimatsuma, T. Nakaya, K. Miura, K. Hirao, J. Appl. Phys. **108**, 073533 (2010)
109. Y. Bellouard, M.-O. Hongler, Opt. Express **19**, 6807 (2011)
110. Y. Liao, Y. Ju, L. Zhang, F. He, Q. Zhang, Y. Shen, D. Chen, Y. Cheng, Z. Xu, K. Sugioka, K. Midorikawa, Opt. Lett. **35**, 3225–3227 (2010)
111. K. Sugioka, Y. Hanada, K. Midorikawa, Laser Photon. Rev. **4**, 386 (2010)
112. N.M. Bulgakova, in *Ultrafast Laser Processing: From Micro- to Nanoscale*, ed. by K. Sugioka, Y. Cheng (Pan Stanford Publishing, 2013), p. 48
113. K. Miura, K. Hirao, Y. Shimotuma, M. Sakakura, in *Ultrafast Laser Processing: From Micro- to Nanoscale*, ed. by K. Sugioka, Y Cheng (Pan Stanford Publishing, 2013), p. 251
114. M. Hörstmann-Jungemann, J. Gottmann, M. Keggenhoff, J. Laser Micro/Nanoeng. **4**, 135 (2009)
115. K. Mishchik, G. Cheng, G. Huo, I.M. Burakov, C. Mauclair, A. Mermillod-Blondin, A. Rosenfeld, Y. Ouerdane, A. Boukenter, O. Parriaux, R. Stoian, Opt. Express **18**, 24809 (2010)
116. S. Richter, M. Heinrich, S. Döring, A. Tünnermann, S. Nolte, J. Laser Appl. **24**, 042008 (2012)
117. M. Beresna, M. Gecevičius, P.G. Kazansky, Opt. Mater. Express **1**, 783 (2011)
118. M. Beresna, M. Gecevičius, P.G. Kazansky, T. Taylor, A. Kavokin, Appl. Phys. Lett. **101**, 053120 (2012)
119. F. Messina, E. Vella, M. Cannas, R. Boscaino, Phys. Rev. Lett. **105**, 116401 (2010)
120. E.S. Efimenko, A.V. Kim, Phys. Rev. E **84**, 036408 (2011)

Chapter 6
Attosecond Pulses for Atomic and Molecular Physics

Francesca Calegari, Giuseppe Sansone and Mauro Nisoli

Abstract The natural temporal scale of ultrafast electronic processes in atoms, molecules, nanostructures and solids is in the attosecond range. In this work we briefly review advances in attosecond science. Particular attention is devoted to the generation, temporal characterization and application of isolated attosecond pulses.

6.1 Introduction

High-order harmonic generation (HHG) in noble gases driven by high-intensity femtosecond laser pulses is a widely used technique for the generation of extreme-ultraviolet (XUV) radiation. The physical processes at the basis of HHG lead to the generation of trains of attosecond pulses, separated by half the optical cycle of the driving radiation, as first theoretically reported in 1992 by Farkas and Toth [1]. In 2001 the first experimental measurements of attosecond pulses have been reported. Trains of attosecond pulses have been measured by Paul et al. [2] by using the reconstruction of attosecond beating by interference of two-photon transitions (RABITT) technique; isolated attosecond pulses have been measured by Hentschel et al. [3] by using the attosecond streak-camera technique [4]. In the last ten years important applications of attosecond pulses have been reported for the investigation of ultrafast electron dynamics in atoms, molecules, solids and nanostructures [5, 6].

In this work we will review progress in attosecond technology and in the application of attosecond pulses to the investigation of electron localization dynamics in diatomic molecules. We will concentrate on the generation, characterization and application of isolated attosecond pulses. Section 6.2 is devoted to a

F. Calegari · G. Sansone · M. Nisoli (✉)
Department of Physics, Politecnico di Milano, Institute of Photonics and Nanotechnologies, IFN-CNR, Piazza L. da Vinci 32 20133 Milan, Italy
e-mail: mauro.nisoli@fisi.polimi.it

M. Castillejo et al. (eds.), *Lasers in Materials Science*,
Springer Series in Materials Science 191, DOI: 10.1007/978-3-319-02898-9_6,

brief review of various techniques for the generation of high-peak-power few-optical-cycle pulses, required for the production of isolated attosecond pulses. A very important characteristic of the driving pulses is the carrier-envelope phase (CEP): active and passive stabilization of the CEP of femtosecond pulses is discussed in Sect. 6.3. Section 6.4 presents various methods implemented for the confinement of the harmonic generation process to a single event, with particular emphasis on the temporal gating schemes. Temporal characterization of isolated attosecond pulses is analysed in Sect. 6.5, while the application of isolated attosecond pulses to the investigation of electron localization dynamics in D_2 molecules is reviewed in Sect. 6.6.

6.2 High-Peak-Power Few-Cycle Pulses

Few-optical-cycle pulses with energy from hundreds of microjoules to tens of millijoules can be generated by implementing various techniques: optical parametric chirped pulse amplification (OPCPA), optical parametric amplification, external compression techniques based on the use of gas-filled hollow fibers. The OPCPA method, first introduced by Piskarskas et al. in 1986 [7] and demonstrated in a proof-of-principle experiment by Dubietis et al. [8] (for a review of OPCPA progress see [9]), offers the possibility to produce ultrashort laser pulses with peak power up to the PW range. 7.9-fs pulses with an energy of 130 mJ (corresponding to a peak power of 16 TW) and a repetition rate of 10 Hz, have been generated in 2009 by using an OPCPA in noncollinear geometry [10]; temporal contrast of 10^{10}, when measured as close as 2 ps to the pulse peak has been recently achieved [11]. The laser technology for routine use of OPCPA systems is still in a development phase and high-energy systems are in operation only in a few laboratories.

6.2.1 Optical Parametric Amplification for the Generation of mJ-Energy Pulses with Stable CEP

Optical parametric amplification has proven to be an excellent technique for the generation of high-energy, few-optical-cycle pulses tunable in broad spectral regions. Sub-20-fs pulses with stable CEP have been produced by using an optical parametric amplifier based on a Ti:sapphire laser system (800 nm, 60 fs, 120 mJ, 10 Hz) [12]. A portion of the laser energy of about 12 mJ was used for the experiment. A first portion of the laser pulse was used for the generation of a broad continuum by filamentation in a cell filled with krypton. The continuum was then compressed by chirped mirrors and used to produce radiation by difference frequency generation (DFG) in a β-barium borate (BBO) crystal. The DF seed was

Fig. 6.1 Sequence of interferograms acquired with a *f*-to-2*f* interferometer, showing the stability of the CEP of the pulses at the output of the near-infrared OPA described in the text. (Adapted from [13])

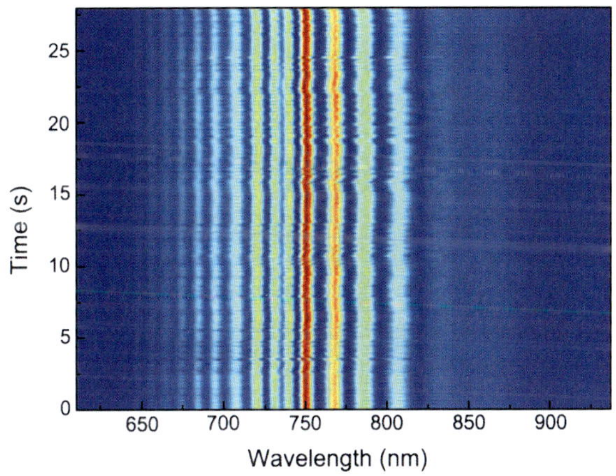

then amplified in a two-stage optical parametric amplifier (OPA) based on BBO in type II configuration pumped by the remaining fraction of the fundamental 800-nm laser source. The amplified DF pulses had an energy of 1.6 mJ and a nearly transform limited pulse duration of about 18 fs [12].

Due to the DFG process, these pulses are intrinsically CEP stabilized. CEP stability has been demonstrated, with a residual rms fluctuation of the order of 200 mrad [13], as displayed in Fig. 6.1. The carrier wavelength of the IR pulses could be tuned, by changing the OPA phase-matching configuration, between 1.4 and 1.6 μm without significantly affecting pulse duration and energy.

6.2.2 Hollow-Fiber Compression Technique

High-repetition rate, few-cycle pulses are routinely generated in several laboratories by using external pulse compression techniques. In particular, a powerful method for the generation of millijoule-level, sub-6-fs light pulses is the hollow fiber compression technique [14], in combination with ultrabroadband dispersion compensation. Such compression technique is based on the propagation of intense laser pulses in dielectric capillaries filled with noble gases. Owing to nonlinear effects occurring during propagation, the pulses undergo spectral broadening; optical compression is then achieved by a dispersive delay line. Wave propagation along hollow fibers can be thought of as occurring by grazing incidence reflections at the dielectric inner surface. Since the losses caused by these reflections greatly discriminate against higher order modes, only the fundamental mode, with large and scalable size, will be transmitted through a sufficiently long fiber [15]. In 1996 the hollow-fiber compression technique was first demonstrated using 140-fs, 0.66-mJ pulses from a Ti:sapphire laser system [14]. Pulses as short as 10 fs, with an energy up to 0.24 mJ were generated. In 1997 20-fs pulses have been compressed to 4.5 fs by using spectral broadening in a krypton-filled capillary and by using a dispersive delay line based on chirped mirrors and two pairs of fused silica prisms of small apex angle [16]. Scaling the input energy to higher values, the generation of sub-10-fs pulses at the sub-terawatt peak power level was also demonstrated. Sub-5-fs pulses with hundreds of microjoules energy have been generated using a tapered hollow fiber (input diameter 0.5 mm, output diameter 0.3 mm) and a set of ultrabroadband chirped mirrors [17]. The hollow-fiber technique has been extended also to the sub-4-fs regime: in 2003 pulses as short as 3.8 fs have been generated by using two gas-filled hollow fibers and ultrabroadband dispersion compensation by using a liquid–crystal spatial light modulator [18].

Recently, 5-fs pulses with an energy up to 5 mJ (1-TW peak power) have been obtained [19] by using a hollow fiber with a pressure gradient [20]. In this case the gas pressure inside the fiber increases from zero at the entrance to the maximum at the output of the fiber. In order to increase the energy throughput of the setup, chirped pulses have been injected into the hollow fiber filled with helium. Due to the excellent spatial beam characteristics at the output of the hollow fiber, such beam has been focused to a nearly diffraction-limited spot size without the need of wavefront correction, with an estimated peak intensity of 5×10^{18} W/cm^2. Circularly polarized pulses have been also used at the input of a gas filled hollow fiber [21]: circular polarization lowers both the ionization rate and the nonlinear refractive index.

6.2.3 High-Energy Pulse Compression by Using Gas Ionization

Another technique recently proposed and implemented for the compression of high-energy femtosecond pulses is based on the use of fast gas ionization in a hollow fiber [22]. 40-fs pulses with an energy of 70 mJ were focused into a 40-cm-long hollow fiber with a 420-μm inner diameter. Upon increasing the pressure of the gas (helium) inside the capillary upto a few millibars, a significant blue-shift and broadening of the spectrum were observed. The use of gas at very low pressure prevents self-focusing limitations. Assuming a one-dimensional geometry, the instantaneous angular frequency of the pulse after propagation through an ionizing gas of length L is given by the following expression [23]:

$$\omega(t) = \omega_0 - \frac{\omega_0}{c} \int_0^L \frac{\partial n(t,z)}{\partial t} dz \tag{6.1}$$

where: ω_0 is the central angular frequency of the input pulse, z is the propagation coordinate, t the time and n is the plasma refractive index:

$$n^2(t,z) = 1 - \frac{n_e(t,z)}{n_c} \tag{6.2}$$

with n_e the free electron density and n_c the critical density. Due to the ultrafast temporal evolution of tunnel ionization and to the large plasma dispersion, large spectral broadening can be obtained at low gas pressure and relatively short interaction lengths. The bandwidth increases up to 107 nm at 8 mbars. The output pulses were compressed by using chirped mirrors with negative dispersion. Pulses as short as 11.4 fs with an energy of 13.7 mJ were obtained.

6.3 Active and Passive Stabilization of the CEP of Femtosecond Pulses

In the case of few-optical-cycle pulses the CEP value is a particularly important parameter, crucial for the generation of isolated attosecond pulses. The pulse-to-pulse variation of the CEP value can be tracked and stabilized by using a *f*-to-2*f* interferometer [24–26]. The spectrum of the pulse is first broadened to a full octave (e.g., by using a photonic crystal fiber) and the heterodyne beat between the frequency doubled low frequency spectral components and the high frequency components, which corresponds to the offset frequency, is detected and used for active phase stabilization. The amplification process and the beam pointing fluctuation introduce a slow CEP drift [27], which can be monitored using a second *f*-to-2*f* nonlinear interferometer. To this purpose a small fraction of the amplified beam is focused on a sapphire plate for the generation of an octave-spanning spectrum by self-phase modulation. The interference between the high-frequency portion of the broadened spectrum and the frequency-doubled light is then measured using an optical multichannel analyzer. The position of the interference fringes depends on the pulse CEP and can be used to extract the CEP drift, using a Fourier-transform-based algorithm. The residual slow CEP drift can be corrected using a second feedback loop.

A different technique has been recently implemented for the active stabilization of the CEP, the feed-forward method [28]. The idea at the basis of this method is simple and effective (see Fig. 6.2). The laser beam is divided into two beams by diffraction off the index grating in an acousto-optic frequency shifter (AOFS). The zero-order beam is used to measure the CEP frequency $f_{CE} = f_r \varphi_{CE}/2\pi$, where f_r is the laser repetition frequency and φ_{CE} is the phase difference, per round-trip, between carrier and envelope inside a femtosecond oscillator. The frequency comb in the Bragg diffracted first-order beam is shifted by the driver frequency f_{RF}. If $f_{RF} = f_{CE}(t)$, the frequency comb in the first-order beam is down-shifted to zero offset, thus corresponding to a train of laser pulses with the same electric field evolution. By using this method a residual phase jitter of 20 mrad has been achieved, corresponding to a timing jitter of only 8 as between carrier and

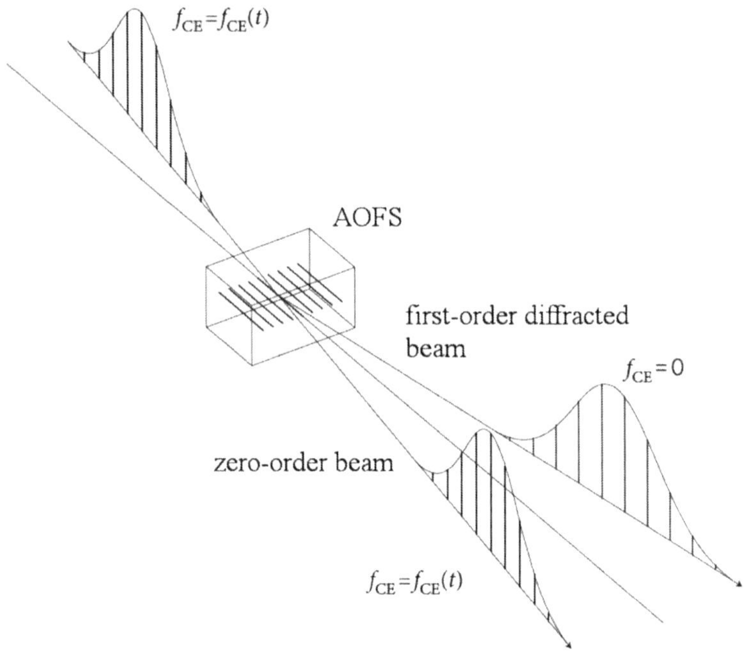

Fig. 6.2 Basic idea of the direct feed-forward method for stabilization of CEP introduced by Koke et al. [28]. The comb modes in the zero-order transmitted beam are unaffected, while each frequency of the diffracted beam is shifted by the input frequency of the acousto-optic frequency shifter (AOFS). (Adapted from [28])

envelope [29], and with a RMS phase error of less than 50 mrad over 10 h of operation [30].

It is also possible to use passive techniques for the stabilization of the CEP, as already mentioned in Sect. 2.1. Phase-stable pulses can be produced by DFG if the two pulses undergoing the DFG process have the same CEP drift ϕ (i.e., $\phi_2 = \phi + c_2, \quad \phi_3 = \phi + c_3$, where c_2 and c_3 represent constant phase contributions). In this case the CEP of the pulse at the difference frequency (DF) is given by $\phi_1 = \phi_3 - \phi_2 - \pi/2 = c_3 - c_2 - \pi/2 =$ const., i.e., the phase fluctuations are automatically cancelled in a passive, all-optical way [31].

6.4 Generation of Isolated Attosecond Pulses

In this section we will briefly review various methods used in the last few years for the generation of isolated attosecond pulses, based on temporal gating of the harmonic generation process [32]. The basic idea is pictorially displayed in Fig. 6.3. A femtosecond driving field generates a train of attosecond pulses,

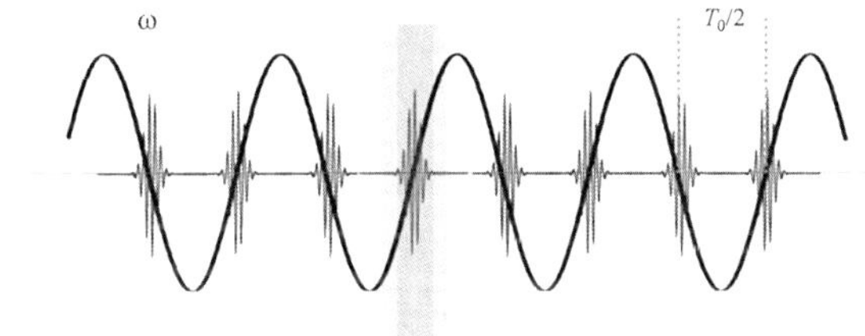

Fig. 6.3 Basic idea of the temporal gating technique for the generation of isolated attosecond pulses. (Adapted from [32])

separated by half an optical cycle, $T_0/2$: isolated attosecond pulses can be generated using a temporal gate, which confines the XUV generation process to a short window, whose temporal extensions depends on the separation between consecutive attosecond pulses.

6.4.1 Polarization Gating

6.4.1.1 One-Color Polarization Gating

The polarization gating method is based on the temporal modulation of the polarization state of the driving pulses isolated attosecond pulses [33–35]. Two birefringent plates can be used to generate a driving field characterized by almost linear polarization in a temporal window shorter than the separation between two consecutive attosecond pulses, as first proposed by Tcherbakoff et al. [36]. A first multiple-order quarter wave-plate introduces a temporal delay, τ, between the ordinary and extraordinary field components. The angle α between the axes of the plate and the incident polarization direction is usually set to $\alpha = 45°$. The second plate is a zero-order quarter wave-plate, with axis forming an angle β with respect to the polarization direction of the incident beam. When $\beta = 0°$ the second plate transforms the linear polarization into circular and vice versa and a short temporal gate of linear polarization can be obtained. The temporal gate of almost linear polarization, defined as the temporal interval during which the ellipticity modulus is smaller than the threshold value ε_{th} (corresponding to the ellipticity value for which harmonic conversion efficiency is decreased by half), can be written as:

$$\tau_g = \frac{\varepsilon_{th}\tau_p^2}{\tau|\cos(2\beta)|\ln 2} \tag{6.3}$$

where τ_p is the pulse duration. By applying this temporal gating scheme, in combination with 5-fs driving pulses with stable CEP, the generation of isolated attosecond pulses at 36-eV photon energy, with a duration of 130 as has been demonstrated in 2006 [35]. In this experiment the first quartz plate was 181 μm-thick, thus splitting the incoming linearly polarized pulse in two orthogonally polarized pulses separated by $\tau = 5.8$ fs, the second one was a zero-order quarter

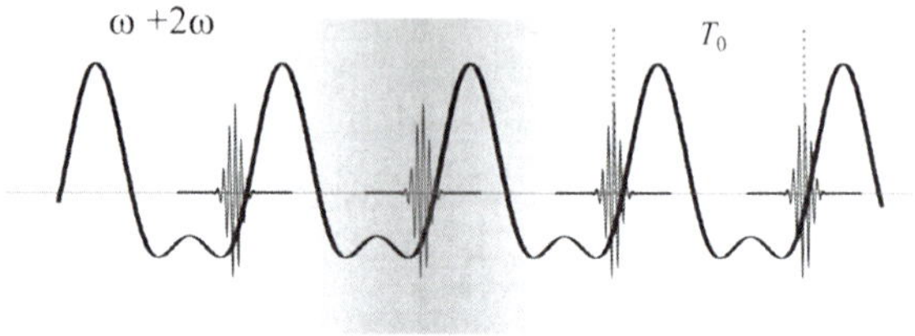

Fig. 6.4 Basic idea of the temporal gating technique for the generation of isolated attosecond pulses by using a two-color driving field. (Adapted from [32])

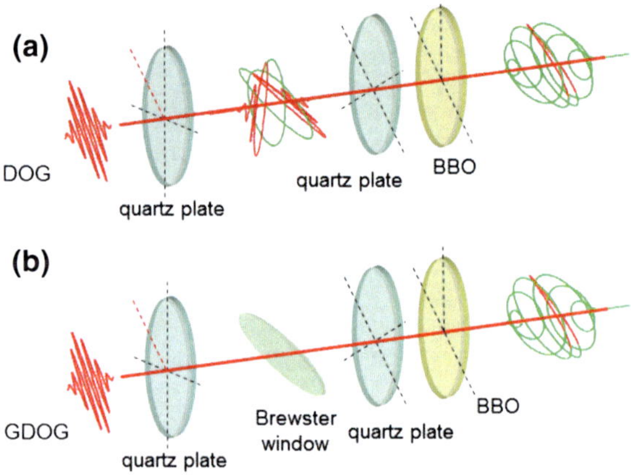

Fig. 6.5 Basic experimental setups for the generation of isolated attosecond pulses using **a** the double-optical gating and **b** the generalized double-optical gating techniques. (Adapted from [32])

wave-plate. It has been demonstrated that the temporal and spectral characteristics of the XUV radiation generated by using the polarization gating scheme can be easily shaped by rotating the birefringent plates used to achieve the temporal gate. In particular, in the case of isolated attosecond pulses it is possible to tune the XUV radiation in a relatively broad spectral range by rotating the first plate [37].

6.4.1.2 Double Optical Gating

The use of a two-color driving field for the generation of high-order harmonics allows one to break the symmetry of the process, so that a train of attosecond pulses separated by a complete optical cycle of the fundamental radiation can be produced, as shown in Fig. 6.4 [38].

The combination of a two-color field, produced by the combination of the fundamental radiation and its second harmonic, with the polarization gating technique (double-optical gating, DOG) allows one to generate isolated attosecond pulses starting from longer driving pulses [39]. The added second harmonic field in the DOG scheme also reduces population depletion caused by the leading edge of the driving field. The collinear configuration of DOG requires two quartz plates and a β-barium borate (BBO) crystal, as shown in Fig. 6.5a: the first quartz plate creates two delayed replica of the incoming pulse with perpendicular polarizations; the optical axis of the second quartz plate and the BBO are aligned and form an angle of 45° with respect to the axis of the first plate. They work together as a quarter wave plate to create a polarization which varies from circular to linear and back to circular. In contrast to the usual polarization gating configuration, the first plate thickness is set to obtain a polarization gate with a duration of one full optical period of the fundamental field. As a result, the longest pulse duration which can be used for generating single attosecond pulses with DOG is almost two times larger compared to the one used by conventional polarization gating. Pulses as short as 107 as have been obtained by using this method [40].

6.4.1.3 Generalized Double Optical Gating

This two-color polarization gating technique can be extended to longer driving pulses by introducing the generalized double optical gating (GDOG) method [41]. The idea is to create a driving field having a polarization which varies from elliptical to linear and back to elliptical with an ellipticity ε. It is possible to demonstrate that in these conditions the width of the temporal gate τ_g is given by:

$$\tau_g = \varepsilon \frac{\varepsilon_{th}\, \tau_p^2}{\tau \ln 2} \tag{6.4}$$

where ε_{th} is the threshold ellipticity, τ is the delay, and τ_p is the input pulse duration. It is possible to demonstrate that the field strength before the gate is lower for GDOG than for DOG, thus allowing a strong reduction of population depletion even for very long laser pulses. The GDOG experimental setup is similar to the DOG one, as shown in Fig. 6.5b: a fused silica Brewster window is placed in between the two plates in order to create the desired ellipticity by rejecting half of the driving pulse. By applying the GDOG method, isolated attosecond pulses with a duration of 148 as have been generated by using 28-fs driving pulses [41].

6.4.2 Ionization Gating

The role of ionization in the generation of attosecond pulses has been theoretically studied in several works [42–44], and experimental investigated in the case of 8-fs [45] and 15-fs [46] driving pulses. In the case of high-intensity driving fields the plasma density rapidly increases on the leading edge of the pulse, thus creating a phase mismatch responsible for the suppression of HHG for all later half-cycles. Even if the excitation intensity is not high enough to create a temporal gate as narrow as required to select a single attosecond pulse, the ionization gating can be used in combination with bandpass filtering to overcome this problem [46, 47]. Recently a different approach based on the combined action of complete population depletion and spatial filtering of the XUV beam has been implemented for the generation of isolated attosecond pulses with duration down to 155 as and

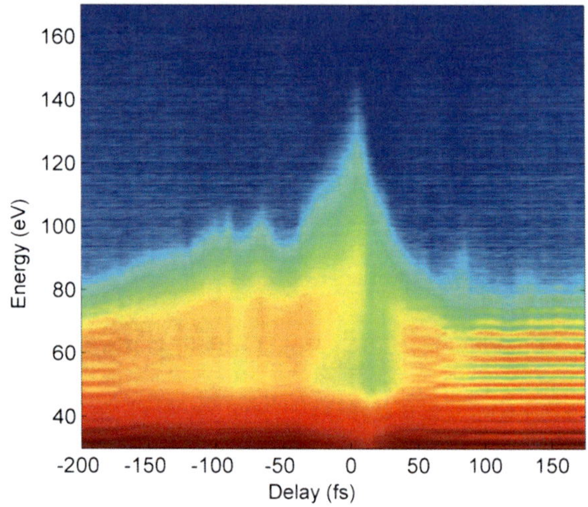

Fig. 6.6 XUV spectra generated in argon as a function of the temporal delay between the 0.8 and 1.45-μm pulses. (Adapted from [55])

energy in the nanojoule range [48]. 5-fs driving pulses with stable CPE and peak intensity of 2.3×10^{15} W/cm^2 have been used for XUV generation in xenon.

The physical mechanisms at the basis of this temporal gating is related to the temporal evolution of the gas ionization. The fast depletion of the neutral atom population induced by the intense and ultrashort driving pulse leads to the confinement of the generation process to a single contribution in correspondence to the leading edge of the excitation pulse. A theoretical calculation based on the Ammosov, Delone and Krainov (ADK) theory [49], clearly indicates a strong dependence of the temporal evolution of the ionization rate on the CEP value of the driving field. This dependence strongly affects the shape of the temporal gate, thus leading to the production of isolated attosecond pulses or a couple of attosecond bursts depending on the CEP value.

6.4.3 Two-Color Gating with Infrared Pulses

The maximum photon energy achievable in the HHG process is given by $h\nu = I_p + 3.17\ U_p$, where I_p is the ionization potential of the neutral atom and U_p is the ponderomotive energy, which turns out to be proportional to $\lambda_0^2 I$, where λ_0 is the wavelength of the driving field and I is the excitation intensity. For this reason parametric sources operating in the IR have been developed to extend the harmonic emission up to keV energies [50–52]. However, the main drawback of using longer wavelength is the reduction of the harmonic yield expected to decrease as λ^{-6} [53, 54].

The first demonstration of an efficient generation of an XUV continuum by using a two-color excitation obtained by mixing 0.8 and 1.45-μm pulses has been reported by Calegari et al. [55]. Figure 6.6 shows the harmonic spectrum generated

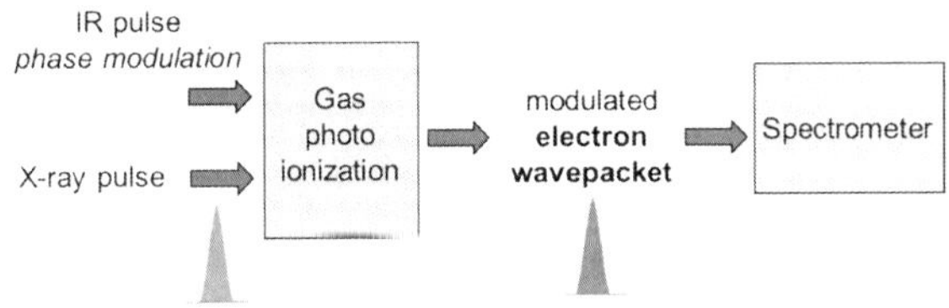

Fig. 6.7 General scheme for the measurement of the duration of attosecond pulses by using a streaking pulse. (Adapted from [57])

in Argon as a function of the delay between the two-color pulses. At zero delay the harmonic emission is dramatically extended and the harmonic spectrum becomes continuous, while when the two pulses do not overlap in time the harmonic spectrum is dominated by the 0.8-μm component. In this experiment the role of the IR pulse is to significantly extend the harmonic emission to higher photon energies, whereas the intense 0.8-μm pulse improves the conversion efficiency of the process. In 2010 Takahashi and co-workers demonstrated that continuous XUV spectra can also be obtained by mixing an 800-nm, 30-fs pulse with a 1300-nm, 40-fs pulse [56]. The idea is to combine long driving pulses in order to relax the constraint on the CEP stabilization of the source. Moreover the synthesized field significantly reduces the ionization probability in the multiple-cycle region, thus allowing to efficiently generate intense attosecond pulses from the neutral medium.

6.5 Attosecond Metrology

A very powerful method for the temporal characterization of attosecond pulses is based on the use of an IR streaking pulse with stable CEP. The basic idea of the measurement, schematically showed in Fig. 6.7 is the following: the attosecond pulse ionizes a gas, by single photon absorption, thus generating an attosecond electron pulse, which, far from any resonance, is a perfect replica of the optical pulse. The conversion of the XUV pulse into an electron wavepacket is obtained in the presence of a streaking IR pulse, whose electric field acts as an ultrafast phase modulator on the generated electron wavepacket. The evolution of the photoelectron spectra as a function of the delay, τ, between the attosecond and the IR pulses allows one to retrieve the temporal intensity profile and phase of the XUV pulses and the electric field of the IR pulse. This technique has been called Frequency Resolved Optical Gating for Complete Reconstruction of Attosecond Bursts (FROG CRAB) [58].

A simple expression of the transition amplitude $a_{\mathbf{v}}$ is the following [58, 59]:

$$a_{\mathbf{v}}(\tau) = -i \int_{-\infty}^{\infty} dt\, e^{i\varphi(t)} \mathbf{d}_{\mathbf{p}(t)}\, \mathbf{E}_X(t-\tau)\, e^{i\left(W+I_p\right)t} \tag{6.5}$$

where $\boldsymbol{E}_X(t)$ is the electric field of the XUV pulse; τ is the temporal delay between XUV and IR pulses, $W = \mathbf{v}^2/2$ is the final kinetic energy of the electron (in atomic units); I_p is the atom ionization potential, $\mathbf{p}(t) = \mathbf{v} + \mathbf{A}(t)$ is the instantaneous kinetic momentum, $\mathbf{A}(t)$ is the vector potential of the IR field, $\mathbf{d}_{\mathbf{p}(t)}$ is the dipole

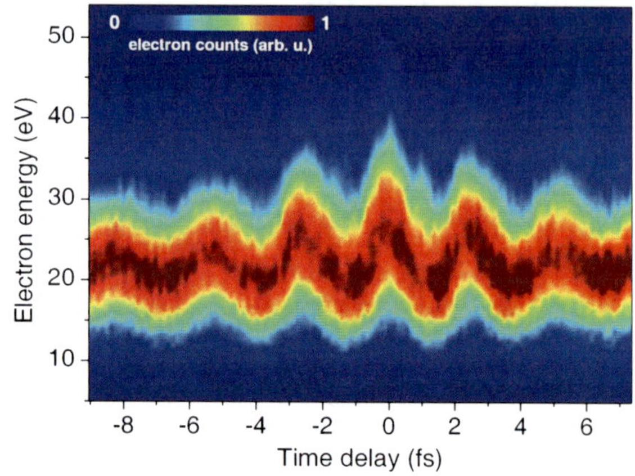

Fig. 6.8 Experimental FROG CRAB trace corresponding to a 130-as isolated pulse. (Adapted from [35])

transition matrix element from the ground state to the continuum with kinetic momentum $\mathbf{p}(t)$ and $\varphi(t)$ is the temporal phase modulation imposed by the IR field to the electron wavepacket, $\mathbf{d} \cdot \mathbf{E}_X(t)$, generated by the attosecond pulse in the continuum. It is possible to demonstrate that the photoelectron spectra $|a_{\mathbf{v}}(\tau)|^2$ as a function of the delay between the XUV and IR pulses contains all the information required for a complete reconstruction of the temporal characteristics of both the attosecond XUV pulse and the IR streaking pulse, and can be seen as a FROG trace. The frequency-resolved optical gating (FROG) is a well known technique used for complete temporal characterization of ultrashort light pulses [60]. In a FROG measurement the pulse to be characterized is divided into temporal slices by using a proper gate pulse. The FROG trace, $S(\omega, \tau)$, is given by the evolution of the spectra of such temporal slices, measured as a function of the delay τ between the pulse to be measured, with electric field $E(t)$, and the gate pulse, $G(t)$, which can be an amplitude or a pure phase gate:

$$S(\omega, \tau) = \left| \int_{-\infty}^{\infty} dt\, G(t)E(t-\tau)\, e^{i\omega t} \right|^2 \tag{6.6}$$

Comparing (6.6) and (6.5) it is evident that the FROG CRAB trace $|a_{\mathbf{v}}(\tau)|^2$ can be considered as a FROG spectrogram of the electron wavepacket generated by the attosecond pulse, with a phase gate $G(t) = \mathrm{e}^{\mathrm{i}\varphi(t)}$. In order to retrieve the temporal phase and intensity profile of the attosecond pulse it is possible to use various iterative algorithms, such as the principal component generalized projection algorithm (PCGPA) [61].

Figure 6.8 shows the FROG CRAB experimental trace corresponding to an isolated attosecond pulse with a duration of 130 as, close to the transform-limited value. The trace follows the temporal evolution of the vector potential of the streaking pulse (5-fs pulse with stable CEP). Due to the almost negligible chirp of the pulse, the bandwidth of the photoelectron spectra is almost constant in the whole range of temporal delays of the measurement [35].

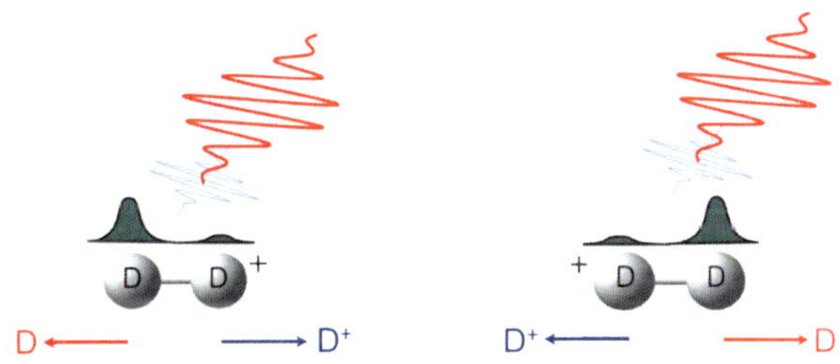

Fig. 6.9 Measurement of electron localization in D_2 molecules after ionization by isolated attosecond pulses

6.6 Application of Isolated Attosecond Pulses to Molecular Physics: Electron Localization in D_2

In various molecular processes the time scale relevant for chemistry is set by electronic motion, which can precede any motion of the nuclei: this is the case when the motion of an electron or a hole along a molecule is the crucial step in the process. Biophysics offers several examples where charge transfer plays a crucial role, such as photosynthesis and various chemical reactions in large peptides and proteins. Theoretical studies have pointed out that very efficient charge transfer can be driven by purely electronic effects, which precede any rearrangement of the nuclear skeleton and which can evolve on a temporal scale ranging from few femtoseconds down to tens of attoseconds [62–64].

This charge dynamics, driven by electron correlations, has been referred to as charge migration. The process is initiated by the generation of a non-stationary electronic state, for example by sudden removal of an electron from a neutral molecule, giving rise to a coherent superposition of electronic states of the resulting cation, which corresponds to the generation of an electron wavepacket migrating along the molecular backbone. A non-stationary state can be generated also by exciting at least two electronic states of the neutral molecule by using a light pulse characterized by a bandwidth wider in energy than the energy difference between the two states. Extremely fast temporal dynamics can be found even in complex bio-molecules.

As a proof-of-principle for the investigation of sub-femtosecond electron dynamics in molecules, the electron localization process in D_2 and H_2 molecules after excitation by isolated attosecond pulses has been investigated [65]. The idea at the basis of the measurement is sketched in Fig. 6.9: the attosecond pump pulse removes one electron from the molecule and the remaining electron undergoes a rearrangement; subsequently the molecule dissociates during the interaction with the IR probe pulse. This allows one to examine how the electron charge distributes itself between the two fragments: since one electron is missing, one fragment will be neutral and the other positively charged. Upon measuring, as a function of the temporal delay between pump and probe pulses, the angular distribution of the ionic fragment it is possible to follow the migration process of the electron remaining on the neutral fragment. The kinetic energy and angular distribution of H^+ and D^+ ionic fragments were measured as a function of the temporal delay

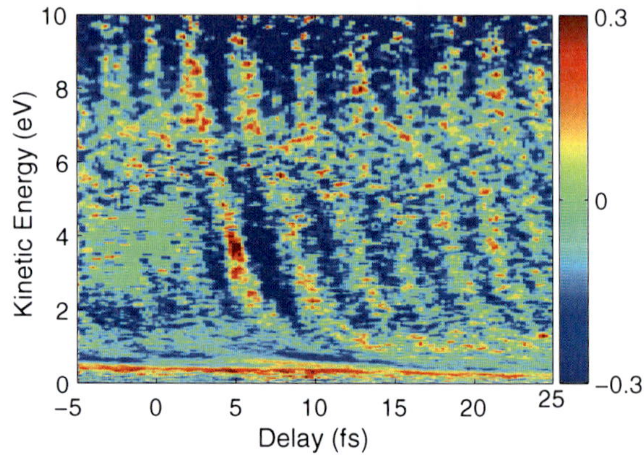

Fig. 6.10 Asymmetry parameter (colour scale) for the formation of D^+ ions after attosecond excitation as a function of the fragment kinetic energy, E_k, and of the temporal delay between pump and probe pulses. (Adapted from [65])

between the XUV pump and a 6-fs, IR pulse with stable CEP. The attosecond excitation leads to dissociative ionization of the molecule, following various pathways [66]. The kinetic and angular distributions of the H^+ and D^+ ions were measured using a velocity map imaging spectrometer for different delays between the attosecond pulse and the synchronized IR pulse. Particularly interesting is the measured temporal evolution of the asymmetry parameter, reported in Fig. 6.10, defined as:

$$A(E_k, \tau) = \frac{N_L(E_k, \tau) - N_R(E_k, \tau)}{N_L(E_k, \tau) + N_R(E_k, \tau)} \tag{6.7}$$

where $N_L(E_k, \tau)$ and $N_R(E_k, \tau)$ are the numbers of ions arriving within 45° of the polarization axis on the left-and, respectively, right-hand sides of the detector. This parameter is directly associated with the electron localization process after the photoionization process initiated by the attosecond excitation. The asymmetry map presented in Fig. 6.10 shows a complex pattern characterized by a dependence on the delay τ and the ion kinetic energy and oscillates with a period given by the optical field period T_0 of the IR pulse.

The measurement demonstrates that the use of CEP-stabilized pulses allows one to control the direction in which the electron is emitted when dissociation is completed. Due to the large bandwidth of the XUV pulses, various mechanisms can lead to the coupling of states of different parity, required for the observation of an asymmetry. In particular two main processes were identified. The first one closely resembles the mechanism discussed in [67], based on IR-induced coupling between the $2p\sigma_u$ and $1s\sigma_g$ states when molecular dissociation occurs. In this mechanism the initial wavepacket dissociates along the $2p\sigma_u$ state and the IR field arriving after the XUV pulse couples this state with the $1s\sigma_g$ giving rise to the electron localization. This mechanism characterizes the asymmetry evolution when the IR pulse arrives after the XUV. The second localization process characterizes the asymmetry around the delay $\tau = 0$ (i.e. for temporally overlapping pulses) and involves the autoionizing states of the Q_1 series. Without IR field the

XUV pulse can ionize the molecule to the $2p\sigma_u$ leading to the emission of an electron characterized by an s-wave. At the same time the attosecond pulse can excite one state of the Q_1 series, triggering the dissociation of the neutral molecule. After few femtoseconds the molecule can autoionize projecting the nuclear wave packet on the $1s\sigma_g$ and ejecting an electron characterized mostly by a p-wave. The IR field redistributes the emitted photoelectron associated to the direct ionization among several angular momentum (including a p-wave contribution); in this way the two paths lead to a coherent superposition of mixed-parity states for the same ion kinetic energy and for the same angular momentum of the outgoing electron.

6.7 Conclusions

Attosecond pulses are unique tools for the investigation of important physical processes in atoms, molecules, mesoscopic systems and solids, where the dynamics of the electron play a crucial role. In this work, various techniques for the generation, temporal characterization and application of isolated attosecond pulses have been discussed.

Acknowledgments The research leading to the results presented in this chapter has received funding from the European Research Council under the European Community's Seventh Framework Programme (FP7/2007–2013)/ERC grant agreement no. 227355—ELYCHE. We acknowledge financial support from the Italian Ministry of Research (FIRB-IDEAS RBID08CRXK) and from MC-RTN ATTOFEL (FP7-238362).

References

1. G. Farkas, C. Toth, Phys. Lett. A **168**, 447 (1992)
2. P.M. Paul, E.S. Toma, P. Breger, G. Mullot, F. Augeé, P. Balcou, H.G. Muller, P. Agostini, Science **292**, 1689 (2001)
3. M. Hentschel, R. Kienberger, C. Spielmann, G.A. Reider, N. Milosevic, T. Brabec, P.B. Corkum, U. Heinzmann, M. Drescher, F. Krausz, Nature **414**, 509 (2001)
4. J. Itatani, F. Quéré, G.L. Yudin, M.Y. Ivanov, F. Krausz, P.B. Corkum, Phys. Rev. Lett. **88**, 173903 (2002)
5. F. Krausz, M. Ivanov, Rev. Mod. Phys. **81**, 163 (2009)
6. M. Nisoli, G. Sansone, Prog. Quantum Electron. **33**, 17 (2009)
7. A. Pirskarskas, A. Stabinis, A. Yankauskas, Sov. Phys. Usp. **29**, 869–879 (1986)
8. A. Dubietis, G. Jonušauskas, A. Pirskarkas, Opt. Commun. **88**, 437–440 (1992)
9. A. Dubietis, R. Butkus, A. Pirskarkas, IEEE J. Sel. Top. Quantum Electron. **12**, 163–172 (2006)
10. D. Herrmann, L. Veisz, R. Tautz, F. Tavella, K. Schmid, V. Pervak, F. Krausz, Opt. Lett. **34**, 2459 (2009)
11. J.M. Mikhailova, A. Buck, A. Borot, K. Schmid, C. Sears, G.D. Tsakiris, F. Krausz, L. Veisz, Opt. Lett. **36**, 3145 (2011)
12. C. Vozzi, F. Calegari, E. Benedetti, S. Gasilov, G. Sansone, G. Cerullo, M. Nisoli, S. De Silvestri, S. Stagira, Opt. Lett. **32**, 2975 (2007)

13. C. Vozzi, C. Manzoni, F. Calegari, E. Benedetti, G. Sansone, G. Cerullo, M. Nisoli, S. De Silvestri, S. Stagira, J. Opt. Soc. Am. B **25**, B112 (2008)
14. M. Nisoli, S. De Silvestri, O. Svelto, Appl. Phys. Lett. **68**, 2793–2795 (1996)
15. S. De Silvestri, M. Nisoli, G. Sansone, S. Stagira, O. Svelto, in *Topics in Applied Physics*, vol. 95, ed. by F. Kärtner (Springer, Heidelberg, 2004), pp. 137–177
16. M. Nisoli, S. De Silvestri, O. Svelto, R. Szipöcs, K. Ferencz, C. Spielmann, S. Sartania, F. Krausz, Opt. Lett. **22**, 522–524 (1997)
17. J.-P. Caumes, G. Sansone, E. Benedetti, M. Pascolini, L. Poletto, P. Villoresi, S. Stagira, C. Vozzi, M. Nisoli, J. Mod. Opt. **53**, 67–74 (2006)
18. B. Schenkel, J. Biegert, U. Keller, C. Vozzi, M. Nisoli, G. Sansone, S. Stagira, S. De Silvestri, O. Svelto, Opt. Lett. **28**, 1987–1989 (2003)
19. S. Bohman, A. Suda, T. Kanai, S. Yamaguchi, K. Midorikawa, Opt. Lett. **35**, 1887 (2010)
20. A. Suda, M. Hatayama, K. Nagasaka, K. Midorikawa, Appl. Phys. Lett. **86**, 111116 (2005)
21. X. Chen, A. Jullien, A. Malvache, L. Canova, A. Borot, A. Trisorio, C.G. Durfee, R. López-Martens, Opt. Lett. **34**, 1588 (2009)
22. C. Fourcade Dutin, A. Dubrouil, S. Petit, E. Mével, E. Constant, D. Descamps, Opt. Lett. **35**, 253 (2010)
23. E. Yablonovitch, Phys. Rev. Lett. **60**, 795 (1988)
24. H.R. Telle, G. Steinmeyer, A.E. Dunlop, J. Stenger, D.H. Sutter, U. Keller, Appl. Phys. B **69**, 327 (1999)
25. D.J. Jones, S.A. Diddams, J.K. Ranka, A. Stentz, R.S. Windeler, J.L. Hall, S.T. Cundiff, Science **288**, 635 (2000)
26. A. Apolonski, A. Poppe, G. Tempea, C. Spielmann, Th Udem, R. Holtzwarth, T.W. Hänsch, F. Krausz, Phys. Rev. Lett. **85**, 740 (2000)
27. A. Baltuška, Th Udem, M. Uiberacker, M. Hentschel, E. Goulielmakis, C. Gohle, R. Holzwarth, V.S. Yakovlev, A. Scrinzi, T.W. Hänsch, F. Krausz, Nature **421**, 611 (2003)
28. S. Koke, C. Grebing, H. Frei, A. Anderson, A. Assion, G. Steinmeyer, Nat. Photon. **4**, 462 (2010)
29. B. Borchers, S. Koke, A. Husakou, J. Herrmann, G. Steinmeyer, Opt. Lett. **36**, 4146 (2011)
30. F. Lücking, A. Assion, A. Apolonski, F. Krausz, G. Steinmeyer, Opt. Lett. **37**, 2076 (2012)
31. A. Baltuška, T. Fuji, T. Kobayashi, Phys. Rev. Lett. **88**, 133901 (2002)
32. G. Sansone, L. Poletto, M. Nisoli, Nat. Photon. **5**, 655 (2011)
33. P.B. Corkum, N.H. Burnett, M.Y. Ivanov, Opt. Lett. **19**, 1870 (1994)
34. I.J. Sola, E. Mével, L. Elouga, E. Constant, V. Strelkov, L. Poletto, P. Villoresi, E. Benedetti, J.-P. Caumes, S. Stagira, C. Vozzi, G. Sansone, M. Nisoli, Nat. Phys. **2**, 319 (2006)
35. G. Sansone, E. Benedetti, F. Calegari, C. Vozzi, L. Avaldi, R. Flammini, L. Poletto, P. Villoresi, C. Altucci, R. Velotta, S. Stagira, S. De Silvestri, M. Nisoli, Science **314**, 443 (2006)
36. O. Tcherbakoff, E. Mével, D. Descamps, J. Plumridge, E. Constant, Phys. Rev. A **68**, 043804 (2003)
37. G. Sansone, E. Benedetti, C. Vozzi, S. Stagira, M. Nisoli. New J. Phys. **10**, 025006 (2008)
38. J. Mauritsson, P. Johnsson, E. Gustafsson, A. L'Huillier, K.J. Schafer, M.B. Gaarde, Phys. Rev. Lett. **97**, 013001 (2006)
39. Z. Chang, Phys. Rev. A **76**, 051403(R) (2007)
40. H. Mashiko, S. Gilbertson, M. Chini, X. Feng, C. Yun, H. Wang, S.D. Khan, S. Chen, Z. Chang, Opt. Lett. **34**, 3337 (2009)
41. X. Feng, S. Gilbertson, H. Mashiko, H. Wang, S.D. Khan, M. Chini, Y. Wu, K. Zhao, Z. Chang, Phys. Rev. Lett. **103**, 183901 (2009)
42. K.T. Kim, C.M. Kim, M.G. Baik, G. Umesh, C.H. Nam, Phys. Rev. A **69**, 051805(R) (2004)
43. W. Cao, P. Lu, P. Lan, X. Wang, G. Yang, Phys. Rev. A **74**, 063821 (2006)
44. M.B. Gaarde, J.L. Tate, K.J. Schafer, J. Phys. B **41**, 132001 (2008)
45. T. Pfeifer, A. Jullien, M.J. Abel, P.M. Nagel, L. Gallmann, D.M. Neumark, S.R. Leone, Opt. Express **15**, 17120 (2007)

46. I. Thomann, A. Bahabad, X. Liu, R. Trebino, M.M. Murnane, H.C. Kapteyn, Opt. Express **17**, 4611 (2009)
47. M.J. Abel, T. Pfeifer, P.M. Nagel, W. Boutu, M.J. Bell, C.P. Steiner, D.M. Neumark, S.R. Leone, Chem. Phys. **366**, 9 (2009)
48. F. Ferrari, F. Calegari, M. Lucchini, C. Vozzi, S. Stagira, G. Sansone, M. Nisoli, Nat. Phot **4**, 875 (2010)
49. M.V. Ammosov, N.B. Delone, V. P. Krainov, Zh. Eksp. Teor. Fiz. **91**, 2008 (1986) [Sov. Phys. JETP **64**, 1191 (1986)]
50. E.J. Takahashi, T. Kanai, Y. Nabekawa, K. Midorikawa, Appl. Phys. Lett. **93**, 041111 (2008)
51. T. Popmintchev, M. Chen, O. Cohen, M.E. Grisham, J.J. Rocca, M.M. Murnane, H.C. Kapteyn, Opt. Lett. **33**, 2128 (2008)
52. T. Popmintchev, M.-C. Chen, D. Popmintchev, P. Arpin, S. Brown, S. Ališauskas, G. Andriukaitis, T. Balčiunas, O.D. Mücke, A. Pugzlys, A. Baltuška, B. Shim, S.E. Schrauth, A. Gaeta, C. Hernández-García, L. Plaja, A. Becker, A. Jaron-Becker, M.M. Murnane, H.C. Kapteyn, Science **336**, 1287 (2012)
53. P. Colosimo, G. Doumy, C.I. Blaga, J. Wheeler, C. Hauri, F. Catoire, J. Tate, R. Chirla, A.M. March, G.G. Paulus, H.G. Muller, P. Agostini, L.F. DiMauro, Nat. Phys. **4**, 386 (2008)
54. A.D. Shiner, C. Trallero-Herrero, N. Kajumba, H.-C. Bandulet, D. Comtois, F. Légaré, M. Giguere, J.-C. Kieffer, P.B. Corkum, D.M. Villeneuve, Phys. Rev. Lett. **103**, 073902 (2009)
55. F. Calegari, C. Vozzi, M. Negro, G. Sansone, F. Frassetto, L. Poletto, P. Villoresi, M. Nisoli, S. De Silvestri, S. Stagira, Opt. Lett. **34**, 3125 (2009)
56. E.J. Takahashi, P. Lan, O.D. Mücke, Y. Nabekawa, K. Midorikawa, Phys. Rev. Lett. **104**, 233901 (2010)
57. F. Quéré, Y. Mairesse, J. Itatani, J. Mod. Opt. **52**, 339 (2005)
58. Y. Mairesse, F. Quéré, Phys. Rev. A **71**, 011401(R) (2005)
59. M. Lewenstein, P. Balcou, M.Y. Ivanov, A. L'Huillier, P.B. Corkum, Phys. Rev. A **49**, 2117–2132 (1994)
60. R. Trebino, *Frequency-Resolved Optical Gating* (Kluwer Academic, Boston, 2000)
61. D. Kane, IEEE J. Quantum Electron. **35**, 421–431 (1999)
62. L.S. Cederbaum, J. Zobeley, Chem. Phys. Lett. **307**, 205 (1999)
63. F. Remacle, R.D. Levine, PNAS **103**, 6793 (2006)
64. H. Hennig, J. Breidbach, L.S. Cederbaum, J. Phys. Chem. A **109**, 409 (2005)
65. G. Sansone, F. Kelkensberg, J.F. Pérez-Torres, F. Morales, M.F. Kling, W. Siu, O. Ghafur, P. Johnsson, M. Swoboda, E. Benedetti, F. Ferrari, F. Lépine, J.L. Sanz-Vicario, S. Zherebtsov, I. Znakovskaya, A. L'Huillier, MYu. Ivanov, M. Nisoli, F. Martín, M.J.J. Vrakking, Nature **465**, 763 (2010)
66. K. Ito, R. Hall, M. Ukai, J. Chem. Phys. **104**, 8449 (1996)
67. M.F. Kling, C. Siedschlag, A.J. Verhoef, J.I. Khan, M. Schultze, T. Uphues, Y. Ni, M. Uiberacker, M. Drescher, F. Krausz, M.J.J. Vrakking, Science **312**, 246 (2006)

Chapter 7
Laser Interactions for the Synthesis and In Situ Diagnostics of Nanomaterials

David B. Geohegan, Alex A. Puretzky, Mina Yoon, Gyula Eres, Chris Rouleau, Kai Xiao, Jeremy Jackson, Jason Readle, Murari Regmi, Norbert Thonnard, Gerd Duscher, Matt Chisholm and Karren More

Abstract Laser interactions have traditionally been at the center of nanomaterials science, providing highly nonequilibrium growth conditions to enable the synthesis of novel new nanoparticles, nanotubes, and nanowires with metastable phases. Simultaneously, lasers provide unique opportunities for the remote characterization of nanomaterial size, structure, and composition through tunable laser spectroscopy, scattering, and imaging. Pulsed lasers offer the opportunity, therefore, to supply the required energy and excitation to both control and understand the growth processes of nanomaterials, providing valuable views of the typically nonequilibrium growth kinetics and intermediates involved. Here we illustrate the key challenges and progress in laser interactions for the synthesis and in situ diagnostics of nanomaterials through recent examples involving primarily carbon nanomaterials, including the pulsed growth of carbon nanotubes and graphene.

7.1 Introduction

The special properties of nanomaterials originate from size-dependent atomic configurations that often result in a departure of electronic band structure, chemical reactivity, and physical properties from their bulk counterparts [1].

D. B. Geohegan (✉) · A. A. Puretzky · M. Yoon · C. Rouleau · K. Xiao · J. Jackson
Center for Nanophase Materials Sciences, Oak Ridge National Laboratory, Oak Ridge, TN 37831, USA
e-mail: geohegandb@ornl.gov

G. Eres · J. Jackson · J. Readle · M. Regmi · M. Chisholm · K. More
Materials Science and Technology Division, Oak Ridge National Laboratory, Oak Ridge, TN 37831, USA

N. Thonnard · G. Duscher
Department of Materials Science and Engineering, University of Tennessee, Knoxville, TN 37996, USA

M. Castillejo et al. (eds.), *Lasers in Materials Science*,
Springer Series in Materials Science 191, DOI: 10.1007/978-3-319-02898-9_7,

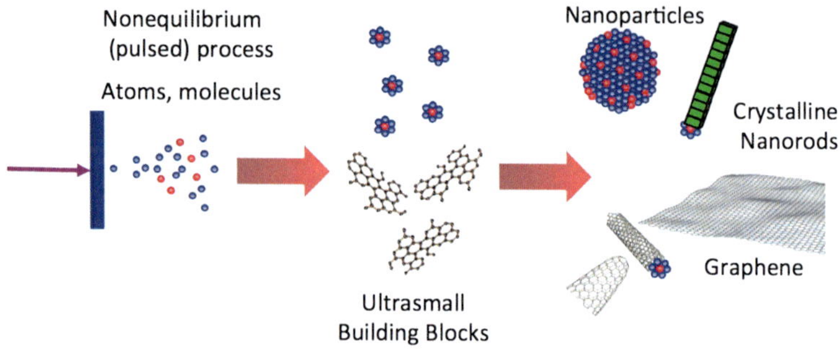

Fig. 7.1 Schematic of the synthesis of nanomaterials from nonequilibrium, pulsed laser-induced process such as laser vaporization of a target into atoms and molecules. A great variety of nanoparticles, crystalline nanorods, and polymorphs (e.g., for carbon, nanohorns, nanotubes, graphene, etc.) are energetically possible, both with and without catalyst-assistance, depending on the nucleation and growth of intermediate, ultrasmall building blocks and subsequent growth kinetics

Under equilibrium synthesis conditions, thermodynamic energy barriers severely constrain the number of configurations and structures that can occur [2]. However, numerous metastable configurations can emerge during synthesis as a result of competition between thermodynamic and kinetic pathways [3]. Non-equilibrium growth methods can overcome the thermodynamic energy barriers for metastable nanostructure formation, and are important tools for materials discovery.

Laser interactions with materials provide remarkably versatile nonequilbrium conditions for nanomaterials synthesis. Laser vaporization of solid targets, for example, typically results in stoichiometric material removal, plasma formation ($T_e \sim$ 1–10 eV), and translational kinetic energies up to $\sim$100 eV [4, 5]. Spatial confinement of these reactants using a background gas can provide the time, temperature, and fluxes of reactants necessary for the formation of nanostructures with metastable phases. In carbon nanomaterials, as shown in Fig. 7.1 for example, a great variety of nanostructures may emerge depending upon the nucleation and growth pathways induced by this spatial confinement. Various allotropes of carbon can be formed depending on the growth times and temperatures provided by the spatial confinement, including fullerenes, [6] carbon nanohorns [7, 8], nanodiamond [9], graphene [10], and carbon nanotubes [11]. For other materials, a great variety of nanoparticles, crystalline nanorods, nanotubes, planar materials, and curved structures can similarly be obtained under such nonequilibrium growth conditions [12]. Metal catalysts introduced into the reactant mixture can greatly alter the product distribution by lowering reaction barriers in chemical reactions. As a result, metal catalyst-assisted chemical vapor deposition is often explored to duplicate a subset of the nanomaterials (such as single wall nanotubes (SWNTs), silicon nanowires (SiNWs), etc.) discovered by laser vaporization.

However, as indicated in Fig. 7.1, understanding and controlling the growth processes that determine the atomic structure of different forms of nanostructures remains a fundamental grand challenge [13] for nanoscience. Can laser interactions reveal the nuclei and ultrasmall building blocks that form the basis for the growth and assembly of larger nanostructures? Can pulsed laser interactions reveal the timescales for their formation? Answering these questions requires real-time diagnostics of the growth environment and nanostructure formation. Through time- and spatially-resolved plasma spectroscopy, the concentrations of atomic and molecular reactants as well as plasma temperatures can be remotely determined in situ (see [4, 5]). Moreover, through optical spectroscopy, the size and structure of many nanostructures can be remotely determined, as well as their evolution in size and concentration. Therefore, valuable kinetics information can be inferred which can be used to model nucleation and growth kinetics, and from which the presence of reactive intermediates may be inferred. A major challenge remaining for laser interactions with materials is the development of laser-based, real-time diagnostics that can identify the ultrasmall building blocks and reactive intermediates in situ.

In this chapter, current challenges in laser interactions with materials for both synthesis and in situ diagnostics of nanomaterials growth processes will be illustrated with an emphasis on carbon nanomaterials. The evidence for "ultrasmall building blocks" in nanomaterials growth will be examined through a review of both high- and low-temperature nanomaterials synthesis processes, including laser vaporization and chemical vapor deposition.

7.2 Cluster and Nanoparticle Growth in Pulsed Laser Vaporization

Pulsed laser vaporization (PLV) of targets using nanosecond lasers typically results in an energetic plasma plume that can penetrate moderate pressure background gases with sufficient deposition rate for the pulsed laser deposition (PLD) of thin films (pressures typically <300 mTorr for reasonable distances of 5–10 cm). As indicated in Fig. 7.2a, atoms and molecules comprise the principal components of the ablation plume arriving at the substrate in vacuum. However, as the background gas pressure is raised, the plasma plume is slowed and confined, leading to the condensation of clusters and nanoparticles as shown in Fig. 7.2b. Under these conditions, the film morphology can completely change.

Measurements of the ion flux penetrating through different background gas pressures reveals that material is scattered from the plume in accordance with Beer's law with an elastic scattering cross section $\sim 1 \times 10^{-16}$ cm^2, with material from the initial shifted center of mass Maxwell–Boltzmann distribution disappearing exponentially with background gas pressure or distance [14]. At particular distances and pressures, some fraction of the initial 'vacuum' distribution of ions can still be detected while a second, delayed component of material arrives much later in time [15]. This situation is illustrated in Fig. 7.2d for the case of ZnTe target ablation into background nitrogen gas [16]. At a distance of 10 cm from the

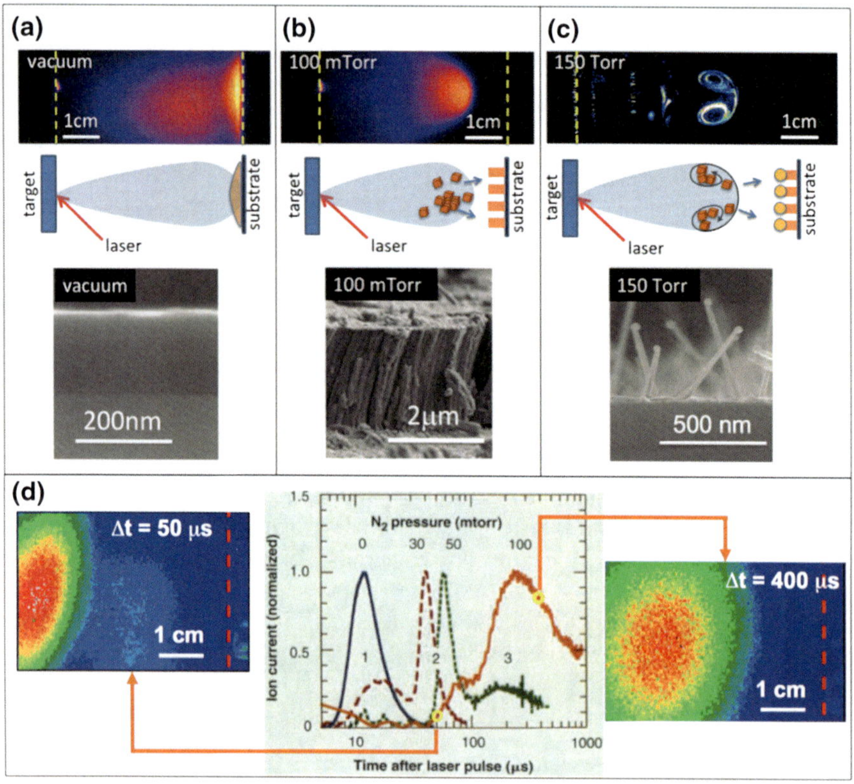

Fig. 7.2 **a** Gated-ICCD images of laser plasma plumes for PLD in vacuum, **b** 100 mTorr, and swirling clouds of aggregated nanoparticles at **c** 150 Torr (Rayleigh scattered light from a second laser sheet). Corresponding schematics show that PLD growth on a substrate changes results from atoms and molecules in vacuum, and an increasing fraction of clusters and nanoparticles in higher pressure background gases. The resulting synthesized structures can be varied from **a** smooth epitaxial films (e.g. TiO_2), to **b** columnar nanoparticle assembles (e.g. TiO_2), to **c** crystalline nanorods (e.g. SnO_2 on Au/Si). **d** Ion probe current pulses (center panel) measured at d = 10 cm from a ZnTe target in background pressures ranging from 0 to 100 mTorr N_2 show corresponding evolution of 3 plume components: (1) atoms and molecules, (2) atoms and molecules + clusters, (3) clusters and nanoparticles. Corresponding images at two times in 100 mTorr show the position of the third component of the plume at Δt = 50 μs (*left*) and at Δt = 400 μs (*right*) after laser ablation relative to d = 10 cm (*red dashed line*) [(**d**) from 16]

target, the initial vacuum distribution of ions arriving to the substrate is labeled "1", while the time-dependence of ions arriving through 30 mTorr N_2 shows two components, a fast component labeled "1" and a slowed component labeled "2". This "plume-splitting" phenomenon is a special circumstance which is difficult to detect using plume imaging alone because the fast 'vacuum' component at the front of the plume represents that material which has penetrated the background gas without collisions. This fast component is far less bright in comparison to the delayed second component, where collision-induced fluorescence resulting from slowed atoms and molecules condensing into a 'shock front' can become very bright. Despite the fact that this highly visible component of the plume is often used to track the apparent propagation of the plume propagation (using, shock, drag, or other models which assume hydrodynamic propagation) the successful description of plume splitting which takes into account both components of the plume involves a scattering model formalism [17]. Plume splitting is a general phenomenon which has been successfully described using a single elastic scattering cross section for both atoms propagating into lighter, or heavier, gases [18].

At longer distances, this second component can be seen to split into yet another, third component labeled "3" in Fig. 7.2d. This component corresponds to larger nanoparticles and nanoparticle aggregates which retain some charge from the plasma. Photoluminescence from neutral nanoparticles and their aggregates can be induced by a second, delayed laser pulse as shown in the two images in Fig. 7.2d [19]. Thus, at low pressures and different distances from the target, mixtures of fast atoms and molecules with clusters can be replaced with a flux composed of nearly entirely nanoparticles and their aggregates. We label this nanoparticle assisted PLD (NA-PLD) in Fig. 7.2b.

At higher pressures, the plume condenses rapidly into nanoparticles, which aggregate and become trapped in sharp vortex rings [20], a cross-section of which is shown in Fig. 7.2c. With metal catalyst-doped targets, a great variety of semiconductor nanowires, nanorods, and nanotubes can be grown while this material is suspended at high temperatures within a tube furnace, although the process is typically referred to as a VLS, or vapor–liquid–solid process [12]. Evidence suggests that small clusters and ultrasmall nanoparticles dominate the "vapor" in such catalyst-assisted growth processes. This can be explicitly demonstrated, as indicated in Fig. 7.2c, by directing preformed nanoparticles as the feedstock for the growth of nanowires on substrates. In Fig. 7.2c SnO_2 nanowires are shown growing from Au catalyst nanoparticles at their tips, demonstrating that the aggregated nanoparticles in 150 Torr background Ar following SnO_2 target ablation can serve as feedstock to nucleate and grow nanowires on a substrate within a tube furnace in a nanoparticle-assisted chemical vapor deposition (CVD) process. Similarly, imaging of ZnO ablation at high pressures was used to demonstrate that crystalline ZnO nanorods can be grown by nanoparticle-assisted PLD without the use of any metal catalyst layer [21]. Hence, the role and dynamics of small nanoparticles as "building blocks" in the assembly of crystalline thin films and nanostructures remains a key question for in situ diagnostic experiments.

7.3 Characterization and Modeling of Ultrasmall Nanoparticle "Building Blocks"

What determines the size, shape, stoichiometry, and crystallinity of nanostructures formed within the laser plume? Spatial confinement of the plume, time, and temperature determine the synthesis conditions for the thermalization and condensation of atoms and molecules into clusters and nanoparticles. This has been modeled for the idealized isentropic expansion of a plume into vacuum [22], however the dynamics of plume expansion into background gases plays a major role in the adjustment of these parameters. As described above, the standard hydrodynamic models for plume expansion do not accurately describe low-pressure plume expansions where a scattering formalism is more appropriate.

Only recently, with the advent of aberration-corrected atomic-resolution transmission electron microscopy, have such views of the atomic structure of ultrasmall nanoparticles (UNPs, diameters <3 nm) become available. Figure 7.3a, b show high resolution TEM and atomic-resolution Z-contrast scanning TEM images of TiO_2 nanoparticle aggregates collected on TEM grids under ablation of a TiO_2 target into low-pressure, room temperature background argon. Such HRTEM and EELS data of TiO_2 UNPs show that the individual nanoparticles are typically crystalline (see Fig. 7.3a and b) with structures that do not match those of known bulk phases. It is quite likely that such nanoparticles, which are formed under highly nonequilibrium conditions, are energetically metastable. In addition, the high fraction of surface atoms leads to undercoordinated bonding and energetically active surfaces. The chemistry induced by such surfaces may facilitate the oriented attachment of the particles [23, 24].

Identifying and classifying unknown phases of ultrasmall metastable nanoparticles is extremely challenging, and requires new computational approaches that are capable of rapidly surveying the energy landscape of a wide range of nanoparticle morphologies and structures. However, most atomistic studies of ultrasmall nanoparticle structure have assumed a known bulk phase as a starting point, which is then permitted to undergo relaxation at different temperatures [25–28]. Even with this limited approach, first principles modeling is currently computationally tractable only for nanoparticles with sizes <3 nm. Hybrid approaches combining force-field potentials and first principles calculations are being developed to rapidly survey configuration space to identify unknown structures.

Figure 7.3c and d show three minimum-energy nanoparticle structures that emerged from numerical simulations that began with 1 nm nanoparticles sculpted from bulk anatase and rutile phases of TiO_2, respectively. The structural stability increases from left to right. Here a hybrid method using Monte Carlo (MC) simulations and conjugate gradient (CG) optimization has been used. The total energy is based on the Matsui–Akaogi force-field potential [29] for its relative accuracy as compared to the ab initio methods. The technique's accuracy was verified against previously published results on small $(TiO_2)_n$ systems (n = 1, 2, and 3) based on ab initio approaches (quantum chemistry and density functional theory) [30].

The results clearly demonstrate that the structural properties and the thermodynamic stabilities of ultrasmall nanoparticles can be very different to their bulk counterparts. For example, localized melting of 2-nm anatase is found to start at just T = 500 K and persists until ~T = 1,000 K, as compared to the bulk melting point of ~2,150 K, with regions of the nanoparticle starting first near the

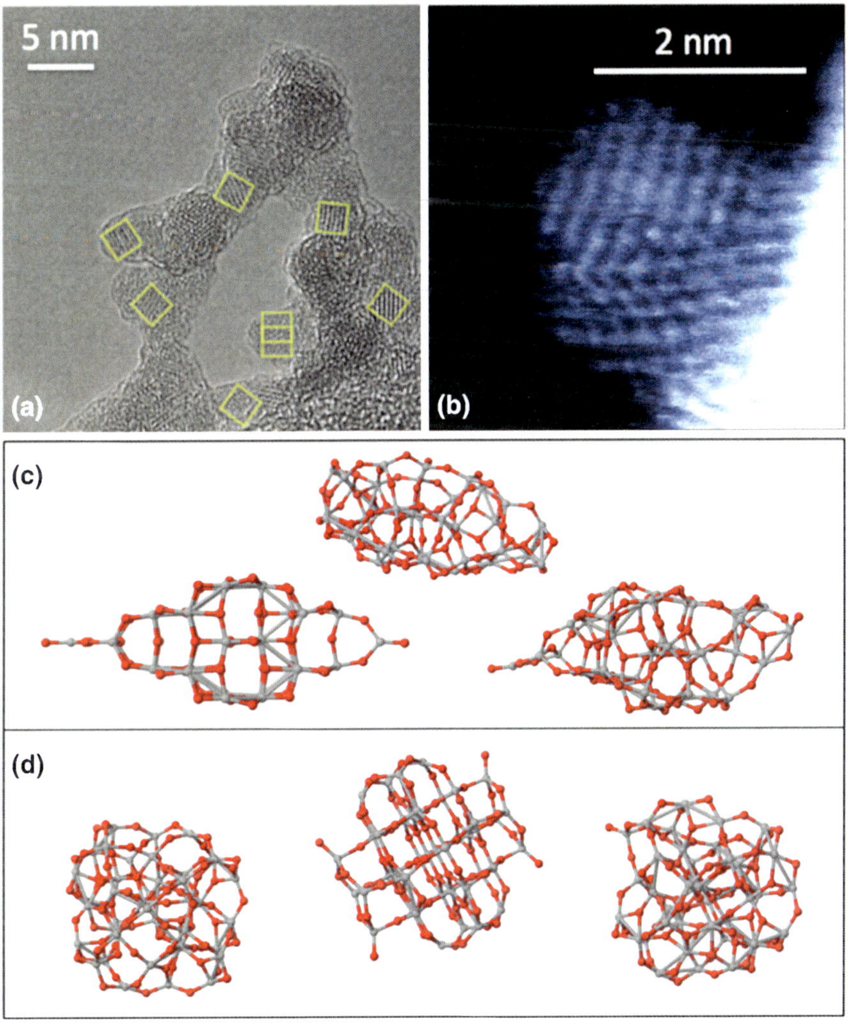

Fig. 7.3 **a** Aggregated ultrasmall nanoparticles of titanium dioxide formed by PLV of TiO_2 into 100 mTorr of O_2. *Yellow boxes* indicate 2–3 nm crystalline domains. 100 mTorr. **b** AR-Z-STEM image of a TiOx nanocrystal of an unknown phase. **c**, **d** Examples of several local minima theoretically predicted for 1-nm **c** anatase and **d** rutile nanocrystals of TiO_2 which have been allowed to relax into metastable nanophases

surface, then the tips. Such calculations provide a qualitative understanding of how UNPs may serve as "building blocks" and integrate into larger nanoparticles or crystalline nanorods under conditions found in Fig. 7.2 for pulsed laser vaporization, NA-PLD, and nanoparticle assisted-CVD (NA-CVD).

7.4 Carbon Nanostructure Synthesis in Laser Vaporization

7.4.1 Fullerenes

Carbon exhibits a remarkable ability to self-assemble into a variety of novel nanostructures within the highly non-equilibrium conditions of laser vaporization plumes. Fullerenes are the early example of carbon clusters with magic numbers that emerged from laser ablation of carbon and condensation in a background gas [31] within specially constructed nozzle sources [32]. Theory and modeling have shown that synthesis temperatures of $\sim$3,000 K are required to induce the curvature necessary for the formation of fullerenes and other curved carbon nanostructures, while synthesis temperatures of $\sim$1,000–2,000 K produce flat carbon chain structures and sheets [33]. Pentagons necessary for fullerene curvature can be viewed as local graphene defects, each costing $\sim$45 kcal/mol energy [34]. However, while the soccer-ball-shaped C_{60} molecule is relatively easy to produce in the highly non-equilibrium conditions provided by laser ablation or electric arc vaporization of pure carbon rods, for many years computer simulation of the formation process failed to produce the structure of the stable fullerenes. For many years, all of the many reaction pathways and models of fullerene formation were based upon intermediate structures that were in thermodynamic equilibrium. However hot carbon plasmas are far from equilibrium, and such equilibrium models could not describe how such perfect, regular structures like C_{60} could emerge in such numbers and with such consistency. (See [35] for a good review.)

More recently, Irle et al., considered the *nonequilibrium* assembly of C_{60} in hot carbon vapor, proposing a dynamic fullerene self-assemby mechanism wherein a great variety of superfullerenes are formed and then whittled down by the elimination of C_2 molecules in a pathway termed the "shrinking hot giant road" [35]. Quantum chemical molecular dynamics (QM/MD) simulations of hot carbon vapor revealed that linear, sp-hybridized carbon polyyne chains nucleate and then assemble into rings, which then condense into curved carbon bowls, and finally, closed irregular giant fullerenes. Following this "size-up" formation of giant fullerenes, the model indicates that the high vibrational temperatures lead to irreversible pop-out processes of C_2 molecules or falloff of the weakly-bonded linear antennae of carbon due to violent wagging and stretch vibrations, thereby reducing the size and shape of the molecules toward ever-more circular and stable molecules with fewer dangling bonds and sp^2-bonded networks with less strain. Thus, the emergence of stable carbon nanostructures from such a chaotic, non-equilibrium environment was modeled to result from the formation and self-assembly of intermediates into a large variety of products, followed by the ejection of primarily C_2 in an annealing period, toward more thermodynamically stable structures [35].

Of course, computational cost require that QM/MD simulations are performed at artificially high temperatures for very short (typically picoseconds to nanoseconds) periods, whereas laser plasmas persist for times of microseconds to milliseconds. What are the actual times of formation of carbon nanostructures in the high-temperature laser plasma plumes? To answer this question, time-resolved in situ diagnostics have been employed and their application to understand the catalyst-assisted synthesis of single-wall carbon nanotubes and the catalyst-free assembly of single-wall carbon nanohorns will be summarized below. Additional details have been summarized in other reviews [36].

7.4.2 Single-Wall Carbon Nanotubes

Figure 7.4 summarizes key aspects of the laser vaporization growth of single-wall carbon nanotubes (SWNTs) and time-resolved diagnostic experiments. First, in Fig. 7.4a a normal photograph taken within the 1,050 °C tube furnace shows the glowing circular 1.0″ carbon (with 1 wt. % each Ni, Co) target and many vortex rings of laser vaporized material floating within the gently-flowing Ar gas, silhouetted against the black background. Each vortex ring results from confinement of the ejecta from one laser pulse (laser running at 10 Hz in these experiments). Confining the reactants into such small volumes for sufficient time for synthesis is crucial for the self assembly of carbon chains and clusters as noted in the modeling simulations above, and also for the catalyst-assisted growth of extended structures like SWNTs, which can reach microns lengths during the 10-s of time these rings remain within the hot growth zone of the furnace.

Time-resolved, in situ imaging and spectroscopy of laser vaporization plumes, schematically illustrated in Fig. 7.4b, utilize a time-delayed sheet beam from a second laser to induce laser-induced fluorescence from atomic or molecular species (e.g., ground state Co atoms in Fig. 7.4c), laser induced blackbody emission from clusters (e.g., C clusters in Fig. 7.4c), or Rayleigh scattering from aggregated clusters (e.g., the vortex rings in Fig. 7.4g) [37, 38]. A time-delayed, gated, intensified CCD-array detector with variable gate (>3 ns) is typically used to capture images or optical spectra during, or at a well-defined delay with respect to, the probe laser pulse. For example, in Fig. 7.4c, images taken through different colored filters at $\Delta t = 1.0$, and 2.0 ms after the initial laser (vaporizing) pulse revealed that molecular carbon disappeared by $\Delta t = 1.0$ ms and blackbody emission from carbon clusters appeared. Within another millisecond, the vortex ring trapping aggregated carbon clusters was well established. However, during the same time period, ground state atomic Co was reaching its maximum population at $\Delta t = 1.0$ ms (following relaxation of the hot laser plasma, not shown) and just began to condense during the next millisecond. These measurements revealed that carbon condenses into clusters *first,* relatively late in time and at extended distances from the target, followed by the condensation of the metal atoms. Thus, the raw feedstock for SWNT growth was *condensed phase* clusters of carbon and

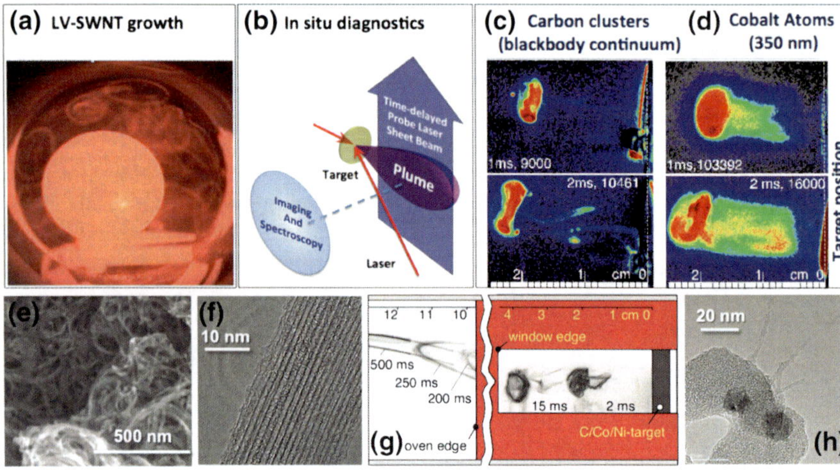

Fig. 7.4 Summary of time-resolved, in situ diagnostic measurements of the laser vaporization growth of SWNTs. **a** Digital photograph looking down a 1,050 °C (2.0″) tube furnace along the laser beam direction, showing a 1.0″ C/Ni/Co target, focused Nd:YAG laser spot, and many vortex rings wherein the ejecta within each laser pulse become trapped in the 500 Torr Ar background pressure. **b** Schematic of the typical in situ diagnostic approach for imaging and spectroscopy, including the option for a second, time-delayed laser sheet beam to probe the plume. **c**, **d** Filtered, gated ICCD images of the expanding laser plume from the target at $\Delta t = 1$, 2 ms. Aggregated carbon clusters are observable in vortex dynamics by laser-induced blackbody continuum radiation, while laser-induced ($\lambda_{ex} = 308$ nm) fluorescence at 350 nm is used to observe the ground-state Co atoms at the same times. **e** SEM and **f** TEM images of profuse bundles of SWNTs collected after ~10 s growth times under these conditions. **g** Gated ICCD images of Rayleigh scattering from the vortex rings at the designated times inside a windowed tube furnace obtained using a second, time-delayed laser. **h** Several individual SWNTs resulting from time-restricted growth to <100 ms, each <100 nm long, protrude from <20 nm NiCo nanoparticles which are embedded in unconvereted carbon. [(**c**) and (**g**) reproduced from [37, 38]

metal. Rayleigh scattering images of the propagating vortex rings (e.g., see Fig. 7.4g) revealed that, depending upon the time spent within the hot furnace, SWNTs grow over extended periods of time at rates of 1–5 μm/s. Stop-growth experiments for times $\Delta t \sim 100$ ms revealed only very short (<250 nm-long) SWNTs as shown in Fig. 7.4h, meaning that the majority of SWNT growth occurs for extended times (hundreds of milliseconds to seconds) [38]. The SWNT products of this *condensed phase* catalyst-assisted growth are remarkably uniform. Large bundles of SWNTs, shown in Fig. 7.4e, f are found to contain individual SWNTs with remarkably narrow diameter distribution (1.2 ± 0.2 nm), despite a large variety of catalyst nanoparticle diameters (2–20 nm). This mystery of nucleation and growth persists, and will be contrasted with the broad diameter distributions from low-temperature chemical vapor deposition growth described below in Sect. 7.5.

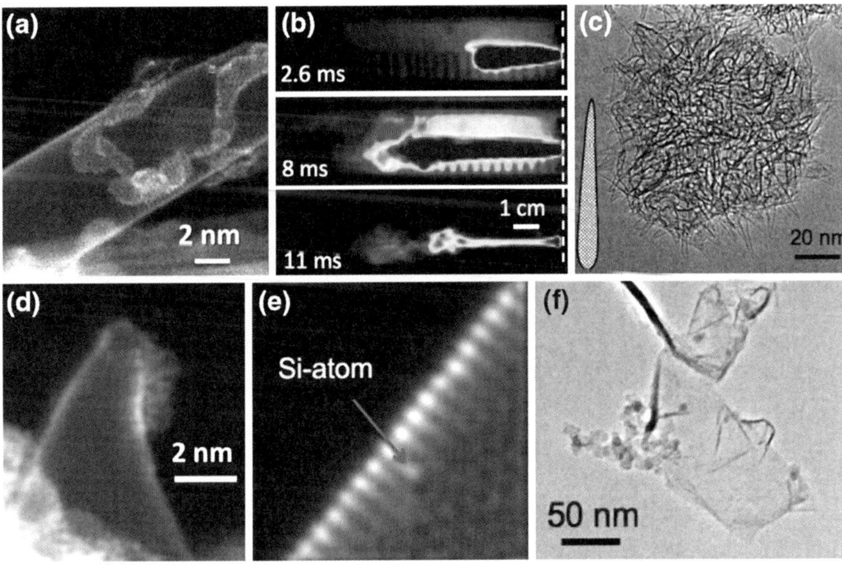

Fig. 7.5 **a** Z-contrast, Atomic resolution STEM images of single-wall carbon nanohorn with attached graphene flakes. **b** Side-on images of plasma plumes used for SWNH growth by 10 ms-long, 100 J pulses from a Nd:YAG (1,064 nm) laser on a C target in 1 atm. flowing Ar. **c** Roughly spherical SWNH aggregate, composed of hundreds of closed nanohorns (subunit depicted as inset). **d** AR-Z-STEM image of a SWNH tip with attached flake. **e** AR-Z-STEM image of a SWNH edge, showing lattice incorporation of a single Si atom. **f** HRTEM image of graphene flake synthesized under similar conditions, although with hydrogen in the background gas

7.4.3 Single-Wall Carbon Nanohorns

Like giant fullerenes, single wall carbon nanohorns (SWNHs) are hollow, single-walled carbon nanostructures that are grown in the highly nonequilibrium conditions of laser plasmas or electric arcs where pure carbon is vaporized into background gases. Discovered in 1999, [7] SWNHs are extended tubular structures with conical, horn-shaped tips as shown in Fig. 7.5. As schematically diagrammed in Fig. 7.5c, the individual SWNH units pack into agglomerates which are roughly spherical in shape, with many protruding ends. They are formed efficiently in high yields by high power laser vaporization into room-temperature background rare-gas ambients. Diagnostics of the synthesis process, as shown in Fig. 7.5b, have been performed to characterize the growth environment [39–41], and it has been shown that SWNHs grow at ~1 nm/ms growth rates (equivalent to the 1 μm/s growth rate of SWNTs with catalyst-assistance) [42].

Recently, atomic resolution Z-contrast scanning transmission electron microscope (AR-Z-STEM) images have revealed the atomic structure of carbon nanotubes, nanohorns, and graphene sheets—making it possible to directly observe defects, both structural and substitutional. For example, in Fig. 7.5a, d, and e it is possible to see the 5–7 Stone–Wales defects responsible for the curvature of the SWNHs. Because Z-contrast STEM utilizes an annular detector, the intensity of the image is proportional to atomic density. For single atoms, such as the single Si atom embedded in the SWNH in Fig. 7.5e, heavier atoms appear brighter. Single-atom electron energy loss spectroscopy (EELS) can be used to identify the atoms and the nature of their bonding using characteristic near-edge structure [43].

However, equally interesting is the atomic structure of the carbon that did not convert to nanotubes or nanohorns. As shown in Fig. 7.5a and e, carbon adhering to the walls and tips of the nanotubes and nanohorns are not 'amorphous' carbon, but in most cases are apparently well-formed or defective small platelets of single-layer graphene. These products are either the results of a competing reaction pathway, or intermediates in the synthesis pathway for nanotubes and nanohorns. It has long been speculated that such small and possibly defective graphene flakes might be the feedstock for the growth of larger structures. In support of this argument, annealing experiments of the carbon soot collected after electric arc and laser vaporization of solid carbon (or carbon/metal) have shown the ability of closed, nanohorn-like structures to grow from pure carbon soot, and much longer SWNTs to grow from short 'seeds' of SWNTs that have already nucleated from metal catalyst nanoparticles that are embedded in carbon soot, such as those shown in Fig. 7.4h [44–46]. Therefore, exploration of the remarkable and facile synthesis of closed single-wall carbon structures such as SWNHs in high-temperature laser plasmas is still in its infancy. Can high quality graphene sheets grow in such non-equilibrium conditions? Preliminary experiments have shown that graphene platelets can been synthesized, as shown in Fig. 7.5f through the addition of small quantities of hydrogen in an attempt to stabilize the dangling bonds thought responsible for the sealing of the graphene platelets during synthesis. Such small sheets can be used as 'building blocks' to construct new nanostructures such as metal-encapsulated "nanooysters"—metal quantum dots within hollow metal shells—by mixing reactive metal nanoparticles and laser annealing [47]. High-surface area novel nanocarbons such as nanotubes, graphene, and nanohorns are highly promising supports for metal nanoparticles in catalysis or hydrogen storage [48], or electrodes in batteries and supercapacitors [49].

7.5 Laser Diagnostics of Single-Wall Carbon Nanotube Growth by Chemical Vapor Deposition

Although laser vaporization is a highly versatile, nonequilibrium method for the discovery of nanomaterials, chemical vapor deposition (CVD) is the method of choice for nanomaterial growth wherever possible in order to take advantage of well-established wafer scale processing fabrication technology. CVD has the potential to deterministically grow well-aligned nanomaterials in precise locations with controlled lengths and properties. As described above for laser vaporization of SWNTs, metal 'catalyst' nanoparticles with an affinity for carbon can grow carbon nanotubes with diameters nearly matching their own. In CVD, the metal nanoparticles typically perform a dual role, first to crack the hydrocarbon gas, and second to assist in the assembly of a carbon nanotube. Two scenarios are considered: (1) vapor–liquid–solid growth (also known as the dissolution/precipitation model) wherein carbon dissolves in the metal nanoparticle (either liquid or solid) and precipitates due to a gradient in the chemical potential, or (2) surface growth, wherein carbon is assembled essentially on the surface of the metal nanoparticle to form sheets, the nanoparticle essentially acting as a form that guides a tubular sock.

Despite the huge body of work on the chemical vapor deposition of carbon nanotubes, many questions remain about nanotube nucleation and growth. The questions generally concern either the factors that control the overall efficiency of the growth process with implications for large scale production, or the factors that govern the exact structure of an individual nanotube with implications for electronic devices. SWNTs have been the focus of effort due to their applicability of semiconducting SWNTs as single electron transistors, and metallic nanotubes as interconnects or transparent conductors. The factors that control the precise crystal structure of SWNTs (their chirality, which determines their electronic structure) at nucleation are not known, however it is clear that once nucleated SWNTs continue to grow with the same chirality up to centimeters in length [50]. The reason that chirality changes do not occur more frequently is that the energy cost to introduce numerous defects simultaneously is prohibitively large. Therefore nanotube nucleation, the initial formation and liftoff of a nanotube "cap", is a crucial process which remains poorly understood, although environmental TEM studies have given tantalizing views of the process [51]. Is it an epitaxial process wherein the carbon atoms precisely register with a crystalline (solid) metal nanoparticle, or are the nanoparticles liquid during nucleation and subject to random thermal fluctuations (thereby implying that the controllable synthesis of nanotubes with prescribed chirality is impossible). Evidence seems to indicate that both solid and liquid nanoparticles can grow carbon nanofibers and nanotubes, so hope is still alive that chirality control is possible through proper control over the synthesis of metallic nanoparticles. In addition, selectivity has been achieved in the preferential growth of metallic vs. semiconducting SWNTs on miscut substrates (beyond their 1/3 : 2/3 ratio expected based upon random fluctuations).

Resonance Raman spectroscopy has quickly become a well-developed method to remotely characterize the diameter and chirality of single SWNTs [52, 53]. In SWNTs, the density of electronic states is bunched close to the van Hove singularities of these one dimensional structures. When the incident laser wavelength is tuned in resonance with van Hove singularities in the valence and conduction bands of a SWNT, the Raman scattering cross section becomes very large as electrons and phonons couple strongly under these conditions. Single SWNTs are observable, including their characteristic radial breathing mode (RBM), G-band (graphite-related optical mode), D-band (defect-induced), and G' band (overtone of the D-band). Metallic and semiconducting SWNTs are easily distinguished by

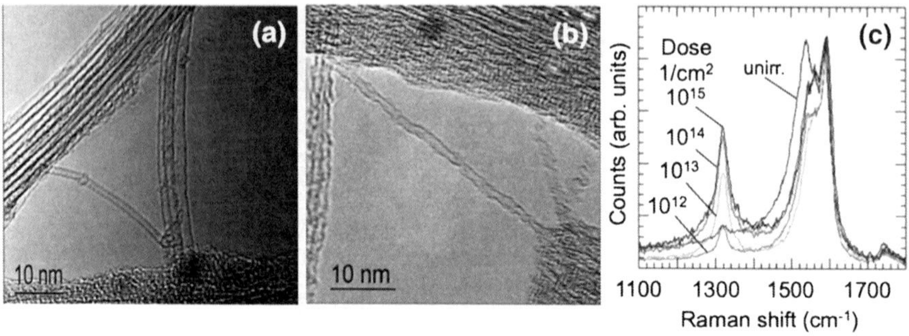

Fig. 7.6 TEM images of SWNTs produced by laser vaporization **a** before and **b** after ion-beam irradiation by Ar^+ (5 keV) with well-defined doses between 10^{12} and 10^{15} cm^{-2}, correlating to an estimated 0.01–10 defects per nanometer of nanotube length. **c** Raman spectroscopy (λ_{ex} = 633 nm) of corresponding samples show the evolution of the D-band ($\sim$1,320 cm^{-1}) with induced defect density. The G-band shows more pronounced quenching of the metallic SWNTs ($\sim$1,540 cm^{-1}) than the semiconducting SWNTs ($\sim$1,590 cm^{-1}) with evolving dose. Raman spectroscopy was also used to observe healing of the damage by heat treatments (not shown). High doses ($>5 \times 10^{13}$ cm^{-2}) lead to irreversible damage as evidenced by HRTEM observation, a broadening of the D-band width approaching that of amorphous carbon, and the inability to reduce the D-band intensity to that of pre-irradiation values

the shape of their G-band (graphite-related optical mode) in the 1,500–1,605 cm^{-1} range. The diameter of the nanotube (in nm) is $\sim$248/(RBM in cm^{-1}). In fact, the exact chiral indices (n, m) of the nanotube may be determined by careful comparison of the resonance Raman condition with theoretical predictions [52, 53]. Very recently, the growth rates of *individual* SWNTs with different chiral indices were measured in situ during chemical vapor deposition using resonance Raman scattering performed at high temperatures [54]. The technique inferred the growth rate of the nanotubes from the increase in the G-band intensity. The measured growth rates agreed with recent theoretical predictions of a chirality-dependent growth rate [55]. Raman spectroscopy is more commonly used with a variety of fixed-wavelength lasers to characterize the overall diameter distribution and level of defects in ensembles of SWNTs.

For example, Fig. 7.6 shows TEM images and Raman spectra of SWNTs (a) before and (b) after the introduction of defects with well-defined doses of Ar^+ (5 keV) ions. With 633-nm laser excitation, resonances exist with both metallic and semiconducting SWNTs within the 1.2–1.4 nm diameter distribution in the sample, as evidenced by the two-peaks in the G-band. As defects are introduced with doses of 10^{12}–10^{15} Ar^+ ions cm^{-2} the D-band is seen to rise and saturate, while the G-band of the metallic SWNTs decreases much more strongly than that of the semiconducting SWNTs. Using this technique to monitor the level of defects in damaged SWNTs it was observed that annealing could reduce the D-band intensity to pre-irradiation values only for Ar^+ doses $<5 \times 10^{13}$ cm^{-2}, above which irreversible damage was incurred.

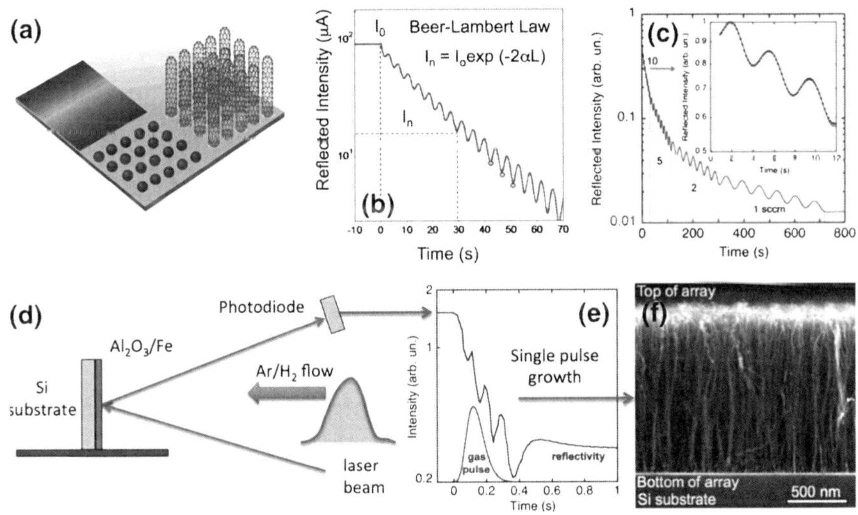

Fig. 7.7 **a** Schematic of vertically-aligned nanotube array (VANTA) growth by CVD. Thin metal films roughen into nanoparticles attached to the substrate, from which aligned arrays of nanotubes grow upon exposure to hydrocarbon gas at high temperature. **b** The reflected intensity of a laser from the growing nanotube array decreases exponentially with length, L, due to the effective extinction coefficient, α. Fringes result from optical interference from light reflected by the substrate and the top surface of the array. **c** In situ monitoring of VANTA growth starting at *10*, *5*, *2*, and *1* sccm acetylene flow. The growth rate can be estimated at 300 nm/fringe. **d** Experimental setup for pulsed CVD of VANTAs. **e** Time dependence of a single 0.2-s CVD gas pulse amplitude (*red*) arriving at the substrate and the corresponding change in laser reflectivity (*blue*) resulting in a 1.3-μm-tall VANTA in (**f**) which shows a side-view cross section of the array grown in ~0.6 s [(**c**) and (**f**) from 65]

One of the most interesting and widely studied ensembles of nanotubes are vertically aligned nanotube arrays (VANTAs, also known as carpets or forests). As schematically shown in Fig. 7.7a, these forests are typically grown by CVD from thin films of metal which dewet into densely-packed nanoparticles. With sufficiently high nucleation density, nanotube forests self-assemble as the individual nanotubes grow into dense, self-supporting, oriented arrays that can be grown to centimeter lengths. The VANTA platform of vertically-oriented, continuous nanotubes is highly promising for a wide range of different applications such as gas sensors [56], flexible electronics and field emission devices, vertical interconnects for microelectronics [57], thermal interface materials [58], heat pipes [59], and unique optical absorbers. The unique morphology of VANTAs has been shown to be especially well suited for "gecko" type adhesives, super-compressible foams, and carpets from which fibers and transparent, conductive sheets [60] can be continuously spun. VANTAs can be synthesized at "super growth" rates [61] such that a millimeter tall forest can grow within ten minutes, thereby enabling mass production of this unique, aligned nanotube architecture. The forests tend to grow in a coordinated growth mode despite the disparity of nanotube diameters and individual growth rates present in the VANTA arising from catalyst nanoparticles of different size and catalytic activity. Typical arrays are only ~5–10 - vol. % dense, corresponding to areal nanotube densities of 5×10^{11} cm^{-2}. However, through reduction of catalyst particle size and spacing, small-diameter SWNTs packed in VANTAs with areal nanotube densities up to 10^{13} cm^{-2} have been reported [62].

Laser-based interferometry and absorption have proven to be invaluable real-time diagnostics of VANTA growth kinetics under actual synthesis conditions [11, 63–67]. As shown in Fig. 7.7d, laser light is reflected from the substrate on which a buffer layer (typically Al_2O_3) and metal catalyst film is deposited. As the nanotubes nucleate and grow upon exposure to the CVD gases at high temperatures, optical absorption and interference occurs as indicated in Fig. 7.7c. Interference oscillations occur due to constructive and destructive interference between light reflected from the top of the growing forest, and light which has penetrated the forest and reflected from the substrate. Depending upon the index of refraction of the array and the laser wavelength employed, each fringe corresponds to a discrete amount of array growth (e.g., ~300 nm/fringe in Fig. 7.7b, c). As indicated in Fig. 7.7c, changes in the growth rate (such as those occurring from changing the feedstock supply) are immediately apparent in the fringe spacing. The absorbance of the array corresponds to an effective extinction coefficient. Since the arrays typically grow from their bottoms where the catalyst nanoparticles are anchored to the substrate, it is important to note that each additional fringe and attenuation of the signal that occurs reflects the incremental growth of a layer at the bottom of the array, providing a step-by-step record of each layer of the array. The diagnostic technique and analysis has been described in detail in [11]. Such measurements form the basis for a complete rate equation-based model of the nanotube CVD process which can predict, from a few measurements at different temperatures during growth test runs with a particular catalyst system, subsequent growth behaviors including: (1) growth rate, (2) number of walls of the nanotubes in the array, (3) the terminal length of the array (4) the growth kinetics versus feedstock flux. The interferometry technique is typically applicable over approximately four orders of magnitude in signal attenuation, which corresponds to roughly the first 20 microns of growth. For growth to millimeter lengths, direct optical imaging is used to measure the kinetics.

Two of the most important parameters governing the properties and applicability of VANTAs are their length and density. The density of VANTAs appears to be integrally linked to their cooperative growth mechanism, which is still not understood despite a number of interesting studies in this area. Due to variations in metal catalyst nanoparticle size and their interactions with the substrate, it is reasonable to expect that the nanoparticles which support nanotube growth have different catalytic activities, with some fraction capable of maintaining growth of a nanotube for a very long time under favorable conditions. Ostwald ripening of metal catalyst nanoparticles [68, 69] during the VANTA growth process can result in a change in nanotube density and diameters, and diffusion of the metal nanoparticles into the substrate can occur [70], leading to further decrease in array density over extended times.

Recently, laser reflectivity and attenuation coupled with fast pulsed gas delivery have revealed the nucleation and growth kinetics of VANTAs on rapid timescales, and have also revealed that the density and diameter of nanotubes in VANTAs are directly related to the feedstock flux [65–67]. As shown in Fig. 7.7d–f, pulses of gas delivered from a pulsed valve within a tube furnace can nucleate, grow, and terminate growth of aligned nanotube arrays within $\sim$0.5 s. By using pulsed gas delivery at low pressure, delays in nucleation can be observed with sub 0.1 s resolution, and the kinetics of the nucleation and growth process can be recorded by the interference oscillations and attenuation described above [67].

Using pulsed CVD, it was discovered that the nucleation time for growth decreased over three orders of magnitude with increasing flux (i.e., partial acetylene pressure) in the pulse [66]. Moreover, it was discovered that the changing flux within each pulse induced significant density changes within the arrays [65]. Figure 7.8 summarizes the results. As shown in the SEM micrograph of Fig. 7.8d, repetitively pulsing the acetylene feedstock results in nanotube arrays with a distinct striped appearance, the contrast variation (shown in yellow) resulting from the fluctuation in density within each layer. As shown in the TEM and Z-contrast STEM micrographs of exfoliated strips of the array representing the first three gas pulses in Fig. 7.8a and b, respectively, the density of the arrays varies dramatically within each gas pulse. The Z-STEM image is directly proportional to carbon density, and the profile of the array is shown in blue in Fig. 7.8c. The density of the array is seen to vary by up to a factor of 1.6. The red trace in Fig. 7.8c is the effective extinction coefficient obtained by analyzing data similar to that in Fig. 7.7e for each pulse. By simultaneously estimating the increase in array length versus time from the interference fringes, a correlation between extinction coefficient and length is obtained. Therefore, the effective extinction coefficient is directly proportional to the density in growing VANTAs, providing a key real-time diagnostic [65].

By analyzing the real-time kinetics of the growth process pulse by pulse, the growth rate and efficiency of the nanotube ensemble is seen to drop on each successive growth event, as shown in Fig. 7.8e. This is also reflected in the decreasing length of each stripe from the top to the bottom of the array. Moreover, detailed analysis of the nanotubes within each array (by both Raman spectroscopy and ex situ HRTEM analysis) revealed that high-flux pulses tended to extinguish the small-diameter nanotubes within the ensemble, while low flux pulses permitted the full spectrum of nanotube diameters to renucleate and regrow on successive pulses.

Two distributions of nanotubes were observed in SEM images of the arrays: continuous nanotubes which continued throughout all of the layers, and disordered nanotubes at the top (high flux portion) of each layer. These smaller-diameter nanotube products apparently grew, terminated growth, and were ripped off and carried off by the continuously-growing, larger-diameter nanotubes. This additional

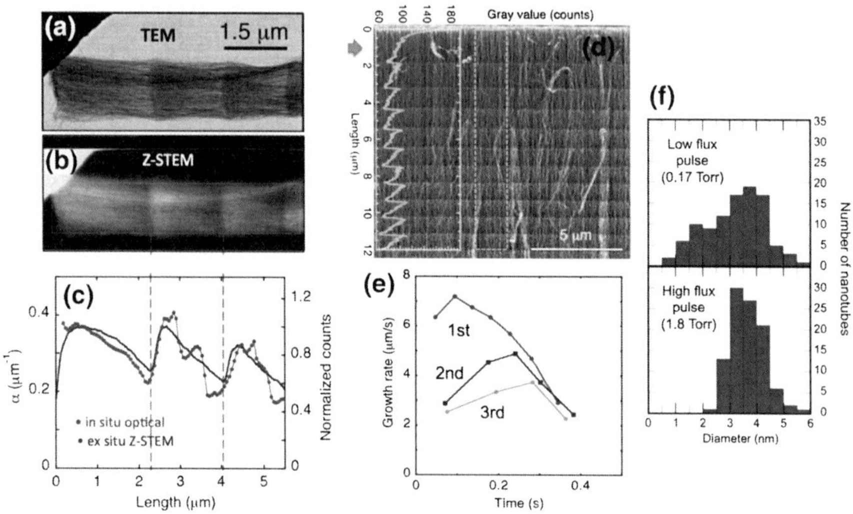

Fig. 7.8 (**a**) TEM and Z-contrast STEM images of a strand of VANTA grown by successive pulses of gas. The Z-STEM image in (**b**), when integrated within the *yellow box* yields the *blue curve* in (**c**) which is directly proportional to the density of the nanotubes in the array, which is seen to vary by a factor of 1.6. The *red curve* in (**c**) is the density in the array inferred from the extinction coefficient α(L) calculated from the measured α(t) and array length L(t). (**d**) Such density variations lead to the striped appearance in SEM images of VANTAs grown by pulsed CVD. The integrated intensity in the dashed zone yields the intensity variations shown at *left*. (**e**) Growth rate changes during the first 3 gas pulses measured from laser interference fringe spacings versus time. (**f**) Nanotube diameter distributions measured by TEM analysis for low- and high-flux gas pulses, showing that high-flux pulses result in the extinction of nanotubes with <2 nm diameter [from 65, 66]

distribution of "crossbar" nanotubes, oriented randomly during the initial growth region of each pulse, produced the observed variation in density.

Through the in situ diagnostics provided by the real-time laser reflectivity, the gas flux could be lowered to produce VANTAs without stripes, density changes, or variations in diameter within the array. Moreover, the dependence of catalyst nanoparticle nucleation efficiency on feedstock flux and the origin of the coordinated growth mode was probed by lowering the flux in each gas pulse. Nanotubes were shown to nucleate and then renucleate repeatedly, to grow "digitally" in incremental fashion on successive gas pulses [67]. Figure 7.9a shows the in situ laser reflectivity curves for two growth runs where clear interference oscillations confirm that the nanotube arrays are growing in coordinated fashion. Each gas pulse causes a segment of the overall growth curve, as shown in the inset to Fig. 7.9a, corresponding to an incremental growth of ~20 nm/pulse. However, at least ~60–100 nm of nanotube growth is required before nanotubes self-align into a vertical array. How can one tell if the nanotube nucleation efficiency is sufficient to produce an aligned array?

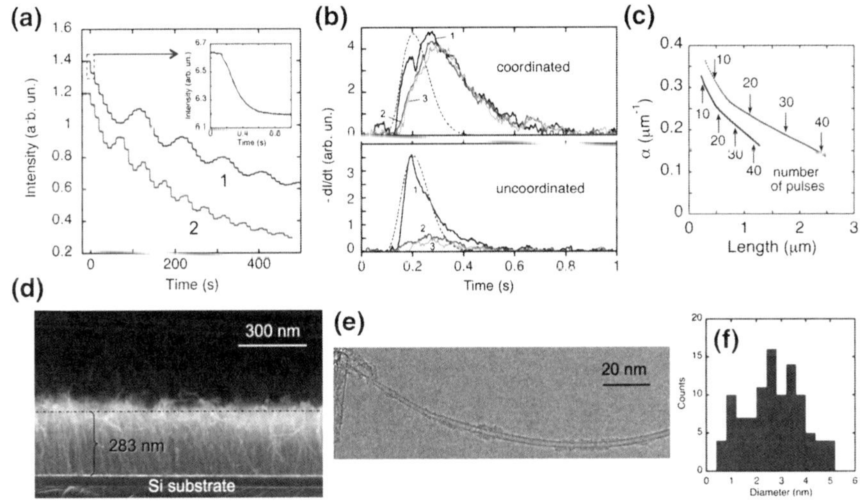

Fig. 7.9 In situ laser reflectivity diagnostics of the incremental, pulsed CVD growth of aligned nanotube arrays. **a** Reflected laser intensity during two different growth runs using multiple gas pulses to multiply stop and start the growth of SWNTs (inset shows a single step, corresponding to growth on the first gas pulse). Fringes indicate aligned array growth, ~20 nm/pulse. **b** Derivatives of laser reflectivity signal for pulses *1*, *2*, *3* as compared to gas pulse (*dashed line*) in cases where growth is (*top*) coordinated and aligned and (*bottom*) uncoordinated. **c** Changes in the density of the nanotubes grown on each pulse as measured by optical extinction coefficient, α, as a function of pulse number and cumulative length of the array. **d** Side-view SEM of a 283-nm aligned array grown incrementally by pulsed CVD. **e** TEM image of SWNT, showing no visible evidence of being grown incrementally in 20-nm steps. **f** Diameter distribution of SWNTs in the array measured by TEM [from 67]

Inspection of the derivative of the reflectivity curves on each pulse, *dI/dt*, yields the instantaneous growth rate. Two behaviors were observed corresponding to coordinated and uncoordinated growth, as illustrated in Fig. 7.9a and b. For arrays which would later exhibit interference oscillations indicative of coordinated, aligned growth, the first exposure to the acetylene pulse produced a double peak—a fast (~50 ms) peak followed by a broader, delayed peak, which extends beyond the tail of the modeled gas pulse. All subsequent pulses show only the second peak. The relative intensities of these two peaks vary considerably, depending on the growth run. In the case of uncoordinated growth, the intensity of the second peak is relatively low and appears as a shoulder on the pronounced first peak of the first gas pulse. Moreover, the total width is comparable to the calculated gas pulse width and is narrower than that for coordinated growth. In these cases, growth typically stops quickly as indicated by the lack of reflectivity changes after a few additional gas pulses.

Practically, these in situ laser reflectivity diagnostics afford the opportunity to grow ultrashort SWNT arrays digitally to prescribed lengths >60 nm with ±20 nm accuracy, as shown for the 283-nm VANTA in Fig. 7.9d [67].

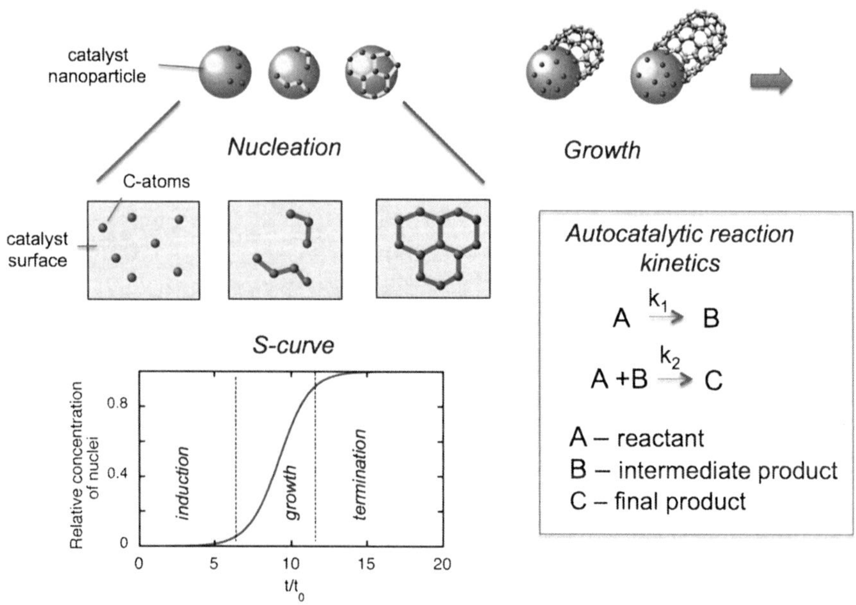

Fig. 7.10 Diagram indicating the kinetic evolution of carbon nanostructures based upon flux-dependent, time-resolved laser diagnostic measurements of the CVD process. Measured kinetics indicate that *S-shaped growth curves* indicative of autocatalytic chemical reactions representing induction, growth, and termination periods. Intermediate products react with reactants to accelerate the formation of closed (graphenic) carbon structures on a catalyst nanoparticle surface. For nanotubes, cap liftoff and continued growth can continue this process over multiple pulses

Importantly, the SWNTs show no evidence of stopping and restarting growth (Fig. 7.8e) with no visible diameter changes or defects on these length scales. Moreover, the entire diameter distribution is preserved (Fig. 7.9f). Such ultrashort SWNT arrays may be useful for thermal interface materials, interconnects, field emitters, sensors, and other applications. It should be noted, however, that as in the pulsed growth of long VANTAs the nanotube density drops with repeated renucleation events, as indicated by the drop in the optical extinction coefficient per pulse in Fig. 7.9c. By comparing the drop in density per pulse it can be estimated that over 98 % of all nanotubes regrow on each successive pulse [67].

The observed double-peak feature in the evolution of the growth rate, *dl/dt*, can be interpreted in the context of autocatalytic kinetics for both nanotube nucleation and growth where in both cases induction delays are explained by the time required to form intermediate species which accelerate chemical conversion of the feedstock gas. Such autocatalytic kinetics result in "S-shaped" growth curves as shown in Fig. 7.10, with a slow induction period during the buildup of the necessary intermediates, a period of rapid and efficient reaction, and finally a period of decline as the chemical reactions terminate due to lack of reactants or available sites. Such "S-shaped" growth kinetics have been observed in carbon fiber [71]

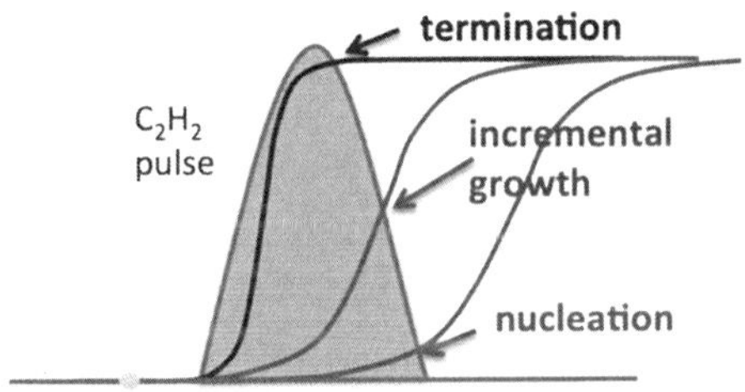

Fig. 7.11 Schematic summarizing the results of the laser reflectivity and pulsed CVD nanotube growth experiments in the context of autocatalytic kinetics model, where growth proceeds in an *S-shaped curve*, with a nucleation period, a growth period, and a termination period due to the requirement for the formation of intermediate reactants. Nanoparticles with different activity catalyze the growth of nanotubes at different rates, leading to either barely nucleation (*red*), rapid growth and termination (*blue*), or repeated nucleation and growth (*green*) during each gas pulse

and nanotube [72] growth experiments for years, and more recently in the growth of carbon nanotube arrays [73–75]. However, separating the kinetics of nucleation and growth has been difficult until now.

Recently, Latorre et al. [76] developed a phenomenological autocatalytic model to describe "S-shaped" kinetics in the growth of nanotube forests measured by in situ Raman spectroscopy. The model used the framework of growth by dissolution-precipitation driven by the gradient of carbon concentrations at the metal catalyst nanoparticle surface. First, however, the chemical reactions required for the nucleation and growth of a stable nanotube "cap" were considered, followed by the processes for cap liftoff and nanotube growth. The termination phase for nanotube nucleation resulted from the lack of available sites on a catalyst nanoparticle, while the collection termination of nanotube growth in the arrays could be due to a variety of different reasons including Ostwald ripening of the catalyst, [69] mechanical stress, [77–80] catalyst particle overcoating [11, 81], but also including chemical reactions, such as those discussed by Eres et al. to explain the preferential growth of nanotube arrays with acetylene as the primary precursor [74] and proposed mechanisms for the self-assembly of nanotubes via acetylene by autocatalytic reactions [75].

This latter model involving autocatalytic chemical reactions explains the pulsed CVD nanotube growth kinetics measured by laser reflectivity in the context of Fig. 7.11. Ensembles of catalyst nanoparticles are likely to contain a range of activities for the decomposition of feedstock gas and the processing of carbon into nanotubes. Our studies of the growth kinetics reveal S-shaped curves which are characteristic of autocatalytic chemical reactions which imply intermediate states that are required to accelerate the rapid growth region after the nucleation period. The S-curves represent a nucleation period, a period of rapid growth, and a termination period where particles may get overcoated with carbon, for example. Higher fluxes tend to activate chemical reactions on each nanoparticle, the most active of which may undergo their entire life cycle within the gas pulse lifetime.

Therefore, at the highest fluxes, a variation in density is explained to yield the arrays in Fig. 7.8, wherein smaller catalyst nanoparticles nucleate, grow, and terminate small nanotubes in the early part of the acetylene gas pulse (blue curve in Fig. 7.11) while larger nanoparticles and their corresponding nanotubes survive the high flux conditions to incrementally grow pulse after pulse. Similarly, at lower fluxes, termination of growth for the particles of both high and low catalytic activity can be avoided to grow continuous nanotubes incrementally over a wide range of nanotube diameters, as shown in Fig. 7.8 [66, 67].

7.6 Graphene and Beyond: Laser Processing for 2D Layered Materials

Two dimensional materials just one to several atomic layers thick, but extending to macroscopic dimensions, are frontier mesoscale materials that promise to enable a wide range of practical applications. Stimulated by the rapid progress in the experimental synthesis and exploration of single layer graphene (SLG) and few layer graphene (FLG), similar efforts are now being applied to obtain ultrathin specimens of other familiar 2D materials such as MoS_2 and other chalcogenides [82–85], h-BN [86], transition metal oxides and hydroxides [87], and compounds such as Bi_2Te_3, and Bi_2Se_3, [3, 88] and GaSe [89]. Here, we will briefly outline the areas of future research where laser interactions with these materials should play key roles in their synthesis, processing, and remote characterization.

The rapid exploration of the unique properties of graphene were stimulated by the transfer of small flakes of that were mechanically exfoliated from highly-oriented pyrolytic graphite (HOPG) [90]. The outstanding electrical conductivity (mobilities up to 15,000 $cm^2\ V^{-1}\ s^{-1}$), zero band gap, and near optical transparency (98 %) of the ultrathin material appear promising as a flexible, transparent, conductive electrodes for a variety of applications including touch panels, displays, and flexible organic electronics such as organic photovoltaics, light emitting diodes, and transistors.

7.6.1 Mechanical and Chemical Exfoliation Methods and Laser Processing

To realize such large-scale applications, methods must be developed to produce large areas of high-mobility nanosheet architectures. Conductivity across junctions between individual nanosheets is a key problem, requiring large individual grains. However, large-scale production of exfoliated material typically involves techniques such as sonication in liquids. 2D nanosheets produced in this way are typically small (<150 nm in diameter) and can suffer damage to their structure and properties [91]. Finding controllable methods to exfoliate layered materials without damage is a major challenge where laser-surface interactions may play a key role. Laser exfoliation of HOPG in liquids for the production of microns-sized flakes of FLG has been demonstrated using 532 nm irradiation at 1.0 J cm^{-2} which was thought to occur through the compression and expansion of the HOPG surface [92]. Similarly, 248-nm laser exfoliation of WS_2 in water was demonstrated to result in microns-sized multilayer flakes, presumably due to laser induced shock waves and the possible intercalation of OH and H radicals created by the two-photon dissociation of water [93]. Recent theoretical predictions indicate that ultrafast excitation of graphite might induce exfoliation of intact graphene layers by inducing spillout of electrons and subsequent Coulomb repulsion between the top layers [94]. The controllable deposition of laser energy may therefore hold unique processing advantages.

Chemical treatments are alternate or auxiliary methods for exfoliation. For example, graphite oxide (GO) is hydrophilic and can be easily exfoliated in water. GO is produced (by oxidizing graphite with sulfuric acid, sodium nitrate, and potassium permanganate) in the Hummers method [95]. The processed material has a disordered, buckled structure with various types of oxygen-containing functional groups (e.g., epoxide, carbonyl, hydroxyl, phenol). The disordered structure of GO varies due to carbon:oxygen ratios variation between 2.1 and 2.9 [95]. Alternatively, GO can be thermally exfoliated by rapidly evolving the oxygen-containing groups such that pressure exceeds the van der Waals forces holding the layers together [96]. In both cases, after exfoliation of GO, functionalized single-layer graphene oxide can be obtained which must be reduced, either chemically [97] or with prolonged thermal treatments, to provide graphene-like layers for applications. Raman spectroscopy and scanning transmission electron microscopy [98] shows that this material is still highly defective, however.

Recently, laser-based reduction of dry GO has been demonstrated using CW and pulsed 532-nm laser irradiation [99] or 248-nm irradiation to directly write graphene lines and patterns [100]. The use of lasers to tunably desorb functional groups alter 2D material stoichiometry will take advantage of pulsed laser vaporization and desorption studies primarily targeted for pulsed laser deposition of thin films.

In addition, it is often desirable to adjust the number of layers in 2D nanosheets. Recently, the controllable laser thinning of WS_2 nanosheets was demonstrated and verified with Raman spectroscopy [101].

7.6.2 Laser Interactions in the Synthesis and Characterization of Graphene and other 2D Nanosheets

The highest quality graphene is synthesized by either top-down or bottom-up processes. Extremely high quality graphene can be obtained by thermally sublimating silicon from silicon carbide substrates in vacuum at temperatures between

1,000 and 1,500 °C. "Mulilayered epitaxial graphene (MEG)" produced in this way can be tailored to produce ribbons along predetermined step edges or grains along the Si or C terminated faces of SiC with mobilities up to 5,000 cm^2/Vs. The material is epitaxially bonded in places to the SiC substrate, forming a unique material which is envisioned to be used for a new type of graphene electronics directly on SiC wafers. Since graphene is a semimetal with no bandgap, new types of ambipolar transistor are being developed [102].

Laser heating and decomposition of SiC to form monolayer, bilayer, and tri-layer graphene was demonstrated with different fluences using 500 shots of KrF-laser irradiation in vacuum [103]. More recently, continuous-wave CO_2 laser irradiation of SiC (0001) was demonstrated to induce epitaxial graphene growth on timescales of ~ 1 s in ambient Ar atmospheres, enabling the writing of graphene patterns without pretreatments [104]. Laser processing therefore has the advantage of rapid heating and cooling rates, controllable and localized energy deposition, and high instantaneous surface temperatures through the choice of pulse width.

By far, the most widely studied technique for the formation of graphene over large areas is CVD, where conditions nearly identical to those used for the growth of carbon nanotubes are employed. Two types of metal foil or film substrates are used: those with low carbon solubility (e.g., Cu) or high carbon solubility (e.g., Ni). On Cu, continuous films of SLG or FLG are formed at 1,000–1,050 °C from methane decomposition at low or high pressure in flowing hydrogen or hydrogen/argon mixtures, very near the evaporation temperature of Cu [105]. Graphene islands nucleate and grow in lobed, hexagonal, or square grains to cover the entire surface in a self-limiting fashion [106]. This graphene can be transferred to arbitrary substrates by attaching a polymer (PMMA) to the graphene surface, etching away the Cu in $FeCl_3$ solution, pressing the graphene to the substrate of choice, and etching away the polymer [107]. This procedure has been demonstrated in roll-to-roll (30″) processing onto PET substrates, achieving 125 Ω/square sheet conductivity [108]. On Cu, large graphene grains can grow across Cu grain boundaries within the foil while on Ni, the grain orientation of the Ni drastically affects the nucleation and growth of the graphene.

However, there is significant disagreement in the literature regarding the growth mechanisms of graphene on metals. Especially for metals with high carbon solubility, it is unclear what fraction of growth occurs isothermally at high temperatures [109] and how much occurs during cool down [110, 111]. Like nanotube growth, the debate centers on growth by surface processes versus that by dissolution/precipitation. Several in situ diagnostics are beginning to provide some insight, though [110]. For example, in situ X-ray photoelectron spectroscopy (XPS) [112], and in situ X-ray diffraction (XRD) [113] have shown recently that graphene can grow isothermally on Ni films, and is not limited to the precipitation mechanism during cool down [110].

To answer these questions, real-time optical diagnostics can be performed in situ during growth [114, 115]. Figure 7.12a shows a setup for confocal laser microRaman scattering spectroscopy, imaging, and reflectivity utilizing a specially-modified growth chamber for CVD of graphene using a pulsed valve to

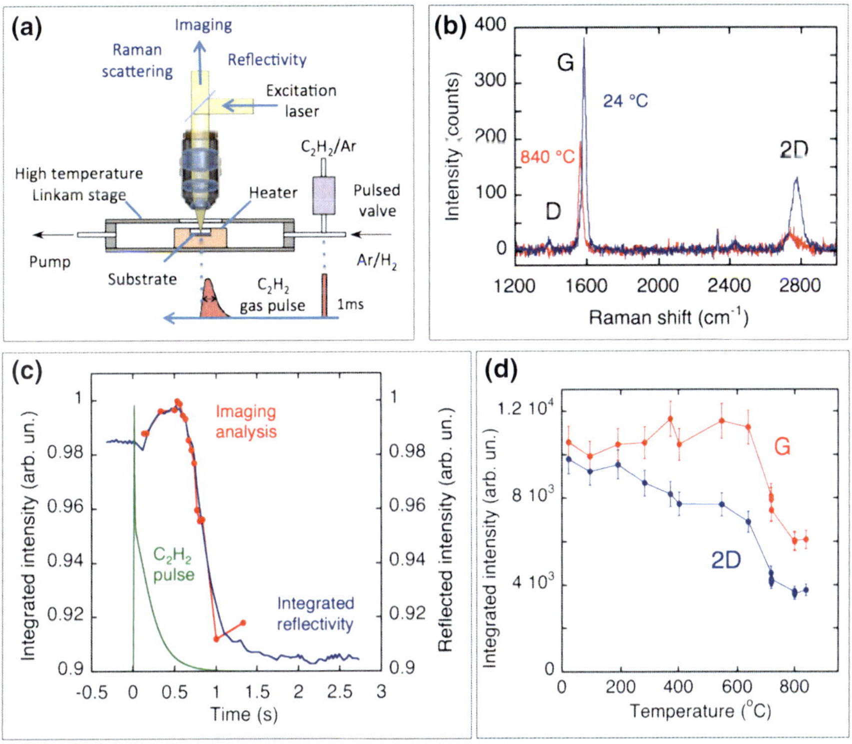

Fig. 7.12 **a** Setup for confocal laser micro Raman spectroscopy, imaging, and reflectivity studies of graphene growth by pulsed CVD. **b** Characteristic Raman spectrum of graphene on Ni (λ_{ex} = 532 nm) showing *D*, *G*, and *2D* bands at 840 and 24 °C. The width and relative *2D*: *G* band ratio can be used to infer the number of graphene layers. **c** Time-resolved kinetics of graphene growth at high temperature (690 °C, C_2H_2 pulse, Ni film substrate) from imaging analysis and integrated reflectivity with a photodiode, in comparison with the ~0.2-s gas-pulse flux (*in green*). **d** Raman *G* and *2D* band integral intensities reveal additional graphene precipitation during cooldown from the growth temperature of 840 °C [115]

introduce pulses of acetylene growth gas within 0.1 s [115]. Raman scattered light is dispersed with a spectrometer to look for the appearance of the characteristic Raman spectrum of graphene, shown in Fig. 7.12b. The Raman shift of the 2D band and the ratio of 2D:G band intensities can be used to infer the number of graphene layers [116]. As shown in Fig. 7.12c after an induction time following the introduction of the hydrocarbon gas at 840 °C, the reflectivity of the sample increases slightly and suddenly drops due to carbon deposition. Imaging through the microscope shows how the patches of graphene nucleate and grow rapidly at this high temperature, with the integrated intensity matching that from the photodiode. Raman spectroscopy with 1-s acquisition time confirms this sub-second growth of graphene, clearly showing that graphene grows isothermally at high temperatures. By monitoring the integrated intensity of the G and 2D bands as the

Fig. 7.13 **a** Aberration corrected Z-STEM image of single-layer graphene (*dark area*) with residues from the synthesis, etching, and transfer process (*bright areas*). **b** AR-Z-STEM image Ga and Se atoms in an exfoliated single-layer GaSe nanosheet. **c** Atomistic simulation of copper pthalocyanine molecules interacting with single-layer graphene [(**c**) from 119]

sample is cooled down, any additional fractional precipitation of graphene from carbon dissolved within the Ni film is revealed by the increase in G-band and 2D-band intensity (Fig. 7.12d). Real-time diagnostics such as these are essential to understand the growth process, and to later serve as real-time diagnostics of graphene nanomanufacturing.

As shown in Fig. 7.12, the growth of graphene can be very rapid on substrates with high carbon affinity. Laser direct writing of graphene patterns on Ni foils has been demonstrated using methane and hydrogen gases where the number of graphene layers is controlled by the scan speed [117, 118].

A major challenge facing graphene and 2D nanosheet materials in general is cleanliness at the atomic level. Many of the applications of graphene as electrodes in organic electronics will require atomically clean surfaces to align and form well-defined interfaces with organic molecules, as shown in Fig. 7.13c [119]. Realizing such surfaces is a huge challenge, especially considering the afore-mentioned etching and polymer transfer processes described above. When imaged by atomic-resolution Z-contrast STEM, such as in Fig. 7.13a and b, it is seen that such processes leave large quantities of residual atoms, molecules, and polymers such that just small regions of atomically clean interface are visible. Cleaning and desorption of these adsorbates from such delicate substrates is a major challenge that lasers may be able to provide.

7.7 Summary

In summary, the role of laser interactions to understand and control nanomaterials synthesis has been briefly examined. Laser interactions provide unique capabilities to remotely deliver excitation to alter, diagnose, and remotely characterize the synthesis and processing of nanomaterials.

Laser vaporization remains a special exploratory tool to create nonequilibrium growth environments that can capture metastable phases and structures of nanomaterials with novel nanoscale properties. Through the development of coordinated in situ laser spectroscopy and imaging diagnostics, the timescales for nanomaterial growth in pulsed laser plasmas and on substrates is revealed. By controlling the growth environment, and with advances in atomic resolution transmission electron microscopy, it is becoming possible to reveal the ultrasmall building blocks that serve as reactive intermediates in growth of loose nanoparticles, nanotubes, nanowires, nanohorns, and nanosheets in laser ablation plumes as well as nanorods and nanostructured thin films deposited on substrates.

Real-time understanding of the kinetics of metal-catalyst-assisted nanotube and graphene growth on substrates during chemical vapor deposition was shown as provided by laser interferometry, attenuation, and Raman scattering. Through such diagnostics, one can remotely characterize the diameter, density, alignment, and kinetics of nanomaterials in real-time. Through the introduction of pulsed gas delivery in CVD, essential nucleation and growth kinetics are revealed, providing invaluable insight into growth processes essential for advanced nanomanufacturing.

The advancements in the understanding of laser interactions for real-time synthesis and characterization are applicable not only for carbon nanomaterials, which have served as the focus for this brief review, but for the great variety of new materials, such as 2D nanosheets of other layered materials, which extend beyond graphene. New challenges for these delicate materials appear very well suited to laser processing as examples above have shown, including: laser thinning and exfoliation, laser healing of defects and methods for their controllable introduction and monitoring, and laser cleaning to produce atomically clean surfaces.

Acknowledgments Synthesis science sponsored by the Materials Sciences and Engineering Division, Office of Basic Energy Sciences, U.S. Department of Energy. Characterization science and explorations of functionality performed at the Center for Nanophase Materials Sciences, and high-resolution electron microscopy was performed in part at the Shared Research Equipment Collaborative Research Center, which are both sponsored at Oak Ridge National Laboratory by the Scientific User Facilities Division, U.S. Department of Energy.

References

1. A.P. Alivisatos, J. Phys. Chem. **100**(31), 13226 (1996)
2. C.C. Chen, A.B. Herhold, C.S. Johnson, A.P. Alivisatos, Science **276**(5311), 398 (1997)
3. P.F. McMillan, Nat. Mater. **1**(1), 19 (2002)
4. D.B. Geohegan, in Pulsed Laser Deposition of Thin Films ed. by D.H. Chrisey, G.K. Hubler (Wiley-VCH, New York, 1994)
5. S. Amoruso, R. Bruzzese, N. Spinelli, R. Velotta, J. Phys. B-at Mol. Opt. **32**(14), R131 (1999)
6. R.E. Smalley, Acc. Chem. Res. **25**(3), 98 (1992)
7. S. Iijima, M. Yudasaka, R. Yamada, S. Bandow, K. Suenaga, F. Kokai, K. Takahashi, Chem. Phys. Lett. **309**(3–4), 165 (1999)

8. A.A. Puretzky, D.J. Styers-Barnett, C.M. Rouleau, H. Hu, B. Zhao, I.N. Ivanov, D.B. Geohegan, Appl. Phys. A –Mater. **93**(4), 849 (2008)
9. G.W. Yang, Prog. Mater. Sci. **52**(4), 648 (2007)
10. K.S. Novoselov, V.I. Fal'ko, L. Colombo, P.R. Gellert, M.G. Schwab, K. Kim, Nature **490**(7419), 192 (2012)
11. A.A. Puretzky, D.B. Geohegan, S. Jesse, I.N. Ivanov, G. Eres, Appl. Phys. A-Mater. **81**(2), 223 (2005)
12. A.M. Morales, C.M. Lieber, Science **279**(5348), 208 (1998)
13. DOE, *New Science for a Secure and Sustainable Energy Future, A Report of a Subcommittee to the Basic Energy Science Advisory Committee.* (U.S. Department of Energy, Dec 2008)
14. D.B. Geohegan, Nato. Adv. Sci. Inst. Se **265**, 165 (1994)
15. D.B. Geohegan, A.A. Puretzky, Appl. Surf. Sci. **96–8**, 131 (1996)
16. D.H. Lowndes, D.B. Geohegan, A.A. Puretzky, D.P. Norton, C.M. Rouleau, Science **273**(5277), 898 (1996)
17. R.F. Wood, K.R. Chen, J.N. Leboeuf, A.A. Puretzky, D.B. Geohegan, Phys. Rev. Lett. **79**(8), 1571 (1997)
18. R.F. Wood, J.N. Leboeuf, D.B. Geohegan, A.A. Puretzky, K.R. Chen, Phys. Rev. B **58**(3), 1533 (1998)
19. D.B. Geohegan, A.A. Puretzky, G. Duscher, S.J. Pennycook, Appl. Phys. Lett. **73**(4), 438 (1998)
20. D.B. Geohegan, A.A. Puretzky, G. Duscher, S.J. Pennycook, Appl. Phys. Lett. **72**(23), 2987 (1998)
21. R.Q. Guo, J. Nishimura, M. Matsumoto, D. Nakamura, T. Okada, Appl. Phys. A-Mater. **93**(4), 843 (2008)
22. Y.B. Zel'dovich, Y.P. Raizer, *Physics of Shock Waves and High-Temperature Hydrodynamic Phenomena* (Academic Press, New York, 1967)
23. A.S. Barnard, L.A. Curtiss, Nano. Lett. **5**(7), 1261 (2005)
24. R.L. Penn, J.F. Banfield, Science **281**(5379), 969 (1998)
25. C.R.A. Catlow, S.T. Bromley, S. Hamad, M. Mora-Fonz, A.A. Sokol, S.M. Woodley, Phys. Chem. Chem. Phys. **12**(4), 786 (2010)
26. D.R. Hummer, J.D. Kubicki, P.R.C. Kent, J.E. Post, P.J. Heaney, J. Phys. Chem. C **113**(11), 4240 (2009)
27. Y. Zhou, K.A. Fichthorn, J. Phys. Chem. C **116**(14), 8314 (2012)
28. V.N. Koparde, P.T. Cummings, J. Phys. Chem. B **109**(51), 24280 (2005)
29. M. Matsui, M. Akaogi, Mol. Simul. **6**, 239 (1991)
30. S. Hamad, C.R.A. Catlow, S.M. Woodley, S. Lago, J.A. Mejias, J. Phys. Chem. B **109**(33), 15741 (2005)
31. H.W. Kroto, J.R. Heath, S.C. Obrien, R.F. Curl, R.E. Smalley, Nature **318**(6042), 162 (1985)
32. T.G. Dietz, M.A. Duncan, D.E. Powers, R.E. Smalley, J. Chem. Phys. **74**(11), 6511 (1981)
33. Y. Yamaguchi, S. Maruyama, Chem. Phys. Lett. **286**(3–4), 336 (1998)
34. B.I. Dunlap, Int. J. Quantum Chem. **64**(2), 193 (1997)
35. S. Irle, G.S. Zheng, Z. Wang, K. Morokuma, J. Phys. Chem. B **110**(30), 14531 (2006)
36. D.B. Geohegan, A. Puretzky, C.M. Rouleau, J.J. Jackson, G. Eres, Z. Liu, D. Styers-Barnett, H. Hu, B. Zhao, I. Ivanov, K.L. More, in *Laser-Surface Interactions for New Materials Production*, vol. 130, ed. by A. Miotello, P.M. Ossi (Springer, Berlin Heidelberg, 2010), p. 1
37. A.A. Puretzky, D.B. Geohegan, X. Fan, S.J. Pennycook, Appl. Phys. A-Mater. **70**(2), 153 (2000)
38. A.A. Puretzky, H. Schittenhelm, X.D. Fan, M.J. Lance, L.F. Allard, D.B. Geohegan, Phys. Rev. B **65**, 24 (2002)
39. F. Kokai, K. Takahashi, D. Kasuya, M. Yudasaka, S. Iijima, Appl. Surf. Sci. **197**, 650 (2002)

40. D. Kasuya, M. Yudasaka, K. Takahashi, F. Kokai, S. Iijima, J. Phys. Chem. B **106**(19), 4947 (2002)
41. M.D. Cheng, D.W. Lee, B. Zhao, H. Hu, D.J. Styers-Barnett, A.A. Puretzky, D.W. DePaoli, D.B. Geohegan, E.A. Ford, P. Angelini, Nanotechnology **18**(18), 185604 (2007)
42. D.B. Geohegan, A.A. Puretzky, D. Styers-Barnett, H. Hu, B. Zhao, H. Cui, C.M. Rouleau, G. Eres, J.J. Jackson, R.F. Wood, S. Pannala, J.C. Wells, Phys. Status Solidi B **244**(11), 3944 (2007)
43. O.L. Krivanek, M.F. Chisholm, V. Nicolosi, T.J. Pennycook, G.J. Corbin, N. Dellby, M.F. Murfitt, C.S. Own, Z.S. Szilagyi, M.P. Oxley, S.T. Pantelides, S.J. Pennycook, Nature **464**(7288), 571 (2010)
44. D.B. Geohegan, H. Schittenhelm, X. Fan, S.J. Pennycook, A.A. Puretzky, M.A. Guillorn, D.A. Blom, D.C. Joy, Appl. Phys. Lett. **78**(21), 3307 (2001)
45. P.J.F. Harris, Carbon **45**(2), 229 (2007)
46. R. Sen, S. Suzuki, H. Kataura, Y. Achiba, Chem. Phys. Lett. **349**(5–6), 383 (2001)
47. K.R.S. Chandrakumar, J.D. Readle, C. Rouleau, A. Puretzky, D.B. Geohegan, K. More, V. Krishnan, M. Tian, G. Duscher, B. Sumpter, Nanoscale **5**(5), 1849 (2013)
48. Y. Liu, C.M. Brown, D.A. Neumann, D.B. Geohegan, A.A. Puretzky, C.M. Rouleau, H. Hu, D. Styers-Barnett, P.O. Krasnov, B.I. Yakobson, Carbon **50**(13), 4953 (2012)
49. A. Izadi-Najafabadi, T. Yamada, D.N. Futaba, M. Yudasaka, H. Takagi, H. Hatori, S. Iijima, K. Hata, ACS Nano **5**(2), 811 (2011)
50. S.K. Doorn, L.X. Zheng, M.J. O'Connell, Y.T. Zhu, S.M. Huang, J. Liu, J. Phys. Chem. B **109**(9), 3751 (2005)
51. S. Hofmann, R. Sharma, C. Ducati, G. Du, C. Mattevi, C. Cepek, M. Cantoro, S. Pisana, A. Parvez, F. Cervantes-Sodi, A.C. Ferrari, R. Dunin-Borkowski, S. Lizzit, L. Petaccia, A. Goldoni, J. Robertson, Nano Lett. **7**(3), 602 (2007)
52. A. Jorio, R. Saito, J.H. Hafner, C.M. Lieber, M. Hunter, T. McClure, G. Dresselhaus, M.S. Dresselhaus, Phys. Rev. Lett. **86**(6), 1118 (2001)
53. M.S. Dresselhaus, G. Dresselhaus, A. Jorio, A.G. Souza, R. Saito, Carbon **40**(12), 2043 (2002)
54. R. Rao, D. Liptak, T. Cherukuri, B.I. Yakobson, B. Maruyama, Nat. Mater. **11**(3), 213 (2012)
55. F. Ding, A.R. Harutyunyan, B.I. Yakobson, P Natl. Acad. Sci. USA **106**(8), 2506 (2009)
56. A. Modi, N. Koratkar, E. Lass, B.Q. Wei, P.M. Ajayan, Nature **424**(6945), 171 (2003)
57. J. Robertson, G. Zhong, H. Telg, C. Thomsen, J.H. Warner, G.A.D. Briggs, U. Dettlaff-Weglikowska, S. Roth, Appl. Phys. Lett. **93**(16), 163111 (2008)
58. H. Huang, C.H. Liu, Y. Wu, S.S. Fan, Adv. Mater. **17**(13), 1652 (2005)
59. I. Ivanov, A. Puretzky, G. Eres, H. Wang, Z.W. Pan, H.T. Cui, R.Y. Jin, J. Howe, D.B. Geohegan, Appl. Phys. Lett. **89**(22), 223110 (2006)
60. C.L. Pint, Y.Q. Xu, M. Pasquali, R.H. Hauge, ACS Nano **2**(9), 1871 (2008)
61. K. Hata, D.N. Futaba, K. Mizuno, T. Namai, M. Yumura, S. Iijima, Science **306**(9), 1362 (2004)
62. S. Esconjauregui, M. Fouquet, B.C. Bayer, C. Ducati, R. Smajda, S. Hofmann, J. Robertson, ACS Nano **4**(12), 7431 (2010)
63. D.B. Geohegan, A.A. Puretzky, I.N. Ivanov, S. Jesse, G. Eres, J.Y. Howe, Appl. Phys. Lett. **83**(9), 1851 (2003)
64. S. Maruyama, Y. Miyauchi, Y. Murakami, S. Chiashi, New J. Phys. **5**, 1 (2003)
65. J.J. Jackson, A.A. Puretzky, K.L. More, C.M. Rouleau, G. Eres, D.B. Geohegan, ACS Nano **4**(12), 7573 (2010)
66. D.B. Geohegan, A.A. Puretzky, J.J. Jackson, C.M. Rouleau, G. Eres, K.L. More, ACS Nano **5**(10), 8311 (2011)
67. A.A. Puretzky, D.B. Geohegan, J.J. Jackson, S. Pannala, G. Eres, C.M. Rouleau, K.L. More, N. Thonnard, J.D. Readle, Small **8**(10), 1534 (2012)
68. P.B. Amama, C.L. Pint, L. McJilton, S.M. Kim, E.A. Stach, P.T. Murray, R.H. Hauge, B. Maruyama, Nano Lett. **9**(1), 44 (2009)

69. S.M. Kim, C.L. Pint, P.B. Amama, R.H. Hauge, B. Maruyama, E.A. Stach, J. Mater. Res. **25**(10), 1875 (2010)
70. A.R. Harutyunyan, G.G. Chen, T.M. Paronyan, E.M. Pigos, O.A. Kuznetsov, K. Hewaparakrama, S.M. Kim, D. Zakharov, E.A. Stach, G.U. Sumanasekera, Science **326**(5949), 116 (2009)
71. R.T.K. Baker, M.A. Barber, R.J. Waite, P.S. Harris, F.S. Feates, J. Catal. **26**(1), 51 (1972)
72. M. Lin, J.P.Y. Tan, C. Boothroyd, K.P. Loh, E.S. Tok, Y.L. Foo, Nano Lett. **6**(3), 449 (2006)
73. A. Li-Pook-Than, J. Lefebvre, P. Finnie, J. Phys. Chem. C **114**(25), 11018 (2010)
74. G. Eres, A.A. Kinkhabwala, H.T. Cui, D.B. Geohegan, A.A. Puretzky, D.H. Lowndes, J. Phys. Chem. B **109**(35), 16684 (2005)
75. G. Eres, C.M. Rouleau, M. Yoon, A.A. Puretzky, J.J. Jackson, D.B. Geohegan, J. Phys. Chem. C **113**(35), 15484 (2009)
76. N. Latorre, E. Romeo, F. Cazana, T. Ubieto, C. Royo, J.J. Villacampa, A. Monzon, J. Phys. Chem. C **114**(11), 4773 (2010)
77. A.A. Puretzky, G. Eres, C.M. Rouleau, I.N. Ivanov, D.B. Geohegan, Nanotechnology **19**, 5 (2008)
78. E.R. Meshot, A.J. Hart, Appl Phys Lett **92**, 11 (2008)
79. P. Vinten, P. Marshall, J. Lefebvre, P. Finnie, Nanotechnology **21**, 3 (2010)
80. M. Bedewy, E.R. Meshot, H.C. Guo, E.A. Verploegen, W. Lu, A.J. Hart, J. Phys. Chem. C **113**(48), 20576 (2009)
81. M. Stadermann, S.P. Sherlock, J.B. In, F. Fornasiero, H.G. Park, A.B. Artyukhin, Y.M. Wang, J.J. De Yoreo, C.P. Grigoropoulos, O. Bakajin, A.A. Chernov, A. Noy, Nano Lett. **9**(2), 738 (2009)
82. Y.H. Lee, X.Q. Zhang, W.J. Zhang, M.T. Chang, C.T. Lin, K.D. Chang, Y.C. Yu, J.T.W. Wang, C.S. Chang, L.J. Li, T.W. Lin, Adv. Mater. **24**(17), 2320 (2012)
83. J.V. Lauritsen, J. Kibsgaard, S. Helveg, H. Topsoe, B.S. Clausen, E. Laegsgaard, F. Besenbacher, Nat. Nanotechnol. **2**(1), 53 (2007)
84. H. Wang, L.L. Yu, Y.H. Lee, Y.M. Shi, A. Hsu, M.L. Chin, L.J. Li, M. Dubey, J. Kong, T. Palacios, Nano Lett. **12**(9), 4674 (2012)
85. Y.J. Zhan, Z. Liu, S. Najmaei, P.M. Ajayan, J. Lou, Small **8**(7), 966 (2012)
86. K.H. Lee, H.J. Shin, J. Lee, I.Y. Lee, G.H. Kim, J.Y. Choi, S.W. Kim, Nano Lett. **12**(2), 714 (2012)
87. R.Z. Ma, T. Sasaki, Adv. Mater. **22**(45), 5082 (2010)
88. K.S. Novoselov, D. Jiang, F. Schedin, T.J. Booth, V.V. Khotkevich, S.V. Morozov, A.K. Geim, P Natl. Acad. Sci. USA **102**(30), 10451 (2005)
89. P.A. Hu, Z.Z. Wen, L.F. Wang, P.H. Tan, K. Xiao, ACS Nano **6**(7), 5988 (2012)
90. K.S. Novoselov, A.K. Geim, S.V. Morozov, D. Jiang, Y. Zhang, S.V. Dubonos, I.V. Grigorieva, A.A. Firsov, Science **306**(5696), 666 (2004)
91. J.N. Coleman, M. Lotya, A. O'Neill, S.D. Bergin, P.J. King, U. Khan, K. Young, A. Gaucher, S. De, R.J. Smith, I.V. Shvets, S.K. Arora, G. Stanton, H.Y. Kim, K. Lee, G.T. Kim, G.S. Duesberg, T. Hallam, J.J. Boland, J.J. Wang, J.F. Donegan, J.C. Grunlan, G. Moriarty, A. Shmeliov, R.J. Nicholls, J.M. Perkins, E.M. Grieveson, K. Theuwissen, D.W. McComb, P.D. Nellist, V. Nicolosi, Science **331**(6017), 568 (2011)
92. M. Qian, Y.S. Zhou, Y. Gao, T. Feng, Z. Sun, L. Jiang, Y.F. Lu, Appl. Surf. Sci. **258**(22), 9092 (2012)
93. J.J. Hu, J.S. Zabinski, J.H. Sanders, J.E. Bultman, A.A. Voevodin, J. Phys. Chem. B **110**(18), 8914 (2006)
94. Y. Miyamoto, H. Zhang, D. Tomanek, Phys. Rev. Lett. **104**, 20 (2010)
95. W.S. Hummers, R.E. Offeman, J. Am. Chem. Soc. **80**(6), 1339 (1958)
96. M.J. McAllister, J.L. Li, D.H. Adamson, H.C. Schniepp, A.A. Abdala, J. Liu, M. Herrera-Alonso, D.L. Milius, R. Car, R.K. Prud'homme, I.A. Aksay, Chem. Mater. **19**(18), 4396 (2007)

97. S. Stankovich, D.A. Dikin, R.D. Piner, K.A. Kohlhaas, A. Kleinhammes, Y. Jia, Y. Wu, S.T. Nguyen, R.S. Ruoff, Carbon **45**(7), 1558 (2007)
98. K.A. Mkhoyan, A.W. Contryman, J. Silcox, D.A. Stewart, G. Eda, C. Mattevi, S. Miller, M. Chhowalla, Nano Lett. **9**(3), 1058 (2009)
99. D.A. Sokolov, K.R. Shepperd, T.M. Orlando, J. Phys. Chem. Lett. **1**(18), 2633 (2010)
100. D.A. Sokolov, C.M. Rouleau, D.B. Geohegan, T.M. Orlando, Carbon **53**, 9 (2013)
101. A. Castellanos-Gomez, M. Barkelid, A.M. Goossens, V.E. Calado, H.S.J. van der Zant, G.A. Steele, Nano Lett. **12**(6), 3187 (2012)
102. W.A. de Heer, arXiv:1012.1644v1 MRS Bulletin (submitted) (2010)
103. S. Lee, M.F. Toney, W. Ko, J.C. Randel, H.J. Jung, K. Munakata, J. Lu, T.H. Geballe, M.R. Beasley, R. Sinclair, H.C. Manoharan, A. Salleo, ACS Nano **4**(12), 7524 (2010)
104. S.N. Yannopoulos, A. Siokou, N.K. Nasikas, V. Dracopoulos, F. Ravani, G.N. Papatheodorou, Adv. Funct. Mater. **22**(1), 113 (2012)
105. S. Bhaviripudi, X.T. Jia, M.S. Dresselhaus, J. Kong, Nano Lett. **10**(10), 4128 (2010)
106. J.M. Wofford, S. Nie, K.F. McCarty, N.C. Bartelt, O.D. Dubon, Nano Lett. **10**(12), 4890 (2010)
107. X.S. Li, Y.W. Zhu, W.W. Cai, M. Borysiak, B.Y. Han, D. Chen, R.D. Piner, L. Colombo, R.S. Ruoff, Nano Lett. **9**(12), 4359 (2009)
108. S. Bae, H. Kim, Y. Lee, X.F. Xu, J.S. Park, Y. Zheng, J. Balakrishnan, T. Lei, H.R. Kim, Y.I. Song, Y.J. Kim, K.S. Kim, B. Ozyilmaz, J.H. Ahn, B.H. Hong, S. Iijima, Nat. Nanotechnol. **5**(8), 574 (2010)
109. R.S. Weatherup, B.C. Bayer, R. Blume, C. Ducati, C. Baehtz, R. Schlogl, S. Hofmann, Nano Lett. **11**(10), 4154 (2011)
110. X.S. Li, W.W. Cai, L. Colombo, R.S. Ruoff, Nano Lett. **9**(12), 4268 (2009)
111. A. Reina, X.T. Jia, J. Ho, D. Nezich, H.B. Son, V. Bulovic, M.S. Dresselhaus, J. Kong, Nano Lett. **9**(1), 30 (2009)
112. A. Gruneis, K. Kummer, D.V. Vyalikh, New J Phys **11**(7), 073050 (2009)
113. K.L. Saenger, J.C. Tsang, A.A. Bol, J.O. Chu, A. Grill, C. Lavoie, Appl. Phys. Lett. **96**, 15 (2010)
114. S. Chiashi, Y. Murakami, Y. Miyauchi, S. Maruyama, Chem. Phys. Lett. **386**(1–3), 89 (2004)
115. A.A. Puretzky, D.B. Geohegan, S. Pannala, C.M. Rouleau, M. Regmi, N. Thonnard, G. Eres, Nanoscale. **5**, 6507 (2013).
116. A.C. Ferrari, J.C. Meyer, V. Scardaci, C. Casiraghi, M. Lazzeri, F. Mauri, S. Piscanec, D. Jiang, K.S. Novoselov, S. Roth, A.K. Geim, Phys. Rev. Lett. **97**, 18 (2006)
117. J.B. Park, W. Xiong, Y. Gao, M. Qian, Z.Q. Xie, M. Mitchell, Y.S. Zhou, G.H. Han, L. Jiang, Y.F. Lu, Appl. Phys. Lett. **98**, 12 (2011)
118. J.B. Park, W. Xiong, Z.Q. Xie, Y. Gao, M. Qian, M. Mitchell, M. Mahjouri-Samani, Y.S. Zhou, L. Jiang, Y.F. Lu, Appl. Phys. Lett. **99**, 5 (2011)
119. K. Xiao, W. Deng, J.K. Keum, M. Yoon, I.V. Vlassiouk, K.W. Clark, A.-P. Li, I.I. Kravchenko, G. Gu, E.A. Payzant, J. Am. Chem. Soc. **135**(9), 8 (2013)

Chapter 8
Laser-Mediated Nanoparticle Synthesis and Self-Assembling

Paolo M. Ossi, Nisha R. Agarwal, Enza Fazio, Fortunato Neri and Sebastiano Trusso

Abstract Elemental and compound nanoparticles (NPs) are increasingly attractive due to their peculiar physico-chemical properties. Any large scale application of NPs requires a strict control on their synthesis and self-assembling. Inherent to the synthesis stage is the control of size, shape, composition, structure of the single NP. When NPs self-assemble on a suitable substrate the morphology and nanostructure of the NP architecture are the key parameters driving the performance of the resulting artificial surface. Pulsed laser ablation allows to pursue the above goals under different conditions including nanosecond and ultra-short femtosecond laser pulses, as well as an ambient fluid, either a gas at high pressure, or a radiation transparent liquid, besides vacuum. In this chapter we offer an outline of the mechanisms underlying NP synthesis in the above environments and of the most popular models currently recognized in the literature to interpret observed experimental trends. Concerning plasma plume propagation through an ambient gas attention is focused on the prediction versus observation of the size of isolated NPs and on a critical discussion of the morphology—properties relationship of noble metal NP arrays, considering their optical properties in the frame of enhanced vibrational spectroscopies (SERS). Ablation in a liquid of a solid target leads to a chemically stable suspension of different nanostructures in a one-step, environment friendly, clean process. For noble metal NPs the effect of liquid layer thickness and

P. M. Ossi (✉)
Dipartimento di Energia, Politecnico di Milano, Via Ponzio 34-3, 20133 Milano, Italy
e-mail: paolo.ossi@polimi.it

N. R. Agarwal
Dipartimento di Chimica, Materiali e Ingegneria Chimica 'Giulio Natta', Politecnico di Milano, Piazza Leonardo da Vinci 32, 20133 Milano, Italy

E. Fazio · F. Neri
Dipartimento di Fisica e di Scienze della Terra, Università di Messina, V.le F. S. d'Alcontres 31= 98158 Messina, Italy

S. Trusso
IPCF-CNR, Istituto per i Processi Chimico-Fisici del CNR, V.le. F. S. d'Alcontres 37, 98158 Messina, Italy

M. Castillejo et al. (eds.), *Lasers in Materials Science*, Springer Series in Materials Science 191, DOI: 10.1007/978-3-319-02898-9_8,

laser spot diameter on the concentration, size distribution and mutual aggregation of the produced NPs is discussed in relation to a more general picture of the process. Irradiation under vacuum with ultra-short fs laser pulses is a clean physical method to synthesize NPs; indeed in the majority of materials, random stackings of NPs, whose size ranges between 10 and 100 nm constitute the deposited film. Selected experiments on NP synthesis upon fs ablation of mainly elemental targets are reviewed focusing mainly on the features of the expanding plasma and on established mechanisms of NP synthesis. Possible lines of future development in the field are envisaged.

8.1 Introduction

Since the first attempts to produce nanoparticles (NPs) and the flourishing of physical and chemical synthesis routes, the interest shifted progressively towards attaining NPs with controlled properties and to their possibly spontaneous assembling in an organized, suitable fashion. During self-assembling, the building bricks organize in an orderly, macroscopic structure: the process goes on via direct interactions, mostly mediated by interparticle forces, or indirectly, usually via an external field, or using a template. To enhance the effectiveness of NP self-assembling for a technological scale-up, a high degree of process direction and control is necessary. The state of the art spans from simpler two-dimensional NP arrays with specific geometrical properties that find application as templates for more complex spatial organizations [1], to more advanced and intricate three dimensional structures, such as cluster-assembled (CA) films.

Particle properties include the size and its distribution, the composition, the chemical and thermal stability, the transport, mechanical, and chemical properties with their specific size dependence. Relevant properties of the artificial structure made of self-assembled NPs are the adhesion to the substrate and the mechanical stability, besides its physico-chemical properties that can differ from those of the constituent NPs.

Particles are found in such different kinds of plasmas as interstellar dust [2] and discharges through low pressure gases [3]. Indeed NPs are spontaneously generated in all plasmas; while usually they constitute a minority plasma population, in plasmas generated by laser ablation a wealth of particles with typical size in the few nanometer range are observed [4, 5]. The technique allows for an independent control of the excitation (laser pulse wavelength, duration) and the irradiation conditions (target nature, surface morphology, orientation, focusing conditions); the possibility to adopt ample ranges of values for the process parameters permits to explore largely different conditions of radiation—matter interaction and to affect the complex dynamics of matter ejected from the irradiated target. Thus, a way to explore and to understand the mechanisms of NP formation and their relative effectiveness is opened.

Here, we consider three different experimental configurations to generate NPs ablating with a laser beam a solid target: the propagation through an ambient gas at high pressure of the ablation plasma produced by a nanosecond (ns) laser pulse, the generation of a colloidal solution of NPs via ablation in a liquid transparent to laser radiation, and ablation in vacuum with ultra-short, femtosecond (fs) pulses. We devote the larger space to the first case and we discuss a selection of results for the remaining two, highlighting established plasma properties and mechanisms of NP formation. In all cases we will refer for simplicity to the synthesis of elemental particles.

Two questions are central to our study: what is the leading mechanism and which is the dominant time scale in NP formation? Searching for answers to these questions requires an *ex post* analysis of the properties of the deposited matter that are usually correlated to changes in values of deposition parameters including target to substrate distance, laser wavelength, power density deposited at the target surface, number of laser pulses [6, 7], ablated mass per pulse, ambient gas nature and pressure [8]. A complementary diagnostics of plasma expansion is essential, as it offers details on the dynamics of the ejected species; the most used techniques are optical emission spectroscopy (OES) [9], optical time-of-flight measurements (TOF) [10], Langmuir probes [11] and fast photography, using an intensified charge-coupled device (ICCD) [12, 13], from which the shape of the expanding plasma, as well as the position and velocity of the plasma front are obtained.

8.2 Propagation of an Ablation Plasma Through an Ambient Gas

After absorbing a ns laser pulse, a solid target undergoes intense fast heating (around 10^{12} Ks^{-1}) in most materials and relevant matter evaporation from the surface occurs. Initially, the high density, highly anisotropic vapor cloud that forms and lies just above the illuminated surface behaves like a hot, collisional fluid, strongly interacting with the laser radiation: an isothermal expanding plasma results up to the end of the pulse; intense ionization of plasma constituents stimulates further plume expansion. Particle ejection from the target surface stops short after the end of the laser pulse. In vacuum and at low ambient gas pressure (up to about 1 Pa) plasma propagation resembles a supersonic free expansion where a linear relation holds between the time delay and the position of the plume front. The weak fluorescence observed close to the target just after the end of the laser pulse is due to collisions among plasma species. The expansion of such a plasma was modeled [14] using gas dynamics equations obtaining the density and pressure of the plume as functions of position and time. Two main assumptions are made, that plasma formation time is much shorter than plasma expansion time and that plasma behaves as an ideal gas with adiabatic index $\gamma = c_p/c_v$ constant. In these conditions, ablation is treated much like thermal desorption, monolayers being evaporated in sequence from the target surface under quasi-equilibrium. Nanoparticles are found in the plume as a minority constituent: their occurrence is explained by a condensation model [15].

When gas pressure increases, plume propagation is slowed down by collisions between plasma species and ambient gas atoms that lead to *shock wave* (SW) formation. The high gas density results in plume braking, more severe in the direction normal to the target with respect to the radial directions. The shape of the propagating plume is observed to rapidly shift from elliptical, strongly forward-peaked, to spherical. At comparatively low (below 10 Pa) ambient gas pressure the *initial* plume expansion is similar to that in vacuum, but at times longer than 1 μs at most, plume front slows down due to the confinement efficiency of the gas [16]. At larger times plume sharpens and its front shows an oscillatory behavior persisting up to gas pressures of a few tens of pascals: at higher gas pressure this stage starts earlier. Plasma oscillations disappear at gas pressure above about 100 Pa. At intermediate gas pressures, between about 30 and 50 Pa, plume sharpening is associated with confinement of plasma emission to plume front; at these pressures braking of plume front begins after a few microseconds and continues until a stationary behavior is achieved. Meanwhile the rear edge of the plume moves backwards to the target. This behavior marks a transition to *diffusion-like* propagation of plume species through the ambient gas typical of longer times, for pressures in the tens of pascals range [17]. A strong interpenetration of plasma species and ambient gas occurs and leads to plume splitting, besides sharpening: the particles of the faster group travel practically at the same velocity as in vacuum and cross the ambient gas nearly collision-less; in turn, the slower, delayed population results from the interaction between ablated species and gas atoms. Splitting into two velocity populations was put into evidence by TOF distribution analysis of the ablated species and affects both ions [18] and neutrals [19]. In coincidence with such a mutual penetration of the laser-generated plasma and ambient gas, a considerable conversion of kinetic energy into thermal energy occurs, so that gas, as well as radiation temperature increase. At pressures around 100 Pa *turbulence* was observed across the slowing down plume front. If gas pressure is further increased, the mutual penetration zone contracts and the plasma front becomes compressed. In Fig. 8.1, we report pictures recorded at the same time delays of silver and gold plasmas produced under identical irradiation conditions and propagating through argon at different pressures.

The above explained evolution experienced by the ablation plume is observed independently of the specific combination target-ambient gas-process conditions. Yet the pressure ranges typical for the different propagation regimes are quite broad and not always all the features of the phenomenology we described are observed. Plume expansion through an ambient gas is a much more complex gas dynamics phenomenon than an expansion in vacuum. The gas affects plume dynamics and the spatial distribution, kinetic energy and kinetic-energy distribution of its constituents, thus influencing NP nucleation and evolution, as well as NP energy distribution in the plume.

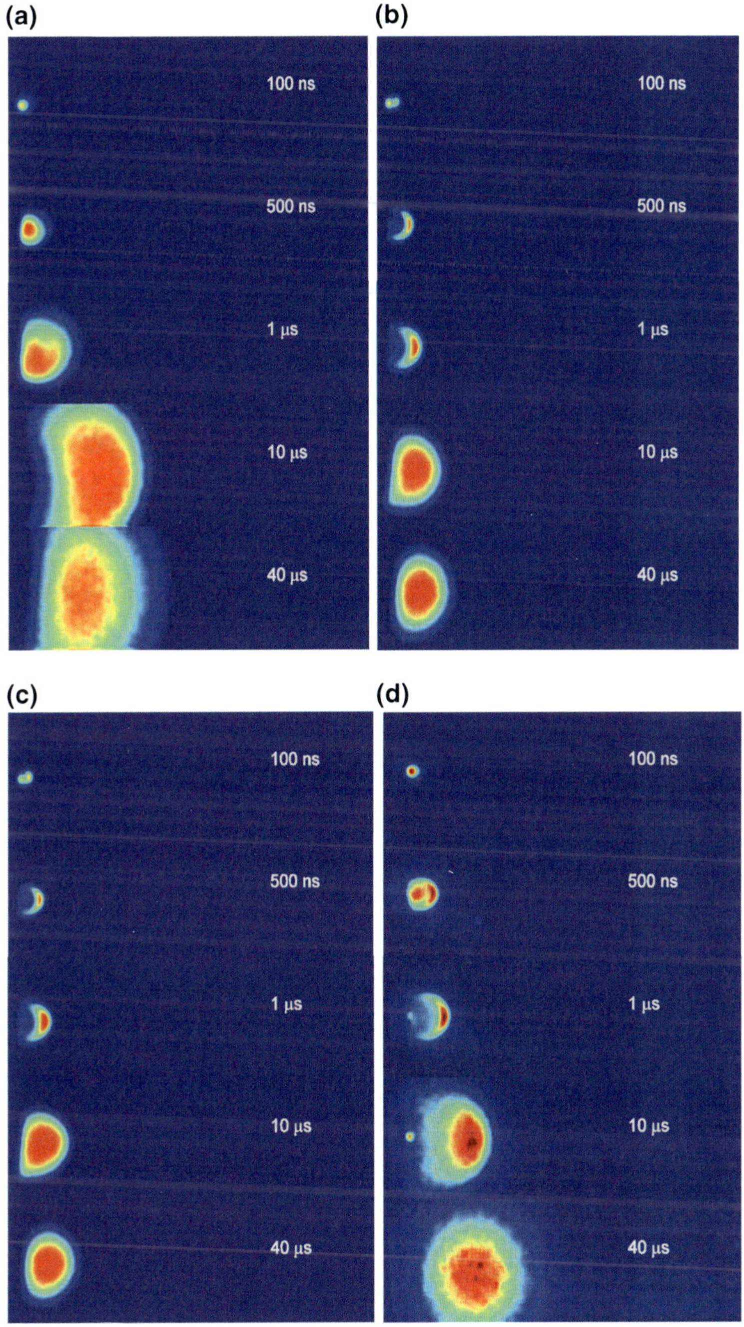

Fig. 8.1 ICCD fast photography pictures of noble metal ablation plasmas propagating through argon at increasing pressure (**a** Ag, 10 Pa; **b** Ag, 40 Pa; **c** Ag, 100 Pa; **d** Au, 70 Pa). The laser beam from a KrF excimer laser (wavelength 248 nm, pulse duration 25 ns, repetition rate 10 Hz, energy density 2 J cm^{-2}) was focused on the rotating target at an angle of 45°

Several phenomenological models were proposed over the years to describe the expansion of a laser-generated plasma through an ambient gas. From the detailed analysis [20] of the most popular drag [21, 22], shock wave [15], diffusion [22, 23] and Arnold [24] models, we must conclude that no one of them gives a satisfactory description of the complete expansion process, so that a detailed understanding of a laser-generated plasma propagating through a background gas is still lacking.

We now introduce with some detail a two-step mixed-propagation model that describes the propagation process over a wide time range for several classes of ablated materials expanding through various ambient gases at different pressures [25]. The model is conceptually simple and is based on modified versions of both drag and diffusion models. In the drag model, the plasma is considered as an ensemble of particles, which expanding experiences a viscous drag force due to the presence of the ambient gas; the position R of the plasma front edge is given by:

$$R(t) = R_0(1 - e^{\beta t}), \tag{8.1}$$

where the stopping distance R_0 is the distance at which the model predicts that the plasma will stop, while β is a slowing down coefficient. In mixed-propagation model drag model predictions are assumed to hold in the initial stage of plasma expansion, provided the phenomenological parameters R_0 and β are replaced by $\tilde{R}_0 = v_0\tilde{D}/\mu$ and $\tilde{\beta} = \mu/\tilde{D}$ respectively. v_0 is the initial plasma velocity, μ is a slowing down coefficient (a diffusion coefficient divided by time) and $\tilde{D}$ a modified diffusion coefficient. Thereafter plasma expansion is described by a diffusion-like equation

$$R(t) = \sqrt{\tilde{D}t}. \tag{8.2}$$

According to gas kinetic theory the diffusion coefficient is $D = (\lambda v)/3$, with $\lambda = (n_g\sigma)^{-1}$ the mean free path, n_g the gas number density, v the relative velocity of the colliding particles and σ the scattering cross section. In mixed propagation model, the diffusion coefficient D is replaced by $\tilde{D} = kv_0(n_g\sigma)^{-1}$ [26] with k an integer number that depends on the atomic mass of the target. Thus, mixed-propagation model describes plasma expansion via a two-step combination of modified versions of drag and diffusion models.

A refinement of mixed propagation model was recently possible [27] after a detailed analysis of the expansion of Ag plasmas in Ar for which experimental data are available over a considerable time interval; besides this the plasma does not come into contact with the substrate, so that an in-depth test of the model was performed. Experimental data are fitted to the equation

$$\begin{aligned} R(t) &= \tilde{R}_0^D(1 - e^{\tilde{\beta} t}), \quad \text{for } R < \tilde{R}_0^D \\ R(t) &= \tilde{R}_0^D + \sqrt{\hat{D}(t - t_0^D)}, \quad \text{for } R > \tilde{R}_0^D \end{aligned} \tag{8.3}$$

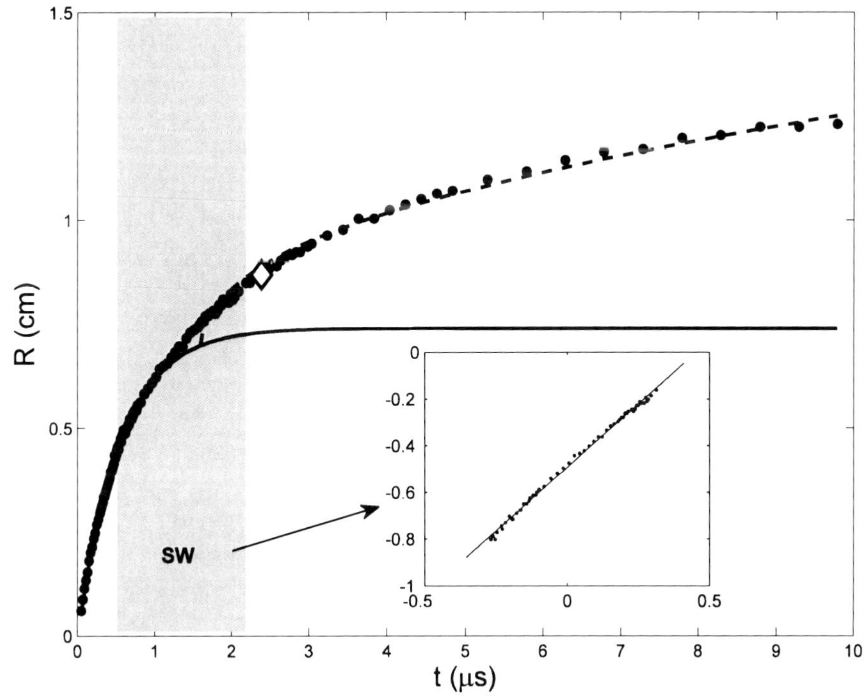

Fig. 8.2 Fit of Ag plasma propagation through argon (40 Pa) by mixed propagation model. Modified drag model is adopted for the initial expansion stage and modified diffusion for the late expansion. The *diamond* marks the change of expansion regime. Although along the *grey area* SW propagation correctly fits the expansion (see the inset), it can be neglected for simplicity

The first term of (8.3) describes the viscous plasma motion in its initial expansion according to modified drag model: this holds until the plasma has traveled, at the time delay t_0^D, the distance $\tilde{R}_0^D$ from the target surface. Such a position is taken as the origin of a subsequent diffusive expansion stage, governed by a diffusion coefficient

$$\hat{D} = \kappa v_0^D (n_g \sigma)^{-1}. \tag{8.4}$$

$\hat{D}$ has the same functional form as $\tilde{D}$, except that the velocity v_0^D, namely plasma velocity at time t_0^D, that is at position R_0^D, is used as the initial velocity. Thus, v_0^D is at the same time the plasma residual velocity at the end of the viscous expansion stage and the initial plasma velocity at the beginning of the modified diffusion stage of expansion. In Fig. 8.2, we report the results of the fitting procedure for Ag plasma propagation in Ar at 40 Pa. Experimental data were fitted using (8.3) with the parameters $\tilde{\beta}$, $\tilde{R}_0^D$, and v_0^D. The dashed line refers to the fit where the first term of (8.3) (modified drag model) holds up to the position $\tilde{R}_0^D$, thereafter, data are fitted with the second term of (8.1) (modified diffusion model)

Table 8.1 Experimental data (p_g, n_g) and parameters from plasma expansion analysis with mixed propagation model (modified drag and modified diffusion) for Ag ablation plasmas propagating through argon at increasing pressure

p_g (Pa)	n_g (cm^{-3})	v_0^D (cm μs^{-1})	$\tilde{R}_0^D$ (cm)	t_0^D (μs)	v_0 (cm μs^{-1})	$\hat{D}$ (cm^2 μs^{-1})
10	2.41×10^{15}	0.09	1.51	5.29	1.56	0.061
40	9.66×10^{15}	0.12	0.82	2.39	1.21	0.020
70	1.69×10^{16}	0.13	0.64	1.67	1.15	0.013
100	2.41×10^{16}	0.20	0.55	0.86	1.18	0.014

using the diffusion coefficient $\hat{D}$ calculated using the v_0^D value as obtained from the fit. In Fig. 8.2, the diamond marks the position at which the velocity v_0^D is imposed to the plasma. Similar results were obtained for expansions in Ar at all the other investigated pressures: remarkably, at 10 Pa a satisfactory fit was obtained up to time delay of 20 μs [28]. The values of the parameters obtained from the fitting procedure are reported in Table 8.1. At the different Ar pressures the values of the initial plasma velocities v_0 deduced from the fitted $\tilde{R}_0^D$ and $\tilde{\beta}$ nearly coincide with those obtained directly from the slope at the origin of the experimental data. v_0 decreases from 1.56 cm μs^{-1} at 10 Pa down to 1.18 cm μs^{-1} at 100 Pa. This is an indication that plasma–Ar interaction results in important deviations from free propagation since the beginning of expansion. The agreement between (8.3) and experiment is notable in particular in the early and the late expansion stages. The reason for this is that $\tilde{R}_0^D$ sets an unphysical *discontinuity* between two different plasma expansion regimes (in the original drag model R_0 indicates the maximum expansion distance of the plasma). Thus, the fit accuracy is poorer over a time interval centered around t_0^D, namely at the transition between the two expansion regimes. The above inaccuracy is due to the (over) simplified structure of mixed propagation model that does not take into account that a SW develops along with plasma propagation. The simplification is acceptable in cases like that of Ag plasma propagation in Ar, where the SW duration is limited to about 1 μs between 70 and 100 Pa and to about 4 μs at 10 Pa [28]. Yet, where SW formation must be taken into account mixed propagation model fails to precisely reproduce plasma dynamics. A notable example is the propagation of gold ablation plumes produced with the same experimental setup and process parameters as for the silver plumes we just discussed. The explored Ar pressures ranged between 10 and 70 Pa and the laser energy density was set at 2.0 J cm^{-2}. We used for fast-photo imaging an iCCD as for the measurements on silver. From Fig. 8.1 already a visual inspection shows the crucial role of SWs to plasma propagation. In Fig. 8.3a, we report the curves for the position of the plasma front edge versus time, as obtained from pictures of the plume taken at increasing time delays with respect to the arrival of the laser pulse at the target. The curves were fitted using the modified version of mixed-propagation model (see (8.3)). Good fits were obtained only for the expansion at 10 Pa: in this case the initial expansion stage was modeled using modified drag model and the subsequent stage by modified diffusion model, like in

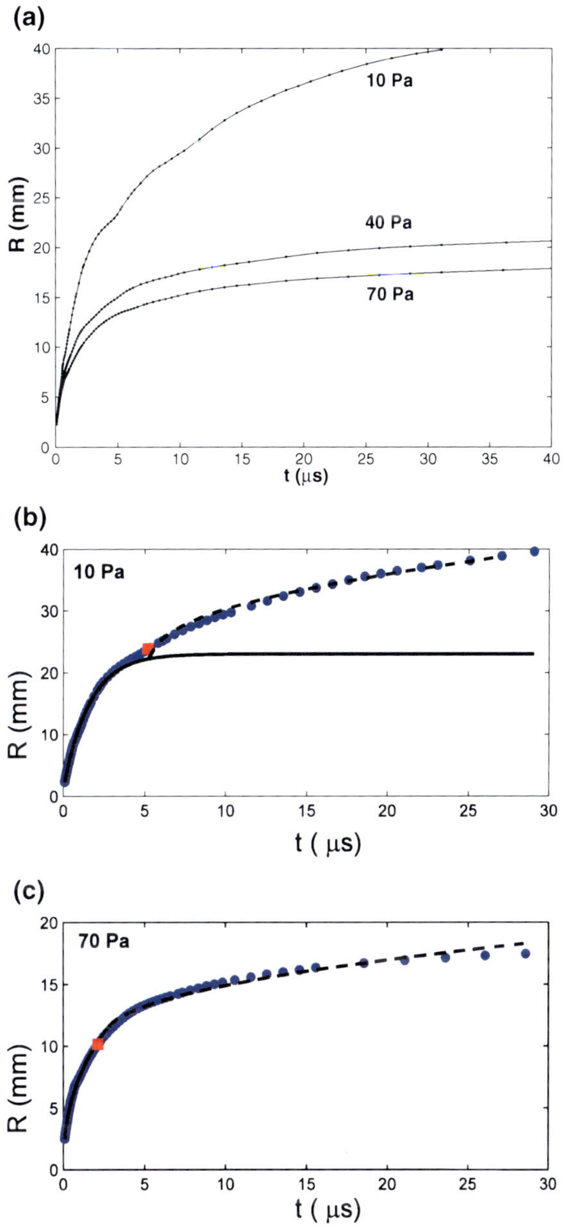

Fig. 8.3 **a** Position of the front edge of Au plasmas versus time at increasing time delays with respect to the arrival time of the laser pulse at the target. The *curves* refer to propagations through argon at increasing pressure (10, 40, 70 Pa). **b** Fit of Au plasma propagation through argon (10 Pa) by mixed propagation model. Modified drag model is adopted for the initial expansion stage and modified diffusion for the late expansion. The *diamond* marks the change of expansion regime. **c** Fit of Au plasma propagation through argon (70 Pa) by mixed propagation model. SW model is adopted for the initial expansion stage and modified diffusion for the late expansion. The *diamond* marks the change of expansion regime

the case of silver. For higher Ar pressures fitting results were poorer. Indeed, the experiments were in all respects identical to those performed on silver, unless that the atomic mass of gold is about *twice* the silver mass: $m_{Au} = 196.97$ and $m_{Ag} = 107.87$ amu. Since plasma dynamics in a gas is dominated by collisions,

Table 8.2 Experimental data (p_g, n_g) and parameters from plasma expansion analysis with mixed propagation model for Au ablation plasmas propagating through argon at increasing pressure. The first column reports the adopted models in the fitting procedure

Model	p_g (Pa)	n_g (cm^{-3})	v_0^D (cm μs^{-1})	$\tilde{R}_0^D$ (cm)	t_0^D (μs)	v_0 (cm μs^{-1})	$\hat{D}$ (cm^2 μs^{-1})
Drag + Diff	10	2.41×10^{15}	0.240	1.55	1.88	1.30	0.221
SW + Diff	40	9.66×10^{15}	0.117	1.31	3.1	0.86	0.027
SW + Diff	70	1.69×10^{16}	0.115	1.10	3.5	0.86	0.015
SW + Diff	100	2.41×10^{16}	0.096	1.4	3.9	0.88	0.009

the mass ratio between plasma species and ambient gas atoms plays a major role that is not considered in any of the previously discussed phenomenological models. In a classical hard-sphere binary collision between an argon atom and a gold atom the kinetic energy loss experienced by the first is expected to be lower than in the case of the lighter silver. Thus, the braking efficiency of the ambient gas is much lower with gold plumes, leading to a significantly different evolution of plasma dynamics with respect to silver. In particular, the time interval during which SW-like expansion is expected to occur will be *longer* for gold than for silver plasmas and thus cannot be neglected as in (8.3). To take this into account, Au plasma expansion data for Ar pressures between 40 and 100 Pa were fitted to the equations:

$$\begin{aligned} R(t) &= \xi \left(\frac{E_0}{\rho_0} \right)^{1/5} t^{2/5}, \quad \text{for } R < \tilde{R}_0^D \\ R(t) &= \tilde{R}_0^D + \sqrt{\hat{D}(t - t_0^D)}, \quad \text{for } R > \tilde{R}_0^D. \end{aligned} \tag{8.5}$$

Differently to (8.3) the initial stage of expansion is reproduced using SW model [15] and the subsequent one using modified diffusion model. In (8.5), the $\tilde{R}_0^D$ position marks an *unphysical* discontinuity: between modified drag and modified diffusion regime (at 10 Pa), as well as between shock wave and modified diffusion regime (at Ar pressures from 40 to 100 Pa). The residual velocity v_0^D at $\tilde{R}_0^D$ represents the plasma *initial* velocity in modified diffusion model. In Fig. 8.3b, c the results of the fitting procedure for two representative cases of Au expansion, in Ar at 10 and 70 Pa, are shown. In Table 8.2, parameter values obtained from the fitting procedures for all the investigated Ar pressures are reported.

8.3 Synthesis of Nanoparticles in the Expanding Plasma

Basically two schemes for NP nucleation and growth are reported in the literature. In the first, exemplified by Cu, NPs grow up to a critical size; coalescence and coarsening follow at large degrees of substrate coverage and drive the growth of nanocrystals with size below 10 nm in nanocomposite films deposited by PLD in argon at low pressure, not higher than 0.66 Pa; at higher gas pressures, up to 13 Pa, again nucleation and growth mainly occur *on* the substrate, but the reduced surface mobility inhibits the coarsening stage, so that highly anisotropic nanocrystals are observed [29]. Although the same scheme of NP growth was reported for Fe/Mo [30] and Fe/Cu (111) [31], it cannot be considered of universal validity. Indeed in the case of silicon NPs with narrow size distribution, plasma spectroscopy indicates that particle nucleation and growth occurs *in* the expanding ablation plume [32], and is strongly affected by ionization processes that occur when plume propagates through the ambient gas, resulting in very high nucleation rates and small cluster critical radii [33]. The synthesis of C NPs in plumes expanding through He and Ar atmospheres up to pressures of 1 kPa was investigated and modeled, considering the leading role of ionization phenomena in the interface region between the SW front and ambient gas [34, 35]. Cluster-assembled W films were deposited in different atmospheres and pressure ranges [36]; the surface morphology, bond coordination and oxidation path of the deposited films, both when exposed to ambient atmosphere and when prepared in dry air, were studied and complemented with a detailed HREM analysis of structure, size and morphology of the deposited NPs [37].

From the bulk of the above investigations a coherent picture of NP formation in the expanding ablation plume follows: small, essentially spherical NPs are synthesized *in* gas phase during plasma expansion and their size depends on plume dynamics; the latter is highly sensitive to ambient gas nature and pressure. Thus, some plasma parameters deduced in the frame of mixed-propagation analysis are useful to model NP *synthesis* in the expanding ablation plume. In particular, we evaluate the average NP asymptotic size, namely the number N of constituent atoms in a particle that reached a steady state during plume propagation.

NPs evolve through the stages of nucleation, growth and cooling [32]. When an ambient gas is present, a set of hydrodynamic equations for plasma expansion, including vapor condensation should be solved. Such an approach is presently impracticable. We assume that the plume, whose initial ionization degree is high, contains a population of tiny clusters [38] and we take that plume evolution is not affected by the mechanisms of NP formation [32]. For a fixed set of ablation conditions, the average asymptotic number N of atoms in a particle that reached a steady state during plume propagation is calculated in the ideal-gas approximation. Although the plume experiences a range of internal pressures and is spatially inhomogeneous, so that at a given time it contains particles at different stages of aggregation, we consider averages over long times; N is given by

$$N = (\langle n_a \rangle \sigma_{a-a} \langle v \rangle t_f)(n_g \sigma_{a-g} \langle v \rangle t_f) \tag{8.6}$$

where $\langle n_a \rangle$ is the average number density of ablated atoms, t_f is the NP growth time (for $t \geq t_f$ we assume that NPs reached their steady size), $\langle v \rangle$ is the mean plasma velocity, σ_{a-a} and σ_{a-g} are the geometric cross sections for ablated particle-ablated particle and for ablated particle-gas atom collisions. If the substrates are placed at a distance smaller than the distance x_{aggr} traveled by the plume during particle growth time, (8.6) should be multiplied by x_{T-S}/x_{aggr}, where x_{T-S} is the target to substrate distance. If x_{T-S} is less than x_{aggr}, NPs with sizes smaller than the maximum achievable are deposited. When distance and gas pressure are expressed in cm and Pa respectively, we have

$$x_{aggr} = kp_g^{-1/5}. \tag{8.7}$$

The equation is an upper limit for x_{aggr}; it was obtained [39] from data on the trend of the intensity versus distance from the target, considering CN [40, 41] and TaO [42] band emissions from plumes ablated by different lasers at quite different energy densities (4–95 J cm^{-2}) over wide ranges of ambient gas pressure (1–10^5 Pa). It is noteworthy that such intensities that are direct indication of plasma reactivity invariably show a peaked behavior: the distance x_M where emission is maximum depends only on gas pressure according to $x_M = k_M p_g^{-1/5}$ with $k_M = 2$ cm $Pa^{1/5}$. From the available data, the maximum distance at which emission is observed is estimated as three times x_M (i.e. $k = 3k_M$): we assume that the same distance coincides with x_{aggr} in the case of a non-reactive species such as NPs.

Time resolved fast photography of plasma expansion allows for a more realistic estimate for x_{aggr} and an evaluation of the physical quantities that are used as input parameters in (8.6), in particular, the average plasma velocity $\langle v \rangle$, the average plasma density $\langle n_a \rangle$, and the aggregation time t_f. We devoted much attention to the formation of noble metal (NM) NPs, in particular Ag and Au, with the aim to prepare artificially roughened substrates with specific optical properties (see Sect. 8.7). We first investigated via time resolved fast photography the synthesis of silver NPs in plasmas expanding in Ar at pressures between 10 and 100 Pa. From the plasma images taken at different time delays with respect to the arrival of the laser pulse, the position of the plasma front (R) as a function of time (t) was determined. According to our discussion in Sect. 8.2, the linear behavior observed in the first expansion stage is followed by a slowing down as a consequence of the collisions with ambient gas atoms at rest, until in the final stage of expansion the plasma stops, or moves very slowly. The time intervals characteristic for the different propagation regimes depend on ambient gas pressure. Increasing the latter both deviation from the linear behavior and plasma stopping are observed earlier. The luminous emission of the plasma, as recorded by the iCCD, is a consequence of relaxation via light emission of collisionally excited neutral or ionized species. In this respect, the intensity of plasma luminous emission is strictly correlated to the collisional rate among plasma species. In Fig. 8.4, we report the trend of emission intensity of Ag plumes at different time delays for a set of Ar pressures, recorded from plasma images taken along the target to substrate direction. Each image was normalized with respect to the integration time and amplification factor of the iCCD. The luminous intensity is highest just after plasma formation, then decreases fast as soon as the plasma starts to expand. We take t_f as the time at which the intensity drops by three orders of magnitude from its initial value: at t_f,

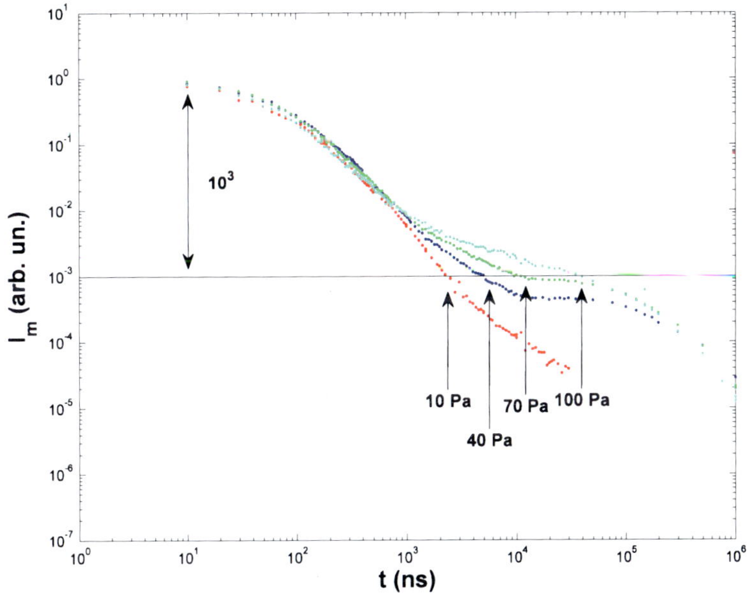

Fig. 8.4 Trend of emission intensity of Ag plasmas at different time delays, for propagations through argon at increasing pressure, as recorded from plasma images collected along the target to substrate direction. The *arrows* put in evidence the time at which each *intensity curve* decayed by three orders of magnitude with respect to the first recorded value

Table 8.3 Experimental (p_g) and model ($\langle n_a \rangle$, $\langle v \rangle$, t_f, x_{aggr}) parameters used to deduce the average number of Ag atoms per NP, $\langle N \rangle$, in plasmas propagating through argon at increasing pressure (adapted from [44])

p_g (Pa)	$\langle v \rangle$ (cm μs^{-1})	$\langle n_a \rangle$ (cm^{-3})	t_f (μs)	x_{aggr} (cm)	d (nm)	$\langle N \rangle$	d_m
10	0.48	7.15×10^{14}	0.79	0.66	0.7	15	–
40	0.28	1.36×10^{15}	3.32	0.49	1.7	158	–
70	0.22	2.24×10^{15}	5.17	0.40	2.8	716	2.9
100	0.21	2.78×10^{15}	7.32	0.39	4.1	2,344	3.8

all collisions that can result in NP growth ended and NPs reached a steady state. The average plasma density $\langle n_a \rangle$ is evaluated from the plasma volume, deduced from plume pictures (at $t = 0$ and at t_f), and from the average number of atoms removed from the target by a single laser pulse (3.14×10^{14}): this value is obtained from measurements of crater volume [43]. The mean plasma velocity $\langle v \rangle$ is calculated as the average between the initial plasma velocity and the plasma velocity at t_f: $\langle v \rangle$ is the impact velocity in a binary collision between a slow gas atom and a fast plasma particle. In Table 8.3, values of model parameters from the analysis of plasma expansion between 10 and 100 Pa are reported together with the estimates for the number of atoms per NP, $\langle N \rangle$. NP size d was calculated for spherical, non-crystalline, close-packed NPs according to indications from

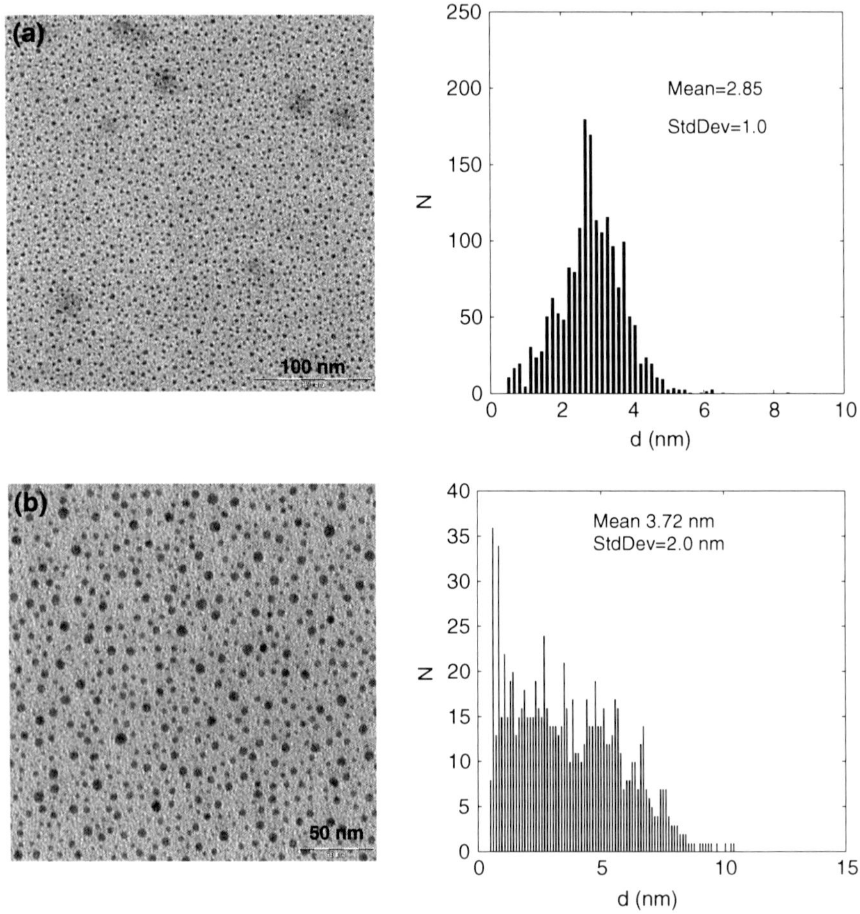

Fig. 8.5 TEM pictures and measured particle size distribution of isolated Ag NPs synthesized in argon at 70 Pa (**a**) and at 100 Pa (**b**) deposited on a-C covered Cu grids

electron diffraction performed during TEM observations (packing coefficient, 0.67). In these depositions the laser pulse number was limited to 10^4 to avoid NP coalescence on the substrate. There is a remarkable agreement between NP size calculated and directly measured from high magnification TEM pictures for films deposited at 70 and 100 Pa as exemplified in Fig. 8.5a, b. The particle size distributions reported in the same figures were obtained after image analysis of the original pictures by a particle counting software [45]. In films grown at 40 and 10 Pa, coalescence among NPs was observed: indeed we find small regions consisting of isolated NPs whose average size is in agreement with model predictions, yet such regions are not large enough to allow a reliable statistical analysis.

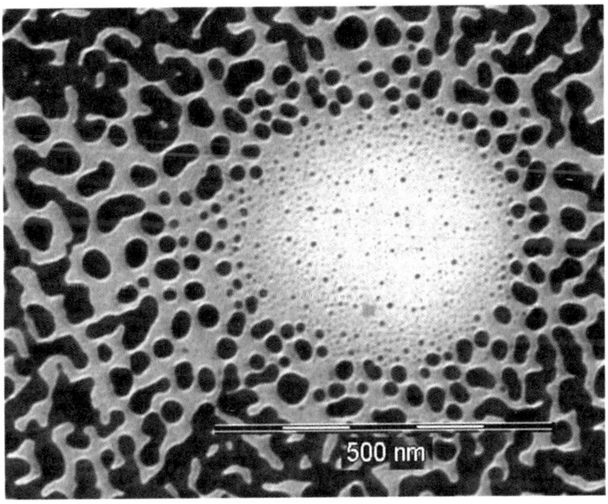

Fig. 8.6 TEM picture of the different stages of Ag film formation, from isolated NPs to coalesced islands, to a percolated structure

8.4 Nanoparticle Self-Assembling on a Substrate and Film Growth

When the ablation plume arrives at the substrate with energy of several eV at^{-1} or above, the nanoparticles it possibly carries either undergo penetration, or pinning [46] resulting in compact, well-adherent films. When the deposition is low energy typically fractions of eV at^{-1}, particles diffuse and aggregate on the substrate surface [47], then they coalesce in larger particles beyond a critical degree of surface coverage. This deposition strategy is adopted to produce cluster-assembled materials that keep memory of the properties of their precursor building blocks.

We discuss the evolution of NP self-assembling on a substrate and the observed stages of film growth, always referring to an initial deposition of small, nearly spherical NPs synthesized in the expanding ablation plasma that land on the substrate with a kinetic energy E_k such that

$$E_k \text{ at}^{-1} < E_{coh} \text{ at}^{-1} \tag{8.8}$$

where E_{coh} is the material cohesive energy; (8.8) gives a necessary condition to keep NP integrity.

In Fig. 8.6 are nicely illustrated all the stages of film formation, in the case of Ag, deposited in Ar at 10 Pa with 10^4 laser pulses. The amorphous C film beneath the Cu grid was not completely covered by particles, yet in the image we clearly identify three distinct regions. A low-density oval area is regularly filled with small prevalently spherical NPs; when the particle size increases the shape deviates from spherical to elliptical. The origin of the oval area is most likely a micrometer sized droplet that accidentally attached itself for a certain time interval to the substrate, shadowing it from NP deposition, then it flew away and NPs re-started to deposit on the bare substrate. Such a region is surrounded by

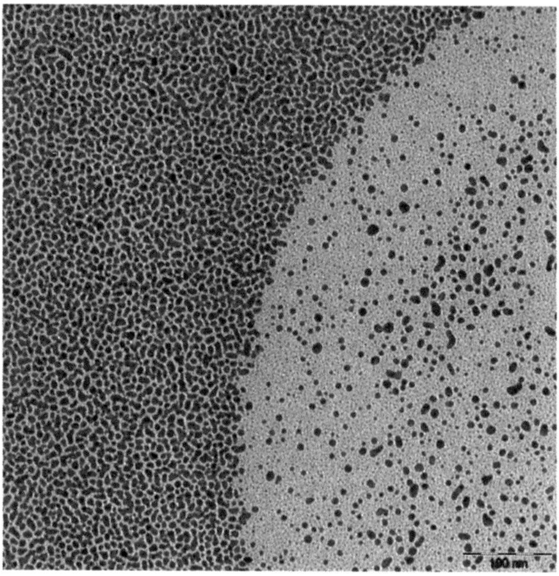

Fig. 8.7 TEM picture of Au NPs synthesized in argon at 100 Pa, deposited on a-C covered Cu grids. The size of some representative NPs is reported

coalesced NPs whose shape is more and more irregular with increasing size, up to big islands interconnected through thin channels. Thus, by chance on the same sample we can observe different time stages of film growth, from the very initial deposition of tiny, isolated NPs, to a nearly-compact, extended film.

The deposition of Au NPs was monitored like that of Ag NPs correlating plasma analysis with TEM observations of isolated, as deposited particles whose size and shape could be individually measured. The experimental set-up and process conditions were kept unaltered with respect to Ag depositions. Figure 8.7 refers to a deposition at 100 Pa of Ar. A nearly circular portion, with micrometer size, populated by a distribution of isolated NPs is surrounded by a film made of NPs at an advanced stage of coalescence into islands interconnected through a network of channels with appreciably regular width. Like in the case we discussed for Ag, the low-coverage region is most likely due to the accidental deposition of a droplet that shadowed the substrate surface from NP deposition for a certain time interval, then detached and flew away. In Fig. 8.7, apart from a minority of very small NPs, the average NP size ranges between 6 and 10 nm, in agreement with predictions from mixed-propagation model (see Table 8.4).

Keeping fixed all the other process parameters, the deposited gold films consist of agglomerated NPs at increasing degree of coalescence with increasing laser pulse number and decreasing ambient gas pressure. This is the very same trend we observed in silver films synthesized in Ar atmosphere. A representative TEM picture, taken across the boundary between two adjacent, well-differentiated regions in a film deposited in Ar at 70 Pa is displayed in Fig. 8.8a. In the left region, similarly to what is observed in films deposited at low pressure, large islands at an advanced stage of reciprocal interconnection are found. The inter-

Table 8.4 Experimental (p_g) and model ($\langle n_a \rangle$, $\langle v \rangle$, t_f, x_{aggr}) parameters used to deduce the average number of Au atoms per NP, $\langle N \rangle$, in plasmas propagating through argon at increasing pressure (adapted from [48])

p_g (Pa)	$\langle v \rangle$ (cm μs^{-1})	$\langle n_a \rangle$ (cm^{-3})	t_f (μs)	x_{aggr} (cm)	d (nm)	N	d_m
10	0.47	3.00×10^{15}	2.3	1.83	2.0	115	–
40	0.41	8.32×10^{14}	5.0	1.50	3.1	478	–
70	0.41	6.80×10^{14}	10.0	1.55	5.6	2,670	–
100	0.37	3.02×10^{14}	35.0	1.51	10.2	16,600	–

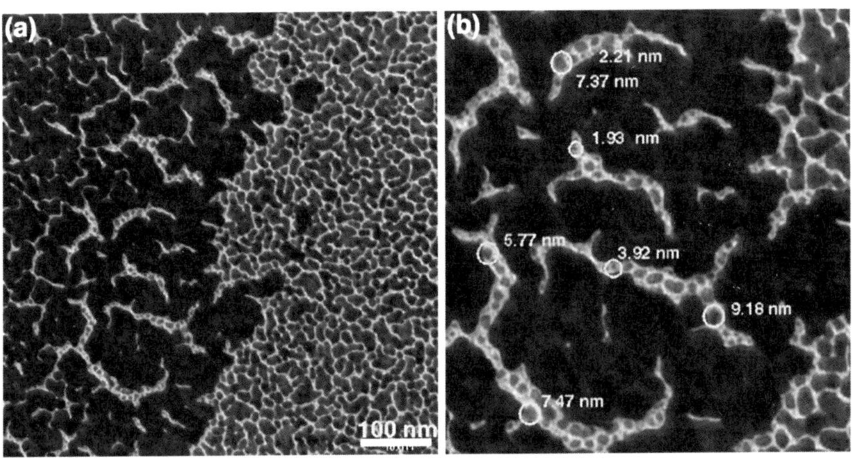

Fig. 8.8 **a** TEM picture of a Au film deposited in argon at 70 Pa. On the *left hand side*, a region at advanced stage of island interconnection. A magnified view of a portion of this region (**b**) highlights the presence of isolated NPs on the *bottom* of inter-island channels at advanced filling stage. On the *right hand side*, coalesced, irregularly shaped islands, with average size between 10 and 20 nm separated by a network of thin channels

island channels are rather short, while their width can be as large as about 20 nm. There is clear evidence of NP deposition on the denuded substrate at channel bottom (Fig. 8.8b). The size of such NPs has to be compatible with the channel width: the shape is spherical for the smallest ones, sized around 2 nm, and progressively distorted to elliptical when coalescence begins. It is noteworthy that such particles most likely had a low kinetic energy, insufficient even to move across the small distance, of the order of channel width, to aggregate themselves to pre-existing islands. As to the region on the right-hand side of Fig. 8.8a, it consists of a dense, uniform coverage of the substrate with coalesced, irregularly shaped, Au islands, sized between about 10 and 20 nm at most, separated by a network of thin channels.

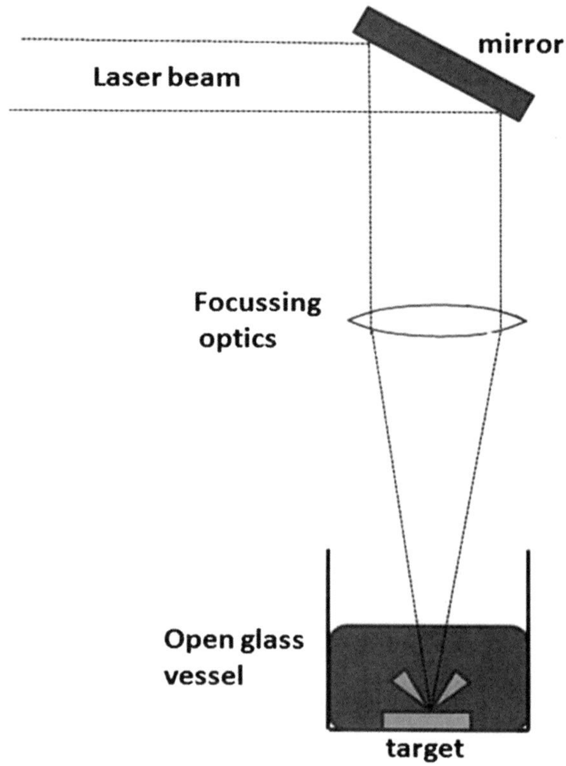

Fig. 8.9 Schematics of the setup for PLA in liquid

8.5 Nanoparticle Production Via Pulsed Laser Ablation in Liquid

Pulsed laser ablation in liquid (PLAL) is a simple, chemically clean and environmental friendly technique to prepare nanoparticles. The essentials of the process can be illustrated referring to Fig. 8.9. A short, high power laser pulse is focused through a lens to a small spot onto a solid target placed in a holder and submerged under variable levels (several millimeters) of a transparent liquid. The liquid-containing vessel may be open, or it may have a window flush with the liquid to prevent splashing. Laser pulses are fired at the target for a time typically ranging from minutes to hours to ablate it. Interaction of light with the target surface causes vaporization of the target material and of a small amount of the surrounding liquid. The spatially confined vapor is rapidly converted to plasma whose initial ionization degree is near unity; such an ambient contains highly excited species, both electronically and translationally [49] so that chemical reactions between the ablated species and molecules of the liquid are favored. The interaction of radiation with a solid target is largely similar when it occurs in vacuum, or at the interface between a solid and a fluid of increasing density, from a diluted gas to a liquid. The qualitative difference concerns plasma expansion that is more and more hindered with increasing fluid density. In particular, even more than in the case of a high-pressure gas that we discussed at length before, the liquid retards plasma expansion, resulting in strong confinement, thus in high plasma pressure and temperature. The high temperatures (of the order of thousands of Kelvin) and pressures (in the GPa range) provide favorable conditions to the synthesis of NPs typically composed of atoms from both the target *and* the liquid, whose composition, structure and morphology belong to exotic, often conventionally unattainable regions of the material phase diagram. Such NPs initially constitute a suspension in the liquid, but after accumulation they result in a colloidal solution. Prolonged interaction with the laser radiation can occur, leading to further changes in NP composition, size, or morphology [50].

The most relevant features of PLAL are: (i) the final product is free from remnants of precursors, intermediate reaction products, reagents, thus no further purification is needed; (ii) the experimental setup is simple and low cost; (iii) the extreme confinement conditions and the resulting high-temperature, high-pressure volume where the process occurs favor the formation of unusual metastable phases; (iv) both the solid target and the liquid are vaporised, so the product can contain atoms from the target material and from the liquid.

These features allow for specific target-liquid combinations to fabricate compound nanostructures with *ad hoc* designed functions. Nevertheless, compared with other NP generation techniques, this laser-based method is limited by low productivities for longtime generation processes, and optimal parameters should be used to draw upon this technique.

Focusing like in the previous Sections on metallic NPs, the main goal is to obtain colloidal solutions with tailored NP concentration and distribution. Thus, it is vital on the one hand to have an efficient control of the ablation process that determines the structural properties of the nanoparticles, on the other hand to know the details of the influence on NP size and size distribution of the surrounding matrix, or of the solvent in which the nanoparticles are dissolved. When laser processing is performed in vacuum or in a gas, the energy density at the target is usually determined by the nominal deposited pulse energy and the ablation spot area. However, when the laser beam propagates from air to a liquid phase and through the liquid, several changes in the incident laser pulse energy and wavelength occur. In particular, ultra short, highly energetic laser pulses undergo self-focusing and filamentation. Thus, when NPs are generated during the ablation process, absorption and nonlinear interactions are enhanced due to the large linear and nonlinear absorption coefficients of the colloid. Such effects limit the ablation efficiency, though they are rarely considered and careful analysis of the results is required [51]. It has been shown that the ablated mass increases with the energy density at the target so that high intensities are desirable to increase the NP production. Yet, when high energies are used, even determining the focal plane in the liquid via the ablation spot is not straightforward. For example, when Au NPs are prepared in water using nanosecond laser pulses, just changing the liquid thickness requires refocusing and repetition of focal plane determination [52]. Both

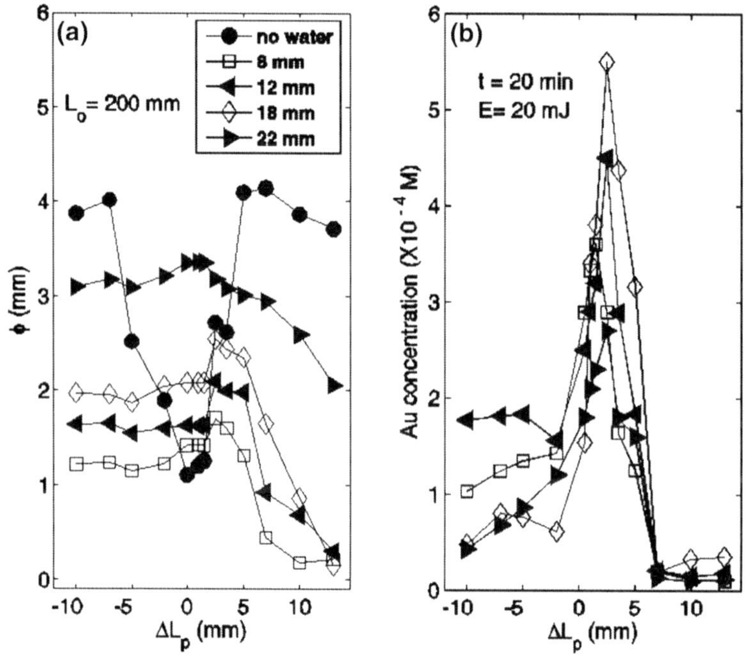

Fig. 8.10 Effect of lens position and water layer thickness on **a** the ablation spot diameter φ and **b** the concentration of Au NPs

the water thickness above the target and the lens position affect the ablation spot diameter that, in turn, determines the operative energy density. From Fig. 8.10a, ablating in air the smallest ablation spot diameter φ is obtained when the distance lens-target L_p coincides with the lens focal length L_o. When the process takes place in water, the ablation spots remain almost unchanged until $\Delta L_p = 3$ mm, being $L_p = L_o + \Delta L_p$, then progressively φ decreases with increasing ΔL_p. A similar trend was observed for different thicknesses of the water layer above the target.

To find the best conditions, in terms of lens position and water layer thickness, to maximize the NP production yield, the ablation process was carried out varying ΔL_p between -10 and $+10$ mm while keeping fixed the laser pulse energy at 20 mJ and the ablation time at 20 min. The Au NP concentration was deduced from the intensity of the absorption band, at about 520 nm, due to the surface plasmon resonance (SPR) of Au NPs in the UV–Visible spectra of the colloidal solutions [53]. As shown in Fig. 8.10b, the largest amount of Au NPs was obtained for $\Delta L_p = 2.5$ mm, when the ablation process was performed with a water layer above the target 18 mm thick. At variance with common observations in vacuum, or in gaseous atmosphere, the largest Au NP concentration is *not* obtained when the laser spot is smallest (in this experiments it was found at about $\Delta L_p = +10$ mm) and it sensitively depends on the water layer thickness. The result can be explained considering that the laser-generated plasma strongly

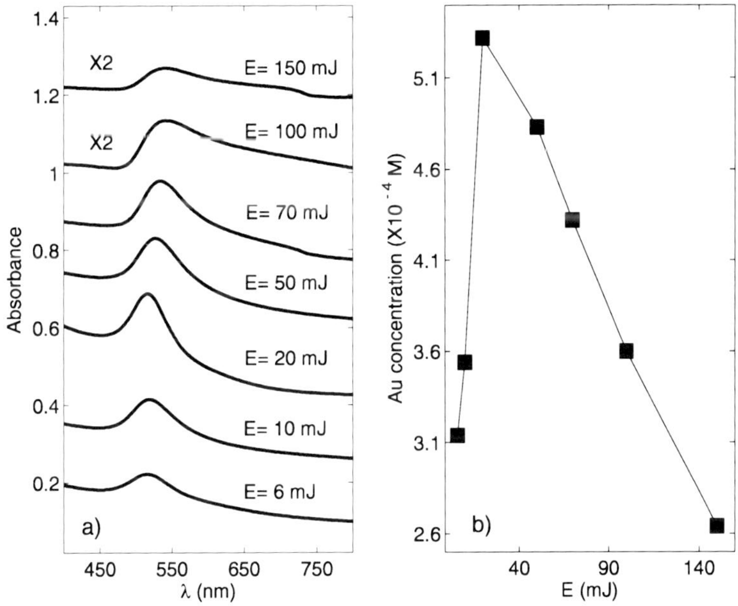

Fig. 8.11 **a** Surface plasmon resonance of Au NP colloids prepared at different laser energies per pulse; the absorbance spectra are shifted for clarity. **b** Au NP concentration versus laser pulse energy E

scatters and absorbs the incoming pulses; only a fraction of each energy pulse reaches the target: thus, the ablation efficiency is drastically reduced [51]. Once the best focusing and water layer conditions were identified, the role of the laser pulse energy to the process was investigated. The ablation parameters for which the highest NP production was found were fixed and ablation was carried out changing the pulse energy between 6 and 150 mJ. The SPR of the colloids so prepared are shown in Fig. 8.11a. They are characterized by the just discussed feature around 520 nm that slightly red shifts when the pulse energy increases; the overall absorbance decreases and a new contribution to absorption around 750 nm is observed, presumably associated to a *clustering* process. From Fig. 8.11b we see that the concentration of Au NPs initially increases up to the value of 5.5×10^{-4} M (corresponding to the laser energy of 20 mJ), then progressively decreases: indeed the extreme pressure and temperature conditions at the interface between the expanding plasma and the confining liquid may favor the fragmentation of existing NPs that will rearrange in clusters [54]. The observed considerable decrease of NP concentration, deduced from optical absorption, reflects such a structural rearrangement.

To gain a direct look at the nanostructuring processes, the morphology of Au products was investigated by TEM imaging. In Fig. 8.12, TEM pictures recorded from all the samples prepared at different laser pulse energies are shown. When the pulse energy is 10 mJ a uniform distribution of nearly spherical NPs smaller than

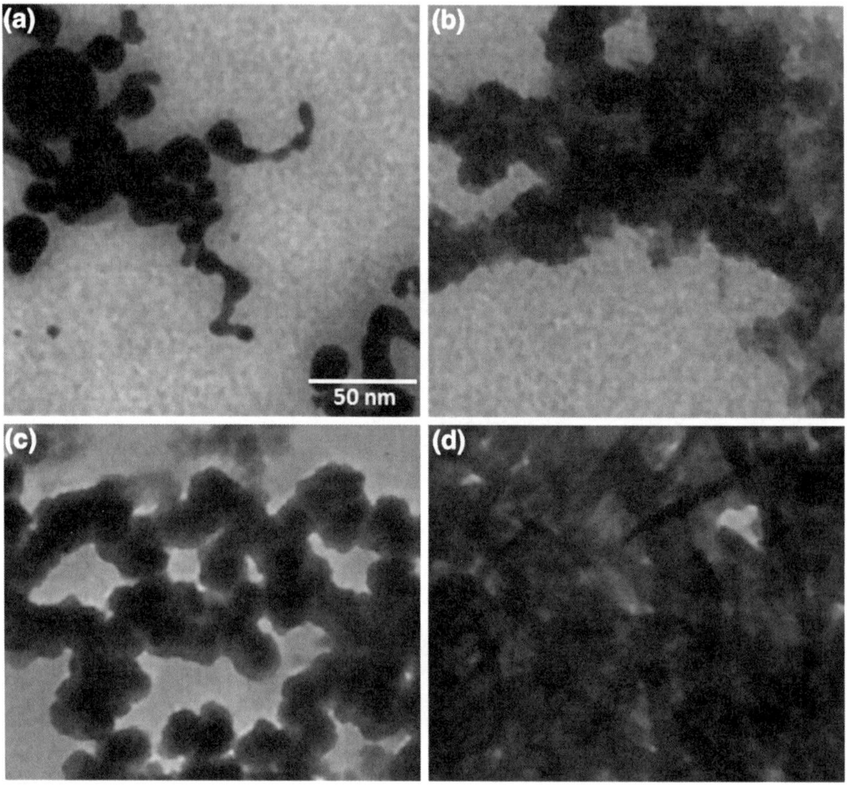

Fig. 8.12 TEM images of samples prepared at different laser pulse energies: **a** 10 mJ; **b** 20 mJ; **c** 50 mJ; **d** 100 mJ

10 nm in diameter is observed, besides some particles larger than 30 nm (Fig. 8.12a). A high density of NPs whose average size reduces to 3–5 nm is found in the sample prepared at 20 mJ (Fig. 8.12b). Increasing the laser energy at 50 mJ (Fig. 8.12c), a re-structuring of spherical NPs sized about 50 nm in chain-like agglomerations is evident. Finally, at the highest energy (100 mJ), some portions of the sample are characterized by columnar-shaped nanostructures (Fig. 8.12d). Overall, the optical response of Au NPs can be tailored by changing the morphological properties of Au nanostructures that, in turn, strongly depend on the ablation parameters, the water layer thickness above the target, the lens position and the laser pulse energy.

Similar trends were observed ablating with femtosecond pulses. In this case, also the focal lens was changed [51]. As previously described, when the thickness of the liquid layer increases, the displacement of the lens with respect to its original position (the position at which the focal plane in air coincides with the target position) should be increased. To explore this point, the laser beam was focused by different focal lengths from 40 mm up to 200 mm: the focusing conditions in the liquid with respect to air differ much and the relative distance between lens and target must be reduced. This is explained considering the refraction at the air–liquid boundary, the liquid vaporization, the self-focusing and the optical breakdown in the liquid. All these phenomena contribute to favor laser-induced welding/sintering which appears to be a promising approach to solder NPs; available examples include Au and Pt [55] as well as Au and TiO_2 NPs [56]: the properties of both hybrid systems are considerably enhanced with respect to those of the elemental constituents.

A comprehensive mechanistic scenario of PLAL and of all the combinatorial library of constituents and interactions between them, including photons, liquid molecules, the solid target, and laser-produced particles was proposed in terms of two main phenomena, shock wave emission and laser-induced bubbles, and their influence on NP formation [57]. After passing through the liquid, the focused laser beam irradiates the solid surface. The solid surface absorbing laser energy emits SWs to relax the excess energy. Two distinct compressive waves are emitted and propagate into both the solid and the liquid. In the solid, a rapid expansion immediately follows the SW. In the liquid, the shock front changes the refractive index, so that the shock front can be observed. The SW carries energy defined as the energy flux cross the area where the SW arrives:

$$E_s = \frac{4\pi r_s^2}{\rho_l} \int_0^{t'} \frac{(P_s - P_l)^2}{U_s - \left(\frac{P_s - P_l}{\rho_l U_s}\right)} dt \tag{8.9}$$

where r_s is the radial distance of the shock front from the origin, P_s is the SW pressure, U_s is the shock front velocity, t' is the time at the tail of the SW and t is set as zero at the shock front. The SW dissipates a considerable fraction of the energy during its propagation through the liquid. Across the SW the liquid is heated. In particular, the SW may induce acoustic cavitation bubbles by travelling across the previously perturbed liquid by photothermal heating and/or photodissociation along the laser path. Besides this, the deposition of photon energy on the target surface gives rise to hot carriers, electrons or electron–hole pairs at a temperature T_e. However, the pathways to ablation change depending on the pulse duration τ_L, the energy density F, and the target properties. At irradiances above a threshold I_p (on the order of about 10^{13} W/cm^2 for metals and semiconductors), the laser pulse induces optical breakdown of the solid and direct solid-plasma transition. Well below I_p, which is the case for most PLA experiments using nanosecond lasers and for some picosecond and femtosecond lasers, both thermal and non-thermal channels may lead to ablation. Two critical time scales are worth mentioning, namely the electron–phonon equilibration time τ_E (about 10^{-12}–10^{-11} s) and the liquid–vapor equilibration time τ_{LV} (about 10^{-9}–10^{-8} s).

8.6 Nanoparticle Synthesis Using fs Laser Pulses

When ablation is performed using fs laser pulses NPs are often observed to form during plasma expansion *in vacuum*: thus the morphology, stoichiometry and nanostructure of the resulting films differ much from those observed with ns laser deposition. Generally the films consist of randomly stacked NPs, typically sized between 10 and 100 nm, and rather narrow size distributions [13]. Some essential features of the interaction between an ultra-short fs pulse and matter are useful to interpret NP formation. The energy released to the target by the pulse is spatially highly concentrated in the immediate surroundings of the laser spot because the interaction time is too short to activate the thermal conduction channel. Consequently a small volume of matter is brought to supercritical conditions while still keeping a density typical of the solid state. The extreme temperature—pressure regions of the phase diagram thus attainable, together with the associated fast quenching associated to material expulsion from the target allow to synthesize otherwise unattainable metastable, exotic phases. Hydrodynamics indicates that the relevant strain underwent at high strain rate during the fast expansion of the high-pressure fluid results in mechanical fragmentation: this mechanism is of basic importance to quantitatively interpret NP formation [58]. Two schemes are available for NP synthesis: besides direct cluster ejection from the target following material disruption by a laser-induced explosion-like process, atom aggregation in the flying ablation plume via a collisional mechanism was proposed [59]. All models and simulations move from the features of fs pulse interaction with the target, an initial ultra-fast heating without changes of matter density, followed by an ultra-rapid expansion and cooling. Hydrodynamic simulations [60] indicate that a thermal wave is followed by an abrupt pressure increase and a propagating shock wave. Molecular dynamics (MD) simulations [61, 62] underline mechanisms including phase explosion, fragmentation, evaporation and mechanical spallation. Two thresholds were found in the model system Ni [62]: spallation follows extended defect formation and it proceeds via void nucleation, growth and coalescence, while phase explosion requires material overheating, after which the material suddenly decomposes in a mixture of vapor and liquid droplets: in Ni the order of magnitude of the time needed before the onset of phase explosion is around 150 ps.

The evolution of a metal film after a localized photoexcitation was simulated depositing a single ultra-short, 200 fs pulse on a small region with Gaussian spatial profile and FWHM of 10 nm, of a Ni slab with diameter of 250 nm and thickness of 20 nm; the corresponding deposited energy density was 3.1 J cm^{-2} [63]. Highly off-equilibrium conditions are set and a sequence of fast phase transformations is observed. First a central nanosized region undergoes a large overheating and a homogeneous transient melting. Such a region protrudes out of the irradiated surface like a bump. The estimated value of the peak pressure within this region reaches up to 8 GPa in 2.5 ps. Relaxation of the temperature gradients via fast two-dimensional electron heat conduction leads to an initial cooling rate of the small melted volume of about 10^{13} K s^{-1}, resulting in deep undercooling, with a liquid–crystal interface that advances at about 80 nm ns^{-1} during the cooling down of the melted region. Around 0.6 T_m (T_m, melting temperature) extended nucleation of crystallites occurs within the undercooled liquid.

Dedicated diagnostics of the ablation plasma produced by ultra-short pulses were essentially performed on elemental targets during irradiation in vacuum with visible to near-IR light at energy densities from intermediate to modest and power densities between a few 10^{12} and a few 10^{14} W cm^{-2}. In metals (Ti) irradiated at energy densities around 0.7 J cm^{-2}, time-of-flight mass spectrometry (TOF–MS) indicates two components of the ion velocity distribution: the kinetic energies of Ti^+, in the keV range, are taken as a marker that Coulomb explosion can significantly contribute to ablation [64].

Since less than 10 years time-resolved optical emission spectroscopy (OES) measurements are coupled to fast imaging of the expanding plasma produced by ultra-fast irradiations. The results collected on several elemental metals yield a consistent picture: at short delay a plume propagates along the normal to the target with narrow angular aperture; the constituents of such a plasma are *atomic* species, both ions and neutrals. Careful velocity measurements show that ions are always faster, with velocities from a few 10^4 to 10^5 ms^{-1}, being strongly coupled to electrons that escape first from the target; atoms are slower by one order of magnitude. We have data for Ti, Zr, Hf [65], Au, Cu [13], Al, Si [66]. We can interpret the strongly directed plasma expansion along the normal to the target surface considering that the pulse duration is shorter than the shortest coupling time between elementary target constituents, electrons and ions (τ_{e-i}), so that plasma forms after laser pulse termination. The relevant initial plasma pressure caused by the strong over-energization drives such a preferential expansion. Notably, this phenomenology was observed also in fs laser produced plasmas from a graphite target, expanding through molecular nitrogen at high pressure ($\sim$53, $\sim$132, $\sim$660 Pa) [67]. In these experiments, besides spectrally integrated plume pictures, also images of C_2 and N were recorded to monitor the evolution of those molecular precursors relevant to the synthesis of carbon nanostructures. C_2 emission is brightest at distances not larger than 1 mm from target surface, at early times, similarly to what is observed when fs ablation is performed in vacuum. It is attributed to direct ablation, or to C_2 formation in three-body-collisions until the plasma is most dense. At longer times C_2 emission is brighter in plume wings, where less fast species are found: indeed the lower temperature in plasma wings favors three-body recombination, whose rate scales with $n_e^2 T_e^{-9/2}$. CN emission intensity, besides at plume wings, is maximum at plume front where its dissociation is less likely to occur than that of C_2, due to the higher dissociation energy of CN (7.72 eV), compared to C_2 (6.27 eV).

In fs ablation of metal targets, a high intensity plume that remains in contact with the target even for times as long as microseconds was studied in detail in the case of Au and Cu irradiations [13]. The expansion of this plasma was attributed to the combined effect of particle size and temperature gradients: smaller, hotter

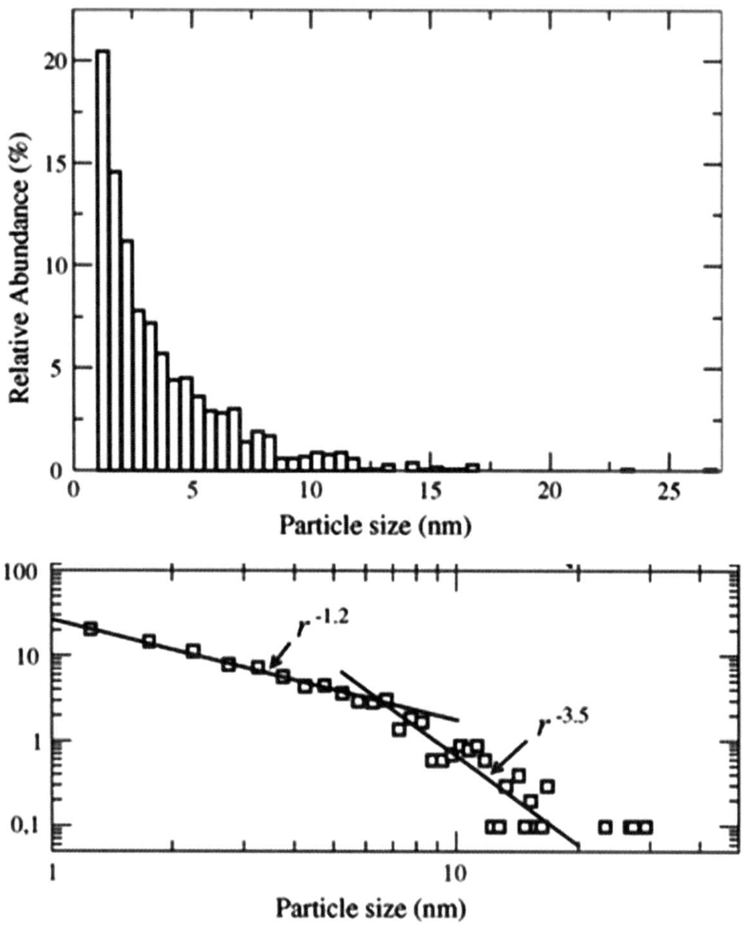

Fig. 8.13 AFM measurements from a deposited surface 2 × 2 μm^2 of size and size distribution of Cu NPs ablated with a Ti:Sapphire laser (wavelength 800 nm, pulse duration 100 fs, repetition rate 1 kHz, energy density 4 J cm^{-2}) (adapted from [13])

particles are grouped in plasma front, similarly to what is observed with nanosecond laser pulses.

In a few cases the size and size distribution of the *dispersed* deposited NPs were determined: as an example, in Fig. 8.13a atomic force microscopy (AFM) measurements on copper are reported. As to the self-structuring of the deposited NPs, whenever the deposited films were analyzed, SEM, or AFM show *random* distributions of NPs [13, 68]; in all these studies sub-monolayers were deposited to avoid NP self-organization on the substrate, most likely resulting in coalescence. Since the above observations are independent of target nature, whether it is a metal or a semiconductor, NP production appears to be a typical phenomenon associated to fs PLD. The velocities just discussed for the different species are nearly constant over quite large distances from the target, from 1 mm to centimeters [68]. The atom/ion to cluster velocity ratio was reproduced in a direct simulation Monte Carlo (DSMC) where molecular dynamics (MD) results were used as input parameters [69]. For both Au and Cu [13] around the threshold energy of ablation E_t, atom velocities increase when measured as a function of the deposited energy density E: the associated mechanism is the increased amount of ablated matter and the concurrent changes in the adiabatic expansion stage. By contrast, NP velocities decrease near E_t, due to changes in their size distribution. Beyond E_t, the velocities of both species are nearly independent of E. Considering atomization as the direct transformation of target material into gas phase, both in Au and in Cu the measured ablation efficiency is maximum at an energy density E_{NP} where atomization is minimum afterwards increasing E atomization slightly increases [13]. Such a strong correlation is attributed to the higher energy cost of atomization, compared to NP production. At energy densities larger than the ablation threshold E_t matter preferentially escapes from deep and cold layers lying more and more below the target surface, so that the amount of energy it absorbs is not enough to lead to complete atomization; instead a mixture of gas–liquid NPs leaves the target. As the ablation efficiency is much higher than thermal vaporization (pure atomization), a relevant fraction of NPs is expected to be *directly* ejected from the target.

The deposited NPs with sphere-like shape belong to *two* distinct populations, what is evident in Fig. 8.13b [13]: the smaller-sized ones are expected to be directly ejected from the target; the size distribution of the larger particles scales as $r^{-3.5}$, r being their radius. This distribution is observed when fragmenting collisions prevail, thus such NPs are likely to result from in-plasma collisional sticking.

Manipulation of NP size and size distribution, operating with delayed fs [70], or ns [71] pulses on the main plasma constituents was studied in a few instances. In particular, irradiating Ni [71] with fs UV pulses ($\lambda = 263$ nm) the average diameters of deposited isolated NPs, as measured by AFM, are nearly half of those obtained irradiating with identical pulses, with doubled wavelength ($\lambda = 527$ nm). In both cases particle size is independent of laser energy density, up to 1 J cm^{-2}. When a further ns-UV pulse intercepts the ablation plasma at different delays, changes are induced in the size distribution of the deposited NPs. At fixed, intermediate (about 0.5 and 0.4 J cm^{-2}, respectively) energy densities of the fs and ns pulse, the average NP size shrinks, this effect being more marked the shorter the delay. Such a result suggests that in the expanding plasma NPs group together as a function of their size, the smaller ones travelling ahead of the bigger ones.

At very long delays, of the order of tens of microseconds, when irradiation is performed at high energy densities (about 10 J cm^{-2} and higher), slow *droplets*, whose velocity is about 1 ms^{-1}, are made visible to the naked eye by their luminous trajectories, to such debris correspond micron-sized particles evident in SEM pictures taken on the surface of the deposited films [72]. The large difference in size and velocity between such species and NPs indicates that droplets are not *fat* NPs, nor the two species are produced by the same mechanism: stress confinement in the target, associated to ultra-short pulses, was proposed as the droplet origin [59].

Whatever the target nature, a scaling among thresholds in laser energy density holds: the thresholds for atomic and nanoparticle emissions, E_a and E_{NP}, are comparable to each other and both are lower than the value for droplet emission, E_d [73]

$$E_d > E_a \approx E_{NP}. \tag{8.10}$$

Relevant experimental results on NP synthesis are correlated with an acceptable degree of agreement to simulations of target behavior under fs pulse irradiation. According to MD [59] at high values of deposited energy density E the leading mechanism of material expulsion from the target is phase explosion of matter heated above the thermodynamical critical temperature T_c. The surface region decomposes into a foam of interconnected liquid regions, containing gas molecules, liquid droplets and small clusters. The liquid fraction increases with increasing depth from the target surface, thus at a progressively lower degree of target overheating, and big droplets form in the tail of the ablation plume. In turn, NPs segregate in different regions of the expanding plasma depending on their size: the smallest ones group together in the front region, the medium-sized ones in the middle. Two different synthesis mechanisms are found: smaller particles are ejected in the explosive decomposition of matter into liquid and vapor, larger ones are the outcome of decomposition and coarsening of transiently interconnected liquid volumes.

Again, two channels of NP production were identified combining smaller-scale MD and larger-scale DSMC techniques [69]: direct ejection, following laser—matter interaction and collisional condensation–evaporation in the flying ablation plume. The processes belonging to the first channel include, besides phase explosion, photomechanical spallation and fragmentation; such volume mechanisms produce NPs *and* atoms/ions. Sticking collisions and evaporation in gas phase are the processes of the second channel: they are similar to those occurring in aggregation sources [74], being favored by the large number of seed molecules and tiny particles in a laser-produced plasma.

8.7 Nanoparticle-Assembled Surfaces with Directed Artificial Roughness: Selected Applications

Moving from the pioneering analysis performed by Mie at the beginning of the twentieth Century on the optical properties of a nanometer sized metallic sphere [75] plasmonics has grown in the last decade as an independent research field, once the possibility to synthesize and assemble together metal NPs supported on suitable substrates became feasible. Within this wide framework we focus on a problem of sensitivity enhancement associated to the morphological peculiarities of surfaces consisting of NP assemblies.

Raman scattering is a powerful, non-destructive, rapid investigation technique commonly adopted to detect molecular species. Indeed each molecule bears its own characteristic spectrum that is a fingerprint of its presence. Unfortunately, since the scattering process is second-order, the typical cross sections are in the range of 10^{-30} cm^2 mol^{-1}; besides this, often fluorescence obscures molecular Raman features. A relatively recent development of the technique is Surface Enhanced Raman Spectroscopy (SERS), based on the giant enhancement of Raman signal intensity, up to a factor of 10^{12}–10^{14} recorded from molecules adsorbed on *roughened* surfaces of some noble and transition metals. Two mechanisms for the enhancement were proposed and are currently accepted. The electromagnetic effect (EM) is due to the strong amplification of the local EM field in the presence of corrugated metal structures, while the chemical effect involves the creation of new electronic states generated by the interaction between the metal and the molecules adsorbed on it. Such new electronic states allow for resonant Raman scattering processes. A first approximation to the EM enhancement Q is the so-called E^4 approximation [76]

$$Q_{SERS} = \left| \frac{E(r, w_0)}{E_0(w_0)} \right|^4, \tag{8.11}$$

where $E(r, w_0)$ is the local electric field at the molecule position and $E_0(w_0)$ is the field associated to the incident radiation. This approximation provides a good estimate of the radiation enhancement, still neglecting polarization and selection rule effects.

Soon after the discovery of SERS effect it was recognized that the morphology details of the substrate play a crucial role to the Raman intensity enhancement. In particular, for *artificially* corrugated surfaces made of NP assemblies the enhancement of spectrum features is highly sensitive on the size, shape and mutual aggregation of NPs. Thus, controlling these parameters when growing nano-structured metallic surfaces is a major issue when pointing at the selective detection of *ultra-low* molecular concentrations, down to the single-molecule identification [77]. Also, it is important that resonance conditions are fulfilled between the exciting radiation and the surface plasmon of the system. It is comparatively much easier·to accomplish this by performing UV-Visible measurements. In the following we discuss the optical properties, in particular the UV-Visible spectra and the SERS performance of Ag and Au artificially roughened substrates obtained depositing on glass metal NPs synthesized in inert gas at different pressures and laser pulse number (see Sects. 8.3, 8.4).

We begin focusing on substrates covered with Ag NPs which were tested for SERS. The SERS activity of such NP assemblies, grown at different deposition conditions, with highly differentiated morphologies was first tested by soaking them for 1 h into aqueous solutions of a reference dye such as rhodamine 6G (Rh6G), at different concentration levels. SERS spectra as shown in Fig. 8.14a were obtained for samples deposited in Ar at 10, 40 and 70 Pa, with laser pulse number fixed at 1.5×10^4, and Rh6G concentration 1×10^{-4} M. Well-defined

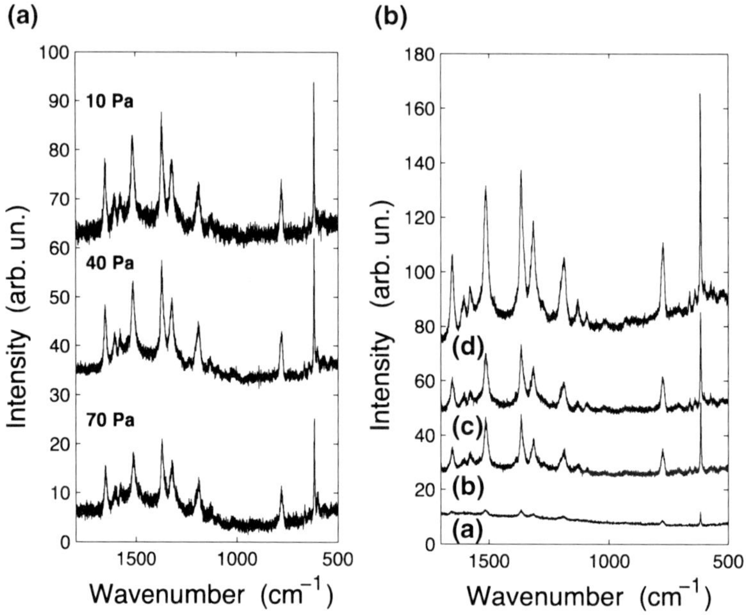

Fig. 8.14 **a**, **b** SERS of Rh6G from Ag substrates **a** deposited in Ar at different pressures with 1.5×10^4 pulses at 1×10^{-4} M conc. **b** deposited in Ar at 70 Pa, with 3×10^4 pulses at *a* 5×10^{-8} M, *b* 5×10^{-7} M, *c* 2×10^{-6} M and *d* 1×10^{-4} M conc

peaks of Rh6G are seen. Looking at this set of samples, the SERS efficiency slightly increases for any Rh6G concentration with decreasing Ar pressure. This inference has been made by taking into account the intensity of the signature (most intense) peak of Rh6G at 615 cm^{-1}. Also, substrates deposited at 10 Pa show highest SERS activity among the set. In Fig. 8.14b, one can observe a second set of SERS spectra which was taken for samples deposited at 70 Pa of Ar, with 3×10^4 laser pulses, at different concentrations of Rh6G i.e. 5×10^{-8}, 5×10^{-7}, 2×10^{-6} and 1×10^{-4} M. The characteristic peaks of Rh6G are evident: though the intensity of the peaks reduces as we move to lower concentrations, the spectrum at 5×10^{-8} M can be identified. The SERS activity of the substrates deposited at various parameter values is strictly related to the surface morphology of the nanostructured films *and* the radiation line used to excite the surface plasmon. Taking these two factors into account, a general explanation can be given for the observed SERS activity. For the case of Fig. 8.14a, as we decrease Ar pressure, there is a progressive increase in the fraction of the substrate surface covered by NPs. At 70 Pa, the surface is covered by isolated (well-separated) spherical Ag NPs and hence the SERS activity is the least. Moreover, a coupling of the excitation line (632.8 nm) with the surface plasmon resonance (SPR) of the substrate deposited at 10 Pa enhances the SERS activity as shown in Fig. 8.15a. Indeed, SPR maximum red shifts and slightly broadens with lowering Ar pressure.

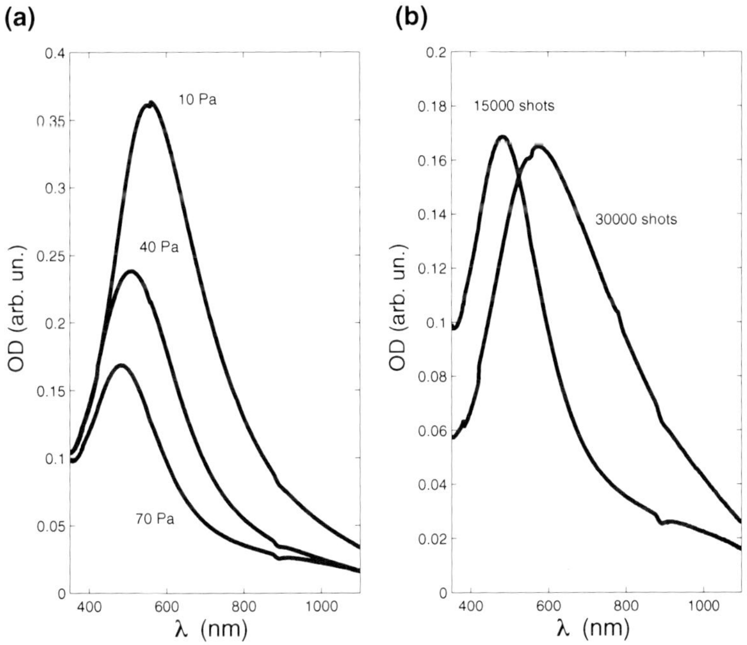

Fig. 8.15 UV-Vis spectra from Ag substrates **a** deposited in Ar at different pressures with 1.5×10^4 pulses; **b** deposited in Ar at 70 Pa, with 1.5×10^4 and 3×10^4 pulses

In the case of Fig. 8.14b, higher SERS activity is observed for the sample deposited at 3×10^4 laser pulses compared to that deposited at 1.5×10^4 pulses since a higher density of NPs is deposited on the substrate in the first case and hence a dense packing of clustered NPs results, still without entering a coalescence stage. The size of the NPs remains the same since they were produced in the same plasma that was confined by the same gas (Ar) at the same pressure (70 Pa). In addition to this, the SPR of the film prepared at 3×10^4 pulses has a larger red shift with respect to the SPR of the film prepared at 1.5×10^4 pulses, which indicates a strong electronic coupling between closely spaced NPs as seen in Fig. 8.15b. A NP in the proximity of another, results in coupling of the plasmon resonance: in such cases the SPR red shifts and the EM effect enhances at the junction between the two adjacent NPs. The effect prevails for closely spaced, clustered, but not yet coalesced NPs: such enhanced region of SPR is called a *hot spot*. In the substrate grown at 70 Pa and 3×10^4 laser pulses, the number of hot spots is large as compared to the substrate grown at 1.5×10^4 laser pulses, hence the SERS activity is enhanced in the former. Moreover in this condition (70 Pa, 3×10^4 laser pulses) the array of NPs is arranged in such a way as to enhance the EM field via dipolar–dipolar interaction as demonstrated in a study performed on the same set of substrates [78].

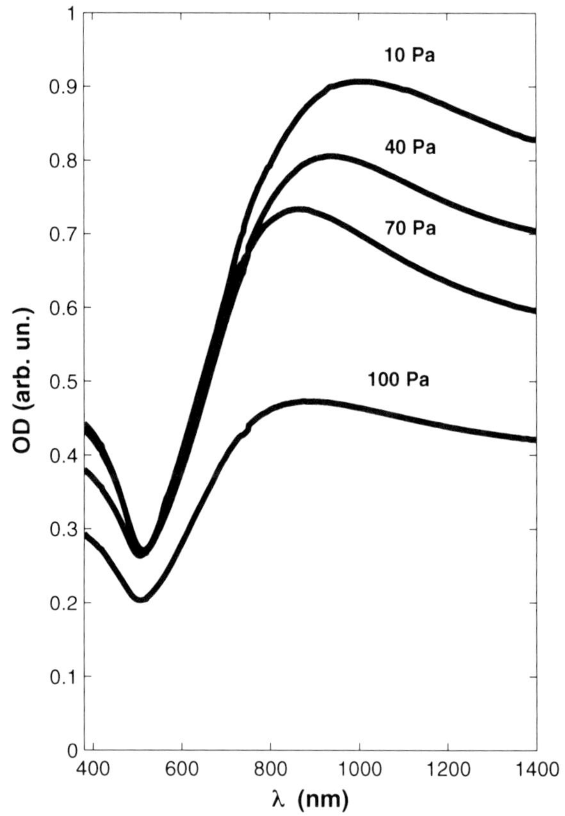

Fig. 8.16 UV-Vis spectra of Au substrates (deposited in Ar, with 3×10^4 pulses)

With an understanding of the SERS activity of substrates made of tailored Ag NPs, we move to the study of SERS activity using Au NPs. Au is chosen since it is non-toxic to blood and hence it offers a possibility for in vivo biological applications. We studied the SERS spectra of apomorphine that is a pharmaceutical drug used as a dopamine agonist in the treatment of Parkinson disease.

We consider Au substrates produced in Ar at 10, 40, 70 and 100 Pa keeping the number of laser pulses fixed at 3×10^4. First we take a look at the trend of UV-Visible SPR absorption peak on increasing Ar pressure (Fig. 8.16): a red shift of the frequency in the absorption curve is observed as expected since substrate morphology changes from coalesced nanoparticles to single, isolated, spherical NPs to generate the EM enhancement at their junctions. For our purpose of studying apomorphine, we have chosen the substrate deposited at 100 Pa since the enhancements are quite large. The plasmon frequency peaks at around 850 nm, hence we use the 785 nm excitation line from a diode laser. This ensures that the surface plasmons are well excited to enhance the weak Raman signals. In Fig. 8.17, we report the SERS spectra obtained by dipping the substrate in 10^{-6}, 10^{-5} and 10^{-4} M of apomorphine solution in distilled water. It is worth noticing

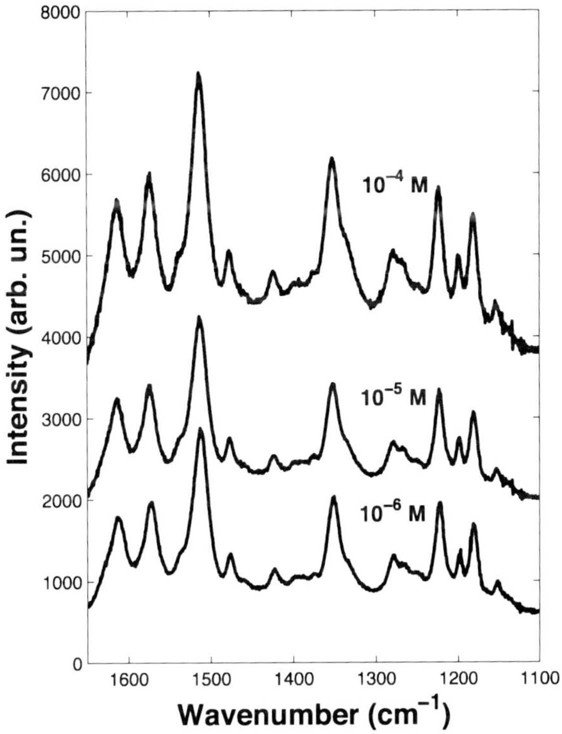

Fig. 8.17 SERS of apomorphine from Au substrates (deposited in Ar at 100 Pa, with 3×10^4 pulses)

that the spectra are reproducible at different concentrations, which is very unlikely when spectra are taken using colloids since apomorphine is quite complicated in its mechanism of protonation in water. Also, it is noteworthy that the control of apomorphine plasma level is not easy and presently it is difficult to reach a concentration of 10^{-6} M [48].

Another field of application of the SERS substrates produced by PLD, is to identify organic dyes in artworks for their dating, conservation and restoration. The main limitations encountered with these dyes when using other characterization techniques is that X-ray fluorescence is missing elemental signature, Fourier Transform Infrared (FT-IR) spectra contain also spectra of other substances present with the dyes i.e. binders and extenders, UV-Visible spectroscopy is often unable to selectively distinguish different chromophores. Thus, micro-Raman spectroscopy is a good candidate to detect and provide the chromophore fingerprint. Yet, due to low Raman scattering cross-sections and a fluorescent background, it is vital to enhance Raman signals and to obtain clear and well-resolved spectra: hence SERS offers a solution to the above difficulties.

As an example, we consider garanza lake: this is a dye having vegetal origins that is extracted from the root of the *Rubea tinctorum* plant. It is constituted of two different chromophores: alizarin and purpurin. UV-Visible spectroscopy of garanza lake shows weak absorption even at highest concentration (5×10^{-3} M) as

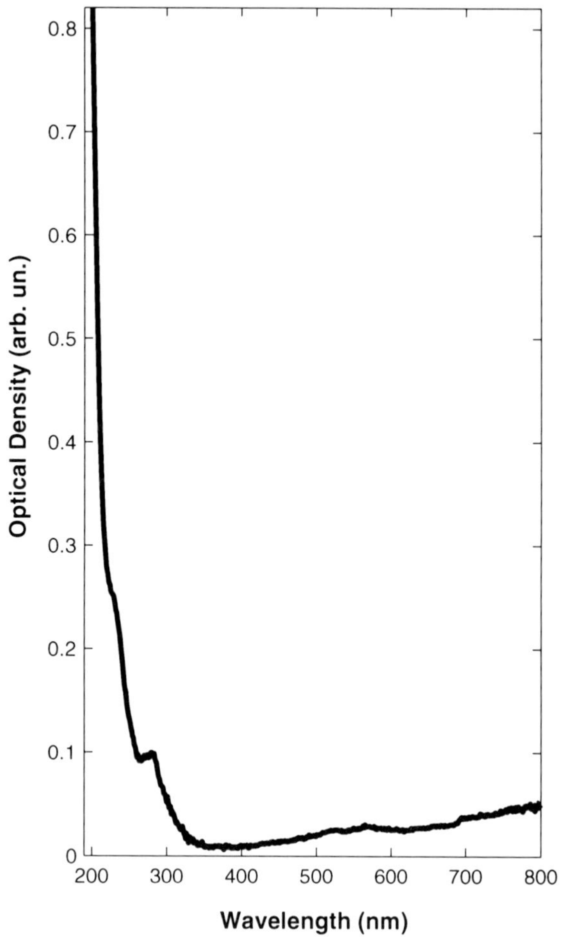

Fig. 8.18 UV-Vis spectrum of garanza lake at 5×10^{-3} M conc. (Ag substrates deposited in Ar at 70 Pa, with 3×10^4 pulses)

shown in Fig. 8.18. From the spectrum, a selective identification of the two chromophores is impossible. Raman spectra were collected using the 632.8 nm excitation line after dipping a bare glass substrate in a garanza lake solution of 5×10^{-3} M concentration: only a fluorescence background is observed in Fig. 8.19a. The SERS substrates are Ag NPs deposited in Ar at 70 Pa and 3×10^4 laser pulses. The SERS spectra obtained by soaking the Ag nanostructured substrates in aqueous solutions at different concentrations starting from 5×10^{-3} M down to 5×10^{-5} M show several peaks in the 600–1,800 cm^{-1} region. The peaks can be assigned to the components of garanza lake, as shown in Fig. 8.19b: alizarin vibrational peaks were attributed at 1,194, 1,286 and 1,559 cm^{-1} which are clearly visible and marked by squares [79, 80]. Peaks marked with an asterisk coincide well with purpurin features reported at 740, 980, 1,027, 1,218 and 1,398 cm^{-1}. Two weak bands are also observed at 582 and 1,007 cm^{-1}.

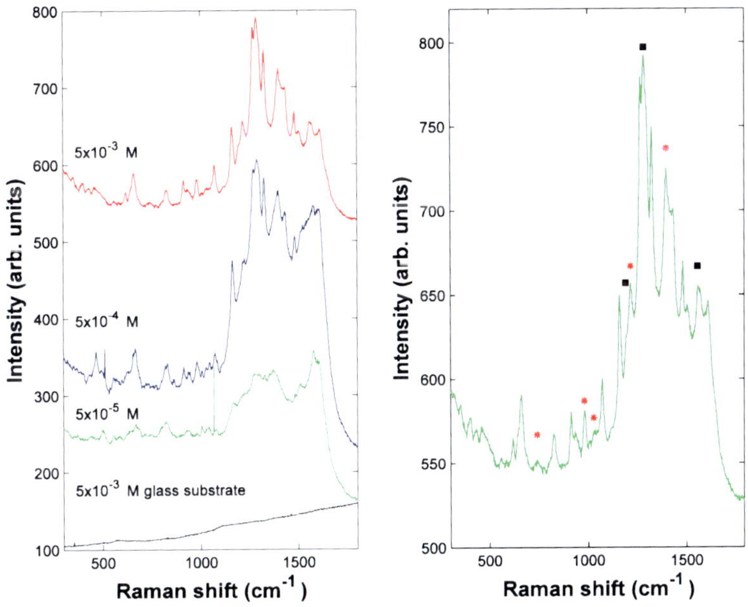

Fig. 8.19 **a** SERS of garanza lake at different concentrations from Ag substrates (deposited in Ar at 70 Pa, with 3 × 10^4 pulses); **b** assignment of peaks to alizarin (*black squares*) and purpurin (*red stars*)

Other kinds of organic dyes, including carmine lake and brazilwood have been studied and their SERS features were unambiguously allocated to their constituents [81], solving the ambiguities inherent in UV-Visible spectroscopy data.

8.8 Conclusions and Perspectives

Nanoparticles continue to attract interest for the possibility to play with their composition, structure and intrinsic properties, as well as for the potentiality to engineer novel materials where NPs act as structural, or functional blocks, or both. In this chapter the role of laser ablation as a powerful tool to synthesize NPs in a controlled way was discussed. Different available experimental setups allow for producing NPs covering a range of structures, morphologies and sizes, often with narrow distributions. The subsequent step of NP self-assembling to give rise to a NP array, up to a thin film is in turn significantly affected by the deposition parameters.

We discussed some selected applications referring specifically to elemental noble metal NP arrays that find application as excellent substrates for surface enhanced Raman spectroscopy (SERS), thus in the field of plasmonics. This is a specific application in the broader field of sensing. We reported on some applications in the field of art conservation, where the potentialities of NP-based substrates result in their performance in terms of high selectivity and sensitivity, two features vital to non-destructive analysis of artworks. Another application we discussed for NP-based substrates is in the recognition of drugs and proteins. We foresee that biomedical applications will become progressively more relevant, besides the established activity of gas sensing.

Also leaching of contaminants and adulterants in food industry is expected to demand high sensitivity, non-invasive small, fast-response sensors, thereby the use of *ad hoc* designed NPs is expected. This is a field where NPs found applications since the beginning of research activity on them.

Still in the field of plasmonics both Si and organic dye based thin film solar cells suffer from comparatively low efficiency. Metallic NPs, specifically Ag, or Au, act as anti-reflection or light-trapping layer, with an adsorption enhancing mechanism able to increase the overall device efficiency.

Metal NPs are largely used in heterogeneous catalysis, where they can act either as the catalysis site, or as a support for the catalytic process. The assessed synergistic effect on catalysis associated to the use of alloy NPs opens a perspective towards bimetallic systems, provided a control over the particle composition and nanostructure is achieved.

In metal matrix composites reinforced with ceramic particles that are used as structural materials in the automotive, aerospace and railway sectors the particle size has strong effect on the strength, ductility and failure mode. Reducing the ceramic particle size down to the nanometer level leads to substantial improvements in the mechanical performance of the material provided the NP distribution throughout the matrix is homogeneous. Such objective is still far from being readily achieved.

Any application of NPs requires a careful control on NP synthesis and self-organization. These goals can be reached by pulsed laser ablation under different conditions, with a separate control over the plasma production and propagation on the one hand and, on the other hand, on NP energy at landing on the substrate.

References

1. M. Grelczak, J. Vermant, E.M. Furst, L.M. Liz-Marzan, ACS Nano **4**, 3591 (2010)
2. P. Biermann, M. Harwitt, Astrophys. J. **241**, 105 (1980)
3. L. Boufendi, J. Hermann, A. Bouchoule, B. Dubreuil, E. Stoffels, W. Stoffels, M.L. De Giorgi, J. Appl. Phys. **76**, 148 (1994)
4. D.B. Geohegan, A.A. Puretzky, G. Duscher, S.J. Pennycook, Appl. Phys. Lett. **72**, 2987 (1998)
5. D. Scuderi, O. Albert, D. Moreau, P.P. Pronko, J. Etchepare, Appl. Phys. Lett. **86**, 071502 (2005)
6. T. Yoshida, S. Takeyama, Y. Yamada, K. Mutoh, Appl. Phys. Lett. **68**, 1772 (1996)
7. S. Eliezer, N. Eliaz, E. Grossman, D. Fisher, I. Gouzman, Z. Henis, S. Pecker, Y. Horovitz, S. Fraenckel, M. Maman, Y. Lereah, Phys. Rev. B **69**, 144119 (2004)
8. S. Amoruso, J. Schou, J.G. Lunney, Appl. Phys. A **92**, 907 (2008)

9. O. Albert, S. Roger, Y. Glinec, J.C. Loulergue, J. Etchepare, C. Boulmer-Leborgne, J. Perriere, E. Millon, Appl. Phys. A **76**, 319 (2003)
10. X.T. Wang, B.Y. Man, G.T. Wang, Z. Zhao, B.Z. Xu, Y.Y. Zia, L.M. Mei, X.Y. Hu, J. Appl. Phys. **80**, 1783 (1996)
11. Z. Zhang, P.A. VanRompay, J.A. Nees, P.P. Pronko, J. Appl. Phys. **92**, 2867 (2002)
12. D.B. Geohegan, A.A. Puretzky, G. Duscher, S.J. Pennycook, Appl. Phys. Lett. **73**, 438 (1998)
13. S. Noel, J. Hermann, T. Itina, Appl. Surf. Sci. **253**, 6310 (2007)
14. S.I. Anisimov, D. Bäuerle, B.S. Luk'yanchuk, Phys. Rev. B **48**, 12076 (1993)
15. Y.B. Zel'dovich, Y.P. Raizer, in *Physics of Shock Waves and High—Temperature Hydrodynamic Phenomena* ed. by W.D. Hayes, R.F. Probstein (Academic Press, New York, 1966)
16. A.K. Sharma, R.K. Thareja, Appl. Surf. Sci. **243**, 68 (2005)
17. T.E. Itina, J. Hermann, P. Delaporte, M. Sentis, Phys. Rev. E **66**, 066406 (2002)
18. R.F. Wood, J.N. Leboeuf, D.B. Geohegan, A.A. Puretzky, K.R. Chen, Phys. Rev. B **58**, 1533 (1998)
19. S. Amoruso, R. Bruzzese, N. Spinelli, R. Velotta, M. Vitiello, X. Wang, Phys. Rev. B **67**, 224503 (2003)
20. F. Neri, P.M. Ossi, S. Trusso, Riv. Nuovo Cimento **34**, 103 (2011)
21. D.B. Geohegan, in *Pulsed Laser Deposition of Thin Films* ed. by D.B. Chrisey, G.K. Hubler (Wiley, New York, 1994) p. 115
22. J. Gonzalo, C.N. Afonso, I. Madariaga, J. Appl. Phys. **81**, 951 (1997)
23. A.V. Rode, E.G. Gamaly, B. Luther–Davies, Appl. Phys. A **70**, 135 (2000)
24. N. Arnold, J. Gruber, J. Heitz, Appl. Phys. A **69**, S87 (1999)
25. A. Bailini, P.M. Ossi, Europhys. Lett. **79**, 35002 (2007)
26. A. Bailini, P.M. Ossi, A. Rivolta, Appl. Surf. Sci. **253**, 7682 (2007)
27. F. Neri, P.M. Ossi, S. Trusso, Laser Part. Beams **28**, 53 (2010)
28. F. Neri, P.M. Ossi, S. Trusso, Rad. Eff. Def. Solids **165**, 559 (2010)
29. C.N. Afonso, J. Gonzalo, R. Serna, J.C.G. de Sandre, C. Ricolleau, C. Grigis, M. Gandais, D.E. Hole, P.D. Townsend, Appl. Phys. A **69**, S201 (1999)
30. P.O. Jubert, O. Fruchart, C. Meyer, Surf. Sci. **522**, 8 (2003)
31. P. Ohresser, J. Shen, J. Barthel, M. Zheng, C.V. Mohan, M. Klaua, J. Kirschner, Phys. Rev. B **59**, 3696 (1999)
32. W. Marine, L. Patrone, B. Luk'yanchuk, M. Sentis, Appl. Surf. Sci. **154–155**, 345 (2000)
33. M.S. Tillack, D.W. Blair, S.S. Harilal, Nanotechnology **15**, 390 (2004)
34. D. Bolgiaghi, A. Miotello, P. Mosaner, P.M. Ossi, G. Radnoczi, Carbon **43**, 2122 (2005)
35. A. Bailini, P.M. Ossi, Carbon **44**, 3049 (2006)
36. F. Di Fonzo, A. Bailini, V. Russo, A. Baserga, C. Cattaneo, M.G. Beghi, P.M. Ossi, C.S. Casari, A. Li Bassi, C.E. Bottani, Catal. Today **116**, 69 (2006)
37. P.M. Ossi, A. Bailini, O. Geszti, G. Radnoczi, Europhys. Lett. **83**, 68005 (2008)
38. D.B. Geohegan, A.A. Puretzky, Appl. Surf. Sci. **96–98**, 131 (1996)
39. P.M. Ossi, A. Bailini, Appl. Phys. A **93**, 645 (2008)
40. R.K. Thareja, R.K. Dwivedi, K. Ebihara, Nucl. Instrum. Methods Phys. Res., Sect. B, Beam Interact. Mater. Atoms **192**, 301 (2002)
41. G. Dinescu, A. Aldea, M.L. De Giorgi, A. Luches, A. Perrone, A. Zocco, Appl. Surf. Sci. **127–129**, 697 (1998)
42. Z.W. Fu, Q.Z. Qin, M.F. Zhou, Appl. Phys. A **65**, 445 (1997)
43. S. Trusso, B. Fazio, E. Fazio, F. Neri, F. Barreca, Thin Solid Films **518**, 5409 (2010)
44. P.M. Ossi, F. Neri, N. Santo, S. Trusso, Appl. Phys. A **104**, 829 (2011)
45. W.S. Rasband, ImageJ (US National Institute of Health, Bethesda, Maryland) 1997–2005
46. M. Di Vece, S. Palomba, R.E. Palmer, Phys. Rev. B **72**, 073407 (2005)
47. J.P. Jensen, Rev. Mod. Phys. **71**, 1695 (1999)
48. N.R. Agarwal, F. Neri, S. Trusso, A. Lucotti, P.M. Ossi, Appl. Surf. Sci. **258**, 9148 (2012)
49. T. Sakka, S. Iwanaga, Y.H. Ogata, A. Matsunawa, T. Takemoto, J. Chem. Phys. **112**, 8645 (2000)

50. A.V. Simakin, V.V. Voronov, N.A. Kirichenko, G.A. Shafeev, Appl. Phys. A **79**, 1127 (2004)
51. A. Menĕndez-Manjŏn, P. Wagener, S. Barcikowski, J. Phys. Chem. C **115**, 5108 (2011)
52. E. Fazio, F. Neri, Appl. Surf. Sci. **272**, 88 (2013)
53. S. Link, Z.L. Wang, M.A. ElSayed, J. Phys. Chem. **103**, 3529 (1999)
54. U. Kreibig, M. Vollmer, *Optical Properties of Metal Clusters* (Springer, Berlin, 1996), p. 265
55. F. Mafuné, J.Y. Kohno, Y. Takeda, T. Kondow, J. Am. Chem. Soc. **125**, 1686 (2003)
56. E. Fazio, P. Calandra, V. Turco Liveri, N. Santo, S. Trusso, Colloids and Surfaces A: Physicochem. Eng. Aspects **392**, 171 (2011)
57. Z. Yan, D.B. Chrisey, J. Photoch, Photobio. C **13**, 204 (2012)
58. B. Holian, D. Grady, Phys. Rev. Lett. **60**, 1355 (1988)
59. L.V. Zhigilei, Appl. Phys. A **76**, 673 (2003)
60. K. Eidmann, J. Meyer-ter-Vehn, T. Schlegel, S. Huller, Phys. Rev. E **62**, 1202 (2000)
61. D. Perez, L. Lewis, Phys. Rev. B **67**, 184102 (2003)
62. L.V. Zhigilei, Z. Lin, D.S. Ivanov, J. Phys. Chem. C **113**, 11892 (2009)
63. D.S. Ivanov, Z. Lin, B. Rethfeld, G.M. O'Connor, T.J. Glynn, L.V. Zhigilei, J. Appl. Phys. **107**, 013519 (2010)
64. M.Q. Ye, C.P. Grigoropulos, J. Appl. Phys. **89**, 5183 (2001)
65. D. Grojo, J. Hermann, A. Perrone, J. Appl. Phys. **97**, 063306 (2005)
66. Y. Okano, K. Oguri, T. Nishikawa, H. Hakano, Appl. Phys. Lett. **89**, 221502 (2006)
67. K.F. Al-Shboul, S.S. Harilal, A. Hassanein, Appl. Phys. Lett. **100**, 221106 (2012)
68. S. Amoruso, B. Toftmann, J. Schou, Phys. Rev. E **69**, 056403 (2004)
69. T.E. Itina, K. Gouriet, L.V. Zhigilei, S. Noel, J. Hermann, M. Sentis, Appl. Surf. Sci. **253**, 7656 (2007)
70. D. Scuderi, O. Albert, D. Moreau, P.P. Pronko, J. Etchepare, Appl. Phys. Lett. **86**, 071502 (2005)
71. S. Amoruso, G. Ausanio, A.C. Barone, R. Bruzzese, C. Campana, X. Wang, Appl. Surf. Sci. **254**, 1012 (2007)
72. C. Boulmer-Leborgne, B. Benzerga, J. Perrier, Proc. SPIE **6261**, 20 (2006)
73. C. Boulmer-Leborgne, in *Laser-Surface Interactions for New Materials Production* ed. by A. Miotello, P.M. Ossi (Springer, Berlin, 2010) p. 125
74. H. Haberland, in *Clusters of Atoms and Molecules* ed. by H. Haberland (Springer, Berlin, 1994) p. 205
75. G. Mie, Ann. Phys. **25**, 377 (1908)
76. M. Moskovits, Rev. Mod. Phys. **57**, 783 (1986)
77. S. Nie, S.R. Emory, Science **275**, 1102 (1997)
78. N. Micali, F. Neri, P.M. Ossi, S. Trusso, J. Phys. Chem. C **117**, 3497 (2013)
79. A.V. Whitney, R.P. Van Duyne, F. Casadio, J. Raman Spectrosc. **37**, 993 (2006)
80. I.T. Shadi, B.Z. Chowdhry, M.J. Snowden, R. Withnall, J. Raman Spectrosc. **35**, 800 (2004)
81. E. Fazio, F. Neri, A. Valenti, P.M. Ossi, S. Trusso, R.C. Ponterio, Appl. Surf. Sci. **278**, 259 (2013)

Chapter 9
Nano-cluster Assembled Films, Produced by Pulsed Laser Deposition, for Catalysis and the Photocatalysis

A. Miotello and N. Patel

Abstract Catalyst architected in form of coating with nano-particles (NPs) are under intense investigation in the catalysis community due to their exceptional activity and selective nature in catalytic processes as compared to the corresponding bulk counterpart, especially because of their large surface-to-volume atomic ratio, size- and shape-dependent properties, and high concentration of low-coordinated active surface sites. Here we report on selected examples to demonstrate how Pulsed Laser Deposition (PLD) technique is able to synthesize NPs in a single step with the required relevant features for catalysis application. Co NPs embedded in B matrix films have been synthesized by PLD technique by taking advantage of the phase explosion process of superheated liquid where a mixture of vapor and liquid droplets leave the irradiated target surface and get deposited on the substrate. Just these NPs of low cost materials (Co–B) exhibit catalytic properties comparable to that of precious metals in hydrogen production by hydrolysis of $NaBH_4$ and NH_3BH_3. The catalytic activity increases further when the Co–B NPs are supported over a porous C films with high surface area synthesized by PLD. PLD was also utilized to produce Co_3O_4 NPs assembled coating by reactive ablation of Co metal in oxygen atmosphere at various substrate temperatures from room temperature to 250 °C. The important characteristics for catalysis such as shape and the size of NPs with narrow size distribution and mixed disordered-nanocrystalline phase were obtained in a single step in Co_3O_4 NPs synthesized by PLD technique. The Co_3O_4 NPs assembled coatings on glass have been tested in degradation of methylene blue dye solution, considered as a water pollutant, via photo-Fenton reaction in presence of H_2O_2. It was observed that the present Co_3O_4 NPs heterogeneous catalyst exhibits significantly better degradation activity for methylene blue solution than that obtained with homogeneous Co^{2+} ions.

A. Miotello (✉) · N. Patel
Department of Physics, University of Trento, 38123 Povo, Trento, Italy
e-mail: miotello@science.unitn.it

M. Castillejo et al. (eds.), *Lasers in Materials Science*,
Springer Series in Materials Science 191, DOI: 10.1007/978-3-319-02898-9_9,

9.1 Introduction

Search of clean fuels is primary objective in research activity to mitigate the problem of the greenhouse-gases while water and air purification still remain urgent tasks. Among clean fuels, hydrogen is a promising choice with zero pollution and medium to store solar energy. However, it is necessary to use renewable energies (solar, wind, biomass, hydro, geothermal) to produce hydrogen in order to avoid pollutant techniques, for example steam reforming of CH_4. Catalysts are key materials to produce H_2 gas, either by water splitting in electrolyzer and photo electrochemical cells or by dissociation of chemical hydrides ($NaBH_4$, NH_3BH_3). In order to replace expensive noble metals (Pt, Pd, etc.), new forms of catalysts with nanostructures should be developed. Coatings made of nanoparticles (NPs) provide the best option for the heterogeneous catalyst with possibility to recover and reuse several times for different reactions. One of the key challenges in this field is to synthesize properly dispersed NPs having small size variance. Thus the main focus of this chapter is on the synthesis of NPs using PLD for applications on catalysis and photocatalysis reactions in the field of energy generation and environmental purification.

9.2 Cobalt NPs Produced by PLD for Hydrolysis of Chemical Hydrides

9.2.1 Co NPs Embedded in B-Matrix Film (Co–NP–B–MA)

To synthesize the Co NPs embedded in B matrix, atomically homogeneous Cobalt-Boride powder was cold pressed in form of cylindrical disks to be used as a target for PLD. KrF excimer laser was used to carry out PLD at the operating wavelength of 248 nm, pulse duration of 25 ns, and repetition rate of 20 Hz. The laser ablation was performed under vacuum condition at room temperature.

Scanning Electron Microscopy (SEM) micrographs of the Co–NP–B–MA films, deposited by PLD using laser fluences of 3 J/cm^2, show NPs-assembled structure with well dispersed spherical particles having size ranging between 50 and 300 nm [1]. However, along with these NPs, by using Transmission Electron Microscopy (TEM) we also observed much smaller Co NPs embedded in B matrix on the film surface. Bright-field TEM micrographs of Co–NP–B–MA film surface (Fig. 9.1a) deposited by PLD show well-dispersed Co–NPs, with average size of 11 ± 4 nm and well defined spherical shape, embedded in the matrix of light element (boron or boron-oxide) as confirmed by energy dispersive X-ray spectroscopy. The histogram of the particle size (D) distribution with standard deviation (σ) value (± 4 nm) are reported in Fig. 9.1b. The particle density (with size $D < 30$ nm) calculated from the histogram is about 320 ± 30 NP/μm^2. The mechanism of direct Co–NP–B–MA formation was attributed to the phase

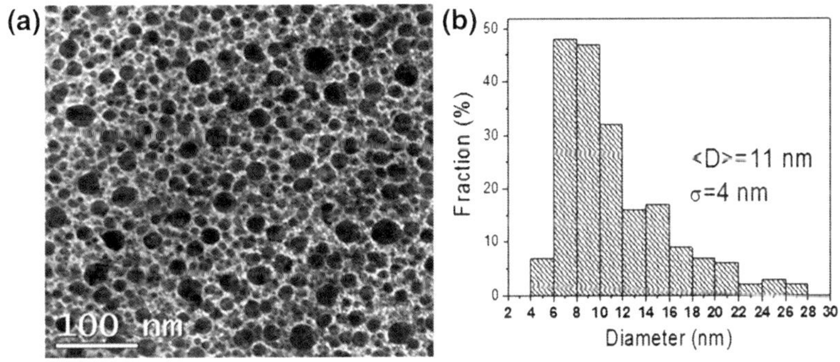

Fig. 9.1 Bright field TEM images (**a**), and particle size histogram (**b**) of Co–NP–B–MA catalyst coating deposited by PLD using energy density of 3 J/cm^2

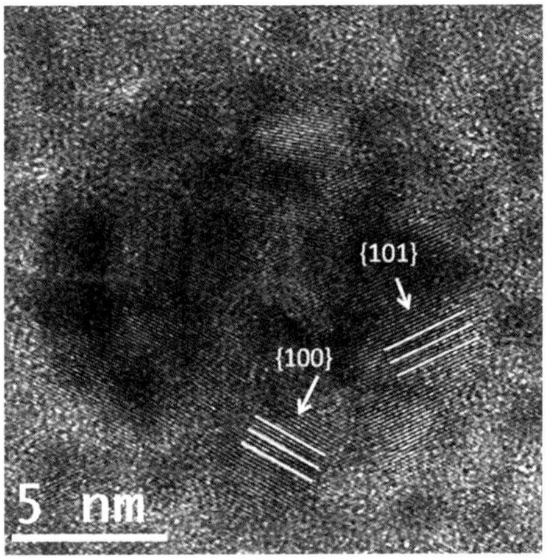

Fig. 9.2 High resolution TEM image of a Co NP deposited with laser fluence of 3 J/cm^2

explosion process that occurs under extreme conditions of high temperature and pressure in target irradiated with laser pulses with energy above the threshold that here is a little lower than 3.0 J/cm^2 [1].

The D value of the Co NPs grows from 11 $\pm$ 4 to 15 $\pm$ 4 nm by increasing the laser fluence from 3 to 9 J/cm^2, while the particle density (with size D $<$ 30 nm) decreases from 320 $\pm$ 30 to 270 $\pm$ 30 NP/μm^2. The ablation performed with laser energy density below 3 J/cm^2 produces mostly boron film with negligible amount of Co NPs on the film surface. Thus, by just varying only the laser fluence in PLD we are able to tune both size and density of Co NPs in B matrix.

High-resolution TEM (HRTEM) of the individual Co NP (Fig. 9.2) shows a polycrystalline structure with three sets of lattice fringes corresponding to Co-hcp phase. Fast Fourier Transform and Selected Area (Electron) Diffraction pattern confirmed the polycrystalline nature of the Co NPs and suggested that each Co NP consists of nano-size domains separated by several grain boundaries. These boundaries, having width of 0.5–1.0 nm contain atoms with lower coordination number as compared to bulk atoms and may act as highly catalytic active sites [2].

Scanning Transmission Electron Microscopy and Atomic Forces Microscopy images of the Co–NP–B–MA films indicate that the Co NPs are only partially embedded in the boron matrix and forms 3-dimensional surface, thus providing high active surface area for the catalysis reaction.

The Co–NP–B–MA coating does not exhibit any major peak in the X-Ray diffraction pattern except single peak of Co with very low intensity, indicating a short-range order but long-range disorder structure of the Co–NP–B–MA film catalyst [3].

Detailed analysis of X-Photoelectron Spectroscopy (XPS) spectra of Co and B binding-energy (BE) shows a positive shift of around 1.2 eV when comparing BE of pure B (187.1 eV) [4] to that of metallic B (188.3 eV) in the present catalyst. This proves electron transfer from alloying B to vacant d-orbital of metallic Co. Just this electron transfer contributes to avoid coalescence and Ostwald ripening of Co NPs, with B acting as an atomic diffusion barrier. In addition, these electron deficient B elements also assist to capture oxygen to protect Co against oxidation. In addition, electron enrichment of the Co active site definitely improves the reactivity of the reactant by providing a pathway for electron transfer in catalysis process.

Co–NP–B–MA coating was utilized as a catalyst to produce hydrogen by hydrolysis of chemical hydrides. The relevance of this catalysis field is motivated by the fact that chemical hydrides, mainly sodium borohydride (SBH, $NaBH_4$) and ammonia borane (AB, NH_3BH_3), with high hydrogen storage capacity and recycling ability, provides pure H_2 to fuel cell at room temperature in presence of heterogeneous catalyst for "on-board" hydrogen fuel cell applications [5–7]. Catalytic hydrolysis reaction of SHB and AB with the present nano-catalyst was performed in an appropriate reaction chamber where the temperature was kept constant within accuracy ±0.1 K (a detailed description of the reactor is reported in [8]). In a typical experiment, catalyst film supported on a glass substrate was introduced in the SBH or AB solution (0.025 M) and the generated H_2 yield was evaluated by accurately measuring the weight of water displaced by the hydrogen volume produced during the reaction course.

Co–NP–B–MA film catalyst was able to generate H_2 with rate 6 times higher than that observed with Co–B target powder for both the hydrolysis reactions (SBH and AB) by using same amount of catalyst (10 mg). Further the catalytic activity of PLD prepared Co–NP–B–MA catalyst (deposited with 3 J/cm^2) was compared with commercially available 5 wt % Pt/carbon black and 10 wt % Pd/carbon black catalyst powders for hydrolysis of SBH and AB solution (0.025 M) and results are presented in Fig. 9.3a, b, respectively.

To make appropriate comparison, the total amount of each catalyst was selected with equal amount (10 mg) of catalytically active metal: Co, Pt, and Pd metals for Co–NP–B–MA, Pt/C, and Pd/C catalysts, respectively. Co–NP–B–MA nano-catalyst was able to produce hydrogen with significantly greater rate than that attained with Pd metal while equivalent to that of Pt metal catalyst. The maximum H_2 generation rate attained with Co–NP–B–MA catalyst is 18.1 and 10.2 L H_2 min^{-1}

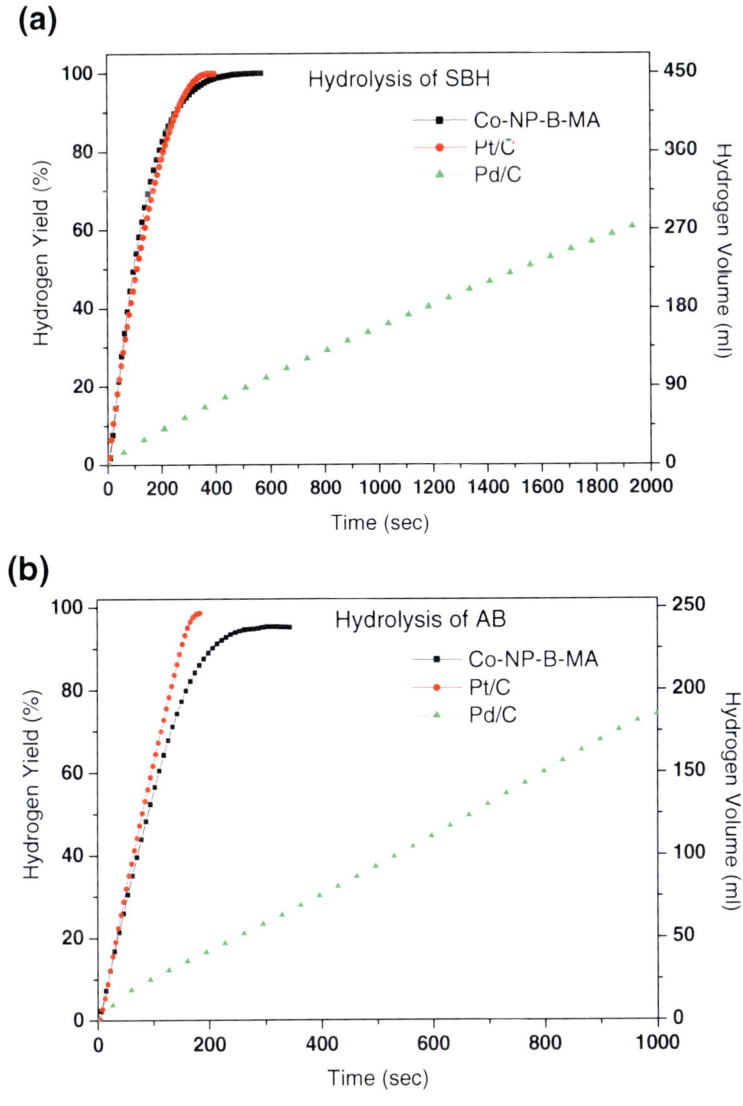

Fig. 9.3 Hydrogen generation yield and volume as a function of reaction time obtained by hydrolysis of **a** $NaBH_4$ and **b** NH_3BH_3 (0.025 M) with: (1) Co–NP–B–MA catalyst deposited by PLD (3 J/cm^2); (2) commercially available 5 wt % Pt/carbon black; and (3) 10 wt % Pd/carbon black catalyst powders. To make appropriate comparison, the total amount of each catalyst was selected with equal amount (10 mg) of catalytically active metal: Co, Pt, and Pd metals for Co–NP–B–MA, Pt/C and Pd/C catalysts, respectively

(g of Co)$^{-1}$ and that with Pt/C powder is 18.2 and 10.5 L H_2 min^{-1} (g of Pt)$^{-1}$ for hydrolysis of SBH and AB respectively. These obtained values are definitely better than that reported for Co-based metal catalyst [9, 10] and comparable to that reported for Pt noble metal catalyst [11, 12]. The experiments performed at different reaction temperatures permitted the evaluation of the activation energy barriers of the rate limiting process: the Co–NP–B–MA catalyst is able to lower the activation energy barrier more than that of the Pt/C and Pd/C catalyst for the hydrolysis reactions [1].

The relevant catalytic efficiency of Co–NP–B–MA nano-catalyst is here attributed to: (a) Co NPs, with spherical shape and average size of a few nanometers (11 nm), that provide large number of surface active atoms with respect to bulk atoms, (b) polycrystalline structure of Co NPs with several grain boundaries which act as active sites for catalytic reaction due to the presence of low coordinated atoms, (c) interaction between surrounding B matrix and Co NPs which plays a vital role not only for better dispersion of NPs but also in protecting against oxidation and stabilizing against coarsening, and (d) electron-enrichment of Co NPs that facilitates optimum interaction with the reactant and product molecules as occurs with Pt metal during hydrolysis reaction. All these features in Co–NP–B–MA nano-catalyst produced by PLD prompts to replace traditional noble metals with low cost catalyst for industrial application: a major breakthrough in catalysis field.

We also studied reusability, stability, and durability of the Co–NP–B–MA nano-catalyst and observed that coating reused several times shows slight decrease in catalytic activity mainly attributed to the detachment of a few weakly bounded NPs from the catalyst coating during the vigorous stirring procedure in the reacting system. However in each cycle the expected amount of H_2 volume is produced (100 % yield). Very severe condition was imposed on our developed catalyst by thermal-treating it in O_2 atmosphere at 623 K for 2 h to check the tolerance against deactivation. The H_2 generation rate for the treated catalyst coating decreased by 10 % and took little more time to initiate the reaction as compared to untreated catalyst. On the other hand, the Co–B catalyst powder (used to prepare target for PLD) treated in similar condition was completely oxidized showing indeed negligible H_2 generation rate. This proves that B matrix efficiently protects Co against oxidation through electron transfer from B to Co. XPS confirms that B is completely oxidized under O_2 gas treatment while protecting Co in catalyst film. The role of B matrix as diffusion barrier for Co NPs was also tested at elevated temperatures where no coarsening of Co NPs was observed at 623 K in Ar atmosphere.

9.2.2 Co–NP–B–MA Nano-catalyst Supported Over Rough Carbon Films

PLD can also be used to produce rough and porous carbon support for NPs catalyst. This C support provides large surface area and better dispersion of the active phase due to its porous nature. Besides that, porous C support: (1) facilitates the diffusion of reactants to the active phase through the pores, (2) improves the dissipation of the reaction heat, (3) contributes in avoiding agglomeration of the active phase, and (4) increases the poison resistance. Thus, PLD presents important advantage of synthesizing in situ both catalyst NPs as well as its support as demonstrated in our previous works [13, 14]. We have deposited C layers having structure ranging from diamond like to highly porous, cluster-assembled coating by using PLD by varying deposition pressure [15]. C-supported Co–NP–B–MA nano-catalyst was synthesized in two steps: (1) depositing C coating by PLD, on a glass substrate, using a laser fluence of 12 J/cm^2 and varying, in the deposition chamber, Ar pressures in the range from 10 to 50 Pa to obtain different surface roughness; (2) depositing Co–NP–B–MA over these carbon coatings in vacuum with laser parameters similar to that used to deposit unsupported Co–NP–B–MA nano-catalyst.

The SEM images of the C coatings deposited under Ar gas pressure of 20, 30, 40 and 50 Pa are illustrated in Fig. 9.4a, c, e and g, while Co–B coatings supported on these carbon substrates are reported in Fig. 9.4b, d, f and h, respectively. Microstructures ranging from flat to highly irregular and to porous are clearly visible. The carbon substrate deposited at low Ar pressure (20 Pa) exhibits columnar structure (observed through cross-section SEM images of the coating, figure not shown) with embedded spherical nodules on the surface. By increasing the pressure (to 30 and 40 Pa) dendritic, highly porous microstructure starts to appear with extremely irregular surface. The nodes in this case appear bigger, loosely packed, and non-spherical. C-substrate adhesion is slightly poor than that reached at low Ar pressure. When using high pressure conditions (50 Pa), the coating appears powder-like, with barely any adhesion to the substrate. The collisional processes imposed by inert gas (Ar) at high pressure in the deposition chamber cause a cooling down of plume along with charges recombination and condensation of carbon clusters in the plume which are deposited on the substrate. In particular, by increasing the gas pressure, cluster-cluster collision may also occur that contributes to formation of bigger clusters which, when assembled, form highly porous and irregular structure on the coating surface. The visible consequence of such a process is the peculiar sequence of morphologies that develop in the growing film. Co–NP–B–MA coating was supported on these carbon substrates and used as catalyst for the H_2 production by hydrolysis of AB. Hydrogen generation yield was measured, as a function of time, by the hydrolysis of AB solution (0.025 M) at 298 K in presence of Co–NP–B–MA catalyst supported on different C substrates deposited with various Ar pressures ranging from 10 to 50 Pa (Fig. 9.5). The inset of Fig. 9.5 shows the maximum H_2 generation rate (R_{max}) as a function of Ar gas pressure used to deposit the carbon substrates. Co–NP–B–MA catalysts supported on C-coatings deposited at low Ar pressures (10 and 20 Pa) show almost similar catalytic activity as un-supported Co–NP–B–MA coating. We attribute this result to the non-porous structure of C-substrate. However, the catalytic activity increases for Co–NP–B–MA catalysts supported on C-coating deposited at higher Ar pressures (30 and 40 Pa). R_{max} reached the maximum for C-coating deposited at 40 Pa. As indicated by SEM images, the roughness and surface area of C-coating increases with the deposition pressure. This trend is also

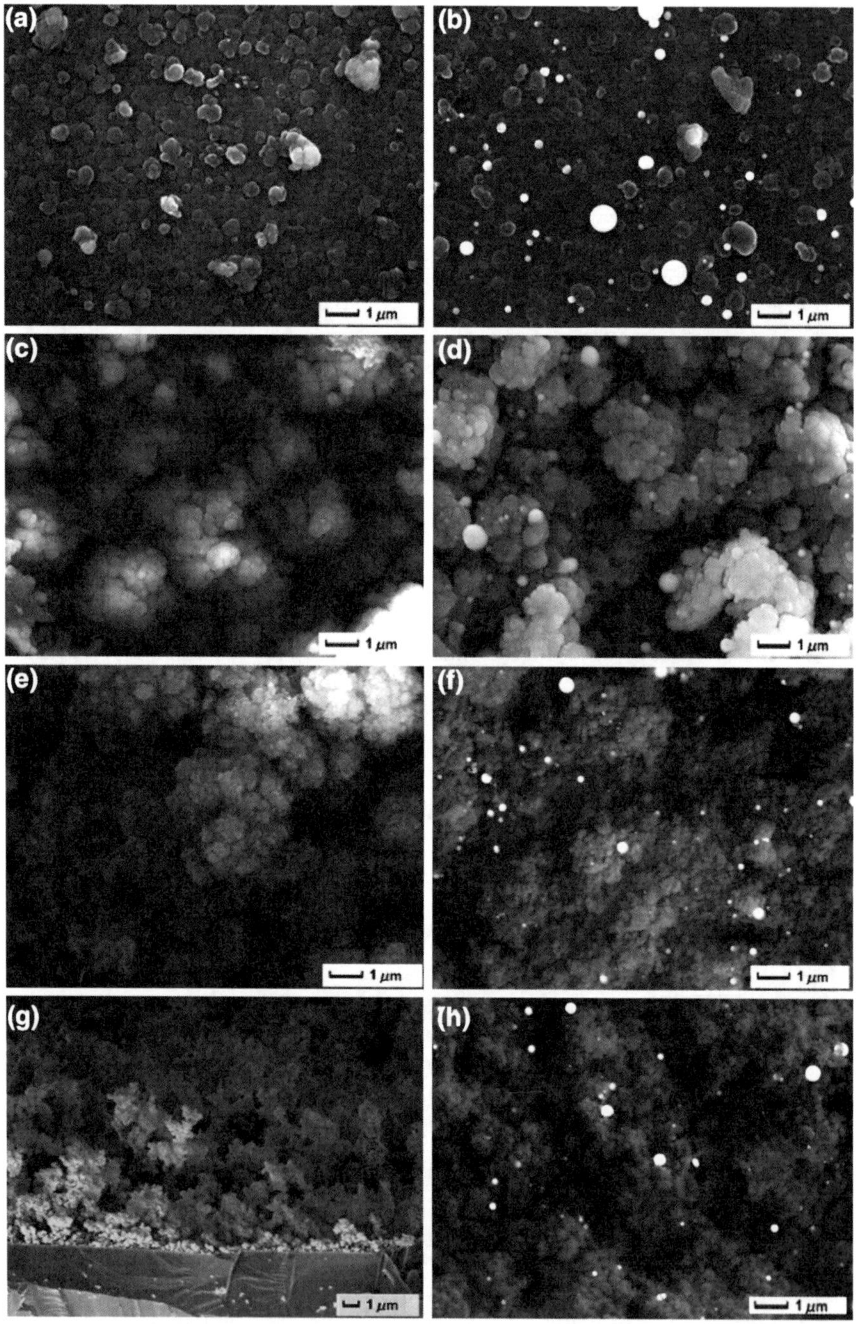

Fig. 9.4 SEM micrographs of the carbon films deposited under different Ar gas pressures **a** 20, **c** 30, **e** 40 and **g** 50 Pa, while **b**, **d**, **f** and **h** are the SEM images of Co–B coatings supported on these carbon substrates, respectively

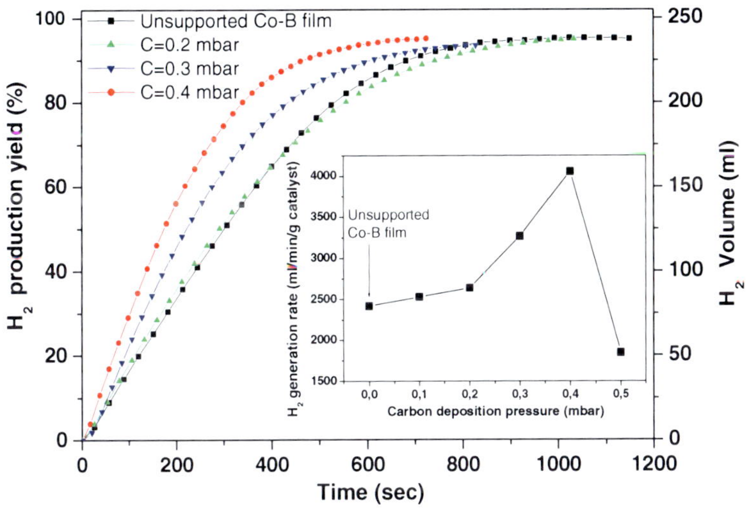

Fig. 9.5 Hydrogen generation yield as a function of reaction time obtained by hydrolysis of NH_3BH_3 (0.025 M) with Co–B catalyst supported on C-coatings deposited with various Ar pressures by PLD. Inset shows the R_{max} as a function of Ar gas pressure used to deposit the carbon coatings

followed by the H_2 generation rate for hydrolysis of AB as demonstrated in the inset of Fig. 9.5. Thus, surface area and roughness of C-coatings play a vital role in the increment of catalytic activity for Co–NP–B–MA film catalyst by providing better dispersion and avoiding aggregation of NPs. On the contrary, Co–NP–B–MA catalyst supported on C-coating deposited at highest pressure of 50 Pa showed drastic decrease in the H_2 generation rate and was not able to complete the hydrolysis reaction of AB. The C-coating deposited at this pressure had a very week adhesion with the substrate and thus under vigorous stirring the coating slowly detached from the substrate in the reactant solution during the AB hydrolysis reaction.

9.3 Co-oxide NPs Produced by PLD for Photocatalysis Application

9.3.1 Co_3O_4 NPs Assembled Coating Photocatalyst

To deposit Co_3O_4 NPs assembled coating, the laser ablation of pure Co target was carried out under O_2 gas pressure of about 4.5×10^{-2} mbar and laser fluence of 3 J/cm^2. The O_2 pressure value was selected, among the other tested values, to ensure the Co oxide stoichiometry. The substrate temperature in the deposition chamber was varied from room temperature to 250 °C.

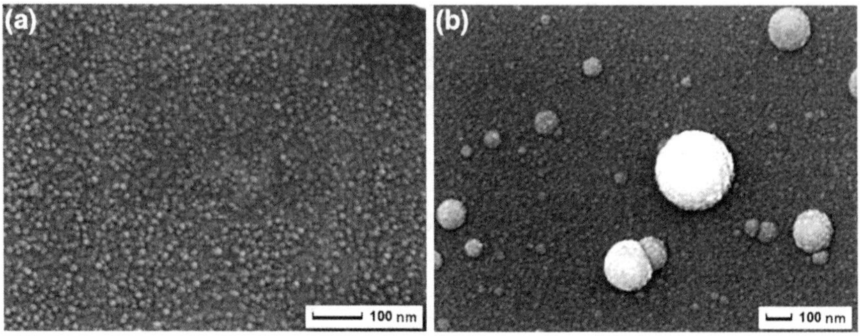

Fig. 9.6 SEM images of Co_3O_4 NPs synthesized by PLD at: **a** 150 °C and **b** 250 °C

SEM images of the Co_3O_4 coatings deposited at RT, 100, 150 (Fig. 9.6a), 200, and 250 °C (Fig. 9.6b) show spherical NPs assembled on the surface. The NPs with average size of about 25 nm with narrow size distribution are densely arranged on the substrate surface after the coating was deposited at RT, 100, and 150 °C (Fig. 9.6a). On the contrary, the surface of the samples prepared at higher temperature, 200 and 250 °C (Fig. 9.6b) shows broad range of NPs from 20 to 200 nm with average size of around 35 and 50 nm, respectively. XRD and Raman analysis showed that Co_3O_4 NPs prepared by PLD at different temperatures are present in form of spinel type cubic structure with Fd3 m space group [16]. Most importantly, it is observed that crystallinity and crystal size increase with increase in substrate temperature. The Co_3O_4 NPs deposited at 150 °C acquire mixed amorphous-nanocyrstalline structure while complete crystallization is seen for the sample prepared at 250 °C with size of 50 nm. These results prove the clear role of the substrate temperature in the observed increase of NP size by favoring adatom diffusion, and also allowing NP to form more ordered structure to increase the crystallite size. However, increasing size of NPs is deleterious in catalysis reaction.

The Co_3O_4 NPs assembled coating was used for the photo-degradation of Methylene Blue (MB) dye solution by advance oxidation process. For photocatalytic studies, the variation of the relative concentration of the remaining MB dye in solution, as function of visible light irradiation time, was measured through optical absorbance measurements at 664 nm.

Figure 9.7 represents the rate of degradation of MB dye in presence of light and H_2O_2 using Co_3O_4 NPs assembled coating deposited with different substrate temperatures in PLD. The best photodegradation activity ($\sim$99 %) was observed for Co_3O_4 NPs prepared by PLD at 150 °C where almost a complete decolourization of MB solution occurred and clear solution was obtained. This enhanced activity at 150 °C is mainly attributed to: (a) the NPs distribution in narrow range of size with average particle size of about 25 nm as compared to Co_3O_4 NPs prepared at 200 and 250 °C, and (b) presence of mixed amorphous-nanocrystalline phase with better crystallinity as compared to Co_3O_4 NPs prepared at RT and

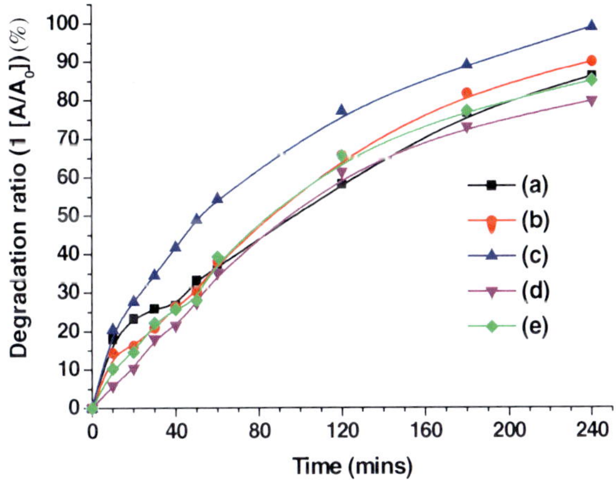

Fig. 9.7 Time dependent photocatalytic degradation ratio of MB solution in presence of H_2O_2 and light irradiation using Co_3O_4 NPs prepared by PLD at: **a** RT, **b** 100 °C, **c** 150 °C, **d** 200 °C, and **e** 250 °C. A_0 is the absorbance at time 0 and A at given time

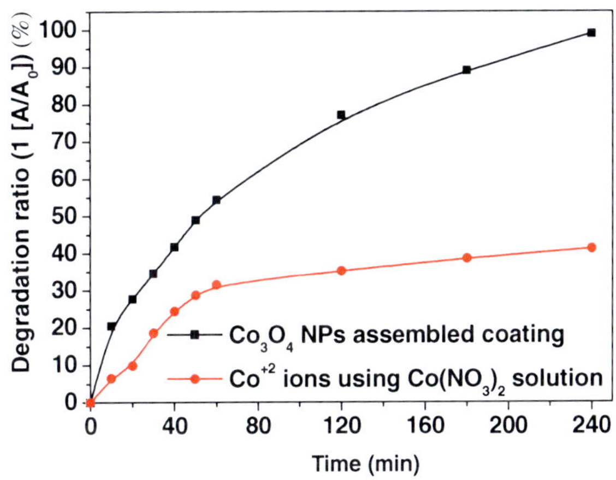

Fig. 9.8 Time dependent photocatalytic degradation ratio of MB solution in presence of H_2O_2 and visible light using heterogeneous Co_3O_4 NPs assembled coating deposited by PLD at 150 °C and homogeneous cobalt nitrate ($Co(NO_3)_2$) powder

100 °C. NPs with average size of 25 nm and narrow size distribution provide large number of surface active atoms over NPs with respect to bulk atoms: a condition which is most favorable to catalytic reaction. The amorphous and nanocrystalline mixed phase are separated by grain boundaries containing atoms with lower coordination number as compared to atoms in the ideal crystallites and therefore these grain-boundaries regions are highly catalytic active sites. This indicates that NPs with smaller size and large number of grain boundaries on the surface produced by PLD present the ideal condition for the photocatalytic reaction of MB.

It is well known that homogeneous catalyst is far more active than the heterogeneous catalyst. To check the effectiveness of laser-synthesized heterogeneous catalysts, the Co_3O_4 NPs were compared with homogeneous Co^{2+} ions for degradation of dye. Figure 9.8 presents the rate of MB degradation in presence of H_2O_2 and visible light using heterogeneous Co_3O_4 NPs assembled coating deposited by PLD at 150 °C and homogeneous cobalt nitrate ($Co(NO_3)_2$) powder.

The amount of Co^{2+} ions in $Co(NO_3)_2$ and Co_3O_4 NPs coating were kept equivalent in the mixture. It is observed that homogeneous catalyst is only able to degrade 40 % of the MB dye in 4 h with very slow reaction rate in comparison with heterogeneous Co_3O_4 NPs catalyst which permits complete decolorization in same amount of time. This indicates that the combined effect of smaller NPs and mixed amorphous-nanocyrstalline structure in present Co_3O_4 NPs coating provides far better activity for MB degradation than that obtained with homogeneous Co^{2+} ions. This result is relevant in this research field where it is important to achieve better performance with heterogeneous catalyst as compared to homogeneous catalyst because the former has biggest advantage of being recovered and reused for different reactions.

9.4 Conclusions

In this chapter we proved the relevance of the PLD synthesis technique to architect NPs catalysts in form of coating having appropriate features to support efficient catalysis process. Co NPs embedded in B matrix films have been synthesized, in a single step by nanosecond PLD, with important characteristic such as narrow size distribution, small average size (11 nm), polycrystalline nature, and electron enriched sites. These features lead the Co NPs to have catalytic performance comparable to that of precious metals (Pt) in hydrogen production by hydrolysis of $NaBH_4$ and NH_3BH_3. The H_2 generation rate was further enhanced by about 30 % by supporting these NPs on rough and porous C coating again synthesized by PLD. Using reactive PLD, Co_3O_4 NPs assembled coatings were also synthesized at various substrate temperatures for the photocatalytic purification of water under visible light. These Co_3O_4 NPs were able to completely degrade the methylene blue dye solution with significantly higher rate as compared to homogeneous Co^{2+} ions. This enhanced photoactivity is attributed to the shape and size of NPs with narrow size distribution and mixed disordered-nanocrystalline phase as produced by PLD technique in a single step.

Acknowledgments The research activity is partially supported by the PAT (Provincia Autonoma di Trento) project ENAM in cooperation with Istituto MCB of CNR (Italy).

References

1. N. Patel, A. Miotello, V. Bello, Appl. Catal. B: Environ. **103**, 31 (2011)
2. V. Johánek, M. Laurin, A.W. Grant, B. Kasemo, C.R. Henry, J. Libuda, Science **304**, 1639 (2004)
3. A. Baiker, Faraday Discuss. Chem. Soc. **87**, 239 (1989)
4. H. Li, H.X. Li, W.L. Dai, Z. Fang, J.F. Deng, Appl. Surf. Sci. **152**, 25 (1999)
5. L. Schlapbach, Nature **460**, 809 (2009)

6. L. Schlapbach, A. Züttel, Nature **414**, 353 (2001)
7. U.B. Demirci, P. Miele, Energy Environ. Sci. **2**, 627 (2009)
8. C. Zanchetta, B. Patton, G. Guella, A. Miotello, Meas. Sci. Technol. **18**, N21 (2007)
9. M. Rakap, S. Özkar, Appl. Catal. B: Environ. **91**, 21 (2009)
10. Q. Xu, M. Chandra, J. Power Sources **163**, 364 (2006)
11. Y. Bai, C. Wu, F. Wu, B. Yi, Mater. Lett. **60**, 2236 (2006)
12. Q. Xu, M. Chandra, J. Alloys Compd. **446**, 729 (2007)
13. N. Patel, R. Fernandes, A. Santini, A. Miotello, Int. J. Hydrogen Energy **37**, 2007 (2012)
14. N. Patel, R. Fernandes, N. Bazzanella, A. Miotello, Catal. Today **170**, 20 (2011)
15. N. Patel, R. Fernandes, G. Guella, A. Kale, A. Miotello, B. Patton, C. Zanchetta, P.M. Ossi, V. Russo, Appl. Sur. Sci. **254**, 1307 (2007)
16. T. Warang, N. Patel, A. Santini, N. Bazzanella, A. Kale, A. Miotello, Appl. Catal. A: Gen. **423–424**, 21 (2012)

Chapter 10
Multifunctional Oxides Obtained by PLD: Applications as Ferroelectric and Piezoelectric Materials

N. D. Scarisoreanu, Maria Dinescu and F. Craciun

Abstract In this chapter we provide an overview of the results obtained on both lead based and lead free ferroelectric thin films deposited by PLD, in relation with the actual scientific and economic tendencies. There is an increasing trend to replace or to reduce the use of toxic elements such as lead, but there is no obvious complete solution for this problem, both types of multifunctional oxides having attractive properties. The perovskite materials, ABO_3 oxides, environmental-friendly or not, will continue to be studied and their properties enhanced through different methods. Moreover, based on these properties (piezoeletric, ferroelectric, transport, optical or magnetic), new functionalities of the material can be added or modified with the help of material nanostructuring techniques. Properties of lead based oxides thin films such as PLZT and PMN-PT were investigated. PLZT thin films with different compositions have interesting dielectric and electro-optic behaviour. The effect of self-polarization in thin PMN-PT relaxor films could be of interest for pyrosensors and other applications based on the pyroelectric effect. As regarding the lead-free oxide materials, the obtaining and the characterisation of SBN and NBT-BT thin films will be presented. For the SBN thin films, the resulting value of the electro-optic coefficient r_{eff} was calculated to be higher than the values reported for $LiNbO_3$. Having a rather high Curie temperature, of 128 °C, compared to the higher Sr content compositions, the SBN:50 thin films can be potentially used in electro-optic devices operating near room temperature. Solid-solution systems (1−x)NBT-xBT based on $Na_{0.5}Bi_{0.5}TiO_3$ (NBT) and $BaTiO_3$ (BT) were investigated: for compositions situated at the morphotropic phase

N. D. Scarisoreanu · M. Dinescu (✉)
NILPRP, National Institute for Lasers, Plasma and Radiation Physics, Bucharest, Romania
e-mail: dinescum@ifin.nipne.ro

N. D. Scarisoreanu
e-mail: snae@nipne.ro

F. Craciun
CNR-ISC, Istituto dei Sistemi Complessi, Area della Ricerca Roma-Tor Vergata, Rome, Italy
e-mail: floriana.craciun@isc.cnr.it

M. Castillejo et al. (eds.), *Lasers in Materials Science*,
Springer Series in Materials Science 191, DOI: 10.1007/978-3-319-02898-9_10,

boundary between rhombohedral and tetragonal phase, high piezoelectric coefficient values and huge electric field-induced strain have been obtained.

10.1 Introduction

The spreading of complex oxide materials in devices used in daily life is becoming more important: from non-volatile ferroelectrics memories, electro-optic devices and different types of sensors to microwave applications area. Great efforts are made to improve ferroelectric properties of such materials and to decrease their production costs. All ferroelectric materials studies have the aim to obtain materials with better properties than those reported to date, with possible applicability in high-performance electronic devices [1, 2]. In order to be used in devices, ferroelectric materials must fulfil a series of conditions, such as:

- Chemical stability and thermal properties are also essential for the performance and reliability of electronic devices in various environments.
- High dielectric constant and low dielectric losses; the impedance of a resonator is given by the thickness of the piezoelectric thin layer, resonator dimensions and dielectric constant, which means that the high dielectric constant value allows the decrease of the resonator dimensions. For example, lead zirconate titanate ($Pb(Zr_xTi_{1-x})O_3$ or PZT) has a dielectric constant, much higher than that of aluminium nitride (AlN), but low dielectric losses represent one of the key properties for electro-acoustic applications and AlN exhibit significantly lower losses with respect to PZT [3, 4].
- High electro-mechanic coupling coefficient; low coupling decreases the quality factor (Q) and limits the bandwidth for the devices used in telecommunications.
- The possibility to change the dielectric properties by applying an electric field is a characteristic of ferroelectric materials (i.e. tunability).

Due to these reasons, thin film technology is in a continuous expansion in the last period; for example, the minimization of the electronic devices is a direct consequence of the progress achieved in thin film technology and in new materials with superior properties. Regarding the qualitative level of the deposited thin films and some practical issues, such as production costs and raw materials or precursor's abundance, the deposition techniques of such materials can be divided in two major classes. The first category of deposition techniques mainly addresses the academic and research areas where the quality of the deposited layers and versatility of the technique are the main arguments. Some of the representative techniques are molecular beam epitaxy (MBE) and pulsed laser deposition (PLD). The second category has the practicality as its main feature, which usually means low costs and large-scale production for ferroelectric thin films and not always *superior properties*. This category is represented by chemical deposition techniques such as chemical vapour deposition (CVD) and sol-gel, but also by physical

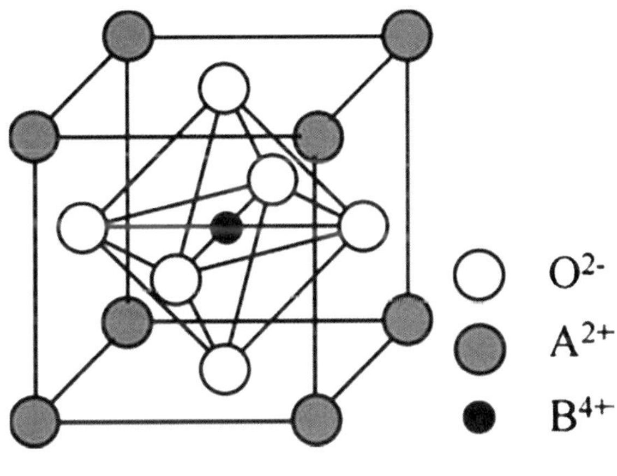

Fig. 10.1 The perovskite structure ABO_3 (from [5])

deposition techniques like sputtering method. Having this in mind, pulsed laser deposition—PLD—has advantages over other methods, which justify the spreading of this technique in research and academic environments. The combination between versatility of this technique and the capability of complex stoichiometry transfer from the bulk target to the thin films, as well as low maintenance costs, makes it very useful for initial material and device studies before moving on to large scale production using a different deposition technique. Therefore, this chapter presents some of the main achievements of ferroelectric thin films deposition by PLD, with emphasis on both lead based and lead free compounds.

Ferroelectric materials have the paramount property to be both piezoelectric and pyroelectric. For this reason, their applicability relates either to one or all of these properties. Consequently, these properties will be, in special cases, discussed together for each material.

For many years, the most important class of ferroelectric materials with widespread industrial applications are perovskite oxide materials, which have the ABO_3 general formula, as can be seen in Fig. 10.1. As a representative ferroelectric material for this class, lead zirconate titanate ($Pb\ (Zr_xTi_{1-x})O_3$ or PZT) is nowadays the most used ferroelectric material. A combination of excellent piezoelectric, pyroelectric and ferroelectric properties of PZT ceramic was found for compositions situated at the so-called morphotropic phase boundary (MPB) region, therefore the connection between the enhanced properties and the MPB compositions being made [6].

The morphotropic phase boundary is situated between the tetragonal (*P4mm*) and rhombohedral (*R3m* and *R3c*) phases (PZT: $x \approx 0.53$). The existence of a monoclinic phase (*Cm*) close to this MPB region was proved to exist and the phase diagram was modified accordingly [7, 8]. However, the origin of this favourable combination of properties for PZT as bulk and, especially, thin films is still under debate [9, 10].

For PZT thin films, as well as for other ferroelectric materials, a fine influence of the small deviations from the stoichiometry or crystalline structure on their physical properties has been demonstrated [10–13]. Compared to metals or

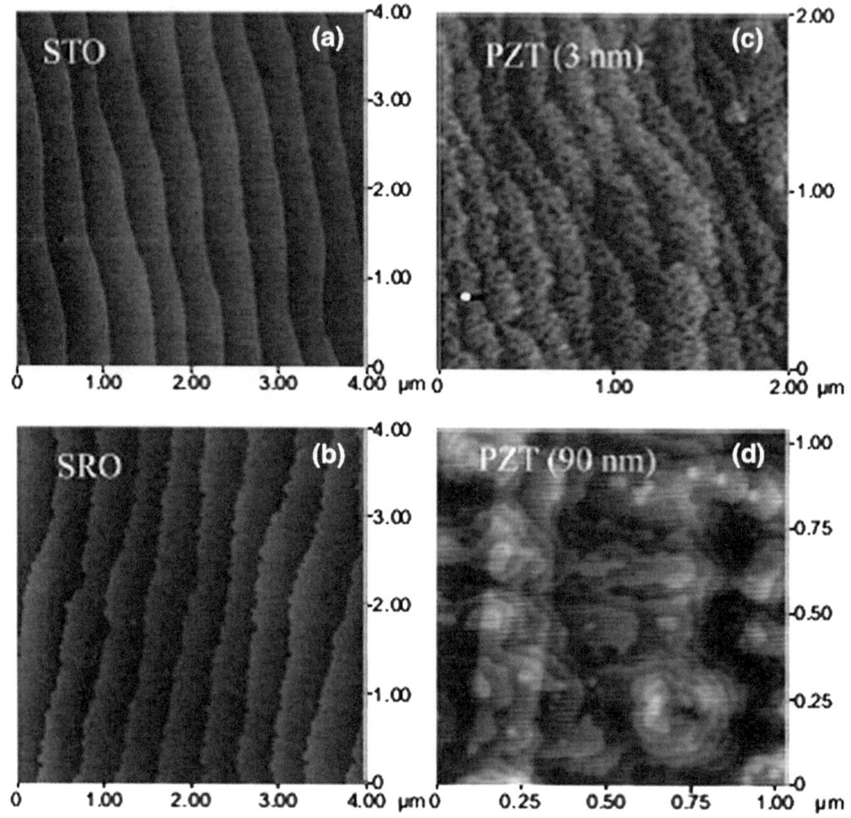

Fig. 10.2 AFM images of morphologies forming on: **a** the vicinal STO substrate; **b** the SRO strained films; **c** 3 nm thick PZT thin films; **d** 90 nm PZT thick films (from [16])

semiconductors, the chemical and structural complexity of most ferroelectric materials has a huge effect on the thin film deposition technique requirements. The obtaining of high quality ferroelectric films implies a thorough understanding of the deposition process and control of certain key parameters, which are responsible for the films properties. To demonstrate these assertions, the case of epitaxial PZT thin films deposition and the degree of sophistication needed to achieve this goal is presented as follows. Using the advantages of PLD [14, 15], epitaxial PZT ($PbZr_{0.2}Ti_{0.8}O_3$ or PZT 20/80) thin films have been obtained and reports on their physical properties have been made [16]. The defect-free, epitaxial PZT 20/80 has been obtained starting from a Pb-enriched PZT 20/80 target using a KrF excimer laser. The role of the substrate proved to be important for achieving the goal of growing epitaxial PZT thin films with no extended structural defects. The vicinal, single-crystalline $SrTiO_3$ (001) (STO) substrate was chemically and thermally treated for achieving the step-flow growth of the strained $SrRuO_3$ (SRO) buffer

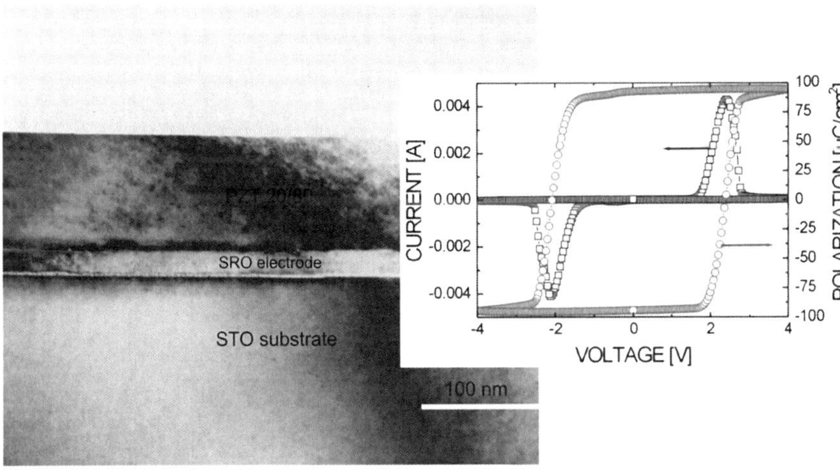

Fig. 10.3 Structural and ferroelectric properties of PZT 20/80 thin films (from [16])

layer. The SRO buffer layer was chosen due to small lattice mismatch with the PZT 20/80 in-plane lattice parameters. The step-flow method of growing strained thin films was reported by Hong et al. [17]. Using atomically flat SRO buffer layers grew on the terraces of the STO vicinal substrates and using special deposition conditions, layer-by-layer PZT films have been obtained by Vrejoiu et al. [16], as can be seen in Fig. 10.2. The ferroelectric and piezoelectric properties of such PZT films are reported to be high, as compared with the previously reported ones (see Fig. 10.3). The remnant polarization value was reported to be $P_r \approx 105\ \mu C/cm^2$, higher than that of bulk material [18].

This properties combination had little competition until now, as compared with other ferroelectric material types. However, due to the toxicity of lead, the use of lead-based ferroelectric materials would be the subject of different restrictions. Banning such materials, however, would require the development of suitable materials with similar properties in order to replace them. Nowadays, the market is still dominated by lead-containing ferroelectric materials, such as PZT, relaxor ferroelectric $Pb(Mg_{1/3}Nb_{2/3})O_3$ (PMN) and its solid solutions with $PbTiO_3$, $Pb(Mg_{1/3}Nb_{2/3})_{1-x}Ti_xO_3$ (PMN-PT) or La-doped lead zirconium titanate $Pb_{1-x}La_x(Zr_yTi_{1-y})O_3$, as can be seen in the latest scientific articles [19, 20].

Efforts are made to find viable replacements for all these materials which contain harmful elements and various studies have been reported on lead-free ferroelectric materials such as $Sr_xBa_{1-x}NbO_6$—strontium and barium niobate, $Na_{0.5}Bi_{0.5}TiO_3$—sodium and bismuth titanate etc. Part of the tungsten-bronze structural family, strontium barium niobate $Sr_xBa_{1-x}NbO_6$ (SBN) is a ferroelectric oxide material with remarkable pyroelectric and electro-optical properties [21]. For both bulk and thin film, SBN was proven to have an impressive electro-optic behavior with very large electro-optic coefficients values, higher than for lanthanum-doped lead zirconate titanate (PLZT) or an actual industrial standard material, lithium niobate

($LiNbO_3$) [21, 22]. Solid-solution systems (1-x)NBT-xBT based on $Na_{0.5}Bi_{0.5}TiO_3$ (NBT) and $BaTiO_3$ (BT), are promising lead-free materials due to their dielectric and piezoelectric properties. For compositions situated at the morphotropic phase boundary between rhombohedral and tetragonal phase, high piezoelectric coefficient values and huge electric field-induced strain have been reported [23].

10.2 Relaxor Ferroelectric PLZT Thin Films

La-doped lead zirconium titanate $Pb_{1-x}La_x(Zr_yTi_{1-y})O_3$ (PLZT) behave like normal ferroelectrics for $x < x_c$ and like relaxors for $x > x_c$, where the critical value of La content, x_c, depends on the Zr content y. Disorder can be more easily introduced in rhombohedral PZT compositions ($y > 0.53$) than in tetragonal PZT. Therefore, in the first case, x_c value is below or around 0.1, whereas in the second x_c can reach 0.25 [24]. Since La^{3+} ions substitute the lower valence Pb^{2+} ions in the A site of the perovskite structure, vacancies must be introduced on the cations A and/or B position to preserve neutrality. This introduces further disorder in these materials.

PLZT compounds are interesting materials for their high dielectric constants and electro-optical linear and quadratic coefficients. In the last years different investigations have been devoted to PLZT thin film growth by different techniques [25–31]. We have selected two relaxor compositions based on rhombohedral and tetragonal PZT, PLZT 9/65/35 and PLZT 22/20/80 (named in the most used notation PLZT x/y/1−y, with x and y expressed as percents). In this section we review some of our results obtained on PLZT 9/65/35 [32–37] and PLZT 22/20/80 [33, 36–42] thin films deposited in different conditions.

10.2.1 PLZT 9/65/35 Thin Films

$Pb_{1-x}La_x(Zr_{0.65}Ti_{0.35})_{1-x/4}O_3$ (PLZT) is one of the most investigated PLZT materials [43, 44]. For $0.07 < x < 0.12$ it shows a relaxor behaviour with a very broad dielectric peak and strong dispersion. Compositions with $x = 0.09$ have a broad dielectric maximum at $T_m \sim 340$ K and high dielectric and electro-optical constants, but the obtaining of high quality thin films is still a difficult problem.

We show results obtained on PLZT 9/65/35 thin films deposited on Pt/Si substrates by PLD and by RF-PLD [32–37]. The RF-PLD technique uses a radiofrequency discharge in oxygen in the PLD deposition chamber for increasing the plasma reactivity and to reduce the oxygen vacancies in films and at the interface with the electrode. The films have been grown in the following conditions: laser wavelength = 300 nm, pulse duration = 10 ns, laser fluence = 3–5 J/cm^2, pulse repetition frequency = 10 Hz, oxygen pressure = 0.3–0.6 mbar, substrate temperature = 650 °C. Different sets of films have been grown by RF-PLD in similar conditions but with an RF discharge power of 100 and 200 W, respectively.

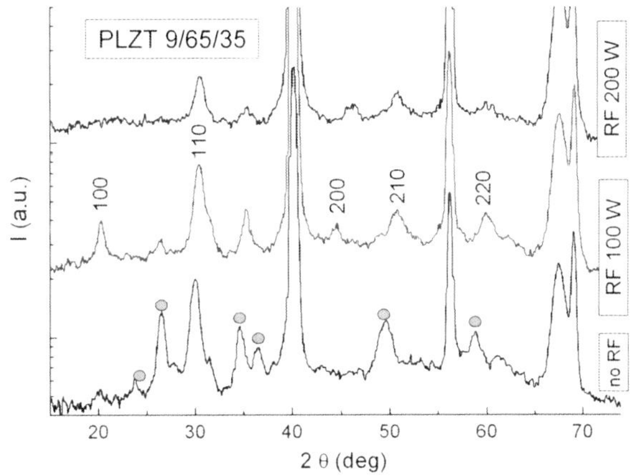

Fig. 10.4 XRD spectra of PLZT 9/65/35 thin films deposited on Pt/Si by PLD (*lower spectrum*) and by RF-PLD (*upper spectra*). Full symbols mark the pyrochlore phase peaks. The peaks at 2θ–40° and 68° correspond to Pt, while those at ~56° and 69° to Si

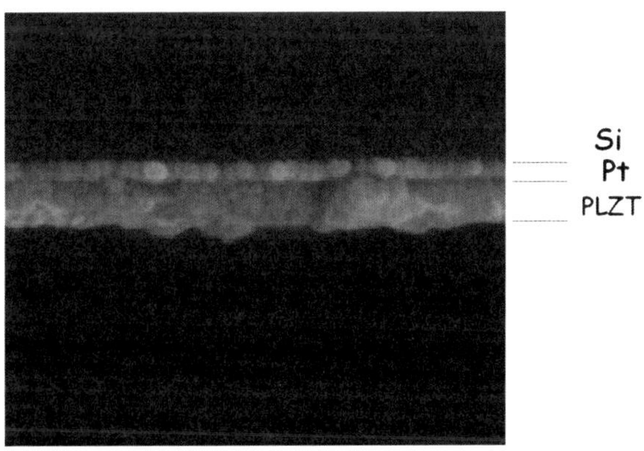

Fig. 10.5 Cross-section SEM image of a PLZT/Pt/Si heterostructure obtained by RF-PLD

Figure 10.4 shows the XRD spectra of PLZT 9/65/35 thin films deposited on Pt/Si by PLD (lower spectrum) and by RF-PLD (upper spectra). The reflection peaks corresponding to the perovskite phase are indexed. Full symbols mark the secondary pyrochlore phase peaks.

It can be observed that films grown by RF-PLD show a lower pyrochlore content. This can be explained by the action of excited and ionized oxygen species in the RF discharge on the lead oxidation and the formation of perovskite phase.

Figure 10.5 shows a cross-section SEM image of a PLZT/Pt/Si heterostructure obtained by RF-PLD. It shows a dense, compact PLZT thin film of ~300 nm thickness over a thin polycrystalline Pt layer. Pt/Si and Pt/PLZT interfaces are also rather smooth.

In Fig. 10.6 AFM images taken over a 40 × 40 μm area (top view) and over a 5 × 5 μm area (3D view) of a PLZT thin film deposited by RF-PLD are shown. The film has a thickness of about 500 nm.

A compact structure formed by uniform grains appears on a larger scale, however details on the magnified image show well-formed submicronic grains

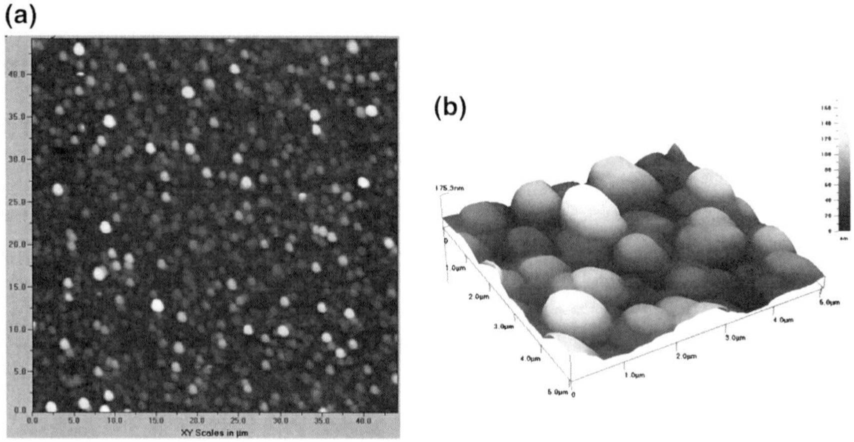

Fig. 10.6 **a** AFM *top view* image (40 × 40 µm) of the surface of a PLZT thin film with ~500 nm thickness deposited by RF-PLD; **b** a 3D detail image on a 5 × 5 µm area

with distinct boundaries and some interspace among few grains. The roughness measured over the surface in Fig. 10.6a was about 50 nm.

Dielectric spectroscopy measurements have been carried out in the 100 Hz–1 MHz frequency range and the 300–570 K temperature range. The variation of capacitance and loss with temperature for RF-PLD deposited films presented a broad peak at $T_m \sim 345$ K and 10 kHz (close to T_m in bulk), which shifted to higher T when frequency increased [32, 35]. However, much higher dispersion was registered in C(ω, T) curves above T_m. It has been shown in [45] that this behaviour is typical for the presence of a low permittivity layer at the interface with the electrode. In order to obtain the true behaviour of the relaxor films we used the correction relationship proposed in [45] for relaxor heterostructures. This is based on the assumption that at T_m the capacitance of the heterostructures is limited by the capacitance of the passive layer. By using this model we have obtained the permittivity curves corresponding to the PLZT film, which are represented in Fig. 10.7.

However, it can be observed that the permittivity values are still much lower than bulk values [46]. This could be caused by the presence of some pyrochlore phase which could not be completely eliminated even in RF-PLD films deposited with RF discharge power of 200 W (Fig. 10.4). A more important detrimental effect could be produced by the porosity and grain boundary layers.

Figure 10.8 shows the frequency dependence of capacitance and loss at room temperature measured on a PLZT 9/65/35 film deposited by RF-PLD (RF power 100 W) on Pt/Si. As it has been shown in [47], the freezing temperature of bulk relaxor PLZT 9/65/35 is $T_f = 320.5$ K. Below this temperature the dynamics of PNRs is significantly slowed and the dispersion decreases. This is reflected also in the small variation with frequency of capacitance and loss values in Fig. 10.8, measured at 300 K. The loss tangent is very small (below 2 %) in the frequency

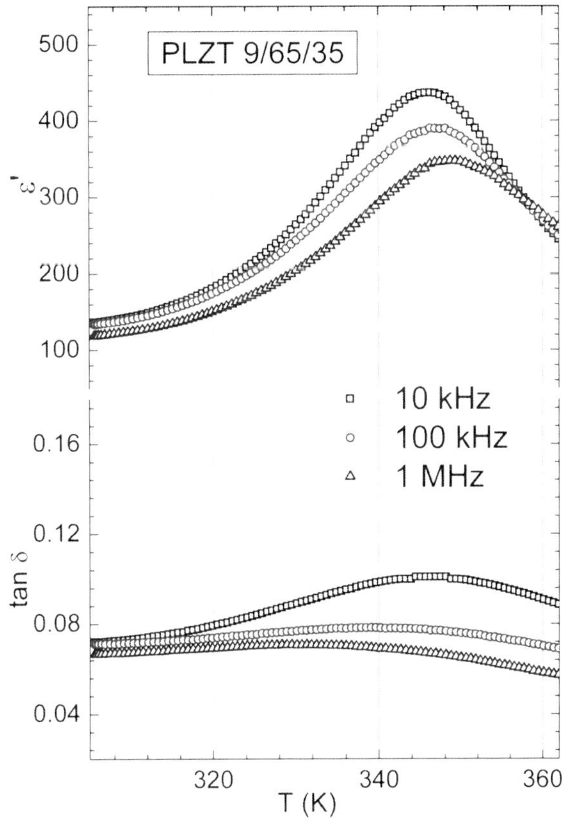

Fig. 10.7 Dielectric permittivity ε′ and loss tan δ versus temperature T, measured on a PLZT 9/65/35 film deposited by RF-PLD (RF power 100 W)

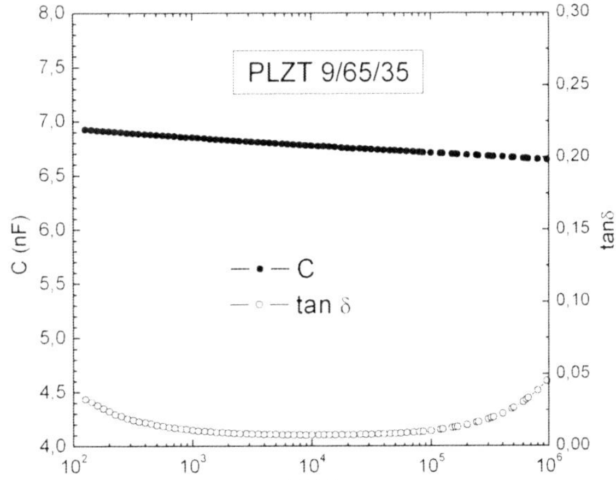

Fig. 10.8 Frequency dependence of capacitance and loss at room temperature measured on a PLZT 9/65/35 film deposited by RF-PLD (RF power 100 W)

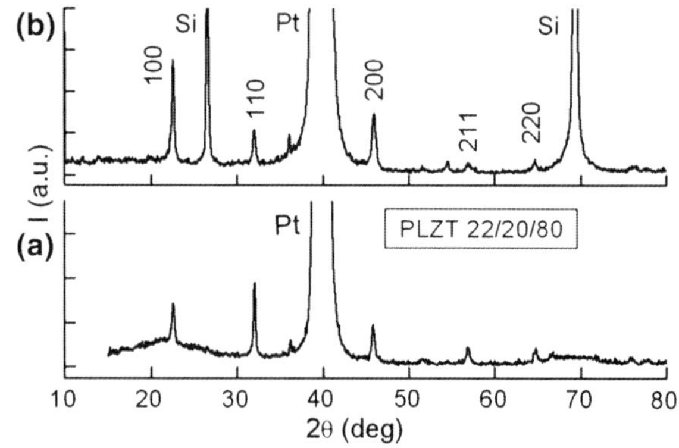

Fig. 10.9 XRD spectra of PLZT 22/20/80 thin films grown on Pt/Si by **a** PLD and **b** RF-PLD

range 1–100 kHz. The increase below 1 kHz is due to the contribution of electrical conductivity.

10.2.2 PLZT 22/20/80 Thin Films

While many investigations focused on the Zr-rich region of the PLZT phase diagram, only few studies have been dedicated to the compositions in the Ti-rich region, despite their high quadratic electro-optic effect and electrostrictive coefficients [48, 49]. We have selected for our investigations the composition $(Pb_{1-3x/2}La_x)(Zr_{0.2}Ti_{0.8})O_3$ with x = 0.22 (PLZT 22/20/80). Details about this material and target preparation can be found in [49]. It has been shown that the tetragonal distortion is decreased with La addition and for the composition PLZT 22/20/80 a relaxor cubic structure is obtained [49].

We have deposited PLZT 22/20/80 thin films on different substrates by PLD and RF-PLD [33, 36–42].

Several sets of films have been deposited on Pt/Si substrates in the following conditions: laser wavelength = 266 nm, pulse duration = 10 ns, pulse repetition frequency = 10 Hz, laser fluence = 2 J/cm^2, oxygen pressure = 0.4 mbar and substrate temperature = 650 °C. The deposition has been performed by PLD or by RF-PLD (RF power = 200 W).

Figure 10.9 shows the XRD spectra of PLZT 22/20/80 thin films grown on Pt/Si by (a) PLD and (b) RF-PLD. It can be observed that films grown by PLD have some amorphous phase, while the growth by RF-PLD produces pure crystalline phase films. These structural differences are similar with those obtained on PLZT 9/65/35 and discussed in the Sect. 10.2.1.

AFM top-view images of PLZT 22/20/80 thin films obtained by PLD (a) and RF-PLD (b) are presented in Fig. 10.10. It can be observed that the film grown by RF-PLD shows larger grains and a more compact structure.\

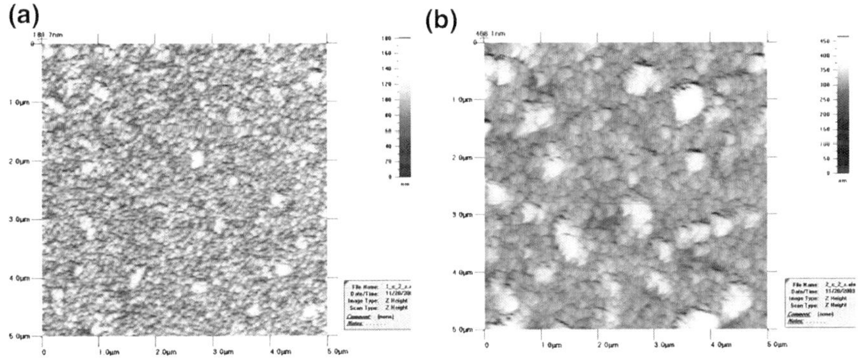

Fig. 10.10 AFM *top-view* images taken on a 5 × 5 μm area of PLZT 22/20/80 thin films obtained by PLD (**a**) and RF-PLD (**b**)

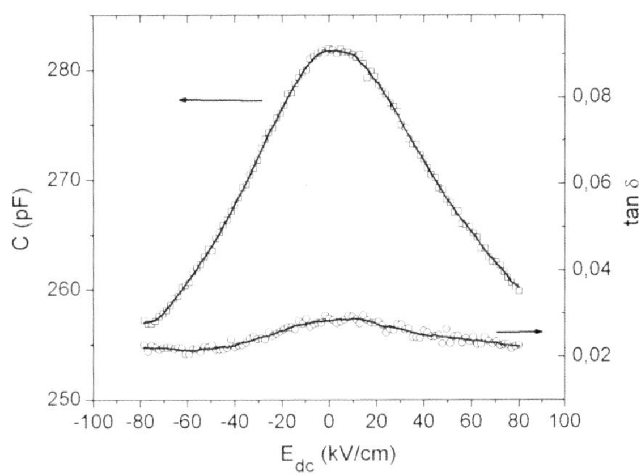

Fig. 10.11 Capacitance-bias electric field dependence C-E_{dc} and tan δ-E_{dc} measured on a PLZT 22/20/80 deposited on Pt/Si by RF-PLD

On these films we have investigated mainly the nonlinear variation of dielectric permittivity with an applied dc electric field. This effect is interesting for tunable dielectric devices. The capacitance tunability is defined as [C(0) − C(E)]/C(0).

The variation of capacitance C and loss tan δ with the bias electric field has been obtained by applying a dc field E_{dc} = 0–80 kV/cm and by measuring the response to a low level ac signal with frequency 1 kHz–20 MHz. Figure 10.11 shows the C-E_{dc} and tan δ-E_{dc} curves measured at 1 kHz on a PLZT 22/20/80 deposited on Pt/Si by RF-PLD.

The capacitance tunability measured at different ac signal frequencies (1 kHz–20 MHz) and dc field E_{dc} = 80 kV/cm varied in the range 10–20 %.

Films deposited in similar conditions but without RF showed a similar dependence but higher values of dielectric loss [38].

Figure 10.12 shows that the variation of capacitance with bias electric field is obtained also at higher electric field (the maximum E_{dc} value applied in our experiments was E_{max} = 150 kV/cm). Moreover it can be observed that the

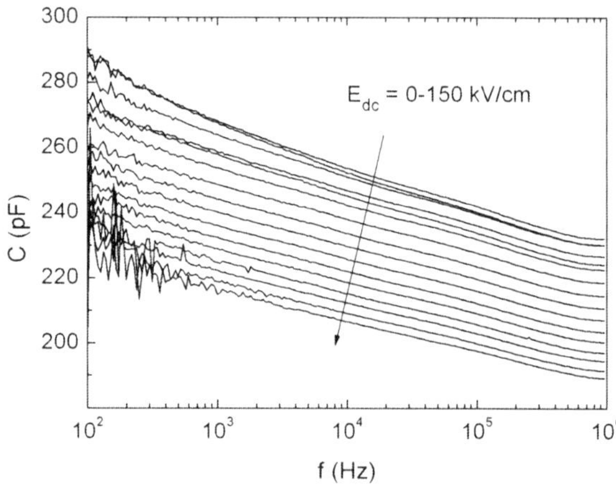

Fig. 10.12 Capacitance versus frequency for different bias field amplitudes in the range 0–150 kV/cm measured on a PLZT 22/20/80 deposited on Pt/Si by RF-PLD

difference C(0) − C(E_{max}) remains almost constant for frequencies up to 1 MHz, while C(0) decreases with frequency, therefore the relative variation [C(0) − C(E_{max})]/C(0) (tunability) increases with frequency.

Complex impedance spectroscopy has been employed to measure the ac conductivity of the PLZT 22/20/80 thin films deposited by PLD and RF-PLD [50]. The ac conductivity has been studied as a function of frequency and dc bias field amplitude. A Jonscher-type dependence [51]:

$\sigma(\omega) = \sigma_0\left[1 + (\omega/\omega_p)^n\right]$ (where σ_0 is the dc conductivity, ω_p is a characteristic relaxation frequency and n is a fractional exponent ($0 < n < 1$)) was followed by the $\sigma(\omega, E)$ curves.

Thus, at high frequencies and/or low field amplitude the frequency dependent term is dominant and it is only slightly dependent on dc field values, while at low frequencies and/or high field amplitude the dc conductivity contribution is dominant. Measurements at bias fields above 100 kV/cm indicated that films grown by RF-PLD have the dc conductivity ($\sigma_0 \sim 10^{-9}\ \Omega^{-1}\mathrm{cm}^{-1}$) decreased by one order of magnitude with respect to PLD-grown films. This is probably due to a minor amount of oxygen vacancies in these films [32]. These results are in agreement with the main improvements which are expected from RF-PLD system, resulting from the effect of RF-plasma species improving oxygen incorporation in the film structure.

Further improvements in the physical properties of PLZT 22/20/80 have been achieved by depositions on other substrates like Nb-doped $SrTiO_3$ (Nb:STO) [41]. Film growth has been performed by RF-PLD, by using an ArF ecimer laser at 193 nm wavelength and 10 Hz pulse repetition frequency, with a laser fluence of 1.8 J/cm^2. The growth has been done on Nb:STO and Pt/Si substrates heated at 600 °C, in an oxygen atmosphere with pressure = 30 Pa.

Figure 10.13 shows XRD patterns of the PLZT thin films deposited on Nb:STO and Pt/Si in the same conditions. It can be observed that while the growth on Pt/Si substrate produces a polycrystalline film with random orientation, the Nb:STO substrate induces a highly c-axis preferred orientation growth. The cubic lattice

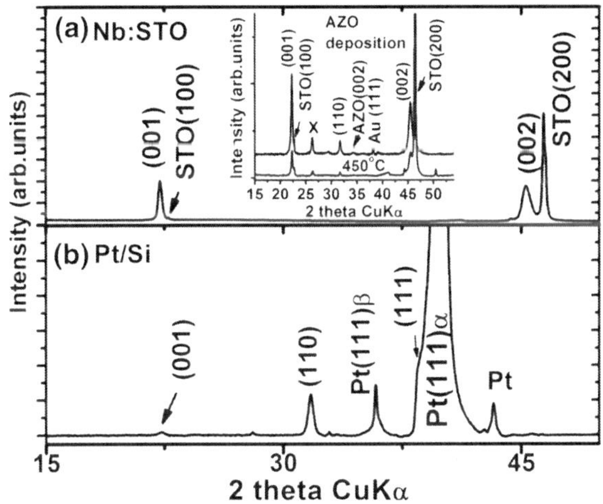

Fig. 10.13 XRD patterns of the PLZT 22/20/80 thin films deposited on Nb:STO (*upper graph*) and Pt/Si (*lower graph*) in the same conditions

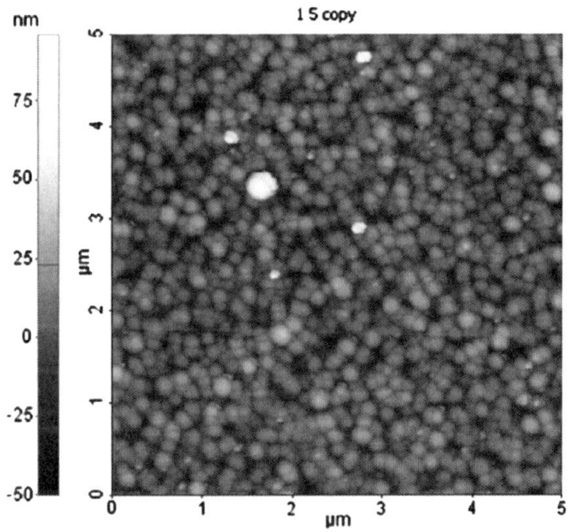

Fig. 10.14 AFM image of a PLZT 22/20/80 deposited on Nb:STO substrate

parameters of the films (3.991 Å for the film on Nb:STO and 3.987 Å for the film on Pt/Si) are close to the bulk value (3.987 Å) which indicates that the PLZT films are relaxed and the stoichiometry is close to the target chemical composition. Upon the deposition of a top transparent conductive electrode of Al-doped ZnO (AZO) on some films (in order to allow electro-optic characterization) and the reheating of the heterostructures to 450 °C, a new peak, besides the AZO (002) reflection, appears in the spectra (inset in Fig. 10.13). This has been assigned to a Pb_3O_4 phase [41] which probably appears due to the reheating at 450 °C.

The AFM image in Fig. 10.14 shows that the deposition on Nb:STO induces growth of a uniform structure, with round, well-defined crystallites of about

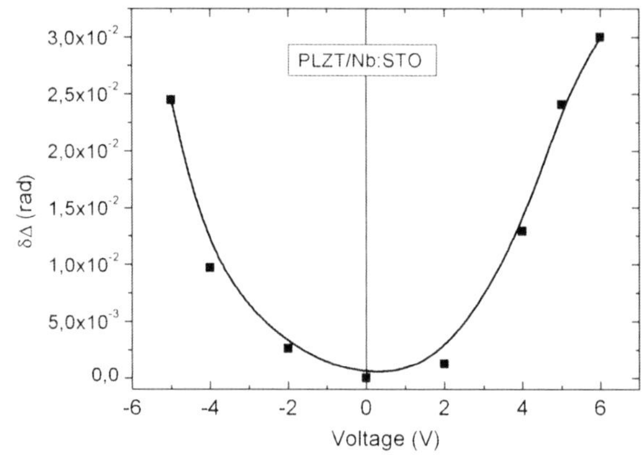

Fig. 10.15 Birefringence shift for the PLZT 22/20/80 thin film deposited on Nb:STO as a function of the applied dc electric field at $\lambda = 540$ nm and a polar angle of incidence of 65°

300 nm. The roughness value is low (around 12 nm). The growth on Pt/Si electrode produces a more compact but less uniform structure, with higher roughness (21 nm) [41].

Electro-optical investigations have been further carried out by using reflection type spectroscopic ellipsometry measurements [41]. The investigations have been performed on AZO/PLZT/Nb:STO heterostructures, due to the superior morphologic and structural properties of PLZT/Nb:STO thin films with respect to those of PLZT/Pt/Si films. In these investigations the variation of the polarization state of the linearly polarized light upon reflection at the surface of the PLZT film is measured as a function of the applied electric field, in order to obtain the electro-optic coefficient.

The experimental phase shift values shown in Fig. 10.15 evidence the quadratic electro-optic behaviour of PLZT 22/20/80 thin films.

The quadratic electro-optic coefficient R has been evaluated from the relationship:

$$R = \delta\Delta \times \lambda / \left(\pi n^3 E_z{}^2 d\right),$$

where $\delta\Delta$ is the relative phase shift in radians, λ is the wavelength, n is the refractive index of the film, E_z is the applied electric field and d is the film thickness [41]. The calculated value is $R = 0.331 \times 10^{-17}$ m^2/V^2, which is relatively small compared to that of bulk [41].

Further improvements in the quality of PLZT thin films have been attempted by substituting the Nb:STO substrate, which does not function as a good electrode with ohmic contact in the heterostructure, by a $SrRuO_3$ (SRO)/STO substrate [42]. Besides good lattice matching with PLZT and STO, SRO also shows good electrical properties. The SRO layer has been deposited on STO (001) substrates at 700 °C and 5 Pa oxygen atmosphere by using an ArF laser working at 193 nm. The PLZT 22/20/80 thin film has been deposited over the SRO layer without opening the deposition chamber. The deposition of PLZT has been performed at temperatures between 600 and 700 °C. For electro-optic investigations an AZO

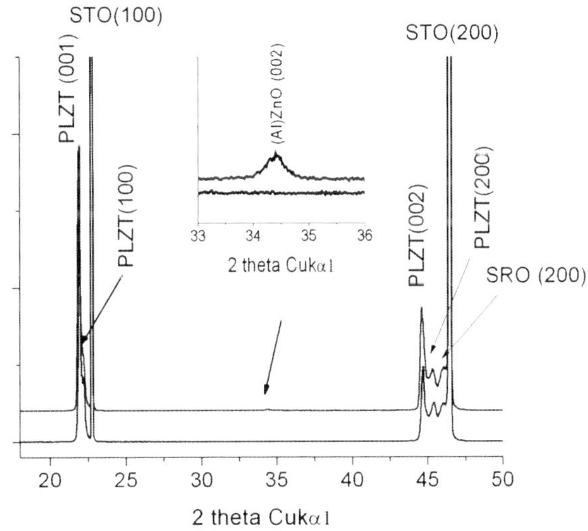

Fig. 10.16 The XRD pattern of the PLZT 22/20/80 deposited on SRO/STO substrate, before and after the AZO deposition

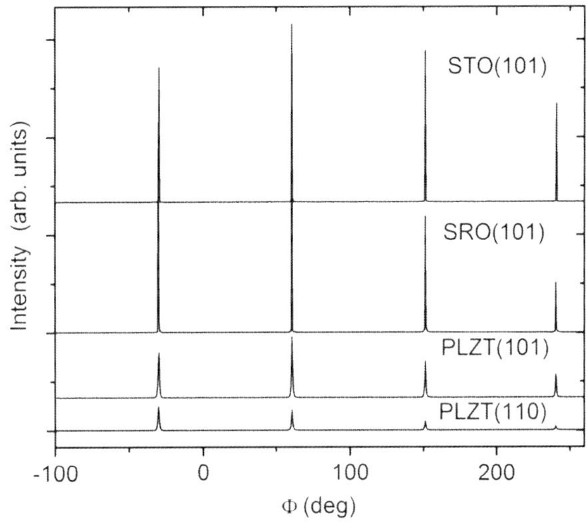

Fig. 10.17 Set of Φ scans of the AZO/PLZT/SRO/STO(100) thin films taken at different reflection peaks

layer has been deposited on top of the PLZT/SRO/STO structures, at temperatures around 250 °C.

Figure 10.16 shows the XRD patterns of PLZT thin films before and after the AZO deposition. The films exhibit only (h00)/(00l) reflections with no evidence of other peaks or pyrochlore phase. After the AZO deposition an additional Zn(Al)O (002) peak is observed, besides the same PLZT structural features.

Φ scans taken at different reflection peaks of the heterostructure confirmed the four-fold in-plane symmetry of the PLZT, SRO and STO and the epitaxial growth of the films (Fig. 10.17).

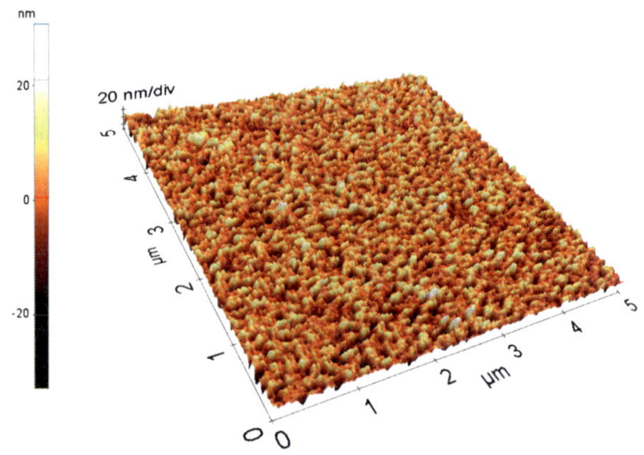

Fig. 10.18 AFM image of PLZT 22/20/80 thin film deposited on SRO/STO substrate

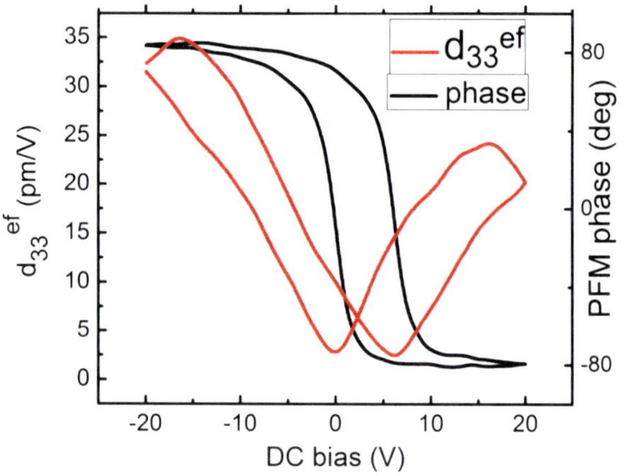

Fig. 10.19 Piezoresponse measurements of the PLZT/SRO/STO thin films (from [42])

The surface features of the PLZT/SRO/STO thin film can be observed in the AFM image shown in Fig. 10.18. The surface of the film has a relatively small roughness (around 7 nm), with no droplets or other defects.

In order to verify the stoichiometry of the samples, the chemical composition of the PLZT thin films deposited on SRO/STO substrate has been investigated by using Sputtered Neutral Mass Spectroscopy (SNMS) technique [42], with special emphasis on the lead content. A good matching between the lead content of the PLZT film and target has been obtained. Moreover, the SNMS depth profile evidenced sharp interfaces between the different layers [42].

Piezoresponse force microscopy (PFM) was used to test the switching properties and the piezoelectric effect in PLZT thin films. During the measurements, a dc bias and a test ac electric field have been applied between the Pt tips of the device and the bottom electrode of the samples. The PLZT thin films show good switching and piezoelectric properties, as shown in Fig. 10.19. The locally measured values of the effective piezoelectric coefficient $d_{33}^{eff} \sim 35$ pm/V are comparable with those measured on PZT 20/80 thin films [42].

Electro-optical investigations carried out by using reflection type spectroscopic ellipsometry measurements evidenced the linear variation of the birefringence shift with the square of the bias electric field, thus the quadratic electro-optic behaviour of PLZT 22/20/80 thin films [42].

10.3 Relaxor Ferroelectric PMN-PT Thin Films

Relaxor ferroelectric $Pb(Mg_{1/3}Nb_{2/3})O_3$ (PMN) and its solid solutions with $PbTiO_3$, $Pb(Mg_{1/3}Nb_{2/3})_{1-x}Ti_xO_3$ (PMN-PT), have been intensively investigated, both in bulk [52, 53] and in thin film form [54–64], due to their excellent dielectric, piezoelectric and electromechanical properties. Unlike normal ferroelectrics, which show a narrow huge dielectric peak at the temperature of para-electric-ferroelectric phase transition, relaxor ferroelectrics are characterized by a broad dielectric maximum at T_m which shifts with frequency increasing toward higher temperatures [52, 53]. The relaxor behaviour in PMN-PT is due to disorder in the occupation of B site of the perovskite structure by Mg^{2+}, Nb^{5+} and Ti^{4+} ions with different valence and size [52, 53]. The ferroelectric order is preserved only on short range, in the so-called polar nano regions (PNRs). Increased correlations between PNRs when temperature decreases, together with freezing of their movement at low temperatures give rise to a huge and broad dielectric anomaly at T_m, generally not far from room temperature [52, 53].

Thus, the dielectric permittivity and piezoelectric coefficients of relaxor ferroelectrics are high in a large temperature range, making them useful for applications.

In this section we review some of our most significant results regarding PMN and PMN-PT ($x = 0.1$) thin film deposition [32, 33, 55–67]. PMN and PMN-PT thin films have been deposited by pulsed laser deposition (PLD) on different substrates and electrodes. For every type of heterostructure the deposition conditions have been optimized in order to obtain good crystalline properties.

Structural investigations demonstrated that the best results have been obtained for films deposited on LSCO electrodes in different heterostructures: LSCO/Si, LSCO/MgO and LSCO/Pt/Ti/MgO. The perovskite oxide LSCO electrodes have been employed due to their high electrical conductivity and good lattice matching with PMN-PT. Moreover, it has been shown that they are also good buffer layers and prevent oxygen vacancies at the interface [34]. Since the LSCO electrode quality is influenced by the substrate type, several combinations have been tested. The LSCO layers have been grown by PLD on the different substrates in the same experimental set-up as the PMN and PMN-PT thin films.

The PMN-PT thin films deposition has been performed in the following optimized conditions: laser wavelength = 266 nm, laser fluence = 2 J/cm^2, oxygen pressure = 0.4 mbar and substrate temperature = 600 °C. In Fig. 10.20a–c the XRD patterns corresponding to PMN-PT films deposited on LSCO/Si, LSCO/MgO and LSCO/PT/Ti/MgO substrates are shown.

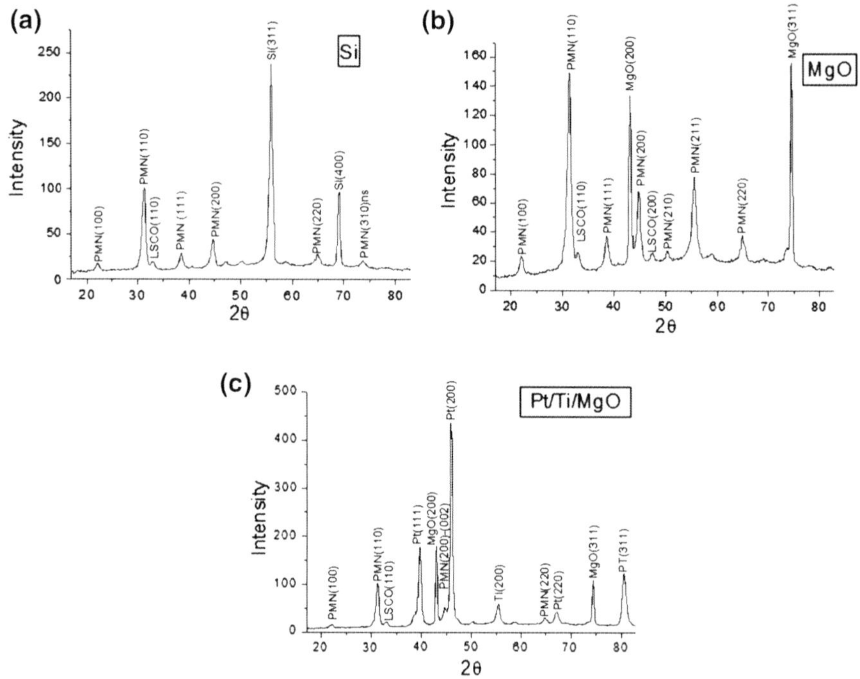

Fig. 10.20 **a** X-ray diffraction spectra of PMN-PT ($x = 0.1$) films deposited on (**a**) LSCO/Si substrates; **b** LSCO/MgO and **c** LSCO/Pt/Ti/MgO

The displayed XRD patterns confirm the formation of a pure perovskite phase and good crystallinity. No pyrochlore phase was detected. The films are policrystalline and no preferential orientation is observed.

SIMS analysis performed on these heterostructures [66] evidenced a uniform distribution of elements and sharp interfaces. The sharp interface was confirmed also by cross section TEM [66]. The growth achieved on LSCO was columnar [66]. Also, the interface with LSCO was smooth, as can be observed in Fig. 10.21, where a detail of cross section TEM is shown. No intermediate layer of different stoichiometry was detected, as demonstrated by SAED analysis, where only PMN and LSCO were identified [66].

PMN and PMN-PT have been deposited also on other technologically important substrates like Au/Pt/NiCr/Glass and Pt/NiCr/Si [65, 67]. It has been found that the deposition of PMN directly on metallic electrodes was accompanied by the formation of minor pyrochlore phase. This is probably due to the interdiffusion of lead from PMN in Pt and Au electrodes and the subsequent formation of a mixed interface layer and a pyrochlore phase in PMN.

In order to avoid this undesired effect, a multitarget system was employed to deposit first an intermediate layer of PZT on the Pt/NiCr/Si structure, followed by the deposition of the PMN layer. In this way the pyrochlore phase was very much reduced [65]. However the interface between PMN and PZT is not so smooth, as

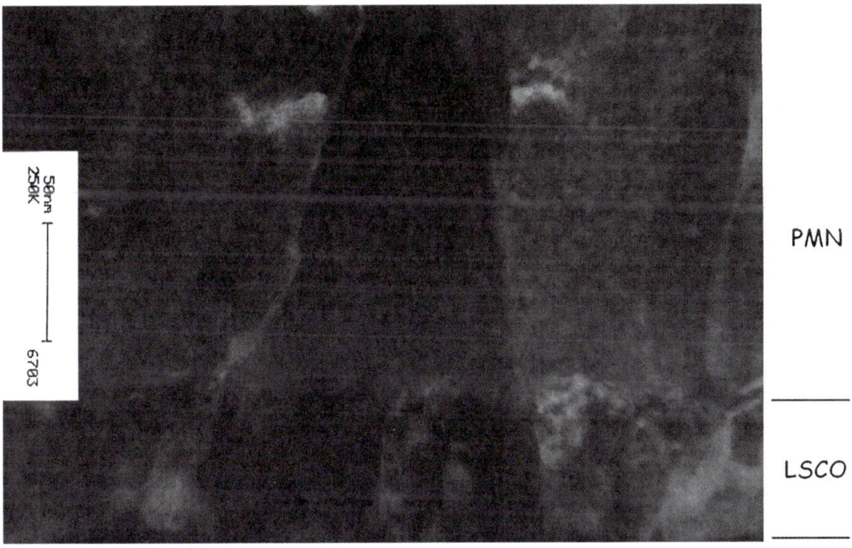

Fig. 10.21 Cross section TEM of a PMN-PT/LSCO heterostructure

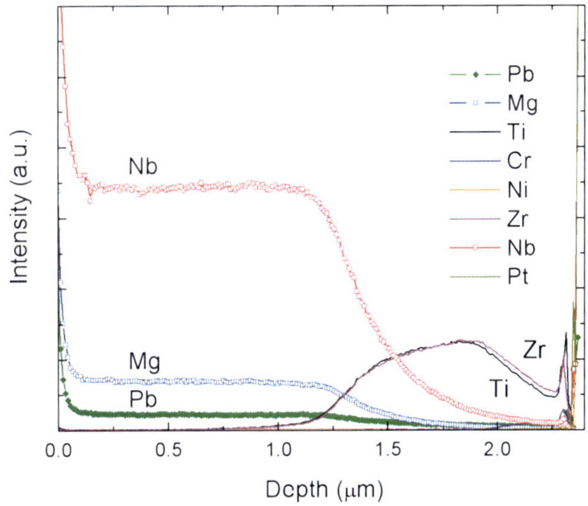

Fig. 10.22 Composition profiles measured by SIMS on a PMN/PZT/Pt/NiCr/Si heterostructure

shown by the SIMS spectra of PMN/PZT/Pt/NiCr/Si heterostructures (Fig. 10.22) which evidenced a uniform distribution of elements in the PMN film, but a strong interdiffusion at PMN/PZT interface [67].

Dielectric measurements (capacitance C and loss tan δ) have been performed at different frequencies and temperatures. The obtained dielectric permittivity varied between 500 and 900. The lower value with respect to bulk is explained by the presence of the interface layer and grain boundaries.

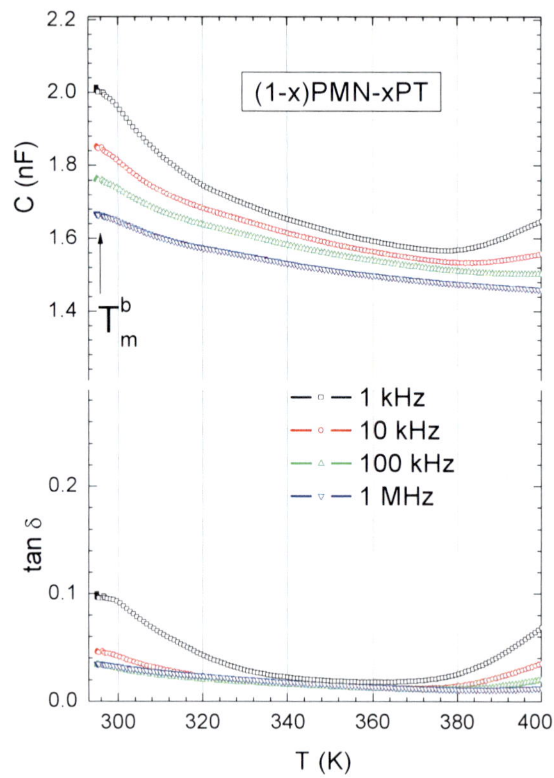

Fig. 10.23 Capacitance C and loss tangent tan δ versus temperature T for a PMN-PT (x = 0.1) thin film deposited on a Pt/NiCr/Si substrate. The *arrow marks* the temperature $T^b{}_m$ of the dielectric maximum for bulk material

Figure 10.23 shows the dependence of capacitance C and loss tangent tan δ with temperature for a PMN-PT (x = 0.1) thin film deposited on a Pt/NiCr/Si substrate. The arrow marks the temperature $T^b{}_m$ of the dielectric maximum for bulk material. It can be seen that, unlike bulk material [68], the dielectric properties of PMN-PT thin films are characterized by a strong dispersion also above the temperature of dielectric maximum T_m, which in PMN-PT (x = 0.1) is about 290 K at 1 kHz and increases to 310 K at 1 MHz [68].

This dispersion (variation of capacitance with the frequency of measurement signal) is explained in relaxors by the existence of a broad distribution of relaxation times, due to polar nanoregions (PNRs) and interactions among them. However, above T_m the interactions among PNRs decrease and at $T \gg T_m$ they behave almost like independent relaxor units. Thus, at $T > T_m$, the dielectric $C(\omega, T)$ curves superpose in normal relaxor ferroelectrics. The existence of frequency dispersion in PMN-PT thin films also above T_m are likely related to the existence of other relaxation mechanisms in this frequency range, like that due to space charge in the interface layer [69].

The piezoelectric properties of the PMN-PT thin films have been characterized by an acoustoelectric probe for d_{33} longitudinal piezoelectric coefficient measurement [70]. This measures the electrical response of the film to a mechanical

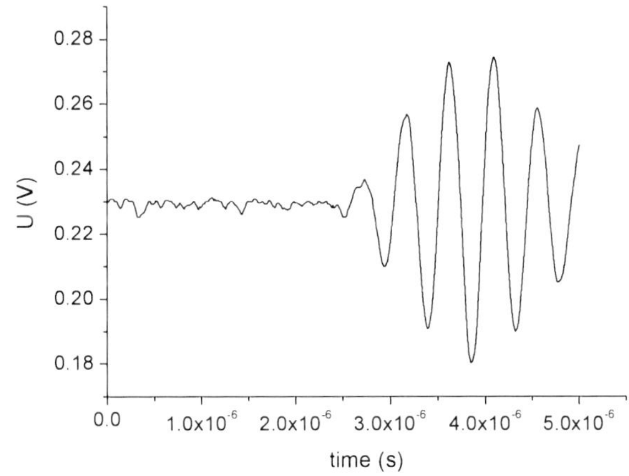

Fig. 10.24 Piezoresponse of an as-grown PMN-PT thin film to a burst acoustical signal working at 2 MHz frequency

stress produced by a 2 MHz acoustic longitudinal wave. An example of a PMN-PT film piezoresponse to a burst acoustical signal of few cycles is shown in Fig. 10.24.

The response signal is compared to that obtained on a standard piezoelectric thin plate in the same conditions, in order to evaluate d_{33} piezoelectric coefficient.

The measurement has been performed on the as-grown PMN and PMN-PT thin films, without any prior polarization treatment. A piezoelectric d_{33} coefficient of 40–60 pC/N has been measured, with some variations between films grown in different conditions.

The presence of a piezoelectric effect in as-grown ferroelectric this films indicates the existence of self-polarization. This effect has been evidenced in other thin films as well [55, 56] and it has been attributed to the presence of oxygen vacancies and structural defects at the interface [56]. The effect of self-polarization in thin relaxor films could be of interest for pyrosensors [56] and other applications based on the pyroelectric effect.

10.4 Lead-Free SBN Thin Films

Strontium barium niobate—$Sr_xBa_{1-x}Nb_2O_6$ (SBN: *x*)—thin films have attracted attention from both research and industry communities due to their interesting electrical and optical properties. Bulk SBN has a relatively complicated tetragonal tungsten-bronze structure ($0.25 \leq x \leq 0.75$), exhibiting a ferroelectric-relaxor behaviour closely related to the value of Sr/Ba ratio [71]. The symmetry at room temperature belongs to the P4bm space group, having a displacive phase transition from (*4mm*) tetragonal ferroelectric phase to (*4/mmm*) paraelectric phase. The NbO_6 octahedral arrangement in the tetragonal tungsten bronze structure (Fig. 10.25) allows the existence of vacant sites in the structure, which means that it is possible to easily modify the SBN material by doping with different dopant elements such as transitional metal species. The bulk SBN has the largest reported linear electro-optic coefficients of all known materials to date ($r_{33} \sim$ 1,410 pmV^{-1} for $Sr_{0.75}Ba_{0.25}Nb_2O_6$ composition). This value of the linear electro-optic coefficient is many times larger than the value exhibited by $LiNbO_3$ ($r_{33} \sim$ 31 pm/V), which is the primary electro-optic material for industrial applications [72].

But there are other intrinsic features of the bulk SBN materials, such as the pyroelectric property, owing to which there is a high demand for SBN thin films in the industry. For tetragonal bulk SBN, the pyroelectric coefficients are higher than those of other well-known ferroelectric materials [73]. The lasing of Yb^{3+}—doped SBN crystals has been revealed for both ferroelectric and paraelectric phases of SBN: 60 crystals, a stable laser radiation at 1,070 nm being observed [74]. Using ultrafast laser inscription technique, buried optical waveguides have been written in multidomain strontium barium niobate with the domain orientated at 180° with respect to the $\pm c$ sample directions [75].

Recently, interesting n-type thermoelectric behaviour of nonstoichiometric tungsten-bronze SBN single crystal has been found by Lee et al. [76]. Another important application for SBN materials was reported to be in photonic crystal structures [77]. Photonic crystals are 2-D or 3-D optical engineered structures which forbid certain wavelengths due to periodic scattering. Photonic crystals can be built with specific photonic band gaps which control light propagation, irrespective to direction or polarization.

For example, photonic crystal waveguides are responsible for light coupling in and out of an integrated circuit, as well as inside the circuit. The final purpose of using photonic crystal structures is to reduce the optical losses and the size of the modulators, switches, etc. Photonic crystals slab (or 2-D tunable photonic crystal structure) using LNO thin films produced by PLD, have been fabricated by employing focused ion beam (FIB) technique [78]. If the thin films-based photonic crystals have limited tunability voltage due to small electro-optic coefficient value, by employing materials with high linear electro-optic coefficient value such as tungsten bronze SBN, the shift induced in the photonic band gap wavelength can be significant [77].

As a result, there is an increasing trend to use high-quality SBN thin films, mainly in electro-optic effect driven devices [73]. However, the transfer of the SBN single crystal properties to the SBN in thin films form is not very successful, mainly because of the SBN complicated stoichiometry and crystallographic structure. The complicated chemistry of this material translates into a major difficulty in controlling stoichiometry of high quality SBN thin films. Works on heteroepitaxial growth of SBN thin films made by techniques such as MOCVD [79, 80], plasma enhanced-CVD [81], sol-gel [82–85] or PLD [22, 86, 87], have reported good optical and electro-optical properties by MOCVD (second harmonic generation) and PLD (linear electro-optic coefficient r_{33} = 844 pm/V). Due to some unique properties with respect to other depositions techniques, PLD and the associated RF-PLD technique seem to be suitable for the deposition of SBN:x thin films. With respect to classical PLD technique, in RF-PLD a supplemental plasma beam of excited and ionized species generated by a RF discharge can be directed towards the substrate during the deposition process [88]. This hybrid technique has some important advantages:

- At the early growth stages, reactive oxygen supplied in the deposition-reaction area will strongly diminish the oxygen vacancies at the interface.
- As the film is growing, the atomic oxygen arriving quasi continuously on the surface (the RF generators works at 13.56 MHz) will contribute to the rearrangement of atoms coming from the laser plasma plume in order to achieve the right stoichiometry and to incorporate the desired amount of oxygen in the structure.
- Decreasing of the deposition temperature and the diminishing the substrate roughness as a result of supplementary energy assured of rf plasma species.

For SBN thin films deposition, we have selected one specific composition, namely $Sr_{0.5}Ba_{0.5}Nb_2O_6$ or SBN: 50, situated at the border between the high Sr content (relaxor-type) and low Sr content (normal ferroelectric behaviour type) compositions. Even though the SBN compositions with high Sr content show high value electro-optical coefficients, they have relatively low Curie temperature. This makes rather critical the use of such material for high speed photonic devices where, due to intense packs of light, high temperatures can be achieved. Until now, SBN thin films with high Sr content (SBN: 60 and SBN: 75) have been investigated, despite the fact that SBN: 50 has the highest Tc temperature (128 °C), as compared to SBN: 60 (78 °C) or SBN: 75 (56 °C). Moreover, SBN: 50 films exhibit high linear electro-optic coefficient $r_{33} = 180$ pm/V for single crystal, as compared to the standard material, lithium niobate ($r_{33} = 31$ pm/V) [21].

SBN thin films have been obtained starting from a ceramic, stoichiometric $Sr_{0.5}Ba_{0.5}Nb_2O_6$ target. The addition of radiofrequency discharge in oxygen makes possible the obtaining of heteroepitaxial SBN: 50 thin films on two different substrate types, namely MgO and conductive Nb:STO single-crystals. A parametric study has been performed to check the influence of different deposition parameters (laser fluence, oxygen pressure, and substrate type or deposition temperatures) and to identify the best deposition conditions to obtain epitaxial SBN: 50 thin films. Both MgO and Nb:STO substrates were heated to deposition temperature (650 and 700 °C) with a ramp of 40 °C/min and then, after deposition, cooled with 5 or 10 °C/min, while the oxygen pressure during deposition was maintained at 0.6 mbar. The cooling process was done in the presence of the radiofrequency discharge. The substrates were placed at a distance of 4.5 cm from the target. The vacuum system allowed obtaining of a base pressure of about 10^{-6}–10^{-7} mbar before deposition. The laser fluence was between 2.0 and 2.3 J/cm^2 and the radiofrequency oxygen plasma beam power was set at 100 W during deposition and 150 W during cooling procedure [89]. Heteroepitaxial SBN: 50 thin films were obtained on magnesium oxide (MgO) substrate, which is the most frequently used substrate due to the high refractive index contrast and small lattice mismatch of MgO with respect to SBN [90–95]. Based on these facts, the SBN/MgO system

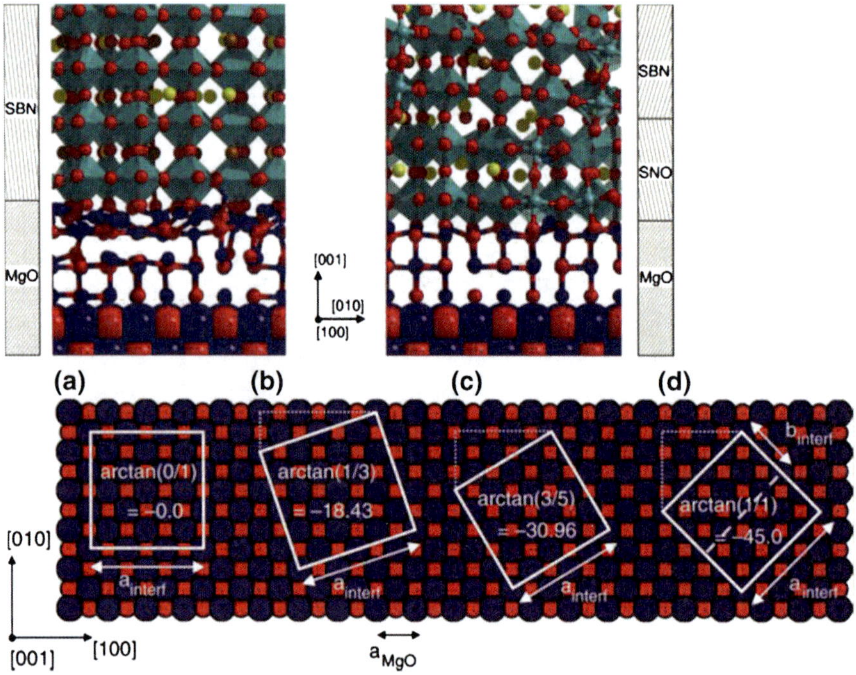

Fig. 10.25 Domains twinned in-plane, with the SBN: x (100) axes parallel to the MgO (310) and (130) equivalent axes [from 90]

has been intensively studied for optical and electro-optical applications of SBN thin films.

Regarding the matching between the lattice parameters, (001) oriented MgO single crystal is an appropriate substrate for the growth of SBN: x thin films because MgO unit cell parameter (0.42 nm) is about one-third of the SBN unit cell parameter in the a–b plane (1.245 nm). The studies reporting successful growth on MgO substrates of highly c-axis oriented SBN thin films have revealed the existence of three in-plane orientations of the crystallites relative to the MgO (100) azimuth: 0°, arctan(1/3) = 18.43°, and arctan(3/5) = 31° [89, 90, 93, 95]. The in-plane orientations of (001) SBN/(100) MgO thin films, with the one at arctan(1/3) = 18.43° reported as present and dominant in many reports, have been explained by S. Thony et al. as a result of the electrostatic interactions between the film's positive species and the substrate Mg^{+} ions, which occurs in the SBN/MgO system due to small lattice mismatching [95]. But this explanation was considered as not being adequate by Willmott et al. [90]. Using molecular dynamics simulations, they showed that the predominance of the arctan(1/3) = 18.43° in-plane orientation of the SBN films with respect to the other two reported (0° and 31°), is due to the formation of a parasitic $SrNb_2O_5$ (SNO) phase at the interface between the film and the substrate, especially for high Sr contents (Fig. 10.25).

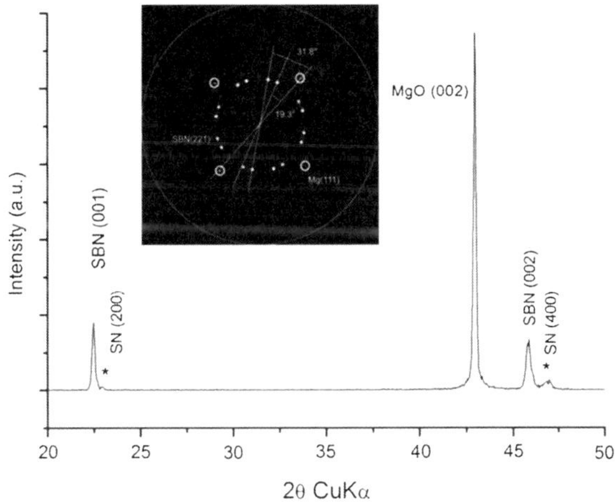

Fig. 10.26 XRD patterns for SBN: 50/MgO thin film, cooled with 10 °C/min from 650 °C to room temperature in oxygen radiofrequency plasma at 150 W. The inset: pole-figure for the SBN:50/ MgO of (221) reflections [from 89]

Indeed, using a fast substrate cooling procedure in the presence of oxygen radiofrequency plasma, highly c-axis oriented SBN/MgO thin films have been obtained. The structural characterization was carried out by X-ray diffraction (XRD). The X-ray analyses were performed using a conventional X-ray (θ–2θ) scan (Bragg-Brentano geometry) and pole figures gathered from the (221) SBN reflection. The XRD spectrum (Fig. 10.26) displays the peaks of a c-axis oriented SBN major phase, along with less intense peaks of a parasitic phase identified as $SrNb_2O_5$ (SNO), also textured.

The X-ray pole figure of this film, measured from the (221) SBN reflection (Fig. 10.26 inset), reveals the occurrence of two in-plane domain orientations relative to the MgO substrate: ±19.3° and ±31.8°, respectively. The in-plane orientations are well defined, equally bright, clearly separated from each other and there is no ring connecting them. The 31.8° orientation is favoured by its lowest lattice mismatch, while the 19.3° orientation is to be associated with the appearance of the observed SNO phase acting as a transitional layer which induces the electrostatic favoured 19.3° orientated domains.

For the SBN: 50/MgO thin films cooled slowly and in the presence of radiofrequency oxygen plasma beam, the X-ray (θ–2θ) scan reveals, besides the substrate MgO (200) peak, only the (001) and (002) reflections of the film, which translates into a c-axis preferred orientation growth (Fig. 10.27). There is no trace of the parasitic SNO phase. The SBN structure has the structure of tetragonal tungsten bronze (TTB) with cell parameters a = 1.246 nm and c = 0.3952 nm respectively, for a $Sr_{0.5}Ba_{0.5}Nb_2O_6$ (SBN50) composition, according to JCPDS-file 39-0265.

The increase of the relative intensity of the (001) peak could be most likely attributed to a desired (001) orientation of the SBN grains or to slight changes in the mode and degree of occupation of the Ba^{2+} and Sr^{2+} sites along the $[001]_{SBN}$ axis [93].

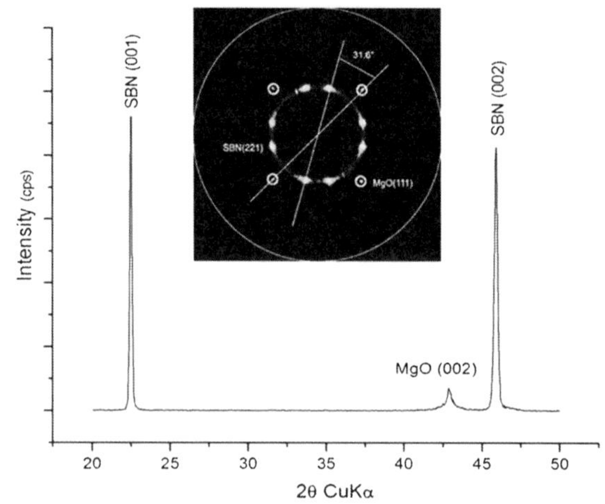

Fig. 10.27 XRD patterns for SBN: 50/MgO thin film, cooled with 5 °C/min from 650 °C to room temperature in oxygen radiofrequency plasma at 150 W. The inset: pole-figure for the SBN: 50/MgO of (221) reflections (from [89])

The values of the full width at half-maximum (FWHM) are 0.17° and 0.25°, respectively, showing a good crystalline quality relative to the mis-orientation from the c-axis, indicating a heteroepitaxial growth of the SBN films onto the MgO substrate. A slight increase of the FWHM for the high angle peak can be noticed, likely due to a strain effect. We applied the treatment of the line broadening used by de Keijser et al. for heteroepitaxial films of lead titanate and strontium titanate grown on MgO substrate, in order to extract the contribution of the strain, w_{strain}.

Therefore, from the total line broadening w:

$w = \sqrt{w_o^2 + w_{strain}^2}$ with w_o being the "natural" broadening, $w_{strain} = 2\frac{\Delta d}{d}\tan\theta$ is the broadening due to strain, with d the lattice spacing, Δd the d variation due to strain and θ the diffraction angle [96]. Using these equations and the Scherrer formula for the crystallite size, one obtains a value of 0.43 % for the strain and of 84 nm for the crystallites size. The X-ray pole figure of the SBN:50 sample shows a dominant 31.6° in-plane orientation of the grains relative to the MgO (100) azimuth, which is the growth-direction (Fig. 10.27 inset). Besides this twin orientation, the presence of a multitude of fine spots connected on the same ring could be observed. They are due to crystalline domains growing perpendicular to the substrate, but with a wide domain of different in-plane orientations relative to the c-axis growth direction. The 31.6° value of the orientation of the dominant crystallites is close to the theoretical value of 30.96° = arctan(3/5), which corresponds to the lowest lattice mismatch between SBN film and MgO substrate of 1.16 %, as Willmott et al. [90] presented. This in-plane orientation was found to be predominant in the literature, as already stated, due to the formation of the very thin, parasitic SNO phase which allows the growth of highly mismatched but electrostatically favoured SBN thin films [90, 91, 93]. There are not any reports yet on the presence of highly c-axis textured crystalline domains, but only with a dominant orientation at 31°, along with a multitude of disordered in-plane orientations relative to the c-axis, as it appeared on these RF-PLD deposited SBN: 50/MgO thin films [89].

These features of the SBN thin films grown on MgO substrates are due to the large tetragonal distortion of the unit cell of SBN, a slow cooling rate being useful to release the thermal strain of the SBN thin films and to produce an appropriate growth base toward highly oriented films. As a result, the growth of SBN films on MgO substrates has to account for processes that minimize the lattice mismatch and those that balance the electrostatic forces. Therefore, deposition conditions, including cooling procedures, are critical in controlling the structural properties of the RF-PLD SBN/MgO films. The presence of oxygen radiofrequency plasma discharge, along with a slow cooling rate, has an important role in the obtaining of single-phase c-axis oriented SBN. The non-uniform residual strain in the films is decreased by the slow cooling rate and by the high oxygen partial pressure combined with the radiofrequency system: it results in the growth of SBN thin films directly on the MgO substrate, displaying good crystallinity, high c-axis preferred orientation, and a dominant in-plane orientation of the grains at 31° relative to the MgO (100) axis.

To investigate the change of the refractive index values (birefringence response) under an applied electric field for the SBN: 50 thin films, some facts must be considered. One of the unique features of ferroelectric materials is that they exhibit favourable properties (electric, electro-optic) along only one crystallographic direction and average properties for the other orientations, which in the case of tetragonal tungsten–bronze SBN materials can translates in excellent ferroelectric and electro-optic properties along c-axis [97]. For this reason, 2 % Al doped ZnO (AZO) transparent conductive oxide has been deposited on SBN/Nb:STO, acting as top electrode in the following heterostructure: AZO/SBN/Nb:STO. Deposition of AZO top layer was done at room temperature and 0.05 mbar oxygen pressure. The electrical resistivity of AZO measured by four point method, has been around 7×10^{-3} Ω cm. Reports on electro-optic behaviour of SBN : x thin films use different techniques to measure electro-optic coefficients (linear r_{33} or transverse r_{51}), like phase modulation (Adachi method) [84, 98, 99] and modulated diffraction methods [86, 87]. The birefringence shift has been studied in our case by using reflection type spectrometric ellipsometry.

From a structural point of view, the SBN/Nb:STO thin films were studied by X-ray diffraction using a Panalytical X'Pert MRD diffractometer, in a parafocusing Bragg–Brentano geometry. Supplementary X-ray diffraction analyses have been performed in 2-axes mode measurements along omega-2 theta direction and omega direction, in order to extract the crystal's mosaicity and peak broadening information from the reciprocal space mapping data. We obtained highly crystalline and c-orientated SBN thin films on orientated Nb:STO (001) substrate, as can be seen from Fig. 10.28. The c-parameter value is similar to the previously deposited SBN/MgO thin films (c = 3.943 Å).

A comparison of structural features of SBN thin films deposited either on MgO or Nb:STO has been also done. We performed 2-axes measurements on two SBN thin films deposited on MgO and Nb:STO along omega-2 theta direction and

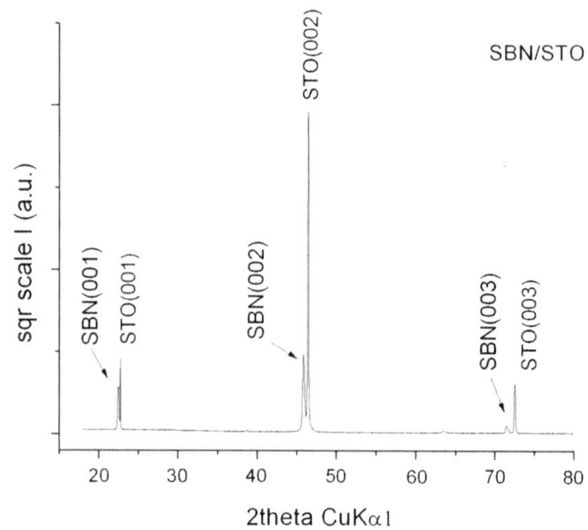

Fig. 10.28 XRD patterns for SBN/Nb:STO thin films

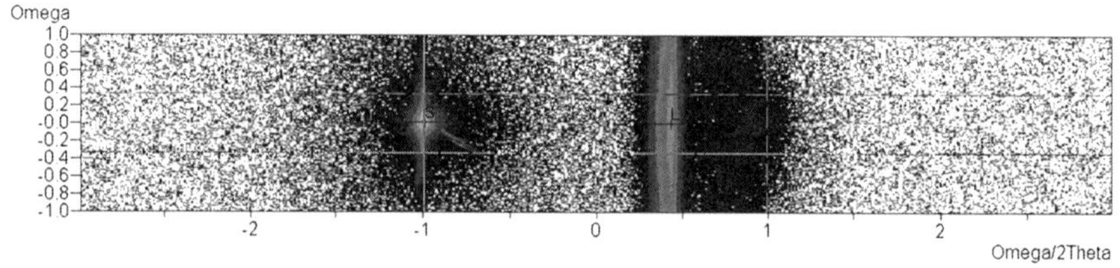

Fig. 10.29 The reciprocal space map for SBN/MgO thin films

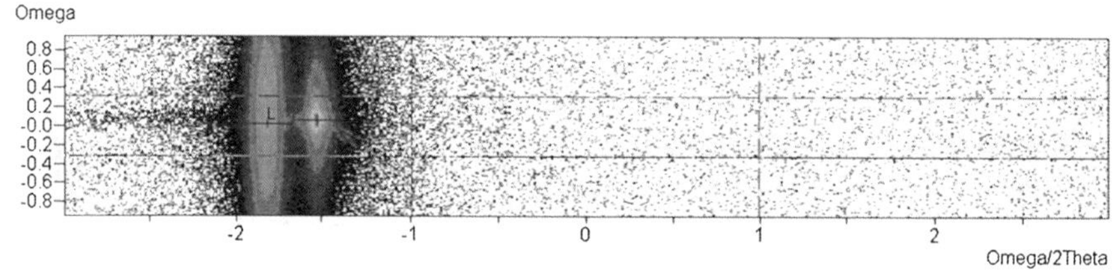

Fig. 10.30 The reciprocal space map for SBN/Nb:STO thin films

omega direction, respectively. The scans were performed in an area around the (002) peak of the SBN film (Figs. 10.29 and 10.30).

The reciprocal space maps reveal similar characteristic (Figs. 10.29 and 10.30). The substrate features, MgO (200) plane and STO (002) plane, respectively, are roughly circular. The features coming from the film are strongly broadened in the omega direction revealing a certain mosaic spread due to the presence of mismatch dislocations. We also observed a broadening along the omega direction due to the thin layer.

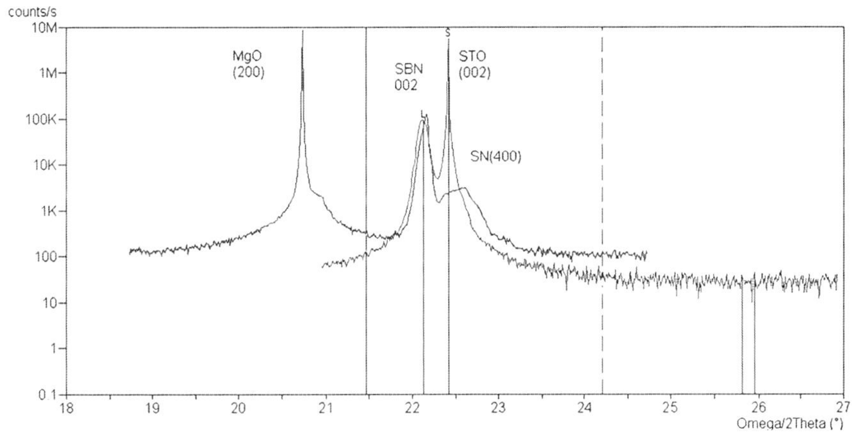

Fig. 10.31 The projection on omega-2theta axis of reciprocal space maps for SBN/MgO and SBN/Nb:STO thin films

For the thin film deposited on MgO there is also an asymmetry along omega-2theta because of a variation in d-spacing due to the presence of the buffer SNO (($SrNb_2O_5$) layer formed between the SBN thin film and the MgO substrate). We select a sample with a higher proportion of this interlayer in order to emphasize the different effect of the two substrates. More clearly, the effect appears in the projection on x (omega-2theta) axis of the RSMs (Fig. 10.31). The intensities are in logarithmic scale. The structural data extracted from the reciprocal space mapping are gathered in the following table:

Sample	c-Parameter (nm)	$FWHM_{x\ (deg)\ (omega\text{-}2theta)}$	$FWHM_{y\ (deg)}$ (omega)
SBN/MgO (1053)	0.3964	0.01356	0.00138
SBN/STO (1051)	0.3954	0.00978	0.00179

The c-parameters are close to the standard $Sr_{0.5}Ba_{0.5}Nb_2O_6$ ($c = 0.39521$ nm), showing the formation of a stoichiometric $Sr_{0.5}Ba_{0.5}Nb_2O_6$ phase in both cases. The FWHM (full width at half maximum) along the x-axis (omega-2theta direction) is higher for the film deposited on MgO due to the formation of the SNO parasitic interlayer phase. The FWHM values along y-axis (omega direction) are relatively low in both cases.

In order to measure the electro-optic behaviour of the SBN: 50/Nb:STO thin films with spectroscopic ellipsometry, a transparent and conductive top electrode (TCO) is necessary to obtain the following heterostructure—TCO/SBN/Nb:STO, presented in Fig. 10.32.

AZO—aluminium doped zinc oxide—was chosen as a transparent and conductive top electrode. The deposition of AZO/SBN/STON heterostructures was

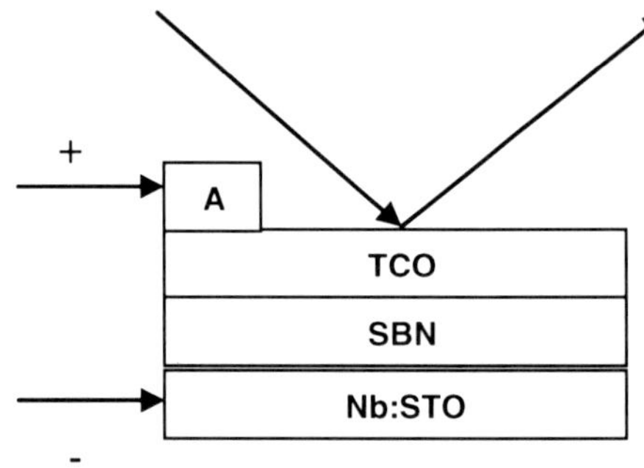

Fig. 10.32 The full oxides heterostructure necessary for electro-optic measurements

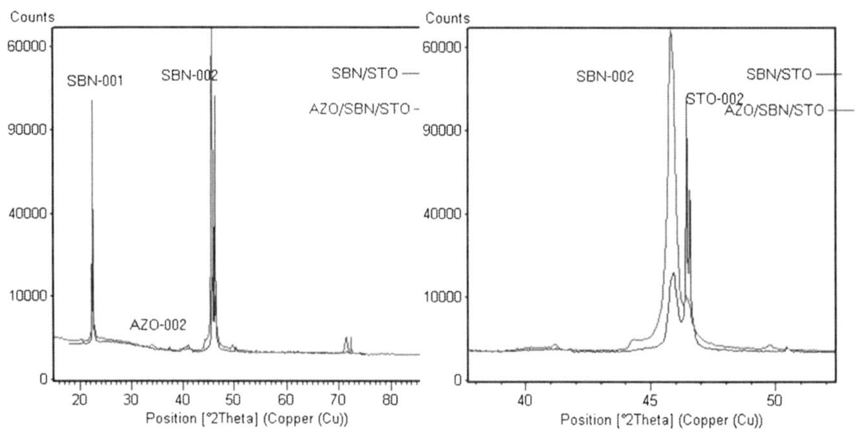

Fig. 10.33 XRD patterns for AZO/SBN/Nb:STO thin films

made by RF-PLD in a single step process, taking advantage of a multitarget system mounted in our experimental set-up.

We check by XRD the preservation of the SBN structure after the AZO deposition procedure. Figure 10.33 presents the XRD spectra of an AZO/SBN/STON sample in comparison with a SBN/STON film. The spectra were recorded in Bragg–Brentano geometry with one degree omega offset in order to diminish the intensity of the highly oriented substrate (STO) peaks.

The preservation of highly c-oriented SBN structure along with the appearance of the (002) peak from the hexagonal AZO (Al doped ZnO) phase can be observed.

The electro-optical (EO) measurements were performed on the SBN thin films by employing the spectroscopic ellipsometry technique. A Woollam Variable Angle Spectroscopic Ellipsometer (VASE) system was used, equipped with a high pressure Xe discharge lamp, which generates light in the 1–5 eV spectral range, from near-IR to UV. The lamp is incorporated in an HS-190 monochromator to enable spectral characterization of the interface. In ellipsometry, the change of the polarization state of linearly polarized light is measured upon reflection at the

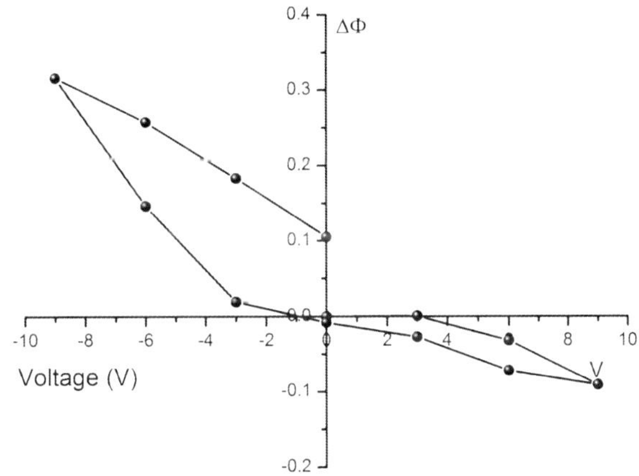

Fig. 10.34 The electro-optic behaviour for SBN:50 thin films in terms of experimental phase shift

surface. Even if the ellipsometry measurement is relatively simple, the analysis of the results is often complicated. An accurate model is required for the system under consideration, which enables simulation or fitting of results.

Measurements were performed from 300 up to 1,200 nm with or without applied electric field, at a fixed angle of incidence. The goal was to obtain the birefringence values (the changes in the refractive index with the applied electric field) by measuring the phase shift, according to the relation:

$$\Delta n = \frac{\Delta\phi \cdot \lambda \cos\theta_r}{4\pi d}$$

where d is film thickness and λ wavelength of incident light.

For $\lambda = 540$ nm and a polar angle of incidence of 60°, the calculated birefringence shift and the experimental phase shift are presented in Fig. 10.34.

For a uniaxial crystal belonging to *4mm* class of symmetry the equation of index ellipsoid can be written:

$$\frac{x^2}{n_o{}^2} + \frac{y^2}{n_o{}^2} + \frac{z^2}{n_e{}^2} = 1 \tag{10.1}$$

and the index ellipsoid is deformed under the effect of an electric field E_z. For an electric field applied along z-axis (c-axis for SBN/Nb:STO), $E_x = E_y = 0$ and the electro-optically induced changes in the refractive indices n_o and n_e, δn_0 and δn_e, can be expressed as:

$$\delta n_o = -n_o^3 r_{13} E_z/2 \tag{10.2}$$

$$\delta n_e = -n_e^3 r_{33} E_z/2 \tag{10.3}$$

The EO change in the phase-shift difference Δ_{sp} under the electric field E_z can be written as:

$$\delta\Delta_{sp} = \frac{\delta\Delta_{sp}}{\delta n_o}\delta n_o + \frac{\delta\Delta_{sp}}{\delta n_e}\delta n_e$$
$$= \left(\frac{2\pi d r_{33} E_z}{\lambda}\right)\left[\begin{array}{l}\frac{n_o n_e \sin^2\theta}{\left(n_e^2-\sin^2\theta\right)^{1/2}} \\ +\frac{r_{13}}{r_{33}}\left(\frac{n_o^3}{n_e}\left(n_e^2-\sin^2\theta\right)^{1/2}-\frac{n_o^4}{\left(n_e^2-\sin^2\theta\right)^{1/2}}\right)\end{array}\right] \quad (10.4)$$

where n_e and n_o are the ordinary and extraordinary refractive indices [100–103]. Assuming $n \sim n_o \sim n_e$, the effective electro-optic coefficient r_{eff} results as $r_{eff} = r_{33} - (n_o/n_e)^3 r_{13} \sim r_{33} - r_{13}$ [104]. Thus, (10.1) can be written as:

$$\delta\Delta = \left(\frac{2\pi d r_{eff} E}{\lambda}\right) \times \left[\frac{n^2 \sin^2\theta}{\left(n^2 - \sin^2\theta\right)^{1/2}}\right] \quad (10.5)$$

where θ is the incident angle of light at the AZO/SBN interface [104].

The resulting value of r_{eff} was calculated to be 75 pm/V, higher than the reported values for $LiNbO_3$. Having a rather high Curie temperature of 128 °C, as compared to the high Sr content compositions, SBN:50 thin films can be potentially used in electro-optic devices operating near room temperature.

10.5 Lead-Free Ferroelectric NBT-BT Thin Films

Ferroelectric materials obtained as solid solutions of $Na_{1/2}Bi_{1/2}O_3$ (NBT) with $BaTiO_3$ (BT), (1−x)NBT-xBT (NBT-BT), are considered among the best lead-free piezoelectric materials [23]. NBT-BT has a morphotropic phase boundary (MPB) at x = 0.06–0.08 where some of the physical properties are most favourable, but other compositions with $0 < x < 0.3$ are also considered interesting for applications. In the last years NBT-BT thin films have been deposited by different techniques, including pulsed laser deposition [105–120].

We have investigated NBT-BT thin films [37, 121–125] with different compositions in the range $0 \leq x \leq 0.08$, grown by PLD on $Pt/TiO_2/SiO_2/Si$ substrates. Target preparation and bulk physical properties have been presented in [126]. The films have been grown by PLD in oxygen atmosphere (pressure = 10–30 Pa) and at substrate temperatures 650–700 °C [124]. The oxygen pressure was optimized in order to avoid the formation of pyrochlore phase [124].

Figure 10.35 shows the XRD pattern of NBT target and NBT thin film deposited on $Pt/TiO_2/SiO_2/Si$ substrate. Both the film and the target display peak reflections characteristic to the rhombohedral symmetry. Moreover, the film pattern evidences a slight (100) orientation.

In Fig. 10.36 XRD patterns of NBT-BT 6 % target and thin film deposited on $Pt/TiO_2/SiO_2/Si$ substrate are shown. The XRD spectrum of the target (lower curve) displays the coexistence of rhombohedral and tetragonal symmetry

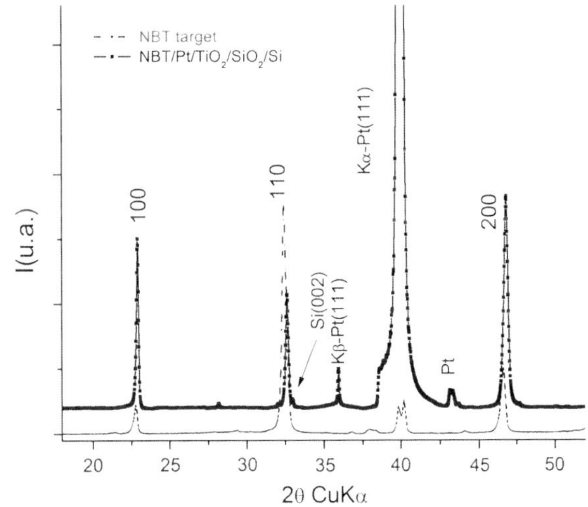

Fig. 10.35 XRD pattern of NBT thin film deposited on $Pt/TiO_2/SiO_2/Si$ substrate. The *lower pattern* corresponds to the NBT target

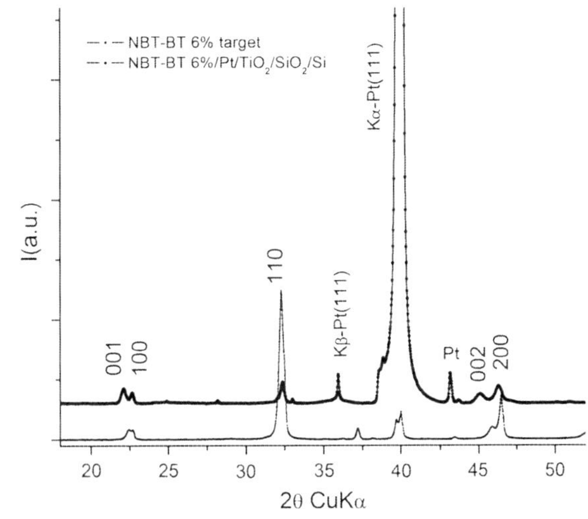

Fig. 10.36 XRD patterns of NBT-BT 6 % target and thin film deposited on $Pt/TiO_2/SiO_2/Si$ substrate

characteristic of the MPB boundary. The tetragonal splitting is also evidenced in the thin film pattern.

XRD patterns of NBT-BT 8 % target and thin films grown on $Pt/TiO_2/SiO_2/Si$ substrate (not shown here) evidenced the reflection peaks characteristic to the tetragonal P4mm symmetry. No particular orientation has been observed for the thin film.

AFM images show that the surface of the deposited films was characterized by well-defined crystallites with size ranging between a few tens of nm up to a few hundreds of nm. A typical image is shown in Fig. 10.37.

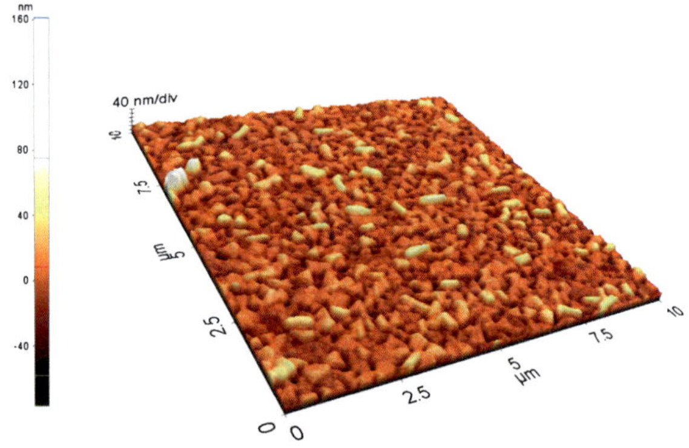

Fig. 10.37 AFM image of NBT-BT 6 % thin films deposited on a Pt/TiO_2/SiO_2/Si substrate at a temperature of 700 °C

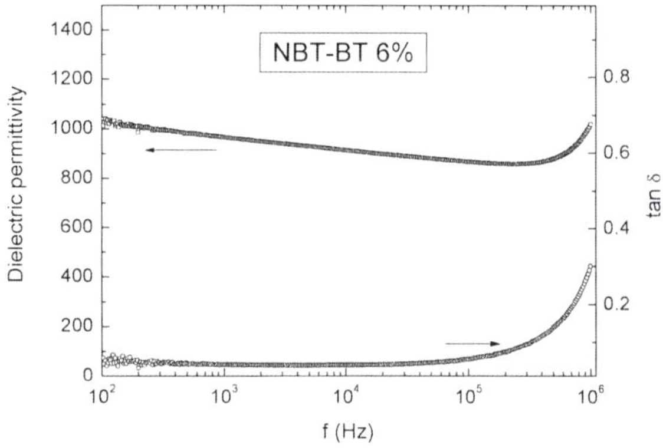

Fig. 10.38 Room temperature dielectric permittivity and loss tangent variation with frequency for NBT-BT 6 % film deposited on Pt/TiO_2/SiO_2/Si substrate

For the electrical measurements gold electrode dots with an area of about 0.22 mm^2 have been evaporated through a mask on the films. Dielectric spectroscopy measurements have been carried out at ambient temperature in the 100 Hz–1 MHz frequency range. The dielectric constant (at 1 kHz and room temperature) obtained for the NBT films was about 1,500, more than three times higher than the bulk value. For NBT-BT 6 and 8 % thin films the obtained room temperature values of the dielectric constant were lower (about 1,000), but close to bulk value (about 1,500).

Figure 10.38 displays the dielectric constant and loss dependence on frequency in the 100 Hz–1 MHz range for NBT-BT 6 % films grown on Pt/TiO_2/SiO_2/Si substrate at 650 °C. A similar dependence has also been measured on NBT and NBT-BT 8 %. A general feature is the relative independence on frequency up to 100 kHz, followed by a rapid increase of both dielectric constant and loss above this frequency. This increase, common for all compositions, is not due to an intrinsic mechanism, but is most probably caused by a relaxation mechanism associated with a space charge layer, which gives contributions to both the real and imaginary parts of dielectric permittivity [69].

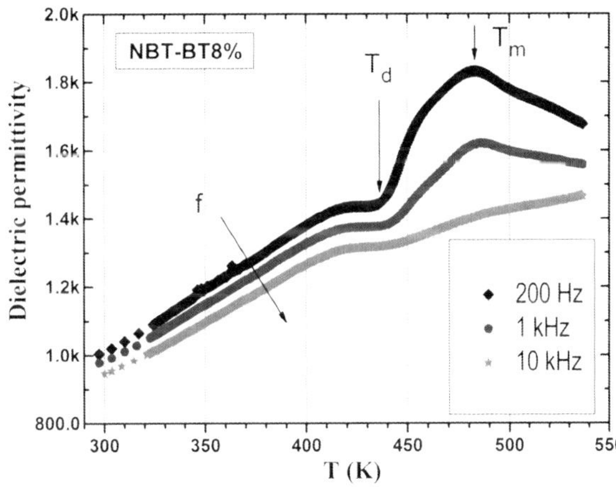

Fig. 10.39 The dielectric temperature variation with temperature measured at different frequencies on a NBT-BT 8 % thin film grown on $Pt/TiO_2/SiO_2/Si$

The variation of dielectric permittivity with temperature has been measured on films with x = 0, 0.06 and 0.08, respectively, deposited on $Pt/TiO_2/SiO_2/Si$. In bulk samples only a broad dielectric peak which shifts with frequency has been evidenced at a temperature T_m in unpoled samples [126]. This behaviour is typical for relaxor ferroelectrics. The relaxor behaviour in NBT-BT compositions is induced by the random occupancy of A-site by ions with different valence and size. In poled bulk samples a ferroelectric phase is induced by polarization. On heating, this induced ferroelectric phase transforms to a relaxor phase at the depolarization temperature T_d.

The variation of dielectric permittivity with temperature on heating is shown in Fig. 10.39 for a NBT-BT 8 % film grown on $Pt/TiO_2/SiO_2/Si$. The arrows mark the T_m and T_d temperatures. Both values are close to the corresponding temperatures measured on bulk poled samples [126].

However, the general behaviour is rather different from bulk samples. A strong difference is related to the peak at T_m which is much narrow than in bulk [126] and displays an almost negligible frequency shift of temperature.

Moreover, although the measurements have been made on the as-grown unpoled films, the depolarization transition is evident on the heating curves. The same behaviour was also observed on NBT-BT 6 % films (Fig. 10.40). It is possible that these changes in the dielectric behaviour are due to the different microstructures of thin films with respect to bulk and to the influence of the nanograin boundaries constraining on the ensemble of polar nanoregions of relaxor phase.

Measurements of polarization hysteresis on NBT-BT 6 % thin films evidenced rather high values of coercive field (about 100 kV/cm), probably due to the pinning of domain walls by defects. The remnant polarization was about 10 $\mu C/cm^2$ and the spontaneous polarization was about 30 $\mu C/cm^2$.

Figure 10.41 displays the results of a polarization fatigue test on a NBT-BT 6 % film, performed by using voltage pulses of ± 5 V. It can be observed that the film shows good resistance to fatigue after having been subjected to 4×10^8 cycles. The polarization hysteresis did not show differences after the test.

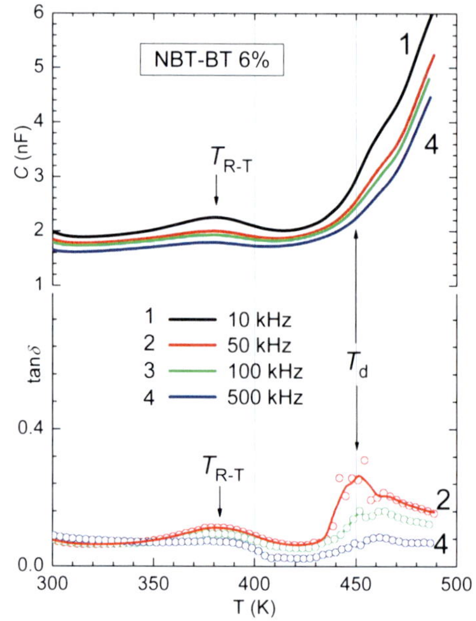

Fig. 10.40 Variation of capacitance C and dielectric loss tan δ with temperature, at different frequencies (10, 50, 100 and 500 kHz, starting from the upper curve) for a NBT-BT 6 % film grown on $Pt/TiO_2/SiO_2/Si$. The arrows mark the rhombohedral-tetragonal transition (T_{R-T}) and depolarization (T_d) temperatures

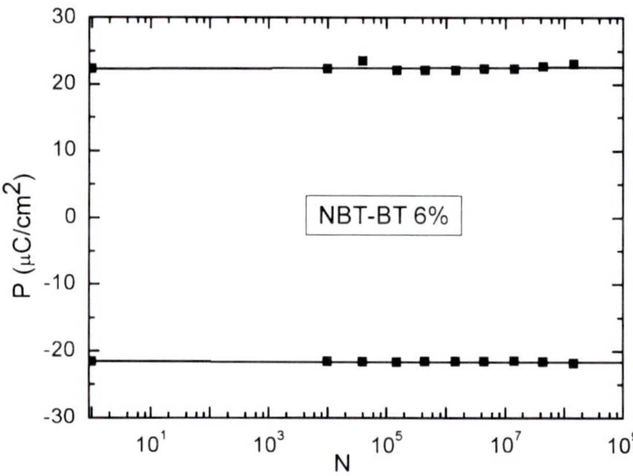

Fig. 10.41 Polarization fatigue determined by using voltage pulses of ±5 V on a NBT-BT 6 % film

Piezoresponse force microscopy (PFM) was used to measure the piezoelectric properties of NBT-BT thin films, for different values of the bias electric field. Typical results are shown in Fig. 10.42 for a NBT-BT 6 % film. They evidence good switching behaviour and piezoelectric hysteresis, confirming the quality of the deposited films. The locally measured values reach a maximum $d_{33}^{eff} \approx 83$ pm/V.

This value is in the range of piezoelectric coefficient values reported in literature for NBT-BT thin films, but is still lower that the corresponding values obtained on ceramic bulk samples, probably due to specific defects and interface problems.

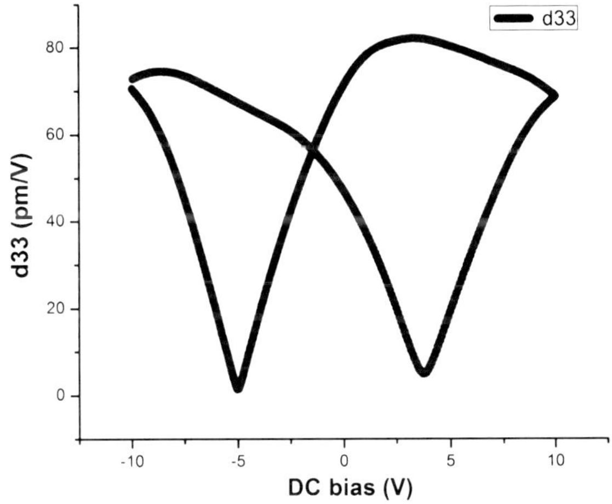

Fig. 10.42 Piezoresponse measurements performed by PFM on a NBT-BT 6 % film

10.6 Conclusions

In summary, we showed some results obtained for ferroelectric thin films grown by PLD, with emphasis on both lead based and lead free compounds, in relation to actual scientific and social tendencies. There is an important trend to replace or to reduce the use of toxic elements such as lead, but there is no obvious complete solution for this problem, both types of multifunctional oxides having interesting properties.

We have investigated and presented properties of lead based oxides thin films such as PLZT and PMN-PT. PLZT thin films with different compositions have interesting dielectric and electro-optic behaviour. Complex impedance spectroscopy has been employed to measure the ac conductivity as a function of frequency and dc bias field amplitude, a Jonscher-type dependence being observed. The electro-optic behaviour of PLZT thin films integrated into a full oxide heterostructure AZO/PLZT/Nb:STO has been revealed by reflection type spectroscopic ellipsometry measurements. Piezoresponse force microscopy (PFM) was used to test the switching properties and the piezoelectric effect in PLZT thin films, the measured values for the effective piezoelectric coefficient $d_{33}^{eff} \approx 83$ pm/V being comparable with those reported for PZT 20/80 thin films. For the ferroelectric relaxor $Pb(Mg_{1/3}Nb_{2/3})_{1-x}Ti_xO_3$ (PMN-PT) thin films the presence of a piezoelectric effect in as-grown ferroelectric this films has been revealed, indicating the existence of self-polarization. This effect has also been evidenced in other thin films. The effect of self-polarization in thin relaxor films could be of interest for pyrosensors and other applications based on the pyroelectric effect.

In the case of lead-free oxide materials, we have presented the obtaining and the characterisation of SBN and NBT-BT thin films. For the SBN thin films, the resulting value of r_{eff} was calculated to be higher than the values reported for

$LiNbO_3$. Having a rather high Curie temperature of 128 °C, compared to the high Sr content compositions, the SBN:50 thin films can be potentially used in electro-optic devices operating near room temperature. Also, we have investigated the NBT-BT thin films properties as a function of different experimental parameters, such as the amount of BT in the target or the deposition temperature. For NBT-BT 6 % thin films, it has been observed that the films show good resistance to fatigue after been subjected to 4×10^8 cycles. The polarization hysteresis did not show differences after the test.

10.7 Perspectives

There are many expectations coming from the multifunctional oxides studies. This is due to their different properties such as magnetic, ferroelectric, multiferroic, optical or transport and due to the potential tuning of these properties. The approach of properties unification has many advantages in production costs, high degree of applicability or miniaturization. Moreover, such functionalities of a material can be added or enhanced with the help of material nanostructuring techniques. The perovskite materials ABO_3 oxides, environmental-friendly or not, will continue to be studied and their properties enhanced through different methods.

Good examples for these facts are bismuth ferrite, a well-known ferromagnetic and ferroelectric material, or PLZT thin films application in photovoltaics, which is an unexpected turn for ferroelectric materials. In conventional photovoltaic applications based on a p–n or Schottky junction, there are some drawbacks like the absorption of only photons with energies above the semiconductor band-gap, and only those photons will contribute to the charge separation (electron-hole) by the internal field generated in a very narrow, micrometer thick interfacial region, called depletion regions. When the band gap value is high (for wide-band gap semiconductors $E_g > 2.5$ eV), visible light will not be efficiently converted to charge carriers. Small photoinduced voltages are produced by this type of conventional photovoltaic devices—equal with the material band-gap and limited by it (typically less than 1 V). In such a situation, the rush for materials with small band gap has never been more intense.

There are other ways to create charge separation in a material, as has been demonstrated for thin films and bulk ferroelectric materials, such as Pb $(Zr,Ti)O_3$, La-doped Pb $(Zr,Ti)O_3$ or $BaTiO_3$ [127–129]. Ferroelectrics are non-centrosymetric materials exhibiting a spontaneous polarization below the Curie temperature. Their polarization vector direction can be switched using an external electric field. In this type of materials, namely non-centrosymetric a new effect was discovered in the second half of the 1960s: the bulk photovoltaic effect [130]. This effect can give rise to large photovoltage, charge carriers separation in an entire homogenous media, not only at a narrow interface region, and to a photocurrent proportional with the polarization degree. Starting from La doped PZT ferroelectric thin films, Qin et al. [128] demonstrated high photovoltaic-power conversion efficiency. The physical mechanism considered at that time was the following: during interaction between a photon, having an energy corresponding to the energy band gap, and an electrically poled ferroelectric, the photon is absorbed by the ferroelectric, resulting in the creation of charge carriers, electrons and holes. The photogenerated charge carriers are driven by the polarization-induced internal electric field in opposite directions, contributing to the photovoltaic output. Therefore, the internal electric field induced by the remnant polarization plays an important role in photovoltaic properties of ferroelectric materials. Unfortunately, for normal ferroelectric perovskites, the band gap value is normally $E_g > 3$ eV, and this is a major limiting factor for visible light absorption. Another drawback results from the composition of actual industrial materials, which are lead-based materials.

In this context, another class of materials started to attract the attention of researchers, namely perovskite materials with small band gap. Among these materials, bismuth ferrite ($BiFeO_3$, BFO) has a major advantage, which is the small value for the band gap [131]. This material is a ceramic material with perovskite structure, high remnant ferroelectric polarization (95 $\mu C/cm^2$) and Curie temperature ($T_c \sim$ 1,103 K), that simultaneously exhibits both ferroelectric and ferromagnetic properties [132, 133]. The band gap value corresponding to maximum absorptivity at visible wavelengths is much smaller than that of most ferroelectric materials. For ferroelectrics with periodic domain structures, the fundamental mechanism for photovoltaic charge separation was discovered by Yang et al. [131], when they studied very thin films of BFO with engineered domain walls in ordered arrays of 71° and 109°. The photovoltaic effect was proven to spontaneously appear at nanoscaled dimension, because of rhombohedrally distorted crystal structure. The domain walls act to separate regions with different ferroelectric and ferromagnetic orientations and have the role of charge separation. Over a distance of 200 μ, Yang and his collaborators succeeded to measure 16 V photoinduced voltages. The photovoltaic effect appears due the polarization vector direction changes at the domain wall, creating steps in the electrostatic potential. On the other hand, these kind of epitaxial films with high control over film properties are very expensive. The alternative is to use polycrystalline BFO thin films which are much cheaper to produce, even if the photovoltaic properties are inferior to those of epitaxial films, and to find a way to improve important parameters like light absorption or surface area of the thin film.

Acknowledgments The authors N.D. Scarisoreanu and M. Dinescu gratefully acknowledge the financial support from Romanian-Swiss Research Programme (RSRP) No. 18 RO-CH, IZERZO-142176/1 (2013–2015); TE 14/2013 and PN 0939/2007–2013 projects.

References

1. Z.-G. Ye, *Handbook of Dielectric, Piezoelectric and Ferroelectric Materials* (Woodhead Publishing Limited, 2008)
2. K.C. Kao, *Dielectric Phenomena in Solids* (Elsevier Academic Press, 2004)
3. N. Izyumskaya, Y.I. Alivov, S.J. Cho, H. Morkoç, H. Lee, Y.S. Kang, Crit. Rev. Solid State Mater. Sci. **32**(3), 111–202 (2007)
4. Y. Yoshino, J. Appl. Phys. **105**, 061623 (2009)
5. M. Dawber, K.M. Rabe, J.F. Scott, Rev. Mod. Phys. **77**, 1083 (2005)
6. B. Jaffe, R. Roth, S. Marzullo, J. Appl. Phys. **52**(6), 809–810 (1954)
7. B. Noheda, D.E. Cox, G. Shirane, J.A. Gonzalo, L.E. Cross, S.E. Park, Appl. Phys. Lett. **74**, 2059 (1999)
8. D.I. Woodward, J. Knudsen, I.M. Reaney, Phys. Rev. B **72**, 104110 (2005)
9. A.M. Glazer, P.A. Thomas, K.Z. Baba-Kishi, G.K.H. Pang, C.W. Tai, Phys. Rev. B **70**, 184123 (2004)
10. N.A. Pertsev, V.G. Kukhar, H. Kohlstedt, R. Waser, Phys. Rev. B **67**, 054107 (2003)
11. D. Damjanovic, Ann. Chim. Sci. Mater. **26**, 99 (2001)
12. M. Tyunina, J. Levoska, S. Leppävuori, J. Appl. Phys. **83**, 5489 (1998)
13. T.J. Zhu, L. Lu, M.O. Lai, Appl. Phys. A **81**, 701 (2005)
14. D.B. Chrisey, G.K. Hubler, *Pulsed Laser Deposition of Thin Films* (Wiley, New York, 1994)
15. D. Bäuerle, *Laser Processing and Chemistry* (Springe, Berlin, 2000)
16. I. Vrejoiu, G.L. Rhun, L. Pintilie, D. Hesse, M. Alexe, U. Gösele, Adv. Mater. **18**, 1657–1661 (2006)
17. W. Hong, H.N. Lee, M. Yoon, H.M. Christen, D.H. Lowndes, Z. Suo, Z. Zhang, PRL **95**, 095501 (2005)
18. M.J. Haun, E. Furman, S.J. Jang, L.E. Cross, Ferroelectrics **99**, 63 (1989)
19. S.H. Baek, M.S. Rzchowski, V.A. Aksyuk, MRS Bull. **37** (2012)
20. C.B. Eom, S.T. McKinstry, MRS Bull, **37** (2012)
21. W.R. Cook, H. Jaff, Electrooptic Coefficients, in *Landolt-Bornstein, New Series*, vol. 11, ed. by K.-H. Hellwage (Springer, Berlin, 1979), pp. 552–651
22. D. Trivedi, P. Tayebati, M. Tabat, Appl. Phys. Lett. **68**, 3227 (1996)
23. T. Takenaka, K. Maruyama, K. Sakata, Jpn. J. Appl. Phys. **30**, 2236 (1991)
24. G.H. Haertling, C.E. Land, J. Am. Ceram. Soc. **54**, 1 (1971)
25. M.H. Yeh, K.S. Liu, I.N. Lin, Thin Solid Films **258**, 82 (1995)
26. S.B. Mah, N.W. Jang, J.H. Park, D.S. Paik, C.Y. Park, Mater. Res. Bull. **35**, 1113 (2000)
27. M. Gaidi, A. Amassian, M. Chaker, M. Kulishov, L. Martinu, Appl. Surf. Sci. **226**, 347 (2004)
28. T. Kim, J.N. Hanson, A. Gruverman, A.I. Kingon, S.K. Streiffer, Appl. Phys. Lett. **88**, 262907 (2006)
29. W. Leng, C. Yang, H. Ji, J. Zhang, J. Tang, H. Chen, L. Gao, J. Mater. Sci. Mater. Electron. **18**, 887 (2007)
30. A. Khodorov, M.J.M. Gomes, Thin Solid Films **515**, 1782 (2006)
31. G. Leclerc, B. Domengès, G. Poullin, R. Bouregba, Appl. Surf. Sci. **253**, 1143 (2006)
32. P. Verardi, F. Craciun, N. Scarisoreanu, G. Epurescu, M. Dinescu, I. Vrejoiu, A. Dauscher, Appl. Phys. A **79**, 1283 (2004)
33. A. Purice, G. Dinescu, N. Scarisoreanu, P. Verardi, F. Craciun, C. Galassi, M. Dinescu, J. Eur. Ceram. Soc. **26**, 2937 (2006)
34. N. Scarisoreanu, F. Craciun, G. Dinescu, P. Verardi, M. Dinescu, Thin Solid Films **453–454**, 399 (2004)
35. F. Craciun, M. Dinescu, P. Verardi, N. Scarisoreanu, C. Galassi, D. Piazza, Ferroelectrics **302**, 313 (2004)

36. P. Verardi, F. Craciun, M. Dinescu, N. Scarisoreanu, A. Moldovan, A. Purice, C. Galassi, Mater. Sci. Eng. B **118**, 39 (2005)
37. N. Scarisoreanu, M. Dinescu, F. Craciun, P. Verardi, A. Moldovan, A. Purice, C. Galassi, Appl. Surf. Sci. **252**, 4553 (2006)
38. F. Craciun, M. Dinescu, P. Verardi, N. Scarisoreanu, A. Moldovan, A. Purice, C. Galassi, Appl. Surf. Sci. **248**, 329 (2005)
39. F. Craciun, M. Dinescu, P. Verardi, N. Scarisoreanu, A. Moldovan, A. Purice, C. Galassi, J. Eur. Ceram. Soc. **25**, 2299 (2005)
40. F. Craciun, M. Dinescu, N.D. Scarisoreanu, C. Capiani, C. Galassi, E. Morintale, IOP Confernce Series: Materials Science and Engineering, vol. 8 (2010), p. 012003
41. N.D. Scarisoreanu, A. Andrei, R. Birjega, R. Pascu, F. Craciun, C. Galassi, D. Raducanu, M. Dinescu, Thin Solid Films **520**, 4568 (2012)
42. N.D. Scarisoreanu, F. Craciun, A. Andrei, V. Ion, R. Birjega, A. Moldovan, M. Dinescu, C. Galassi, Thin Solid Films (to be published)
43. G.A. Samara, Ferroelectricity revisited: advances in materials and physics. Solid State Phys. **56**, in ed. by H. Ehrenreich, F. Spaepen (Academic Press, San Diego, 2001), pp. 239–458
44. A.A. Bokov, Z.G. Ye, J. Mater. Sci. **41**, 31 (2006)
45. M. Tyunina, J. Levoska, Phys. Rev. B **63**, 224102 (2001)
46. F. Cordero, F. Craciun, A. Franco, C. Galassi, Ferroelectrics **302**, 221 (2004)
47. F. Cordero, F. Craciun, A. Franco, D. Piazza, C. Galassi, Phys. Rev. Lett. **93**, 097601 (2004)
48. X. Dai, Z. Xu, D. Viehland, J. Appl. Phys. **79**, 1021 (1996)
49. F. Cordero, M. Corti, F. Craciun, C. Galassi, D. Piazza, F. Tabak, Phys. Rev. B **71**, 094112 (2005)
50. F. Craciun, M. Dinescu, N.D. Scarisoreanu, C. Capiani, C. Galassi, E. Morintale, IOP Conference Series: Materials Science and Engineering, vol. 8 (2010), p. 012003
51. A.K. Jonscher, *Dielectric Relaxation in Solids* (Chelsea Dielectric Press, London, 1983)
52. S.H. Baek, H.W. Jang, C.M. Folkman, Y.L. Li, B. Winchester, J.X. Zhang, Q. He, Y.H. Chu, C.T. Nelson, M.S. Rzchowski, X.Q. Pan, R. Ramesh, L.Q. Chen, C.B. Eom, Nat. Mater. **9**, 309 (2010)
53. B.S.E. Park, T.R. Shrout, IEEE Trans. Ultrason. Ferroelectr. Freq. Control **44**, 1140 (1997)
54. J.P. Maria, W. Hackenberger, S.T. McKinstry, J. Appl. Phys. **84**, 5147 (1998)
55. Z. Kighelman, D. Damjanovic, N. Setter, J. Appl. Phys. **89**, 1393 (2001)
56. J. Wang, K.H. Wong, H.L.W. Chan, C.L. Choy, Appl. Phys. A **79**, 551 (2004)
57. N.J. Donelly, R.M. Bowman, J.M. Gregg, Phys. Rev. B, 064110 (2006)
58. F. Wu, X.M. Li, W.D. Yu, X.D. Gao, X. Zhang, Appl. Phys. A **88**, 781 (2007)
59. J. Shang, H. Zhang, Y. Li, X. Zhou, P. Zhang, J. Cryst. Growth **312**, 1925 (2010)
60. S.H. Baek, J. Park, D.M. Kim, V.A. Aksyuk, R.R. Das, S.D. Bu, D.A. Felker, J. Lettieri, V. Vaithyanathan, S.S.N. Bharadwaja, N. Bassiri-Gharb, Y.B. Chen, H.P. Sun, C.M. Folkman, H.W. Jang, D.J. Kreft, S.K. Streiffer, R. Ramesh, X.Q. Pan, S. Trolier-McKinstry, D.G. Schlom, M.S. Rzchowski, R.H. Blick, C.B. Eom, Science **334**, 958 (2011)
61. J. Jiang, H.H. Hwang, W.J. Lee, S.G. Yoon, Sens. Actuators B **155**, 854 (2011)
62. J. Jiang, S.G. Yoon, J. Alloy. Compd. **509**, 3065 (2011)
63. X. Zhu, E. Defay, G. Le Rhun, M. Aïd, Y. Xu, Q. Zhang, Y. Xiao, H. Gao, D. Liang, Jiliang Zhu, Jianguo Zhu, D. Xiao, J. Appl. Phys. **112**, 054105 (2012)
64. Y. Tang, D. Zhou, Y. Tian, X. Li, F. Wang, D. Sun, W. Shi, L. Tian, J. Sun, X. Meng, J. Chu, J. Am. Ceram. Soc. **95**, 1367 (2012)
65. P. Verardi, M. Dinescu, F. Craciun, R. Dinu, I. Vrejoiu, Appl. Surf. Sci. **168**, 340 (2000)
66. P. Verardi, F. Craciun, N. Scarisoreanu, M. Dinescu, C. Grigoriu, C. Galassi, A.L. Costa, Ferroelectrics **293**, 189 (2003)
67. F. Craciun, P. Verardi, D. Brodoceanu, M. Morar, C. Galassi, C. Grigoriu, M. Dinescu, Mater. Sci. Semicond. Process. **5**, 227 (2003)
68. F. Cordero, F. Craciun, P. Verardi, Ferroelectrics **290**, 141 (2003)
69. A.K. Jonscher, *Dielectric Relaxation in Solids* (Chelsea Dielectric Press, London, 1983)
70. P. Verardi, F. Craciun, Rev. Sci. Instrum. **74**, 4453 (2003)

71. P.B. Jamieson, S.C. Abrahams, J.L. Bernstein, J. Chem. Phys. **48**, 5048 (1968)
72. T. Volk, M. Wohlecke, *Lithium Niobate: Defects, Photorefraction and Ferroelectric Switching* (Springer, 2008), pp. 1–9
73. L. Hesselink, M.C. Bashaw, Opt. Quantum. Electron. **25**, S 611, (1993)
74. E.M. Rodrigo, D. JAque, J.G. Sole, R. Pankrath, Appl. Phys. Lett. **92**, 181107 (2008)
75. D. Jaque, N.D. Psaila, R.R. Thomson, F. Chen, L.M. Maestro, A. Ródenas, D.T. Reid, A.K. Kar, Appl. Phys. Lett. **96**, 191104 (2010)
76. S. Lee, R.H.T. Wilke, S. Trolier-McKinstry, S. Zhang, C.A. Randall, Appl. Phys. Lett. **96**, 031910 (2010)
77. J. Li, M.H. Lu, F. Liang, X.P. Liu, Y.F. Chen, J. Appl. Phys. **101**, 013516 (2007)
78. S. Dizian et al., Appl. Phys. Lett. **95**, 101103 (2009)
79. Z. Lu, S. Feigelson, R.K. Route, R. Hiskes, S.A. Dicarolis, Mater. Res. Soc. Symp. Proc. **335**, 59 (1994)
80. M.J. Nystrom, B.W. Wessels, W.P. Lin, G.K. Wong, D.A. Neumayer, T.J. Marks, Appl. Phys. Lett. **66**, 1726 (1995)
81. L.D. Zhu., Appl. Phys. Lett. **67**, 1836 (1995)
82. W. Sakamoto, T. Yogo, K. Kikuta, A. Kawase, S. Hirano, J. Am. Chem. Soc. **79**, 2283 (1996)
83. J. Koo, J.H. Jang, B.-S. Bae, J. Am. Chem. Soc. **84**, 193 (2001)
84. Z. Shen, Thin Solid Films **488**, 40–44 (2005)
85. J. Ho, Appl. Phys **100**, 083524 (2006)
86. P. Tayebati, D. Trivedi, M. Tabat, Appl. Phys. Lett. **69**, 1023 (1996)
87. A. Infortuna, P. Muralt, J. Appl. Phys. **100**, 104110 (2006)
88. B. Mitu, G. Dinescu, M. Dinescu, A. Ferrari, M. Balucani, G. Lamedica, A.P. Dementiev, K.I. Maslakov, Thin Solid Films **382**, 230–234 (2001)
89. N.D. Scarisoreanu, G. Dinescu, R. Birjega, M. Dinescu, D. Pantelica, G. Velisa, N. Scintee, A.C. Galca, Appl. Phys. A **93**, 795–800 (2008)
90. P.R. Willmott, R. Herger, B.D. Patterson, R. Windiks, Phys. Rev. B **71**, 144114 (2005)
91. S. Schwyn Thöny, K.E. Youden, J.S. Harris, L. Hesselink, Appl. Phys. Lett. **65**, 2018 (1994)
92. K. Tanaka, O. Nakagawara, M. Nakano, T. Shimuta, H. Tabata, T. Kawai, Jpn. J. Appl. Phys. Part **1**(37), 6142 (1998)
93. M. Cuniot-Ponsard, J.M. Desvignes, B. Ea-Kim, E. Leroy, J. Appl. Phys. **93**, 1718 (2003)
94. C.M. Rouleau, G.E. Jellison, D.B. Beach, Appl. Phys. Lett. **82**, 2990 (2003)
95. E.L. Venturini, E.G. Spencer, P.V. Lenzo, A.A. Ballman, J. Appl. Phys. **39**, 343 (1967)
96. Keisjer, Thin Solid Films **266**, 157–167 (1995)
97. A.M. Prokhorov, *Ferroelectric Crystals for Laser Radiation Control*, (Adam Hilger, New York 1990), p. 81
98. H. Adachi, T. Kawaguchi, K. Setsune, K. Ohji, K. Wasa, Appl. Phys. Lett **42**, 867 (1983)
99. Y.H. Xu, H.C. Chen, S.T. Liu, Jpn. J. Appl. Phys. **24**, 278 (1985)
100. S. Ducharme, J. Feinberg, R.R. Neurgaonkar, IEEE J.Quantum Electron. QE. **23**, 12, (1987)
101. Y. Shuto, M. Amano, J. Appl. Phys. **77**(9) (1995)
102. M. Aillerie, N. Theofanous, M.D. Fontana, Appl. Phys. B **70**, 317 (2000)
103. M. Goulkov, T. Granzow, U. Dorfler, Th Woike, M. Imlau, R. Pankrath, W. Kleemann, Opt. Commun. **218**, 173 (2003)
104. J. Koo, C. Lee, J.H. Jang, K. No, B. Bae, Appl. Phys. Lett. **76**, 2671 (2000)
105. H.-W. Cheng, X.-J. Zhang, S.-T. Zhang, Y. Feng, Z.-G. Liu, Appl. Phys. Lett. **85**, 2319 (2004)
106. M. Abazari, A. Safari, S.S.N. Bharadwaja, S. Trolier-McKinstry, Appl. Phys. Lett. **96**, 082903 (2010)
107. T. Harigai, Y. Tanaka, H. Adachi, E. Fujii, Appl. Phys. Express **3**, 111501 (2010)
108. D. Alonso-Sanjosé, R. Jiménez, I. Bretos, M.L. Calzada, J. Amer. Ceram. Soc. **92**, 2218 (2009)
109. M. Bousquet, J.-R. Duclère, C. Champeaux, A. Boulle, P. Marchet, A. Catherinot, A. Wu, C. Bachelet, J. Appl. Phys. **107**, 034102 (2010)

110. D.Y. Wang, N.Y. Chan, S. Li, S.H. Choy, H.Y. Tian, H.L.W. Chan, Appl. Phys. Lett. **97**, 212901 (2010)
111. H. Adachi, Y. Tanaka, T. Harigai, M. Ueda, E. Fujii, Appl. Phys. Express **4**, 051501 (2011)
112. M. Cernea, A.C. Galca, M.C. Cioangher, C. Dragoi, G. Ioncea, J. Mater. Sci. **46**, 5621 (2011)
113. C. Dragoi, M. Cernea, L. Trupina, Appl. Surf. Sci. **257**, 9600 (2011)
114. I. Bretos, D. Alonso-Sanjosé, R. Jiménez, J. Ricote, M. Lourdes Calzada, Mater. Lett. **65**, 2714 (2011)
115. J. Schwarzkopf, M. Schmidbauer, A. Duk, A. Kwasniewski, S.B. Anooz, G. Wagner, A. Devi, R. Fornari, Thin Solid Films **520**, 239 (2011)
116. C.-C. Diao, C.-F. Yang, J.-J. Lin, J. Nanosci. Nanotechnol. **11**, 10562 (2011)
117. Y.-H. Lin, P.-S. Cheng, C.-C. Wu, T.-P. Sun, J.-J. Lin, C.-F. Yang, Ceram. Int. **37**, 3765 (2011)
118. M.M. Hejazi, A. Safari, J. Appl. Phys. **110**, 103710 (2011)
119. M. Cernea, L. Trupina, C. Dragoi, B.S. Vasile, R. Trusca, J. Alloy. Compd. **515**, 166 (2012)
120. M. Bousquet, J.-R. Duclère, B. Gautier, A. Boulle, A. Wu, S. Députier, D. Fasquelle, F. Rémondière, D. Albertini, C. Champeaux, P. Marchet, M. Guilloux-Viry, P. Vilarinho, J. Appl. Phys. **111**, 104106 (2012)
121. M. Dinescu, F. Craciun, N. Scarisoreanu, P. Verardi, A. Moldovan, A. Purice, A. Sanson, C. Galassi, J. Phys. IV France **128**, 77 (2005)
122. N. Scarisoreanu, F. Craciun, V. Ion, S. Birjega, M. Dinescu, Appl. Surf. Sci. **254**, 1292 (2007)
123. N.D. Scarisoreanu, F. Craciun, A. Chis, R. Birjega, A. Moldovan, C. Galassi, M. Dinescu, Appl. Phys. A **101**, 747 (2010)
124. A. Andrei, N.D. Scarisoreanu, R. Birjega, M. Dinescu, G. Stanciu, F. Craciun, C. Galassi, Appl. Surf. Sci. **278**, 162 (2013)
125. N.D. Scarisoreanu, F. Craciun, A. Andrei, V. Ion, R. Birjega, A. Moldovan, M. Dinescu, C. Galassi, Thin Solid Films **541**, 127 (2013)
126. F. Cordero, F. Craciun, F. Trequattrini, E. Mercadelli, C. Galassi, Phys. Rev. B **81**, 144124 (2010)
127. M. Qin, K. Yao, Y.C. Liang, Appl. Phys. Lett. **95**, 022912 (2009)
128. M. Qin, K. Yao, Y.C. Liang, Appl. Phys. Lett. **93**, 122904 (2008)
129. J.F. Li, K. Tatagi, B.P. Zhang, J. Mater. Sci. **39**, 2879 (2004)
130. Y.S. Yang, S.J. Lee, S. Yi, M.S. Jang, Appl. Phys. Lett. **76**, 774 (2000)
131. S.Y. Yang, J. Seidel, S.J. Byrnes, P.R. Ramesh, Nat. Nanotechnol. **5**, 143 (2010)
132. W. Ji, K. Yao, Y.C. Liang, Adv. Mater. **22**, 1763 (2010)
133. J. Wang, J.B. Neaton, H. Zheng, V. Nagarajan, R. Ramesh, Science **299**, 1719 (2003)

Chapter 11
Biomaterial Thin Films by Soft Pulsed Laser Technologies for Biomedical Applications

Ion N. Mihailescu, Adriana Bigi, Eniko Gyorgy, Carmen Ristoscu, Felix Sima and Ebru Toksoy Oner

Abstract We review recent results on biomaterial nanostructured layers transferred by matrix-assisted pulsed laser evaporation (MAPLE). The chapter is organized according to three main applications of these nanostructures: drug delivery systems, biosensing and biomimetic coating of metallic implants. The synthesized layers were optimized based upon the results of investigations performed by physical–chemical methods. Biocompatibility and bioactivity were assessed by dedicated in vitro tests. From the first category we chose the composite alendronate–hydroxyapatite (HA). The coating of metallic implants with these layers demonstrated to enhance human osteoblasts proliferation and differentiation, while inhibiting osteoclasts growth, with benefic effects for the treatment of osteoporosis. Enzyme ribonuclease A (RNase A) immobilized on solid supports has applications in control of the enzymatic reaction, and improved stability as compared to the free enzyme. The results by reversed-phase high-performance liquid chromatography showed that immobilization process does not affect the RNase A behavior. The transfer of pure levan and oxidized levan was obtained by MAPLE without any addition of plasticizers or pigments. The nanostructures exhibited high specific surface areas fully compatible with their potential use in drug delivery systems. For the second application, we refer to the transfer and immobilization of IgG molecules. We investigated the effect of the lipid addition in the initial solution upon the protein

I. N. Mihailescu (✉) · E. Gyorgy · C. Ristoscu · F. Sima
Lasers Department, National Institute for Lasers, Plasma and Radiations Physics, 409 Atomistilor, MG-54 077125 Magurele–Ilfov, Romania
e-mail: ion.mihailescu@inflpr.ro

A. Bigi
Department of Chemistry "Giacomo Ciamician", University of Bologna, 40126 Bologna, Italy

E. Gyorgy
Consejo Superior de Investigaciones Científicas, Barcelona, Spain

E. T. Oner
Industrial Biotechnology and Systems Biology (IBSB) Research Group, Department of Bioengineering, Marmara University, Istanbul, Turkey

M. Castillejo et al. (eds.), *Lasers in Materials Science*, Springer Series in Materials Science 191, DOI: 10.1007/978-3-319-02898-9_11,

thin films adhesion to substrates. From the third class, we selected magnesium substituted octocalcium phosphate (OCP) and strontium substituted OCP deposited by MAPLE on Ti substrates which proved to enhance osteoblast activity and differentiation. We conclude that under optimized conditions, the thin films obtained by MAPLE were similar in composition, morphology and structure with the base material, and most likely preserved their functionality and biological performances.

11.1 Introduction

According to the definition given by the US National Institutes of Health, a *biomaterial* is "any substance (other than a drug) or combination of substances, synthetic or natural in origin, which can be used for any period of time, as a whole or as a part of a system which treats, augments, or replaces any tissue, organ, or function of the body" [1]. Merriam-Webster medical dictionary also defines the *biomaterial* as a "natural or synthetic material (as a polymer or metal) that is suitable for introduction into living tissue especially as part of a medical device" [2].

Due to general increased tendency of the human life expectancy (up to more than 70 years) throughout the world, an augmented demand of biomaterials was recorded in many developed economies during the last decades. Significant research interest was focused on materials with bioactive properties. Unfortunately, reliable biomaterials are expensive, often fragile and brittle in bulk. An alternative solution could be their application in form of thin films, as only the outer layer of a conventional biomedical system comes into intimate contact with the vivid tissue. Nevertheless, the assembling of adequate thin biomaterial layers is not reachable by most of the recognized deposition techniques, because of the incompatibilities related to their irreversible degradation by heat, pressing or chemical action. Recently, lasers opened new doors towards the safe transfer and deposition of very complicated high molecular weight compounds, the most of them of interest for biomedical applications [3].

One of the notorious laser techniques, Pulsed Laser Deposition (PLD), was widely applied to inorganic thin biofilm synthesis [4, 5] but also to some polymers [6]. In the latter case, it was suggested that the UV absorption and consequently the laser ablation is producing polymer chain scission to monomers which subsequently repolymerize once arriving onto deposition substrate. This is however limited to the case of polytetrafluoroethylene (PTFE) and Poly-methyl methacrylate (PMMA), where the laser radiation induces photothermal depolymerization only, which is next followed by the reversible unzipping of the polymer chains.

Matrix-Assisted Pulsed Laser Evaporation (MAPLE) was therefore developed as an alternative to PLD suitable for the transfer and deposition of organic materials. It emerged and developed as a potential replacement method to spray coatings [7]. The technique ensures a soft pulsed laser transfer of "delicate" (organic and/or biologic) biomaterials avoiding the risks of photonic and thermal damage under intense laser irradiation [8]. It demonstrated characteristic advantages such as the accurate control of coating thickness, the relatively uniform spreading of material over rather large areas and the possibility of extension to the synthesis of multistructures. Thus, MAPLE is a single-step fabrication method of complex constructions of multilayers and multistructures due to the potential of transferring and depositing both organic and inorganic materials. Multi-component targets can be easily prepared and utilized for synthesis of layers with superior stability and adhesion to substrate.

In MAPLE, the delicate active substance (solute) is dissolved or suspended in a liquid (solvent) which is frozen. The solvent material acts as a highly protective shield against the action of the laser radiation during evaporation and subsequent transfer of the material of interest onto the substrate surface. During the MAPLE transfer which is supposed to be dry, solvent implication is minimized or even avoided, whilst the synthesis of multilayer structures is rather easy.

The method is a non-contact technique which proved quite versatile and challenging in comparison with other laser based techniques. Additionally, using appropriate masks one can manufacture microsized samples (single or multilayered) for microarray chip applications [9].

MAPLE was applied to a large class of organic compounds for various applications [7] such as thin films of polysaccharides [10–14] for drug delivery, polyfluorene and polythiophene copolymers for metal-insulator-semiconductor diodes and filed-effect transistors structures [15], chemoselective polymers for chemicalsensors [16] or proteins with applications in tissue engineering [17] or biosensing [18], to mention just a few among many others. Recently, MAPLE was used to fabricate nanostructured ultra-stable glassy polymer coatings with good thermal and kinetic stability [19].

11.2 MAPLE Set-up

The preparation of a homogenous solution which consists of the material of interest—solute (0.25–10 % wt) dissolved or suspended in an appropriate solvent stands for the first step of the MAPLE technique. The solvent is characterized by chemical inertness towards the solute, volatility, and high absorbance at the wavelength of the employed laser radiation. Next, the solution is frozen (usually in liquid N_2 or with Peltier elements). The obtained solid cryogenic target is mounted on a cooler in a reaction chamber and kept solid during the deposition process by a continuous liquid N_2 replenishment.

The substrates used to collect the transferred substance (solute) are selected for particular uses or dedicated investigations. Their cleaning is made in baths of acetone, ethanol and/or deionized water. When introduced in the reaction chamber, they are placed at an optimized separation distance facing the cryogenic target. They can be gently heated below the temperature of the solute chemical decomposition.

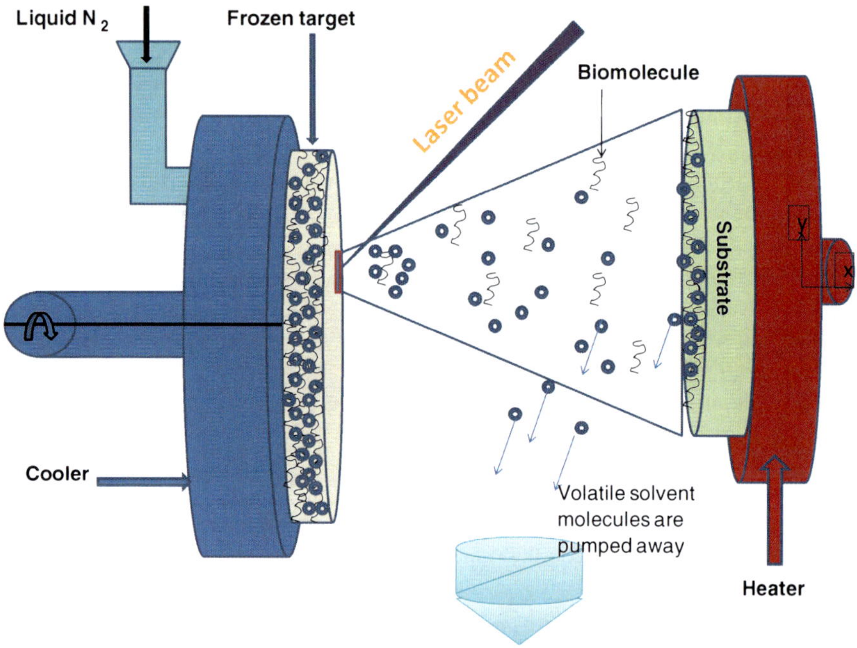

Fig. 11.1 Typical MAPLE set-up. A laser beam (*center top*) irradiates the cryogenic target (*left*) while the dislocated material is transferred and deposited on the facing substrate (*right*)

As an effect of laser irradiation, the solvent is vaporized and directs the solute molecules to the substrates where they are deposited (Fig. 11.1). The key experimental parameters in MAPLE are laser fluence, laser pulse repetition rate, solute concentration, substrate temperature and target—collector separation distance, which are optimized following recurring tests for each pair solute–solvent. A dynamic pressure of <10 Pa is typically kept in the reaction chamber during the laser transfer experiment. Hundreds to thousands laser pulses are applied to get the desired material thickness on the substrate surface. The average deposition rate is generally not higher than 1 Å/pulse.

The laser systems employed for targets evaporation are operating in UV generating pulses of ns duration, i.e. excimer or 4th harmonic of Nd:YAG, with fluences which generally vary from 0.05 to 1 J/cm^2 [20].

We review in this chapter recent MAPLE results obtained with a KrF* excimer laser source (248 nm, 25 ns) for deposition of nanostructures and thin films for biomedical applications. Examples are given with potential utilization as drug delivery systems (DDS), biosensors (BS) and biomimetic coatings for advanced implants (BCI). From the first category we refer to bisphosphonates (BPs) doped hydroxyapatite (HA), ribonuclease A (RNase A) and levan coatings; from the second to IgG structures with or without lipids and from the third to Mg and Sr doped octocalcium phosphate (OCP) layers. We note that BPs-modified HA can be also considered a prospective alternative to BCI.

Scheme 1 Molecular structure of two amino-biphosphonates. Reproduced with permission from [48]

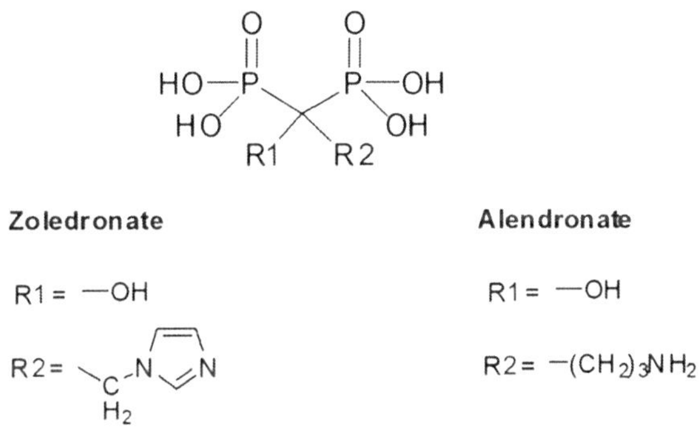

11.3 MAPLE Layers for DDS

11.3.1 Bisphosphonate–Hydroxyapatite Thin Films

The synthesis of coatings constituted of calcium phosphates (CaPs) modified with bioactive molecules allows combining the excellent bioactivity of CaPs with the local availability of the biologically active molecule. A very interesting example is provided by HA modified with BPs. BPs are a class of compounds widely employed for the treatment of disorders of bone metabolism characterized by increased bone resorption, including osteoporosis, dysplasia, myeloma and bone metastases [21, 22]. A common feature of BPs is the presence of the P–C–P bridge, where each P is a phosphonate group, which is resistant both to chemical and to enzymatic hydrolysis [23]. The two covalently bonded side chains attached to the central carbon atom are termed R_1 and R_2 (Scheme 11.1). The presence of hydroxyl (OH) or amino (NH_2) groups as R_1 substituents provides enhanced bone binding capability, most likely via tridentate binding to calcium [23, 24]. The R_2 side group mainly controls the antiresorptive potency of the bisphosphonates. The presence of nitrogen atoms in the R_2 side group is associated with the ability of an individual bisphosphonate to inhibit farnesyl pyrophosphate synthase (FPP) enzyme within the mevalonate pathway in osteoclasts [25–28]. Moreover, it can also influence overall bone affinity [29–32].

The usefulness of BPs in managing specific disorders of bone resorption is widely recognized; furthermore, they are also employed as antitumoral and anti-vascular agents [33–35]. On the other hand, their prolonged use can cause over-suppression of bone metabolism [36–38]. In order to overcome this shortcoming, a number of studies have been recently addressed to the development of strategies aimed to combine these drugs with CaPs for BPs local administration [39–45]. The high affinity of BPs for HA makes the synthesis of BPs-modified HA an uneasy task [44], therefore most of the attempts to prepare HA modified with these compounds have been performed via chemisorption from solution [39, 40, 45–47]. However, HA nanocrystals modified with two potent amino-BPs, alendronate and zoledronate, were recently successfully synthesized in aqueous medium [44, 48].

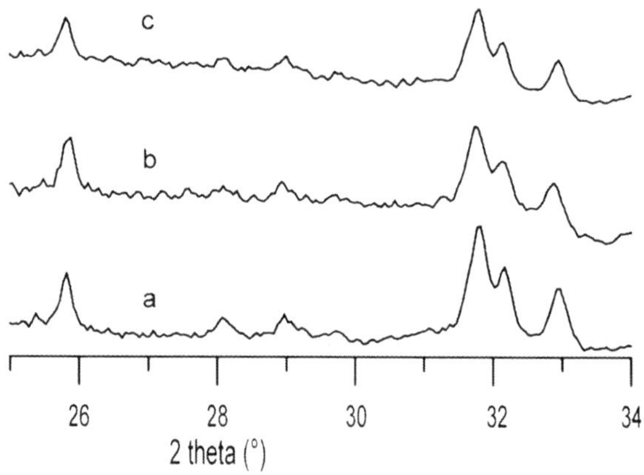

Fig. 11.2 XRD patterns of HA (**a**), HA-AL7 (**b**) and HA-AL28 films (**c**) respectively. Reproduced with permission from [50]

The results of in vitro studies showed that these BPs are able to inhibit osteoclast activity and promote osteoblast growth and differentiation even when included in the HA nanocrystals [48, 49].

The mild conditions characteristic of MAPLE were exploited to synthesize Al:HA thin coatings [50]. The coatings were synthesized using HA powders at increasing alendronate content [0, 3.9 (sample code HA-AL7), 7.1 % wt (sample code HA-AL28)] as starting materials. The powders were suspended in deionised water, frozen at liquid nitrogen temperature and used as targets for MAPLE experiments. The depositions were conducted with a KrF* excimer laser source in mild conditions of temperature and pressure.

The results of X-ray diffraction (Fig. 11.2) and SEM (Fig. 11.3a–c) investigations indicate that the films are constituted of crystalline HA and exhibit a porous-like structure, with pores dimension of 2–4 μm. Neither the morphology, nor the roughness parameters of the coatings, evaluated from the AFM images (Fig. 11.3d), show significant differences depending on the alendronate content.

Bone cells response was evaluated using osteoblast-like MG63 cells and human osteoclasts, which were cultured on the thin films up to 14 days. Osteoblasts show a higher rate of proliferation (Fig. 11.4a) and earlier differentiation in the presence of alendronate, in particular the results indicate that the BP enhances the deposition of collagen type I (CICP) (Fig. 11.4b), as early differentiation marker, and the subsequent production of osteocalcin (OC) (Fig. 11.4c), as later mineralization marker [51]. Conversely, alendronate negatively influenced osteoclast proliferation and differentiation, and it induced osteoclast apoptosis, as revealed by Caspase 3 results (Fig. 11.5a, b) [52]. Moreover, coatings synthesized from HA at relatively high bisphosphonate content (7.1 % wt—HA-AL28) displayed a reduced production of Tumour Necrosis Factor alpha (TNF-α), a pleiotropic cytokine that plays a key role in both inflammation and apoptosis [53], and Interleukin 6 (IL-6), which is involved in the mediation of the inflammatory and immune responses initiated by infections or injuries, suggesting a down-regulatory role of alendronate on the inflammatory reaction.

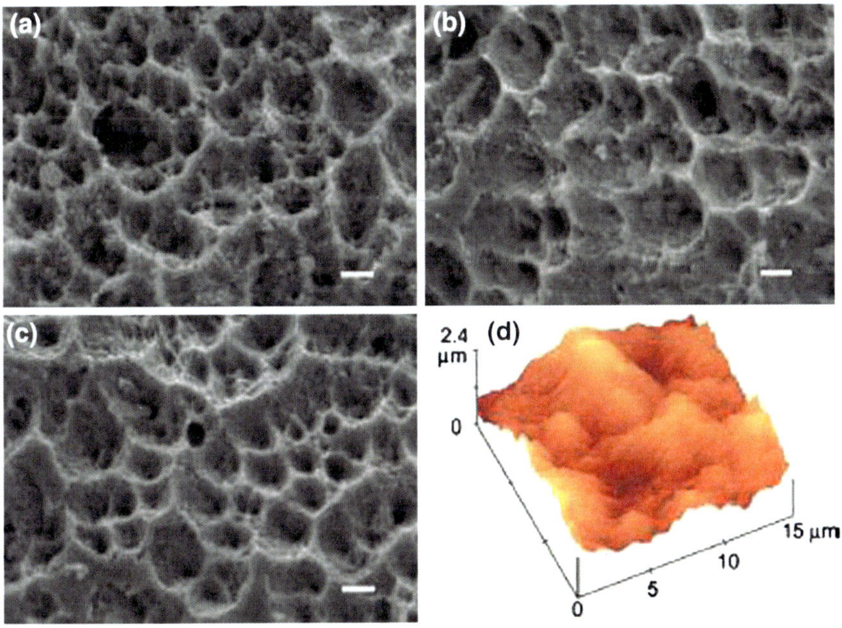

Fig. 11.3 SEM micrographs of thin films deposited from (**a**) HA, (**b**) HA-AL7, (**c**) HA-AL28. Bars = 2 micrometers. **d** AFM image of the surface of a thin film deposited from HA. Reproduced with permission from [50]

The data indicate that the use of MAPLE allows to the successful deposition of Al:HA thin films with enhanced bioactivity, able to promote osteoblast differentiation and to inhibit osteoclast proliferation.

11.3.2 RNase A

RNase A is a single subunit enzyme with a molecular weight of 13.7 ± 0.3 kDa. An important application of RNase A is the removal of RNA from DNA or during other protein preparations [54–57]. Demand for RNA free DNA and protein samples increases with the higher production of DNA vaccines. However, RNase A has to be removed from the DNA or protein solutions after RNA degradation. The alternative is to use RNase A immobilized on solid supports. This approach can offer additional advantages as reuse, control of the enzymatic reaction, or better stability as compared to the free enzyme [58, 59]. RNase A is also recognized for its cytotoxic and cytostatic effects on cancer cells [60–62]. Entrapment of RNase in porous biocompatible three-dimensional (3D) scaffolds can stand for new solutions in cancer therapy aiming for controlled drug release [63–65].

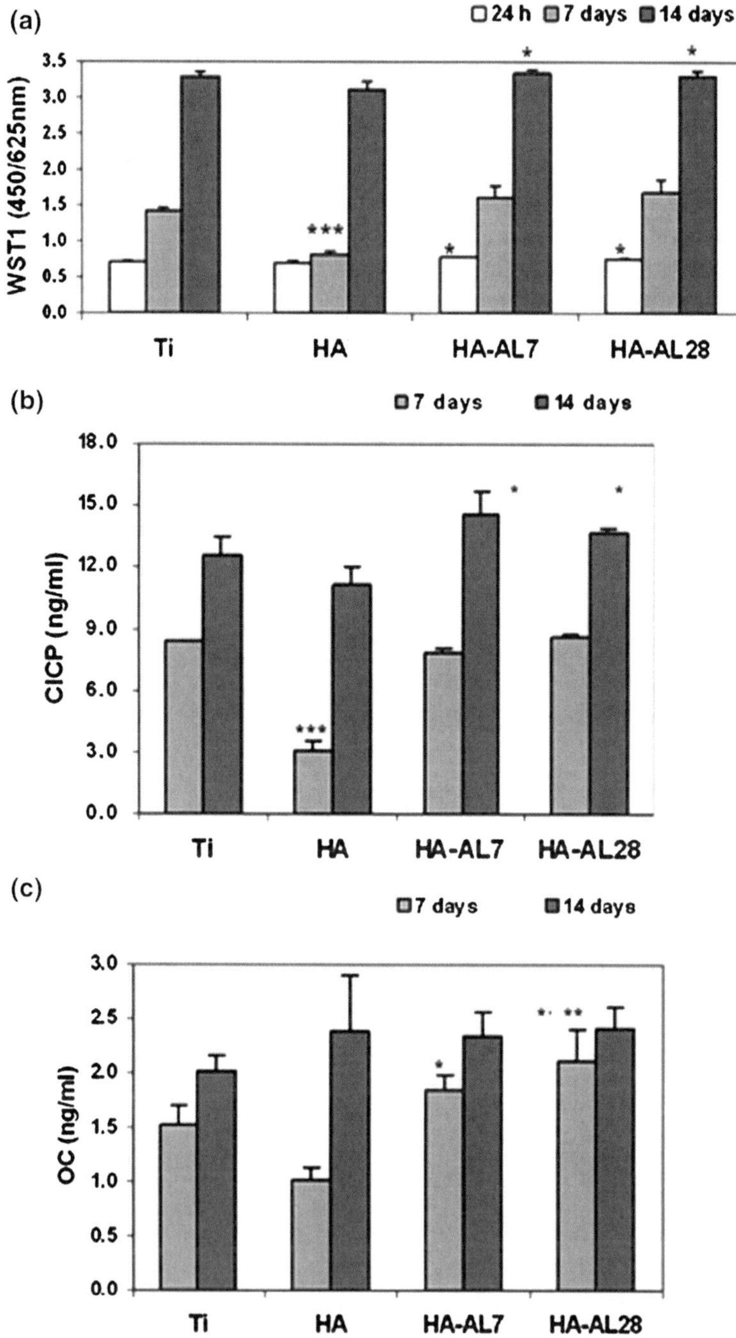
(a)
24 h 7 days 14 days
WST1 (450/625nm)
3.5
3.0
2.5
2.0
1.5
1.0
0.5
0.0
Ti
HA
HA-AL7
HA-AL28
(b)
7 days 14 days
CICP (ng/ml)
18.0
15.0
12.0
9.0
6.0
3.0
0.0
Ti
HA
HA-AL7
HA-AL28
(c)
7 days 14 days
OC (ng/ml)
3.0
2.5
2.0
1.5
1.0
0.5
0.0
Ti
HA
HA-AL7
HA-AL28

◀ **Fig. 11.4** **a** Proliferation of MG63 (WST1 tests) after 1, 7 and 14 days of culture on samples of Ti, HA, HA-AL7 and HA-AL28. Mean ± SD, n = 3. (* = $p < 0.05$; ** = $p < 0.005$; *** = $p < 0.0001$); 1 day: *HA-AL7 versus Ti, HA; *HA-AL28 versus HA; 7 days: ***HA versus Ti, HA-AL7, HA-AL28; 14 days: *HA versus HA-AL7, HA-AL28. **b**, **c** Differentiation and synthetic activity of MG63 after 7 and 14 days of culture on samples of Ti, HA, HA-AL7 and HA-AL28. Mean ± SD, n = 3. (* = $p < 0.05$; ** = $p < 0.005$; *** = $p < 0.0001$):(**b**) CICP. 7 days: ***HA versus Ti, HA-AL7, HA-AL28; 14 days: *HA-AL7, HA-AL28 versus HA. (**c**) OC. 7 days: *HA-AL7 versus HA; *HA-AL28 versus Ti; **HA-AL28 versus HA; 14 days: ns. Modified with permission from [50]

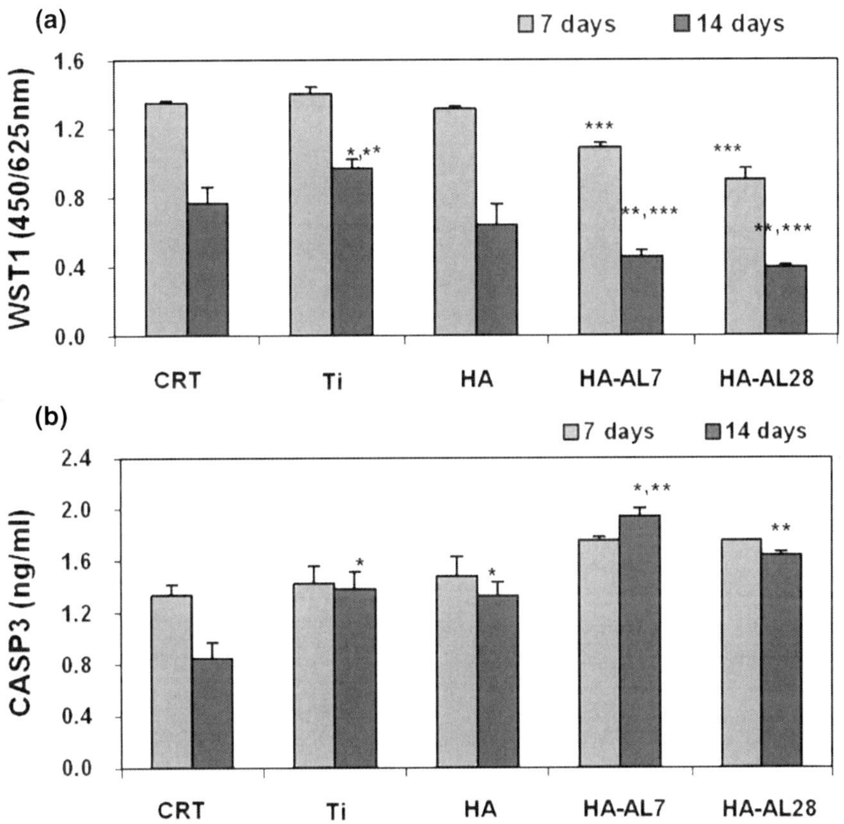

Fig. 11.5 **a** Osteoclasts proliferation after 7 and 14 days of culture on samples of Ti, HA, HA-AL7 and HA-AL28. Mean ± SD, n = 3. (* = $p < 0.05$; ** = $p < 0.005$; *** = $p < 0.0001$). 7 days: ***HA-AL7 versus CTR, Ti, HA; ***HA-AL28 versus CTR, Ti, HA, HA.AL7; 14 dyas: *Ti versus CTR; **Ti versus HA; HA-AL7 versus CTR; HA-AL28 versus HA; ***HA-AL7 versus Ti; HA-AL28 versus CTR, Ti. **b** Caspase 3 values of osteoclasts culture for 14 days on samples of Ti, HA, HA-AL7 and HA-AL28. Mean ± SD, n = 3. (* = $p < 0.05$; ** = $p < 0.005$; *** = $p < 0.0001$). 7 days: ns; 14 days: *CRT versus Ti, HA; HA-AL7 versus Ti, HA; **HA-AL7, HA-AL28 versus CTR. Modified with permission from [50]

Fig. 11.6 Double-wall target holder used for the growth of the RNase A nanostructures

For immobilization of RNase structures, thin film growth experiments were carried out inside a stainless steel irradiation chamber. The composite targets were prepared by dissolving RNase A in distilled water or buffer hepes solvents, at 0.25 wt % RNase A concentration. The prepared solutions were frozen in liquid nitrogen using a double-wall holder mounted inside the irradiation chamber (Fig. 11.6) and kept frozen during the multipulse laser irradiation experiments.

For the deposition of each film 40,000 subsequent laser pulses were applied. The laser fluence value incident on the target surface was chosen in the range (0.25–0.5) J/cm^2.

Before each irradiation run, the chamber was evacuated down to a residual pressure of 10^{-2} Pa. This pressure value was maintained during the RNase A thin film growth experiments. The SiO_2 glass collectors were placed parallel to the target at a separation distance of 5 cm. During the irradiation, the substrates were kept at room temperature.

The surface morphology and growth mode of the deposited RNase A thin films were investigated by atomic force microscopy (AFM) in acoustic (dynamic) configuration with a PicoSPM Molecular Imaging apparatus. The deposited thin films chemical composition and bonding states between elements were studied by Fourier-transform infrared spectroscopy (FT-IR) in the wavenumber range 4,000–500 cm^{-1} with a 4 cm^{-1} resolution.

The surface morphology of the immobilized material changes with the increase of the incident laser fluence value. At 0.5 J/cm^2 laser fluence the transferred RNase A has a uniform surface morphology (Fig. 11.7a) consisting of nanoparticles with dimensions up to a few tens of nm (Fig. 11.7b). Large, several hundreds of μm aggregates formed by nanoparticles can be also observed.

The obtained RNase A structures were characterized by FT-IR spectroscopy. The results were compared with the spectra of the base material used for the preparation of the composite MAPLE targets (Fig. 11.8a). The spectra of the RNase A thin film obtained at 0.5 J/cm^2 laser fluence is identical to that of the base

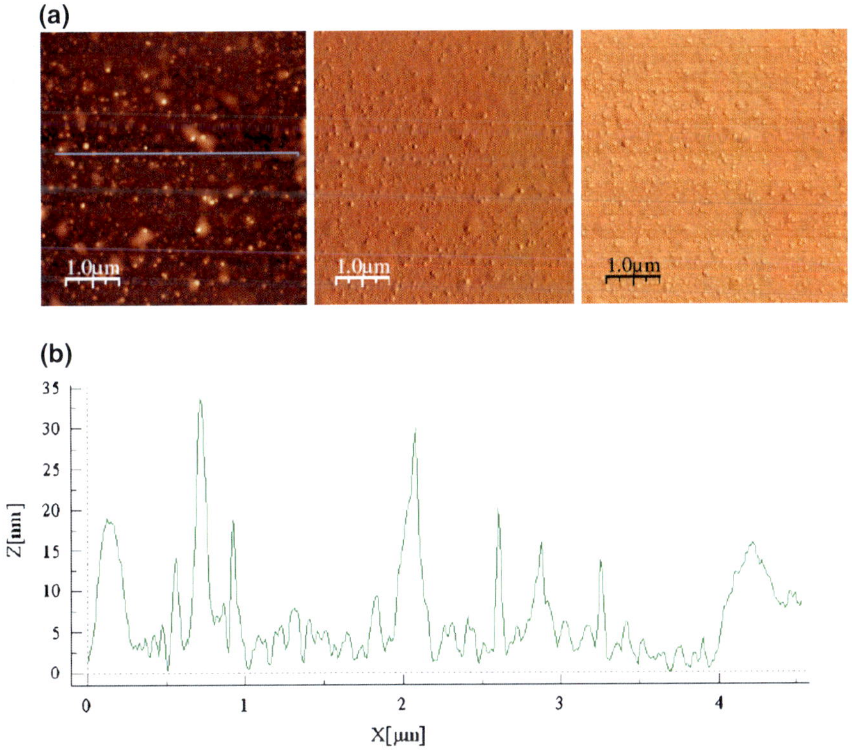

Fig. 11.7 **a** AFM images of RNase A nanostructures obtained through the irradiation of frozen composite targets at an incident fluence value of 0.50 J/cm^2 and 4×10^4 subsequent laser pulses; **b** Surface profile corresponding to the line marked in the AFM image presented in (**a**)

material (Fig. 11.8b). The band at wavenumbers lower than 1,200 cm^{-1} corresponds to the glass substrate.

The enzymatic activity of the obtained structures was analyzed using the synthetic substrate polycytidylic acid [poly(C)] [66]. Poly(C) can be considered to be a synthetic analog of single-stranded RNA and is commonly used for the determination of RNase enzymatic activity [67]. Cleavage of poly(C) by the immobilized enzyme and the products formation were analyzed by reversed-phase high-performance liquid chromatography. The results showed that immobilization process does not affect the RNase A behavior. The product formation pattern was similar to the ones obtained by the non-irradiated base RNase A powder after immersion in solution.

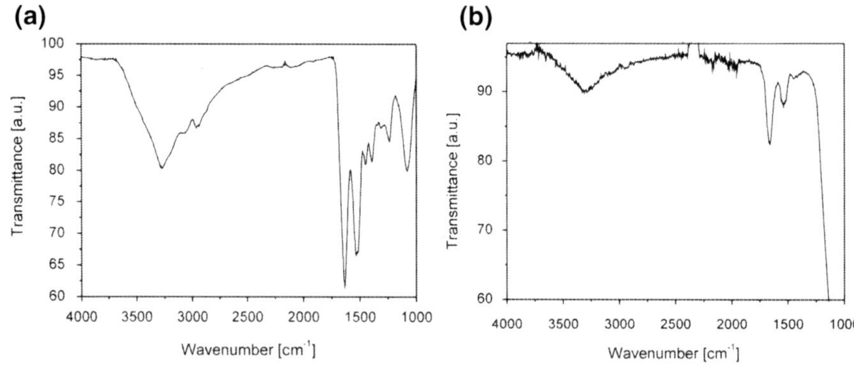

Fig. 11.8 FT-IR spectra of **a** base RNase A material used for the preparation of the MAPLE targets and **b** RNase A thin film obtained using 0.5 J/cm^2 incident laser fluence and 4×10^4 subsequent laser pulses

11.3.3 Levan

Polysaccharides and their conjugates play a key role in various biological mechanisms such as immune response, adhesion, infection and signal transduction. As natural biomaterials, they are highly stable, safe, non-toxic, hydrophilic and biodegradable. For sustainable and economical production of bioactive polysaccharides at industrial scale, rather than plants and algae, microbial sources are preferred since they enable fast and high yielding production processes under fully controlled fermentation conditions [68].

Levan is a naturally fructan homopolysaccharide that is composed of β (2–6) linked fructose units and has many valuable properties like high solubility in oil and water, strong adhesion, good biocompatibility and film-forming ability [69]. It has a great potential as a novel functional biopolymer in foods, feeds, cosmetics, pharmaceutical and chemical industries with many potential uses as emulsifier, stabilizer and thickener, encapsulating agent, osmoregulator or cryoprotector in addition to its uses in medicine as plasma substitute, prolongation of drug activity, radio protector, antitumor and antihyperlipidemic agent [69–71]. In fact, a recent literature analysis on microbial exopolysaccharides clasified levan together with xanthan, curdlan and pullulan as the most promising polysaccharides for various industrial sectors [68]. Although various applications of dextran, gellan, chitosan and pullulan can be found in biomedical field, polysaccharides like curdlan and levan still remained unexploited, mainly due to the limited information related to their biocompatibility and toxicity [72]. Microbial levan is produced as an exopolysaccharide (EPS) from sucrose-based substrates by a variety of microorganisms, including *Acetobacter*, *Aerobacter*, *Azotobacter*, *Bacillus*, *Corynebacterium*, *Erwinia*, *Gluconobacter*, *Mycobacterium*, *Pseudomonas*, *Streptococcus* and *Zymomonas*. In addition, halophilic bacterium *Halomonas* sp. has also been reported as a high level levan producer extremophile [73]. Further research on the potential use of levan by *Halomonas* sp. as a bioflocculating agent [74], its suitability for peptide and protein-based drug nanocarrier systems [71] and an economical production scheme via cheap carbon resources [75] were reported.

Levan by *Zymomonas mobilis* was shown to have anti-tumour [76], immunomodulatory [77], antiobesity and lipid lowering [69] activities, whereas anti-tumour and immunostimulatory activities of levan produced by *Aerobacter levanicum* [78] and by *Microbacterium laevaniformans* [79] were also demonstrated. Further studies on biological activity of levan include cholesterol lowering activity of synthetic levan [80], immunostimulatory and antihistaminic activities of levan by *Bacillus subtilis* [81] and antioxidant role of levan by *Paenibacillus polymyxa* [82]. Moreover, levan by *Halomonas* sp. was not only suggested to be used as an anticytotoxic agent [73] but also credited with high biocompatibility and affinity with HeLa human epithelial cervical cancer and L929 mouse fibroblast cell lines [75]. Recently, levan produced by different *Bacillus* sp. strains showed antiviral activity against pathogenic avian influenza HPAI, H5N1 and adenovirus type 40 [83].

Chemically modified forms of levan polymer have also found wide interest. Sulphate, phosphate and acetate modified forms were prepared and potential applications were patented [84]. Moreover, C3–C4 region of the ß-D-fructofuranozic unit in the levan molecule was suggested to be one of the essential elements for immunochemical binding. Its modification by [85] periodate oxidation and consequent aldehyde activation was found to result in improved bioactivity [86, 87].

Unmodified levan by *Bacillus* sp. was blended with ethylcellulose which is a hydrophobic polymer widely used in coating pharmaceutical dosage forms. Levan/ethylcellulose blends were casted into films and characterized thermally. By extrusion and molding techniques, thermally processed levan films were prepared by adding glycerol as a plasticizer [88]. Thickness of the obtained semi-transparent and cohesive films ranged from 7 to 22 mm. Nevertheless, such thick films are known to hold high risks of poor adhesion, cracking or easy peeling [89]. Thin films with nanoscale thickness are required because of the increased specific surface area, better controlled film dissolution and lower production cost [87]. Thin nanostructured coatings are expected to find numerous applications in drug delivery uses.

Recently, MAPLE deposition of levan (L) and oxidized levan (OL) (Scheme 11.2) has been reported [87]. Frozen solid pellets of levan (or oxidized levan)/dimethyl sulfoxide (DMSO) solutions were used as targets. 0.05 g of L or OL was dissolved in 10 ml DMSO to obtain a homogenous solution. The evaporated material was collected and assembled onto glass slides and Si wafers.

Levan thin films were compact, with good adhesion to substrate and a homogeneous nanostructured surface (Fig. 11.9). The surface features which are confined in volume, were indicative for the starting of film assembling from the bottom. It was suggested that this nanostructured assembling could be due to evaporation of the solvent (DMSO) molecules that accompany levan to the substrate heated at 100 °C. The forming levan layer is drying along a characteristic (evacuation) direction, and a large-scale orientation is boosted. This phenomenon

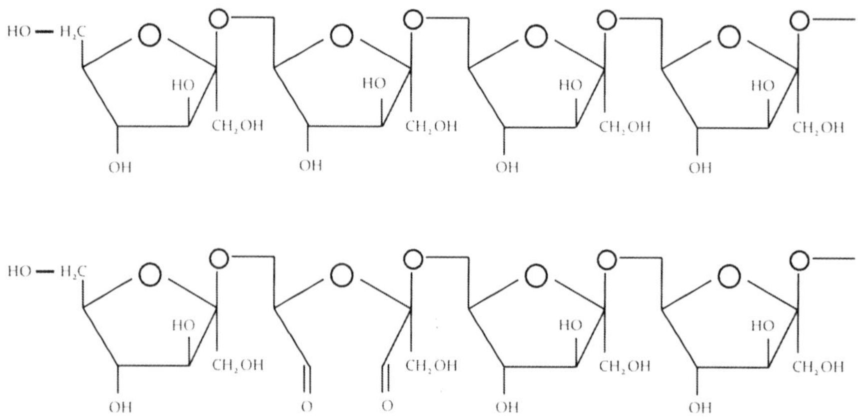

Scheme 2 Chemical Structures of Levan (top) and oxidized Levan (bottom)

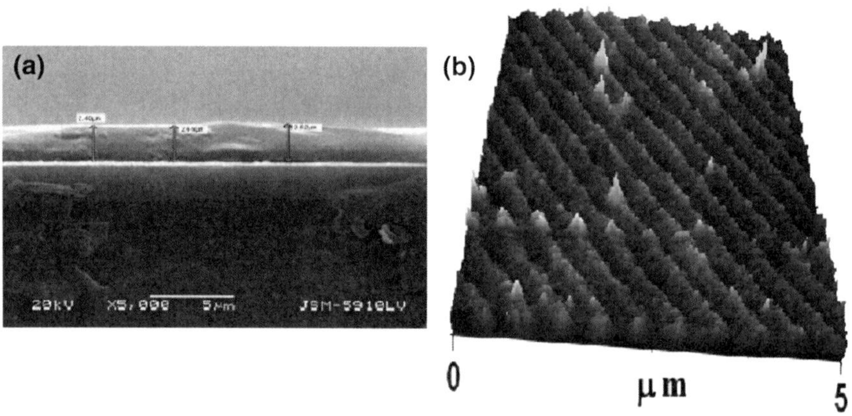

Fig. 11.9 Typical XSEM (**a**) and AFM (**b**) images of oxidized levan thin films obtained by MAPLE (reproduced with permission from [87])

is described as "evaporation-induced self-assembly" (EISA) [90, 91]. The surface assembling of levan was confirmed by AFM images (Fig. 11.9b), with a neat spatial orientation, the most probably due to the collective effect of evaporation-induced assembly with the specific linkages between the linear structures of polysaccharides. It has to be noted that the nanostructured assembling increases the specific surface area, which could boost the surface properties.

The contact angle studies evidenced a higher hydrophilic behavior for oxidized levan structures due to the presence of acidic aldehyde–hydrogen bonds. Moreover, high specific surface areas and biocompatibility of the films supported their potential use in DDS.

11.4 MAPLE Layers for BS

11.4.1 IgG

The application of MAPLE was extended to the transfer and immobilization of IgG molecules on different collectors. As known, five important classes of antibodies are present in human body: immunoglobulin A (IgA), immunoglobulin G (IgG), immunoglobulin M (IgM), immunoglobulin E (IgE) and immunglobulin D (IgD) [92]. Among them, IgG is the smallest one but most abundant which helps to fight with bacterial and viral infections [93]. Clinically, a quantitative antibody test is essential for patients presenting symptoms of autoimmune diseases or allergies and recurring infections. On the other hand, in research, antibodies are used in flow cytometry studies to differentiate cell types [94], in immunoprecipitation to separate proteins [95], or Western blot analyses to identify proteins [96]. Thus, IgGs are considered the main protein category of biological detection elements for biosensor applications.

In the case of protein immobilization by MAPLE, a compromise between the stability in the solution of the material and its safe transfer should be achieved. To this end, solutions of organic and inorganic salts are usually utilized for both preserving the protein stability and increasing the solution laser absorption. The collector temperature during the protein deposition is kept to a maximum of 30–35 °C to prevent the decomposition.

We studied the influence of the laser fluence, as well as the effect of the lipid addition in the target initial solution upon the protein thin films adhesion to substrate [97]. Optical examination and AFM studies showed important differences of the IgG and IgG containing lipid transferred structures in their organization on substrate depending on incident laser fluences. A spot-like arrangement in the case of the IgG protein transferred at the lowest fluences (0.33 J/cm^2) was found probably due to small quantity of transferred material. When increasing the fluence (0.5 J/cm^2), some larger round structures were transferred onto the substrates, although with a proportional fine distribution. At the highest used fluences (0.67 J/cm^2), as expected, more material was deposited but with non-uniform configuration, most probably because of the considerable amount of the salts present in the initial IgG buffered solution.

A similar growth of the deposited IgG containing lipid structures function of incident transfer fluence was revealed. The distribution from spot-like to well protected, encapsulate-like IgG material and finally to irregular structures could be observed when increasing the laser fluence. This reveals a "window" in laser fluence at 0.5 J/cm^2, for which the preservation of structure of IgG after expulsion and deposition is possible, the morphology was well controlled at the same time as a significant quantity of material was transferred [97]. The "window" does not depend on lipid introduction in the mere buffered solution. When dropping the laser fluence one gets substantial loss of deposited protein, while growing it significantly increases the surface roughness due to the removal and transfer of an important amount of salts.

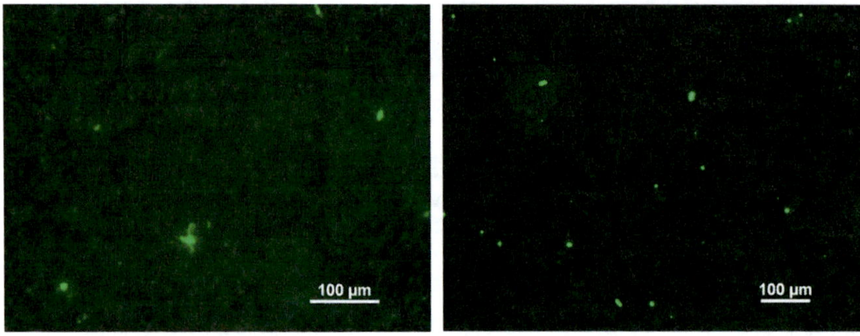

Fig. 11.10 Fluorescence microscopy images of lipid containing (*left*) and lipid free (*right*) samples after incubation with Alexa Fluor conjugated secondary antibody. Permission from [97]

The immunoresponse of the transferred protein was demonstrated by optical labeling methods such as fluorescence microscopy and spectrofluorimetry for the detection of low amounts of antigen (IgG). When comparing samples obtained from lipid-free and lipid containing solutions respectively, while keeping a constant fluence of 0.5 J/cm^2, a tendency of lipid containing deposited sample to emit a higher fluorescence was noticed. No uniform coating but rather a Schweitzer-like structure was found in case of lipid-free structures, due most probably to immobilization deficiencies of the film when immersed in the antibody solution. Conversely, a homogeneous coating distribution with enhanced fluorescent IgG regions was observed after MAPLE transfer from solutions containing lipids. We presume that the areas of increased fluorescence intensity are owing to protein aggregates with superior adhesion at substrate. One can thus conclude that by adding a small fraction of appropriate lipid in an initial protein solution, a homogenous film can be obtained by MAPLE with good adherence to the substrate in comparison with structures deposited from solutions without lipids (Fig. 11.10).

We note that MAPLE deposited IgG coatings could be used as immunosensors for the recognition of specific antigens in research, as well as in clinical investigations. It represents an essential step for the development of personalized, miniaturized biosensors.

11.5 MAPLE Layers for BCI

11.5.1 Magnesium and Strontium Doped Octacalcium Phosphate Thin Films

OCP, $Ca_8H_2(PO_4)_6{\cdot}5H_2O$, is considered the precursor phase of biological apatites. In vitro OCP crystallizes as wide {100} blades, elongated along the *c*-axis [98] (Fig. 11.11). This characteristic morphology is related to OCP structure, where

Fig. 11.11 SEM image of OCP crystals. Modified with permission from [104]

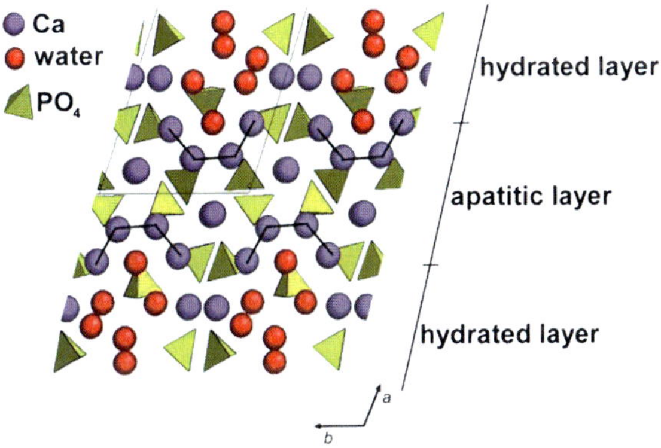

Fig. 11.12 A view of the OCP structure down the c-axis. The hydrated and apatitic layers are highlighted. The positions of Ca atoms connected by black lines and of the phosphate groups of the apatitic layer are very close to those found in the HA structure. Reproduced with permission from [104]

two different layers alternate parallel to the [100] face: cations and anions arrangement is similar to that in HA structure in the "apatitic layer", whereas calcium and phosphate ions are more widely spaced in the "hydrated layer", which contains also interdispersed water molecules (Fig. 11.12). The close similarity between the two structures justifies the easy transformation of OCP into HA by hydrolysis or thermal dehydration [99–101]. As a matter of fact, the relevant stimulatory effect of OCP towards new bone formation has been related to its rapid conversion into apatite in a biological setting [102, 103].

Thanks to its high stability and flexibility, the structure of HA can host a great variety of ionic substitutions [104]. At variance, despite its structural similarity with HA and the results of theoretical studies predicting an easier incorporation of some foreign ions into OCP than into HA lattice [105, 106], experimental data indicate that the presence of foreign ions inhibits OCP crystal growth [107, 108]. Substituted OCP crystals display ill-defined edges and reduced mean dimension, as well as a decreased thermal stability with respect to pure OCP. In particular, although it was recently found that Sr^{2+} and Mg^{2+} ions can be incorporated into OCP structure, the maximum achievable substitutions—7.4 at % for Sr^{2+} and 1

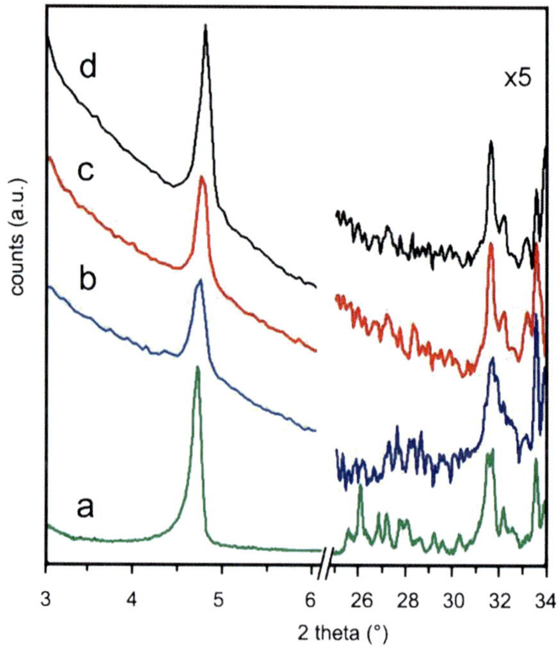

Fig. 11.13 Powder X-ray diffraction patterns of (**a**) OCP powder and of the thin films deposited on Ti substrates from: (**b**) Mg-OCP, (**c**) Sr-OCP, (**d**) OCP. In each plot, the intensities of the 25–34° 2θ range are magnified by 5. Reproduced with permission from [122]

at % for Mg^{2+} [108]—are much smaller than into HA structure—100 at % in the case of Sr^{2+} [109] and 10 at % for Mg^{2+} [110, 111].

Ion-substitution and ion-doping with biologically relevant ions can be used to improve the biological response to CaPs based materials [104]. The positive influence of ion-substitution/doping in HA coatings of metallic substrates has been demonstrated for several ions, such as carbonate, magnesium, strontium and manganese [112–117]. The congruent deposition of ion-substituted or ion-doped OCP is a more difficult task because of the complexity of its structure [118–120]. It was shown that OCP can be successfully deposited on titanium substrates by means of PLD [118]. The deposited material was constituted by nano and microcrystalline domains embedded in a poorly crystalline structure [118, 121].

In order to get better crystallized deposits and to synthesize ion-doped OCP, namely Sr-OCP (Sr^{2+} content: 5.5 at %) and Mg-OCP (Mg^{2+} content: 0.6 at %), we have applied MAPLE that allows for milder working conditions.

The suspension consisting of 0.12 g of OCP powder in 5 ml distilled water was ground with agate balls for 60 min at 360 rot/min in a Retsch centrifugal ball mill suitable for wet grinding. Next, the mixture was collected in a Berzelius glass, mechanically stirred on a Velp Scientifica Vortex agitator, ultrasonicated for 5 min and stirred again. Finally, it was poured into a copper cup which was immediately immersed in liquid nitrogen to serve as a solid target. The target was then mounted on a cooling device inside the reaction chamber and rotated during irradiation to avoid local overheating and laser beam drilling and consequently to circumvent

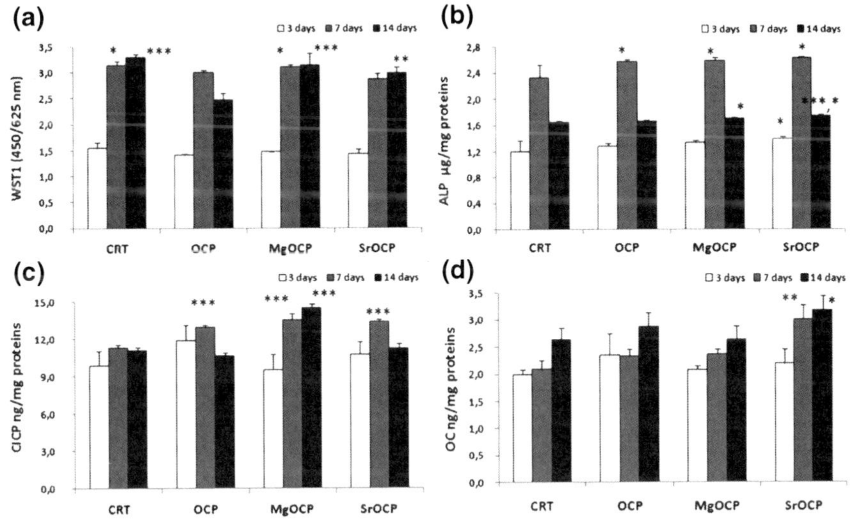

Fig. 11.14 Proliferation, differentiation and synthetic activity of osteoblasts after 3, 7 and 14 days of culture on samples of CTR, OCP, Mg-OCP, and Sr-OCP. Mean ± SD, n = 6. (* = pb0.05; ** = pb0.005; *** = pb0.0001). (**a**) WST1. 7 days: *CTR, Mg-OCP versus Sr-OCP; 14 days: ***CTR, Mg-OCP versus OCP; **Sr-OCP versus OCP. (**b**) ALP. 3 days: *Sr-OCP versus Mg-OCP; 7 days: *OCP, Mg-OCP, Sr-OCP versus CTR; 14 days: ***Sr-OCP versus CTR, OCP; *Sr-OCP versus MgOCP; *Mg-OCP versus CTR, OCP. (**c**) CICP. 7 days: ***OCP, Mg-OCP, Sr-OCP versus CTR; 14 days: ***Mg-OCP versus CTR, OCP, Sr-OCP. (**d**) OC. 7 days: **Sr-OCP versus CTR, OCP, Mg-OCP; 14 days: *Sr-OCP versus CTR, Mg-OCP. Reproduced with permission from [122]

the transfer of non-uniform films [122]. The depositions were performed using a dynamic pressure of 13.33 Pa. The collector was placed parallel to the target at a separation distance of 4 cm and maintained at a temperature of 150 °C during deposition using a conventional 2 inch diameter heater. A pulsed KrF* laser source operating at 7 Hz was used for the multipulse irradiation of the targets. 20,000 subsequent pulses were applied at a laser fluence of 0.75 J/cm^2 for the synthesis of each coating.

Synthesized thin films consisted of crystal fragments, together with cauliflower-like aggregates and droplets, with a homogeneous distribution of magnesium and strontium on the surface of the thin films, as it was proved by the EDS maps [122]. The X-ray diffraction patterns of the thin films (Fig. 11.13), characterized by the strong low angle 100 reflection of OCP at 4.7° in 2θ, demonstrate the good degree of crystallinity of the deposits.

The results of in vitro tests performed using osteoblast-like MG63 cells indicate that the presence of the foreign ions enhanced cells proliferation, activity and differentiation (Fig. 11.14a–d). In particular, Mg-OCP displays a major role on collagen type I (CICP) production, an early marker of osteoblast differentiation, while Sr-OCP has a greater stimulatory effect on Alkaline Phosphatase (ALP) activity and Osteocalcin (OC) synthesis. These features demonstrate that ion doping improves the beneficial effect of OCP on bone cells. The comparison of the results obtained for the differentiation markers on the different coatings indicates that the degree of cell differentiation on the coatings follows the order OCP < Mg–OCP < Sr–OCP.

11.6 Conclusions

We consider that MAPLE technique clearly evidenced a potential to be applied for different delicate biomaterials for DDS, BS or BCI. In particular, it was proved that innovative materials such as Al:HA, RNase A, levan, IgG, as well as pure or doped OCP could be safely transferred and deposited in form of nanostructures and thin film, with good adhesion to facing collectors of interest. Various solvents as well as laser parameters were tested to identify the optimum experimental conditions, with the view of finding the best compromise between the film bioactivity and its morphology. Using appropriate pairs of solute–solvent and laser deposition conditions, MAPLE technique allowed for the transfer and immobilization on the substrate surface of the material of interest with chemical composition, structure, morphology, and most likely functionality resembling the base material, as proved by physical–chemical characterization and in vitro testing.

Acknowledgments INM, CR and FS acknowledge with thanks the support of this work by UEFISCDI under the contracts ID 304/2011, TE 82/2011 and PD 101/2012 and of European Social Fund POSDRU 2007–2013 through the contract POSDRU/89/1.5/S/60746. ETO, INM, CR and FS gratefully acknowledge the support of TUBITAK through project 112M330 and of UEFISCDI under the contract 597/2013.

References

1. Clinical Applications of Biomaterials. NIH Consens Statement **4**, 1 (1982)
2. http://www.merriam-webster.com/medlineplus/biomaterial
3. R.W. Eason, *Pulsed laser deposition of thin films: applications-led growth of functional materials* (Wiley, Hoboken, 2007)
4. I.N. Mihailescu, C. Ristoscu, A. Bigi, I. Mayer, in *Laser-Surface Interactions for New Materials Production Tailoring Structure and Properties*, ed. by A. Miotello, P.M. Ossi (2010), p. 235
5. V. Nelea, M. Jelinek, I.N. Mihailescu, in *Pulsed Laser Deposition of Thin Film*, ed. by R. Eason (Wiley, New Jersey, 2007), p. 63
6. G.B. Blanchet, S.I. Shah, Appl. Phys. Lett. **62**, 1026 (1993)
7. A. Pique, Appl. Phys. A-Mater. Sci. Process. **105**, 517 (2011)
8. A. Pique, in *Pulsed Laser Deposition of Thin Film*, ed. by R.W. Eason (Wiley, New Jersey, 2007), p. 63
9. P.K. Wu, B.R. Ringeisen, J. Callahan, M. Brooks, D.M. Bubb, H.D. Wu, A. Piqué, B. Spargo, R.A. McGill, D.B. Chrisey, Thin Solid Films **398–399**, 607 (2001)

10. R. Cristescu, I. Stamatin, D.E. Mihaiescu, C. Ghica, M. Albulescu, I.N. Mihailescu, D.B. Chrisey, Thin Solid Films **453–454**, 262 (2004)
11. R. Cristescu, C. Popescu, A.C. Popescu, G. Socol, I. Mihailescu, G. Caraene, R. Albulescu, T. Buruiana, D. Chrisey, in *Technological Innovations in Sensing and Detection of Chemical, Biological, Radiological, Nuclear Threats and Ecological Terrorism*, ed. by A. Vaseashta, E. Braman, P. Susmann (Springer Netherlands, 2012), p. 231
12. R. Cristescu, G. Dorcioman, C. Ristoscu, E. Axente, S. Grigorescu, A. Moldovan, I.N. Mihailescu, T. Kocourek, M. Jelinek, M. Albulescu, T. Buruiana, D. Mihaiescu, I. Stamatin, D.B. Chrisey, Appl. Surf. Sci. **252**, 4647 (2006)
13. R. Cristescu, M. Jelinek, T. Kocourek, E. Axente, S. Grigorescu, A. Moldovan, D.E. Mihaiescu, M. Albulescu, T. Buruiana, J. Dybal, I. Stamatin, I.N. Mihailescu, D.B. Chrisey, J. Phys: Conf. Ser. **59**, 144 (2007)
14. M. Jelinek, R. Cristescu, E. Axente, T. Kocourek, J. Dybal, J. Remsa, J. Plestil, D. Mihaiescu, M. Albulescu, T. Buruiana, I. Stamatin, I.N. Mihailescu, D.B. Chrisey, Appl. Surf. Sci. **253**, 7755 (2007)
15. S. Guha, D. Adil, N.B. Ukah, R.K. Gupta, K. Ghosh, Appl. Phys. Mater. Sci. Process. **105**, 547 (2011)
16. A. Palla-Papavlu, V. Dinca, M. Dinescu, F. Pietrantonio, D. Cannatà, M. Benetti, E. Verona, Appl. Phys. A **105**, 651 (2011)
17. F. Sima, P. Davidson, E. Pauthe, L.E. Sima, O. Gallet, I.N. Mihailescu, K. Anselme, Acta Biomater. **7**, 3780 (2011)
18. A. Purice, J. Schou, P. Kingshott, N. Pryds, M. Dinescu, Appl. Surf. Sci. **253**, 6451 (2007)
19. Y. Guo, A. Morozov, D. Schneider, J.W. Chung, C. Zhang, M. Waldmann, N. Yao, G. Fytas, C.B. Arnold, R.D. Priestley, Nat. Mater. **11**, 337 (2012)
20. M. Jelinek, T. Kocourek, J. Remsa, R. Cristescu, I. Mihailescu, D. Chrisey, Laser Phys. **17**, 66 (2007)
21. H. Fleisch, Endocr. Rev. **19**, 80 (1998)
22. R.G. Russell, Bone **49**, 2 (2011)
23. R.G. Russell, N.B. Watts, F.H. Ebetino, M.J. Rogers, Osteoporos. Int. **19**, 733 (2008)
24. M.J. Rogers, Curr. Pharm. Des. **9**, 2643 (2003)
25. J.E. Dunford, K. Thompson, F.P. Coxon, S.P. Luckman, F.M. Hahn, C.D. Poulter, F.H. Ebetino, M.J. Rogers, J. Pharmacol. Exp. Ther. **296**, 235 (2001)
26. B.C. van der Eerden, E.F. Gevers, C.W. Lowik, M. Karperien, J.M. Wit, Bone **30**, 478 (2002)
27. F.H. Ebetino, A.-M.L. Hogan, S. Sun, M.K. Tsoumpra, X. Duan, J.T. Triffitt, A.A. Kwaasi, J.E. Dunford, B.L. Barnett, U. Oppermann, M.W. Lundy, A. Boyde, B.A. Kashemirov, C.E. McKenna, R.G.G. Russell, Bone **49**, 20 (2011)
28. M.J. Rogers, J.C. Crockett, F.P. Coxon, J. Monkkonen, Bone **49**, 34 (2011)
29. R.G. Russell, M.J. Rogers, Bone **25**, 97 (1999)
30. J.E. Compston, Brit. Med. J. **309**, 711 (1994)
31. T.J. Martin, V. Grill, Aust. Prescriber **23**, 130 (2000)
32. G.H. Nancollas, R. Tang, R.J. Phipps, Z. Henneman, S. Gulde, W. Wu, A. Mangood, R.G. Russell, F.H. Ebetino, Bone **38**, 617 (2006)
33. J.R. Green, Oncologist **9**(Suppl 4), 3 (2004)
34. P. Fournier, S. Boissier, S. Filleur, J. Guglielmi, F. Cabon, M. Colombel, P. Clezardin, Cancer Res. **62**, 6538 (2002)
35. J. Wood, K. Bonjean, S. Ruetz, A. Bellahcene, L. Devy, J.M. Foidart, V. Castronovo, J.R. Green, J. Pharmacol. Exp. Ther. **302**, 1055 (2002)
36. S.M. Ott, J. Clin. Endocrinol. Metab. **86**, 1835 (2001)
37. G. Favia, G.P. Pilolli, E. Maiorano, Bone **45**, 406 (2009)
38. M. Pazianas, B. Abrahamsen, Bone **49**, 103 (2011)
39. S. Josse, C. Faucheux, A. Soueidan, G. Grimandi, D. Massiot, B. Alonso, P. Janvier, S. Laïb, P. Pilet, O. Gauthier, G. Daculsi, J. Guicheux, B. Bujoli, J.-M. Bouler, Biomaterials **26**, 2073 (2005)

40. H. Roussière, G. Montavon, S. Laïb, P. Janvier, B. Alonso, F. Fayon, M. Petit, D. Massiot, J.-M. Bouler, B. Bujoli, J. Mater. Chem. **15**, 3869 (2005)
41. H. Roussière, F. Fayon, B. Alonso, T. Rouillon, V. Schnitzler, E. Verron, J. Guicheux, M. Petit, D. Massiot, P. Janvier, J.-M. Bouler, B. Bujoli, Chem. Mater. **20**, 182 (2007)
42. S. Panzavolta, P. Torricelli, B. Bracci, M. Fini, A. Bigi, J. Inorg. Biochem. **103**, 101 (2009)
43. S. Panzavolta, P. Torricelli, B. Bracci, M. Fini, A. Bigi, J. Inorg. Biochem. **104**, 1099 (2010)
44. E. Boanini, M. Gazzano, K. Rubini, A. Bigi, Adv. Mater. **19**, 2499 (2007)
45. B. Peter, D.P. Pioletti, S. Laïb, B. Bujoli, P. Pilet, P. Janvier, J. Guicheux, P.-Y. Zambelli, J.-M. Bouler, O. Gauthier, Bone **36**, 52 (2005)
46. E. Verron, O. Gauthier, P. Janvier, P. Pilet, J. Lesoeur, B. Bujoli, J. Guicheux, J.-M. Bouler, Biomaterials **31**, 7776 (2010)
47. H. Seshima, M. Yoshinari, S. Takemoto, M. Hattori, E. Kawada, T. Inoue, Y. Oda, J. Biomed. Mater. Res. B, Appl. Biomat. **78**, 215 (2006)
48. E. Boanini, P. Torricelli, M. Gazzano, M. Fini, A. Bigi, Biomaterials **33**, 722 (2012)
49. E. Boanini, P. Torricelli, M. Gazzano, R. Giardino, A. Bigi, Biomaterials **29**, 790 (2008)
50. A. Bigi, E. Boanini, C. Capuccini, M. Fini, I.N. Mihailescu, C. Ristoscu, F. Sima, P. Torricelli, Biomaterials **30**, 6168 (2009)
51. G.S. Stein, J.B. Lian, Endocr. Rev. **14**, 424 (1993)
52. D.E. Hughes, K.R. Wright, H.L. Uy, A. Sasaki, T. Yoneda, G.D. Roodman, G.R. Mundy, B.F. Boyce, J. Bone Miner. Res. **10**, 1478 (1995)
53. D.J. MacEwan, Br. J. Pharmacol. **135**, 855 (2002)
54. C. Frazao, C.E. McVey, M. Amblar, A. Barbas, C. Vonrhein, C.M. Arraiano, M.A. Carrondo, Nature **443**, 110 (2006)
55. A. Barbas, R.G. Matos, M. Amblar, E. Lopez-Vinas, P. Gomez-Puertas, C.M. Arraiano, J. Biol. Chem. **283**, 13070 (2008)
56. D.B.E. Battistel, G. Rialdi, Pure Appl. Chem. **63**, 1483 (1991)
57. M. Benčina, J. Babič, A. Podgornik, J. Chromatogr. A **1144**, 135 (2007)
58. F.J. Xu, Q.J. Cai, Y.L. Li, E.T. Kang, K.G. Neoh, Biomacromolecules **6**, 1012 (2005)
59. V. Grazu, O. Abian, C. Mateo, F. Batista-Viera, R. Fernandez-Lafuente, J.M. Guisan, Biotechnol. Bioeng. **90**, 597 (2005)
60. S.M. Rybak, D.L. Newton, Exp. Cell Res. **253**, 325 (1999)
61. U. Arnold, Curr. Pharm. Biotechnol. **9**, 161 (2008)
62. P.A. Leland, K.E. Staniszewski, B.M. Kim, R.T. Raines, J. Biol. Chem. **276**, 43095 (2001)
63. D. Teoli, L. Parisi, N. Realdon, M. Guglielmi, A. Rosato, M. Morpurgo, J. Controlled Release **116**, 295 (2006)
64. M.A. Brook, Y. Chen, K. Guo, Z. Zhang, J.D. Brennan, J. Mater. Chem. **14**, 1469 (2004)
65. S.P. Cullen, X. Liu, I.C. Mandel, F.J. Himpsel, P. Gopalan, Langmuir **24**, 913 (2008)
66. C. Popescu, J. Roqueta, A. Pérez del Pino, M. Moussaoui, M.V. Nogués, E. György, J. Mater. Res. **26**, 815 (2011)
67. M. Moussaoui, A. Guasch, E. Boix, C.M. Cuchillo, M.V. Nogués, The role of non-catalytic binding subsites in the endonuclease activity of bovine pancreatic ribonuclease A. J. Biol. Chem. **271**, 4687 (1996)
68. F. Donot, A. Fontana, J.C. Baccou, S. Schorr-Galindo, Carbohydr. Polym. **87**, 951 (2012)
69. S.A. Kang, J.-W. Seo, K.H. Kim, Y.H. Kim, D. Rairakhwada, M.Y. Seo, J.O. Lee, S.D. Ha, C.-H. Kim, S.-K. Rhee, in *Microbial production of biopolymers and polymer precursors*, ed. by R. BHA (Caister Academic Press, UK, 2009)
70. F. Freitas, V.D. Alves, M.A. Reis, Trends Biotechnol. **29**, 388 (2011)
71. A.D. Sezer, H. Kazak, E.T. Oner, J. Akbuga, Carbohydr. Polym. **84**, 358 (2011)
72. J.F. Mano, H.S. Azevedo, P.B. Malafaya, R.A. Sousa, S.S. Silva, L.F. Boesel, J.M. Oliveira, T.C. Santos, A.P. Marques, N.M. Neves, R.L. Reis, J. R. Soc. Interface **4**, 999 (2007)
73. A. Poli, H. Kazak, B. Gürleyendağ, G. Tommonaro, G. Pieretti, E. Toksoy Öner, B. Nicolaus, Carbohydr. Polym. **78**, 651 (2009)
74. S. Sam, F. Kucukasik, O. Yenigun, B. Nicolaus, E.T. Oner, M.A. Yukselen, Bioresour. Technol. **102**, 1788 (2011)

75. F. Kucukasik, H. Kazak, D. Guney, I. Finore, A. Poli, O. Yenigun, B. Nicolaus, E.T. Oner, Appl. Microbiol. Biotechnol. **89**, 1729 (2011)
76. G.M.T. Calazans, R.C. Lima, F.P. de Franca, C.E. Lopes, Int. J. Biol. Macromol. **27**, 245 (2000)
77. S. Park, K.H. Jang, M.H. Kim, J.D. Lim, E.T. Han, S.A. Jang, K. Kim, S. Pyo, E.H. Sohn, J. Nutr. Food Sci. **13**, 1 (2008)
78. J. Leibovici, A. Siegal, S. Kopel, G. Davidai, H. Yavetz, Int. J. Immunopharmacol. **11**, 133 (1989)
79. E.J. Yoon, S.H. Yoo, J. Cha, H. Gyu Lee, Int. J. Biol. Macromol. **34**, 191 (2004)
80. Y. Yamamoto, Y. Takahashi, M. Kawano, M. Iizuka, T. Matsumoto, S. Saeki, H. Yamaguchi, J. Nutr. Biochem. **10**, 13 (1999)
81. Q. Xu, T. Yajima, W. Li, K. Saito, Y. Ohshima, Y. Yoshikai, J. Brit. Soc. Allergy Clinical Immunol. **36**, 94 (2006)
82. J. Liu, J. Luo, H. Ye, Y. Sun, Z. Lu, X. Zeng, Carbohydr. Polym. **82**, 1278 (2010)
83. M.A. Esawy, E.F. Ahmed, W.A. Helmy, N.M. Mansour, W.M. El-Senousy, M.M. El-Safty, Carbohydr. Polym. **86**, 823 (2011)
84. E.J. Roberts, P.J.Garegg, ed. by W.I.P. Organization, 1998), Vol. WO98/03184
85. O.A. Cavalcanti, B. Petenuci, A.C. Bedin, E.A.G. Pineda, A.A.W. Hechenleitner, Acta Farmaceutica Bonaerense **23**, 53 (2004)
86. S. Gonta, O. Neilands, I. Vina, J. Mol. Struct. (Thoechem) **710**, 61 (2004)
87. F. Sima, E.C. Mutlu, M.S. Eroglu, L.E. Sima, N. Serban, C. Ristoscu, S.M. Petrescu, E.T. Oner, I.N. Mihailescu, Biomacromolecules **12**, 2251 (2011)
88. J.R. Barone, M. Medynets, Carbohydr. Polym. **69**, 554 (2007)
89. S.S. Kim, J.C. Hyun, in *Handbook of solvents*, ed. by G. Wypych (William Andrew and Chem. Tec. Publishing, Ontario, 2001), p. 386
90. C.J. Brinker, Y. Lu, A. Sellinger, H. Fan, Adv. Mater. **11**, 579 (1999)
91. T. Brezesinski, M. Groenewolt, A. Gibaud, N. Pinna, M. Antonietti, B. Smarsly, Adv. Mater. **18**, 2260 (2006)
92. G.W. Litman, M.J. Shamblott, R.N. Haire, M. Hulst, W. Roess, R.T. Litman, K.R. Hinds-Frey, A. Zilch, C.T. Amemiya, Mol. Biol. Evol. **10**, 60 (1993)
93. G.B. Pier, J.B. Lyczak, L.M. Wetzler, *Immunology, Infection, and Immunity* (ASM Press, Washington, 2004)
94. B.F. Brehm-Stecher, E.A. Johnson, Microbiology and molecular biology reviews : MMBR **68**, 538 (2004)
95. N.E. Williams, Methods Cell Biol. **62**, 449 (2000)
96. B.T. Kurien, R.H. Scofield, Methods **38**, 283 (2006)
97. F. Sima, E. Axente, C. Ristoscu, I.N. Mihailescu, T.V. Kononenko, I.A. Nagovitsin, G. Chudinova, V.I. Konov, M. Socol, I. Enculescu, L.E. Sima, S.M. Petrescu, J. Biomed. Mater. Res. A **96**, 384 (2011)
98. R.A. Terpstra, P. Bennema, J. Cryst. Growth **82**, 416 (1987)
99. B.B. Tomazic, M.S. Tung, T.M. Gregory, W.E. Brown, Scanning Microsc. **3**, 119 (1989)
100. A. Bigi, E. Boanini, B. Bracci, G. Falini, K. Rubini, J. Inorg. Biochem. **95**, 291 (2003)
101. Y.-H. Tseng, C.-Y. Mou, J.C.C. Chan, J. Am. Chem. Soc. **128**, 6909 (2006)
102. F. Blanchard, L. Duplomb, M. Baud'huin, B. Brounais, Cytokine Growth Factor Rev. **20**, 19 (2009)
103. A. Bigi, E. Boanini, M. Borghi, G. Cojazzi, S. Panzavolta, N. Roveri, J. Inorg. Biochem. **75**, 145 (1999)
104. E. Boanini, M. Gazzano, A. Bigi, Acta Biomater. **6**, 1882 (2010)
105. K. Matsunaga, J. Chem. Phys. **128**, 245101 (2008)
106. K. Matsunaga, H. Murata, J Phys Chem B **113**, 3584 (2009)
107. M. Iijima, H. Kamemizu, N. Wakamatsu, T. Goto, Y. Doi, Y. Moriwaki, J. Cryst. Growth **135**, 229 (1994)
108. E. Boanini, M. Gazzano, K. Rubini, A. Bigi, Cryst. Growth Des. **10**, 3612 (2010)
109. A. Bigi, E. Boanini, C. Capuccini, M. Gazzano, Inorg. Chim. Acta **360**, 1009 (2007)

110. A. Bigi, G. Falini, E. Foresti, M. Gazzano, A. Ripamonti, N. Roveri, Acta Crystallographica Sect. B **52**, 87 (1996)
111. D. Laurencin, N. Almora-Barrios, N.H. de Leeuw, C. Gervais, C. Bonhomme, F. Mauri, W. Chrzanowski, J.C. Knowles, R.J. Newport, A. Wong, Z. Gan, M.E. Smith, Biomaterials **32**, 1826 (2011)
112. F. Barrere, C.A. van Blitterswijk, K. de Groot, P. Layrolle, Biomaterials **23**, 2211 (2002)
113. B. Bracci, P. Torricelli, S. Panzavolta, E. Boanini, R. Giardino, A. Bigi, J. Inorg. Biochem. **103**, 1666 (2009)
114. X. Wang, A. Ito, Y. Sogo, X. Li, A. Oyane, Acta Biomater. **6**, 962 (2010)
115. W. Xia, C. Lindahl, J. Lausmaa, P. Borchardt, A. Ballo, P. Thomsen, H. Engqvist, Acta Biomater. **6**, 1591 (2010)
116. G. Munir, G. Koller, L. Di Silvio, M.J. Edirisinghe, W. Bonfield, J. Huang, J. Royal Soc. Interface/Royal Soc. **8**, 678 (2011)
117. C. Capuccini, P. Torricelli, F. Sima, E. Boanini, C. Ristoscu, B. Bracci, G. Socol,M. Fini, I.N. Mihailescu, A. Bigi, Acta Biomaterialia **4**, 1885 (2008)
118. G. Socol, P. Torricelli, B. Bracci, M. Iliescu, F. Miroiu, A. Bigi, J. Werckmann, I.N. Mihailescu, Biomaterials **25**, 2539 (2004)
119. P. Habibovic, J. Li, C.M. van der Valk, G. Meijer, P. Layrolle, C.A. van Blitterswijk, K. de Groot, Biomaterials **26**, 23 (2005)
120. X. Lu, Y. Leng, Q. Zhang, Surf. Coat. Technol. **202**, 3142 (2008)
121. M. Iliescu, V. Nelea, J. Werckmann, I.N. Mihailescu, G. Socol, A. Bigi, B. Bracci, Thin Solid Films **453–454**, 157 (2004)
122. E. Boanini, P. Torricelli, M. Fini, F. Sima, N. Serban, I.N. Mihailescu, A. Bigi, J. Inorg. Biochem. **107**, 65 (2012)

Chapter 12
MAPLE and MALDI: Theory and Experiments

Anna Paola Caricato

Abstract Laser induced breakdown spectroscopy (LIBS) and pulsed laser deposition (PLD) are important techniques for the analysis of materials and the fabrication of thin films (metals, alloys and inorganic compounds). These techniques are not applicable to polymers, organic and biomaterials, mostly destroyed by the energetic laser pulses. To overcome this drawback, matrix assisted laser techniques were introduced: matrix assisted laser desorption ionization (MALDI) and matrix assisted pulsed laser evaporation (MAPLE), for mass spectroscopy and thin film deposition, respectively. They offer an efficient mechanism to transfer easy-to-be-decomposed materials from the condensed phase into the vapor phase. The material of interest (polymers, biological cells, proteins…) is diluted in a volatile solvent, with a typical concentration of a few wt%, to form the target to be irradiated with a pulsed laser beam. The laser energy is principally absorbed by the solvent and converted to thermal energy, allowing the solvent to vaporize. The molecules of the material of interest receive enough kinetic energy through collective collisions with the evaporating solvent to be transferred in the gas phase and finally analyzed or deposited on a suitable substrate. Here, important results of MALDI and MAPLE are reported and their working mechanisms are discussed.

12.1 Introduction

Application of laser to materials processing started immediately after the introduction of the first laser, a ruby laser, by Maiman [1]. The first laser applications involved cutting [2], drilling [3] and welding [4] of metals. In 1965, using the different absorptivities of paper and ink, Schawlow introduced the "laser

A. P. Caricato (✉)
Dipartimento di Matematica e Fisica Ennio De Giorgi, Università del Salento,
Via Arnesano 73100 Lecce, Italy
e-mail: Annapaola.Caricato@le.infn.it

M. Castillejo et al. (eds.), *Lasers in Materials Science*,
Springer Series in Materials Science 191, DOI: 10.1007/978-3-319-02898-9_12,

cleaning", having proposed and demonstrated the first "laser eraser" to remove the ink without damaging the underlying paper [5].

Metal evaporation for thin film deposition was demonstrated as early as in 1965 [6], but the process did not attract much attention, due to the existence of simpler and more efficient methods, like electron beam evaporation. During the 1970s and early 1980s, there was a fast development of the pulsed laser ablation deposition (PLD) process, mostly due to the understanding of an important peculiarity of pulsed laser ablation process: the congruent ablation of complex materials [7]. Laser ablation of polymers was first reported in 1982, nearly simultaneous by two groups [8, 9]. In 1987 thin films of high-temperature superconductors were deposited, soon after their introduction by Bednorz and Müller [10].

A wide range of applications was then offered by the PLD technique. It emerged in the last two decades as a versatile technique for the deposition of thin films. Ultra-violet laser light is commonly used, since it efficiently couples with most materials, at energy density (fluence) of the order of 1 J/cm^2, pulse duration of the order of 10 ns and pulse repetition rate of the order of 10 pulses/s (pps). The ablated material deposits on a suitable substrate placed a few cm downstream. High deposition rate, of the order of 0.1 nm/s are usually obtained. PLD was successfully applied to a wide range of materials, like semiconductors [11], metals [12], alloys [13] and compounds [14]. The ability to deposit thin films with complex compositional profiles on substrates at room or at low temperatures is the most significant advantage of this technique with respect to the traditional ones, like for instance electron beam evaporation. The ablation can also be performed in low-pressure atmosphere (of the order of 1 Pa, or less) to promote a chemical reaction between the ablated material and the environment elements. Chemical reactions can be exploited to attain compound formation (nitrides [15], carbides [16] and oxides [17] of the ablated elements), or to restore the exact target stoichiometry eventually lost during the ablation process [18].

Besides preservation of stoichiometry, PLD presents many advantages for thin film growth: monolayer thickness control, good film-to-substrate adhesion, minimum material consumption, among others. In contrast, it is not well suited for the evaporation and deposition of large delicate molecules like polymers and biomaterials. In fact, the powerful laser pulse causes a strong heating at the laser beam spot on the target, breaking molecular bonds in polymers and burning biological materials. There are only a few exceptions, since polymers like Teflon (PTFE) [19], polymethylmethacrylate (PMMA) [20], polypernaphthalene (PPN) [21] and maybe some other polymer films have been successfully PLD deposited on appropriate substrates.

Besides the wide spread of the PLD technique for the fabrication of thin films of simple and complex materials, laser ablation became also an important tool in the field of analytical chemistry, with the introduction of two very important techniques like the laser induced breakdown spectroscopy (LIBS) for the detection of atomic and molecular species [22] (used also to take spectra from Mars soil by the Curiosity rover, 2012) in irradiated targets detecting their emission spectra and the matrix assisted laser deposition ionization (MALDI) for the evaporation of very large molecules and subsequent mass spectroscopy investigations [23, 24]. Here, the material of interest is dissolved in an appropriate solvent, which absorbs most of the laser pulse energy, with the consequent fast evaporation of the solvent and consequent evaporation/ionization of the large biomolecules dispersed in it.

Taking inspiration from MALDI, a new technique for the deposition of thin films of soft materials, like polymers, organic and biomaterials was introduced: the Matrix Assisted Pulsed Laser Evaporation (MAPLE) [25–27]. MAPLE is an evolution of the PLD technique. It is supposed to produce a more "gentle" mechanism to transfer small and large molecular-weight species from the condensed phase into the vapor phase. The main difference with PLD is the target structure: the material of interest (solute) is diluted in a volatile solvent matrix to form a homogeneous solution (solute concentrations typically of the order of 1 wt%). The solution is frozen at liquid nitrogen (LN) temperature and then rapidly placed into a vacuum chamber to act as a target for laser assisted deposition. The frozen target is irradiated with a pulsed laser beam, like in the PLD process. But now the laser pulse energy is principally absorbed by the solvent and converted to thermal energy, allowing the solvent to vaporize. By collective collisions with the evaporating solvent, the molecules of the material of interest receive enough kinetic energy to be transferred in the gas phase, thus covering a suitable substrate where they adhere as a thin film. A highly volatile solvent is chosen, to be pumped away during the flight from the target to the substrate. As a consequence, the deposited film should be composed only of the solute material. Since most of the laser energy is absorbed by the volatile matrix rather than by the solute molecules, their photochemical decomposition can be minimized. Moreover, the ablation onset in MAPLE is defined by the thermodynamic parameters of the volatile solvent, rather than the ones of the solute material. It means that deposition can proceed at much lower fluences (0.05–0.5 J/cm^2), as compared to conventional PLD (a few J/cm^2). Using low fluences, thermal damage or decomposition of the solute molecules is of course prevented or greatly reduced.

The matrix-assisted technique in both MALDI and MAPLE allows evaporation of large and delicate molecules (polymers, proteins, biomaterials, etc.), which would be degraded or destroyed when trying their evaporation using PLD. The working principle is almost the same, but applications are very different: vaporization and ionization for mass spectroscopy investigation for MALDI, film deposition for MAPLE.

Here, an overview of the basic principles of MALDI and MAPLE will be given, together with a short review of the most interesting applications.

12.2 MALDI: Basic Principles and Applications

MALDI was introduced in 1985 by Tanaka [23] and Hillenkamp and Karas [24]. MALDI is a special ionization technique that allows the desorption and ionization of large molecular species ($\sim$10–1,000 kDa). When used in combination with mass spectrometry (MALDI-MS), it allows the analysis of biomolecules and large organic molecules, increasing by far the sensitivity of conventional mass spectrometry instruments (100–240 Da).

MALDI is based on the discovery that dissolving biomolecules (like enzyme, proteins, DNA, etc.) within a particular matrix, specifically chosen to absorb at the irradiation wavelength, can lead to their ejection, upon irradiation, into the gas phase and ionization, with the consequent possibility to "identify" the biomolecules with mass spectrometry (typically time-of-flight, MALDI-TOF). The most important advantage of the method is that the dissolved biomolecules can be ejected in the gas-phase with minimal or little degradation. The characterization of the structure (i.e., sub-group identification) of biopolymers was demonstrated [28].

Different steps take place in the MALDI process:

1. Sample preparation (dilution of the analyte molecules in a matrix with appropriate properties);
2. Irradiation, by means of a pulsed laser beam, of the sample and evaporation of part of the solution;
3. Generation and separation of charges and ionization (protonation or deprotonation) of analyte molecules;
4. Extraction and separation according to mass-to-charge ratio of the ions in the mass spectrometer;
5. Detection of the analyte.

The technique is mainly empirical and it is most frequently based on a trial and error approach on finding appropriate molecular matrices for the various types of biomolecules to be studied. However, molecular dynamic simulations have recently given a big help in understanding the different processes involved in the MALDI technique [29].

At the moment, it is well understood that a successful matrix should exhibit the following criteria:

- It has to isolate the analyte molecules by dilution, to prevent analyte aggregation. Moreover, the dilution of the analyte must be such to minimize its thermal degradation. The critical analyte/matrix ratio increases with molecular weight (for peptide in the low kDa range it is of about 1–5 wt%);
- It has to absorb the laser energy via electronic (UV-MALDI) or vibrational (IR-MALDI) excitation to achieve transition from the condensed to the vapor phase, without excessive destructive heating of the embedded analyte molecules;
- It has to be acidic in order to act as a proton source to promote ionization of the analyte. At present, it is assumed that the matrix plays an important role in the protonation process.

There are many different matrices that can be used for MALDI-TOF. Some of the most common include:

- sinapinic acid (SA) for protein samples;
- alpha-cyano-4-hydroxycinnamic acid (ACH) for peptide samples;
- 9:1 mixture of 2,5-dihydroxybenzoic acid and 2-hydroxy-5-methoxybenzoic acid (sDHB) for carbohydrate and sometimes protein samples.

A solution of one of these molecules is often made in a mixture with highly purified water and an organic solvent, usually acetonitrile (ACN) or ethanol Trifluoroacetic (TFA) acid may also be added. The role of water and organic solvent is to allow both hydrophobic and water-soluble (hydrophilic) molecules to be dissolved into the solution. Then, the matrix solution is mixed with the analyte (e.g., protein-sample). This solution is spotted onto a MALDI plate (usually a metal plate designed for this purpose). The solvents vaporize, leaving only the recrystallized matrix, but now with analyte molecules embedded into matrix crystals. The matrix and the analyte are said to be co-crystallized.

The first MALDI spectrum of a protein (b-D-galactosidase) was presented during the International Mass Spectrometry Conference in Bordeaux in 1988 [30]. The protein was dissolved in nicotinic acid and the solution was irradiated with a Nd:YAG laser ($\lambda = 266\,\text{nm}$). After that, very fast progresses were made, highly enhancing resolution. Later, the direct analysis of viruses, bacteria, fungi and spores was demonstrated [31–34].

As regards the laser characteristics, fast excitation is necessary to avoid destructive thermal excitation of the analyte. It means that the laser pulse duration must be shorter than 25 ns. Most frequently, 0.5–10 ns pulse lengths are used. Wavelengths range from 193 to 430 nm (photon energy of 6.5 and 2.9 eV, respectively), with penetration depths of 50–200 nm. The typical laser fluences are in the range 30–600 J/m^2 for laser spot diameters of $\sim$100 μm [35]. A threshold fluence for ion detection is reported, which strongly depends on the laser spot dimension. It is also evident that the ion signal increases with the laser fluence. Mean ion velocities are in the range 200–1,000 m/s, while the mean neutral velocity is of $\sim$500 m/s. The ion/neutral ratio is of $\sim 10^{-5}$–10^{-3}, depending mainly on the laser fluence.

A unique model for the laser-induced desorption mechanism does not exist. It depends on the laser pulse characteristics, like wavelength and the amount of deposited energy. The desorption mechanism using UV lasers, which means excitation of samples, the phase change and the dynamics of the material plume expansion were studied by Dreisewerd [35]. Primary (matrix) and secondary (analyte) excitation and ionization mechanisms are reviewed in the two closely related papers by Karas and Krüger [36] and Knochenmuss and Zenobi [37].

12.3 MAPLE: Basic Principles

Modern technology requirements offer a great deal of applications for thin-film polymer coatings, mostly in the areas of microelectronics, optoelectronics and miniature chemical and biological sensors. In many cases, the compositional and structural complexity and the anisotropy of the material properties preclude the processing of many of these polymers by conventional physical or chemical vapor deposition methods. Laser techniques seem to offer better opportunities. As mentioned above, some attempts to deposit thin films of polymeric materials were first made by PLD. Several types of polymers (polyethylene, polycarbonate, polyimide and PMMA) were ablated using UV lasers at energies near the ablation threshold [19–21]. A decrease of the molecular weight of the polymers forming the films was observed most of the time. Film formation was supposed to occur via pyrolytic decomposition, followed by repolymerization. Since repolymerization can be incomplete, PLD clearly cannot be used in general for complex polymers.

MAPLE was expressly developed by McGill and Chrisey [38] to overcome the above difficulties presented by PLD of polymers and soft materials. This novel technique was used for the deposition of polymers and other complex materials, as ultrathin and uniform coatings. Polymers were and are now among the most used materials in MAPLE deposition, with excellent structural fidelity of the films with respect to the target materials. High quality, uniform and adherent polymer coatings are produced using the MAPLE technique, paying attention to optimize the laser parameters to the polymer chemical characteristics. With appropriate choice of the experimental parameters, the physicochemical properties of the coating are mostly unaffected by the process.

The MAPLE deposition hardware does not substantially differ from the ones commonly used in PLD. Excimer lasers (or Nd:YAG, third harmonic at 335 nm) are mostly used, since UV radiation efficiently couples with almost any target material. In some particular cases, infrared laser sources are used to selectively dissociate solvent molecules (RIR-MAPLE). Pulse repetition rates are usually in the 1–20 Hz range, with fluences from 0.01 to 1 J/cm^2. The main difference with respect to PLD is the target and target holder, since the target must be kept at very low temperatures during depositions. It means that a liquid nitrogen reservoir must be connected to the target holder. It is usually made of high-conductivity oxygen-free cooper, crossed by a stem of the same material supporting the target holder. The target must rotate (3–10 Hz), like in PLD, to allow smooth erosion of the frozen solution. Feedthroughs and connectors have to be accurately designed, with properly chosen gaskets, to allow rotation at low temperature without seizing problems. A schematic diagram of a deposition system, very similar to the ones used for PLD, except for the target and target holder, is shown in Fig. 12.1. An image of the cooled target holder and substrate configuration is shown in Fig. 12.2. The MAPLE technique is analogous to the MALDI-MS analytical technique, but in MAPLE the target is generally refrigerated at LN temperature, while in MALDI the matrix is usually solid at room temperature.

MAPLE preserves most of the advantages of PLD, like the quite simple procedures, the many independent deposition parameters, the good control of film thickness, the possibility of film deposition on non-planar substrates with good uniformity and the possibility to deposit multilayers and composite films in a single-step process.

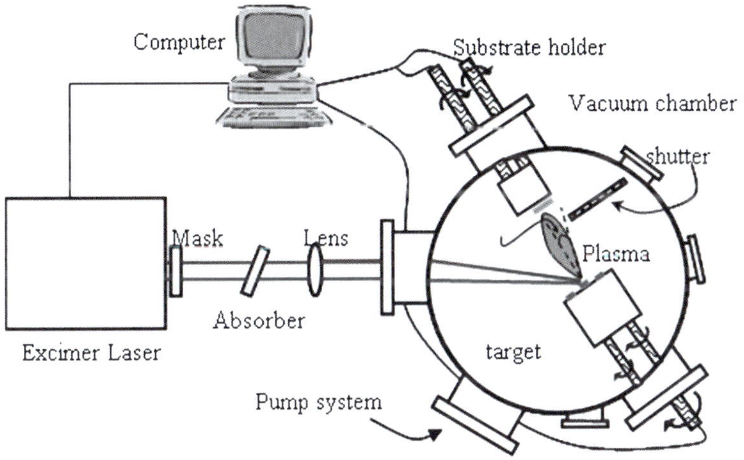

Fig. 12.1 Schematic diagram of a typical MAPLE deposition system, very similar to the ones used for PLD, except for the target and target holder

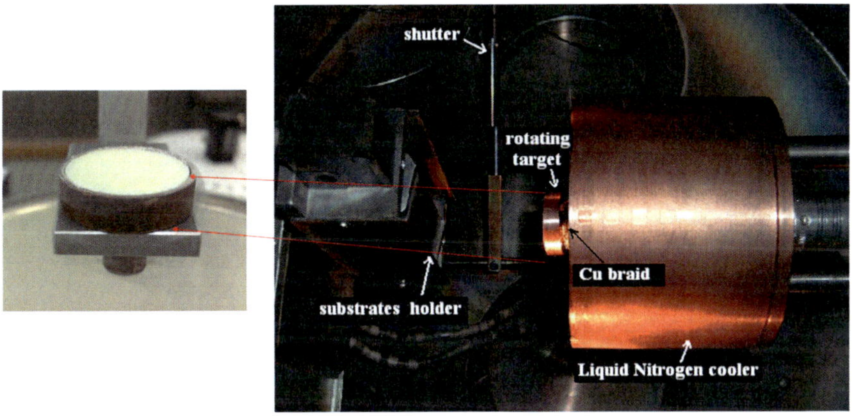

Fig. 12.2 Cooled target holder and substrate holder (MAPLE system at the University of Salento, Italy)

The assumed evaporation mechanism is considered as a photothermal process, where most of the laser energy is absorbed by the volatile solvent (matrix) and converted into thermal energy, producing the vaporization of a surface solvent layer and only a moderate heating of the solute. The collective motion of the many solvent molecules, evaporated by a single laser pulse, carries the few solute molecules present in the evaporated layer to the substrate. The volatile solvent molecules are pumped out from the deposition chamber during the time of flight. Solvent must not react with the solute molecules even under laser irradiation. The most used solvents are isopropanol, deionized water, toluene, acetone,

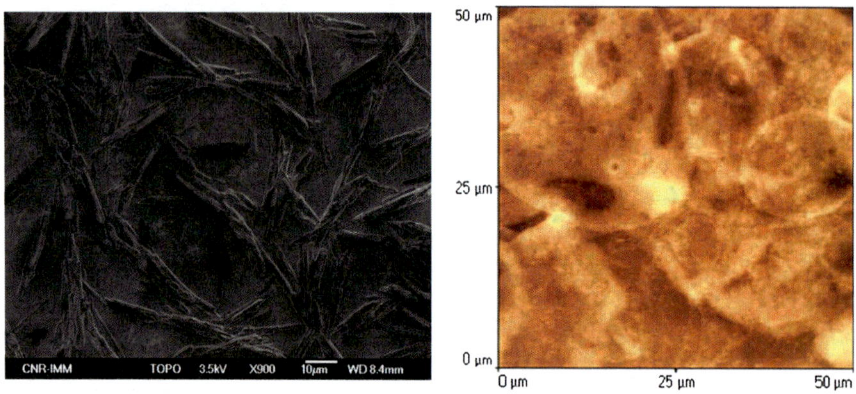

Fig. 12.3 Typical surface structures of a MAPLE-deposited polymer (Ge(TPC)OCH_3) film (*right*) [43] and of an un-evaporated layer of trioctylphosphine [66]

tetrahydrofuran, ter-butyl alcohol, dimethyl sulfoxide, ethyl acetate, chlorobenzene, methanol, dimethoxyethane, phosphate buffer solution and chloroform. Chloroform ($CHCl_3$) and chlorobenzene (C_6H_5Cl) gave frequently origin to formation of reactive radicals, due to photochemical generation of free chlorine atoms, which can react with the solute materials [39].

It has to be noticed that low deposition rates, one order of magnitude lower than those for PLD, are characteristics of MAPLE, due to the lower laser fluences and lower concentration of the material to be deposited. Another frequent drawback is that the scanning electron microscopy (SEM) and atomic force microscopy (AFM) images of MAPLE deposited polymer films show significant surface roughness, aggregates or clusters with sizes ranging from tens of nanometers to tens of microns (Fig. 12.3). The formation of large polymer features is rather unexpected since the polymer concentration in the target is low, polymer molecules are dissolved in the matrix down to the molecular level and the expanding plasma plume should not provide a suitable environment for condensation of polymer clusters. Mercado et al. [40] observed that the geometric patterns present in their MAPLE-deposited poly(lactide-co-glycolide) (PLGA) films were solvent evaporation patterns, consistent with the ones normally associated with the spin casting process. This phenomenon may be due to the ejection of large slices of the target surface layer, explosive evaporation, or spallation of layers onto the substrate, where they melt, leading to evaporation of the solvent. A high ablation rate in their experiment supports this scenario, suggesting that large volumes of the target were ejected by each laser pulse, rather than entrainment transport of individual molecules.

This kind of experimental results stimulated research, which changed the simple model of the MAPLE process. In particular, Leveugle and Zhigilei [41] observed that the initial picture of the ejection and transport of individual polymer molecules in MAPLE introduced by Chrisey et al. [27] cannot explain the results of high-resolution SEM and AFM imaging of many MAPLE deposited films, where significant surface roughness, with well-defined aggregates with characteristic sizes ranging from tens of nanometers to tens of microns, can be observed. Starting from these experimental evidences, Leveugle and Zhigilei formulated a computational model to get a better understanding of the relation between the basic mechanisms of laser interaction with the target material, nonequilibrium processes caused by the fast deposition of laser energy, parameters of the ejected plasma plume, and the resulting morphological characteristics of the growing film. They observed that, even at concentrations of polymer molecules in MAPLE experiments of 0.1–5 wt%, the collective behavior of multiple polymer molecules may play an important role in defining the mechanisms of molecular ejection and the morphological characteristics of the deposited films. To take into account this collective behavior, the laser-induced molecular ejection from a MAPLE target was studied by a coarse-grained molecular dynamics (MD) model. Simulations were performed for MAPLE targets with concentrations of polymer molecules of 1, 3 and 6 wt%, as well as for pure matrix. Irradiation at a wavelength of 337 nm (3.68 eV), with pulse duration of 50 ps, was simulated. The laser fluences (3–9 mJ/cm^2) were chosen to cover the range from below the ablation threshold (3.5 mJ/cm^2 for pure matrix) up to more than twice the ablation threshold. In this irradiation regime, the heat conduction does not contribute to the energy redistribution during the laser pulse and the thermal energy is largely confined within the absorbing region. The conditions of thermal confinement are also characteristic for the majority of MAPLE experiments performed with nanosecond laser pulses. Thus, although the length and time scales of the simulations are very different from the ones in a typical MAPLE experiment, the fact that in the simulations and experiments the MAPLE process takes place under the same physical regime of thermal confinement suggests that the ejection mechanisms are similar, even if at larger time and length scales.

From the simulation, at the early stage of the laser-target interaction, it resulted that below a threshold fluence the process is an evaporative process with the thermal evaporation of the only matrix molecules. No polymer molecules ejection was observed. Above the threshold fluence an explosive process takes place with prompt ejection of small cluster and liquid droplets of matrix-polymer structures as well as of matrix molecules. The polymer molecules are only ejected as part of matrix-polymer clusters. While in the simulation for pure matrix the liquid emerging from the "phase explosion" quickly transforms into spherical droplets as soon as they separate from the target, the presence of polymer chains determines the formation of complex matrix-polymer liquid structures elongated in the direction of the ablation plume expansion. As the polymer concentration increases, the chains become more entangled with the formation of intricate elongated structures, which can reach the substrate, resulting in the formation of complex surface morphology. Evaporation of matrix molecules from the surface of the droplets contributes to the formation of polymer rich surface layer hampering, at first, the escape of the remaining matrix molecules which then escape from polymer voids.

Moreover, Leveugle and Zhigilei observed that MAPLE film depositions are always performed in a multipulse laser irradiation regime and significant structural, morphological, and compositional changes may accumulate in the surface region of a target irradiated by multiple laser pulses. Snapshots of the MAPLE target surfaces, taken at the ends of the simulations, revealed general characteristics of the new surface regions left behind by the ablation process. In addition to the formation of rough target surface morphology, simulations predicted that the composition of the surface region of the target can be significantly altered by the ablation process due to matrix evaporation. One can expect that the effect of the increasing polymer concentration in the target may accumulate during multipulse irradiation, especially at low laser fluences, close to the ablation threshold, and for targets with low initial polymer loading. In addition to the direct effect on the mechanisms of molecular ejection, the compositional and morphological changes in the surface region can have implications on optical properties of the surface, thermodynamic properties of the laser-modified target material, and even heat transfer mechanisms in the heat-affected region of the target.

It can be observed that the simulations performed by Leveugle and Zhigilei support the experimental evidence of minimal chemical modification of polymer molecules in MAPLE film deposition. Indeed, no photothermal bond scission events were detected in any of the simulations performed for polymer concentrations up to 6 wt% and laser fluences up to more than twice the ablation threshold. But simulations also showed that polymer molecules are always ejected as parts of matrix-polymer clusters with a broad cluster size distribution. The ejection of molecular clusters and droplets seems to be inherently connected to the basic mechanism of laser ablation-explosive decomposition of a surface region of the target overheated up to the limit of its thermodynamic stability. Cluster formation decreases with decreasing of polymer concentration in the target. But, the consequence is the decrease of the efficiency of the MAPLE technique for polymer film fabrication.

12.4 MAPLE: Applications and Influence of Deposition Parameters

The main applications of MAPLE are for deposition of polymer thin films, of active proteins and other bioactive thin films. More recent applications are related to the deposition of colloidal nanoparticles and nanorods. Evolutions of the technique are IR-MAPLE, where the energy deposition is achieved via absorption by vibrational modes of the matrix molecules (mainly O–H bending and C–H stretch vibrations at 10 μm, C=O stretch vibrations between 5.5 and 6.5 μm and

O–H or N–H stretch vibrations at ~3 μm) and MAPLE-Direct Write (MAPLE-DW) for the pattering of biomaterials with a spatial resolution of ~5 μm.

12.4.1 Polymer Film Deposition

The first MAPLE trials were performed to deposit films of a chemioselective polymer (a hydrogen bond acid functionalized polysiloxane known as SXFA) onto surface acoustic wave (SAW) devices. The MAPLE deposited films (10–50 nm thick and highly uniform across the whole area) showed higher sensitivity and faster response times to various chemical vapors than analogous films deposited by spray coating technique [25].

After, the MAPLE technique was used to deposit thin films of electrically conductive polymers, like polypyrrole [26]. The MAPLE-deposited polypyrrole films had electrical conductivities similar to polypyrrole films deposited by other techniques. Films of electroluminescent polymers (like poly[2-methoxy-5-(2′-ethylhex-yloxy)-1,4-phenylene vinylene]—MEH-PPV) [42], photoluminescent polymers (e.g. poly(9,9-dioctylfluorene), Ge(TPC)OCH_3) [43] and electro-optic polymers (N-(4-nirophenyl)-(L)prolinol (NPP), tris-(8-hydroxyquinoline) aluminum-Alq3) [26] were also successfully deposited by MAPLE. The deposited polymers presented similar properties to the ones of the bulk counterpart.

Functionalized copolymers thin films prepared on silicon and quartz substrates by MAPLE method were also studied [44]. Two polymeric structures were synthesised by the copolymerization of maleic anhydride and methyl methacrylate, respectively, maleic anhydride and vinyl benzyl chloride, which were subsequently functionalized with 2,4-dinitroaniline. UV–visible, FTIR, Raman and photoluminescence spectroscopy were used to investigate the influence on the properties of the films of different substrate temperatures (150 and 250 °C) during depositions, background N_2 pressure (5–30 Pa) and polymer concentration into the target (2 and 3 %). The authors evidenced that the MAPLE deposition process does not damage the chemical structure of this kind of polymers. SEM investigations revealed the droplets type morphology of the polymeric films, with thickness between 41 and 105 nm, as calculated from ellipsometric measurements.

Paun et al. [45] demonstrated that MAPLE can be used for fabricating films of polymer blends, as well as for incorporating drugs in the polymeric films. The polymers [polyethylene glycol, PEG (1,450 Da), poly(lactide-coglycolide), PLGA (40–75 kDa)] were separately dissolved in chloroform (1 wt%). Then, the two solutions were mixed together, frozen at LN temperature, placed in a vacuum chamber and submitted to laser irradiation (Nd:YAG laser, $\lambda = 266$ nm, $F = 0.1$–1.2 J/cm^2, 10 Hz). Films showed continuous surfaces, with uniform covering of the substrates. Irradiation at low fluences produced films with rms roughness of 20–100 nm. In contrast, irradiation at $F \geq 1$ J/cm^2 led to increased roughness up to 200 nm. FTIR spectroscopy showed that both polymers were transferred preserving their chemical structure at $F < 1$ J/cm^2. Then indomethacin

(INC), an antiinflammatory drug, was incorporated in the polymeric solution. Films of PEG:PLGA:INC blends were MAPLE-deposited. The presence of INC in the films was confirmed by the specific absorption of the drug at 319 nm.

Many other polymers were MAPLE-deposited. For an updated review of MAPLE deposition of polymer films see [46].

12.4.2 Maple Deposition of Bilayer Polymeric Structures

Thin films of polymers are normally prepared by solvent-based methods, such as spin coating, dip-coating and drop-casting, which are quick and inexpensive but not effective in finely controlling the homogeneity on the deposited area and tailoring the final thickness. But severe limitations exist in the case of multilayer deposition, even for the simple bilayer structure. Indeed, most of the conjugated organic molecules are soluble in similar solvents, consequently the deposition of the upper polymer layer tends to dissolve the underlying one, causing intermixing at the polymer/polymer interface [47]. For this reason, when possible, orthogonal solvents are used, i.e. the solvent of the polymer of the upper layer does not affect the morphology of the first polymer layer. However, this stratagem frequently does not work, because it is difficult to find orthogonal solvent since, as said before, the polymers of interest are often dissolved by the same type of solvent.

Few attempts of MAPLE deposition of polymeric multilayers, like e.g. double layers of polyethylene glycol and 3-(3,4-dihydroxyphenyl)-2-methyl-L-alanine thin films, are reported in the literature [48–50]. But, they were always fabricated using two-steps processes, which require breaking the vacuum conditions inside the deposition chamber, resulting in problems of contamination, oxidation and waste of time. This procedure is due to the high difficulty in realizing a rotating multi-target holder working at LN temperature.

Caricato et al. [51] recently reported the first single-step MAPLE (ss-MAPLE) deposition of a bilayer structure consisting of the most-studied donor–acceptor combination for polymer solar cells, namely the electron-donating regioregular polymer, poly(3-hexylthiophene) (P_3HT), and the electron-accepting fullerene, [6.6]-penyl-C_{61}-butyric-acid-methyl-ester (PCBM). The same solvent was used for both polymers, thus overcoming the typical drawback of re-dissolution of the bottom layer occurring in the conventional solution-based deposition techniques.

The two polymers, P_3HT and PCBM, were both dissolved in toluene with the concentration of 0.3 wt%. The obtained solutions, after sonication, were poured in a double-section target holder, fabricated to allow the switch of the laser beam between the two frozen solutions under the same vacuum deposition conditions. The target holder is a copper cup consisting of two concentric cylindrical sections, separated each other, thus allowing both solutions to be frozen simultaneously at the liquid nitrogen temperature (77 K) without intermixing. MAPLE depositions of each material were performed using a KrF excimer laser ($\lambda = 248$nm, $\tau = 20$ ns, pulse rate = 10 Hz) at the fluence of 250 mJ/cm^2. Cross-sectional and

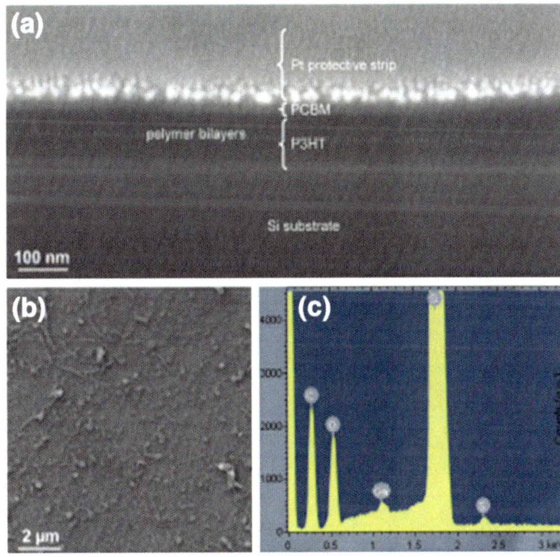

Fig. 12.4 Cross-section (**a**), *top* view (**b**), and EDS spectrum (**c**) of the P_3HT/PCBM bilayer with a ~68 nm thick layer of P_3HT and a ~25 nm thick layer of PCBM, deposited using the ss-MAPLE technique

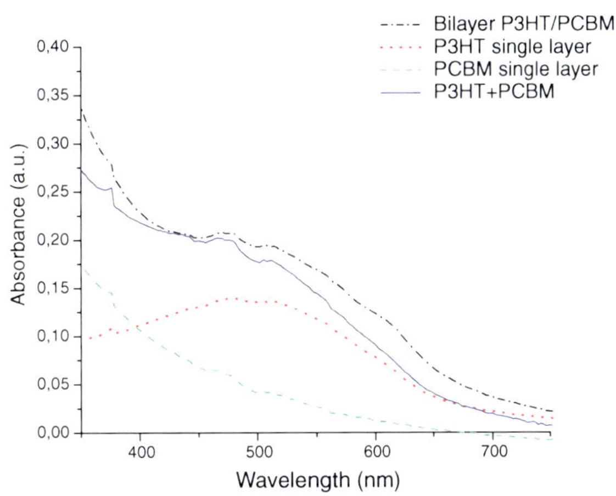

Fig. 12.5 UV-visible absorption spectra of the P_3HT single layer (*dotted line*), PCBM single layer (*dashed line*), and P_3HT/PCBM bilayer (*full line*). The arithmetical sum of the absorption spectra of the two single layer polymers is displayed, too (*dot-dashed line*)

top view images together with the energy dispersive spectrometry, EDS, spectrum of the deposited bilayer are reported in Fig. 12.4a–c respectively. The rms roughness values for P_3HT and PCBM single layers as well as for P_3HT/PCBM bilayer are of about 17, 19 and 50 nm, respectively, as obtained from AFM measurements on a scanned area of 5×5 µm.

The UV–Vis absorption spectrum (Fig. 12.5) of the bilayer P_3HT/PCBM is well fitted by the arithmetical sum of the single layer spectra, P_3HT+PCBM, indicating that both polymeric structures resulted well reproduced and separated in the P_3HT/PCBM sample, thus excluding intermixing or damaging phenomena related to the bilayer deposition.

This paragraph demonstrated the potentiality of the ss-MAPLE technique in depositing polymer bilayer structures in a single step under vacuum conditions, using the same solvent for the two considered polymers. The first polymer bilayer deposited using this technique was used to realize a prototype of an organic solar-cell device.

12.4.3 Organic Materials, Active Protein and Bioactive Thin Films

The processing of organic and biomaterials in the form of thin film requires preserving their molecular function and size. MAPLE was successfully used for this purpose.

Thin films of various carbohydrate molecules such as glucose (molecular weight (mw): 180.2 a.m.u.), sucrose (mw: 342.3 a.m.u.) and dextran (average mw: 148,000 a.m.u.) [52] were deposited from frozen targets made by aqueous solutions of each sugar using an ArF excimer laser. The chemical structure and the molecular weight of the film materials were identical to that of bulk material, except for dextran which presented an about halved molecular weight. This last finding suggests a possible limit on the size of the molecule that can be deposited.

Another kind of sugar, pullulan (average mw: 200,000 a.m.u.) was MAPLE-deposited by Cristescu et al. [53]. The best results were obtained using a solution of pullulan (2 wt%) in dimethyl sulfoxide—$(CH_3)_2SO$. The frozen target was irradiated with a KrF laser at $F = 230$ mJ/cm^2. Trials to deposit pullulan films using PLD were not successful.

Many other biocompatible polymers have been successfully deposited. Polyethylene glycol (PEG) has many biomedical applications. This is why a large number of papers continue to be published on PEG thin films. Thin films of PEG of average molecular weight 1,400 a.m.u. were MAPLE-deposited by Bubb et al. [39]. The deposition was carried out with an ArF laser at $F = 220–230$ mJ/cm^2. Deionized water and chloroform were used as matrices. As already told, the reduction in molecular weight and structural modification of the film deposited using chloroform was attributed to the presence of the highly reactive chlorine atoms, while there was little evidence for photochemical modification when using water. PEG films were also MAPLE-deposited by Toftmann et al. [54]. Water was used as a matrix for the guest material, at concentrations from 0.5 to 4 wt%. The target was irradiated with 6 ns laser pulses at 355 nm at high fluences (2.5–12 J/cm^2). The authors claim that, even at the highest fluence, FTIR spectra indicated a chemical structure of the deposit close to that of the un-irradiated PEG. In a further chapter [55] the laser wavelength was changed from Nd:YAG third harmonic ($\lambda = 355$ nm) to the fundamental one ($\lambda = 1{,}064$ nm). Matrices of 1 wt% PEG frozen solutions were used and emission of PEG molecules and ions from PEG-doped water ice targets was investigated. Even though linear absorption in defect-free water ice is two orders of magnitude larger at 1,064 nm than at 355 nm, the deposition rate and ion current density were found much smaller for IR than for UV laser light.

The MAPLE technique was also used by Bloisi et al. [56] to deposit PEG films. Here a *Q*-switched Nd:YAG laser was used, which was operated at wavelengths ranging from IR to UV to optimize the deposition parameters. The results showed that visible (532 nm) radiation gives better results with respect to UV (355 nm) radiation in preserving the PEG integrity.

As reported above, copolymer of PEG and PLA—poly(d,l-lactide-co-glycolide) films were MAPLE-deposited by Paun et al. [45].

Cristescu et al. [57] report on the successful MAPLE deposition of high quality type I fibrilar collagen thin films. Thin film deposition was performed in a N_2 ambient (20 Pa) using a KrF laser source with energy in the range of 20–35 mJ. They demonstrated that the deposited thin films were composed of collagen, with no impurities.

Active proteins were also deposited in thin film forms. For instance, thin films of horseradish peroxidase (HRP) and insulin were deposited on a variety of substrates, such as Si, NaCl and gold and platinum coated Si [58]. IR spectra and solvent-phase activity test of the HRP films indicated that most of the transferred proteins retained their chemical and physical structure, as well as biological activity. Analyses on insulin films demonstrated near-intact transfer of this protein, with little or no photoinitiated decomposition. These results represented the first demonstration that pure films of active biomolecules can be deposited using a vapor-deposition technique. Additional studies were immediately after performed by the same authors on other biomaterials such as 50/50 Poly(D,L-lactide-*co*-glycolide), biotinylated bovine serum albumin (BSA), and phospholipid polymers, which also maintained function and chemical structure in the MAPLE-deposited thin films [59].

A detailed study of BSA deposition was later made by Martino et al. [60]. Deionized water and phosphate buffered saline (PBS) were used as solvents. The BSA concentration (1 and 2 wt%) and the ArF laser fluence (F = 75–500 mJ/cm^2) were varied, to find the best deposition conditions. FTIR spectra showed that the major BSA absorption bands at 1,653 and 1,550 cm^{-1} were present in the MAPLE-deposited BSA films, without red shifts. This result evidenced the absence of protein denaturation. Biological tests were also performed to check the protein integrity after MAPLE deposition, using the sodium dodecyl sulfate polyacrylamide gel electrophoresis (SDS-PAGE). The band relative to the entire BSA protein was detected in the film prepared at F = 500 mJ/cm^2, indicating that this fluence is high enough to produce a reasonable deposition rate, without damaging the solute.

Stamatin et al. [61] reported the successful MAPLE deposition of fibrinogen blood protein thin films using a KrF excimer laser (F = 440–700 mJ/cm^2) in N_2 atmosphere (15–20 Pa). FTIR spectroscopy demonstrated that the deposited thin films were composed of fibrinogen and fibrin, maintaining their chemical structures. It resulted from FTIR spectra and atomic force microscopy (AFM) micrographs that fibrinogen concentration depend on the laser fluence, the best results being obtained using low fluences.

12.4.4 *Influence of Deposition Parameters*

A correct choice of the deposition parameters is very important to obtain good quality thin films. The most important ones are:

- Laser parameters (wavelength, fluence…);
- Solvent physical and chemical properties (volatility, absorption properties at the laser wavelength, capacity to dissolve the solute, chemical stability…).

An example can clarify these items.

Tunno et al. [62] studied the MAPLE-deposition of Poly(9,9-dioctylfluorene)—(PFO), a material of great technological interest for its high optical gain in the visible range, high charge-carrier mobility and efficient blue emission, using a KrF excimer laser ($\tau = 20$ ns, 10 Hz). The influence of the laser fluence (50–500 mJ/cm^2) and the nature of the solvent on the films properties were carefully investigated. The target was prepared from different solutions of PFO in chloroform ($CHCl_3$), toluene (C_7H_8) and tetrahydrofuran (THF-C_4H_8O). A constant concentration of 0.5 wt% was used in all the depositions. The chemical composition of the deposited films was investigated by FTIR spectroscopy in the range 700–3,600 cm^{-1}. To evaluate the influence of the solvents on the structural properties of the deposited films, the infrared transmission spectra were acquired for films deposited at the same laser fluence (200 mJ/cm^2). The spectra of the MAPLE-deposited films are shown in Fig. 12.6, together with the spectrum of a reference spin coated film. From the comparison, it is clear that chemical decomposition of PFO takes place when chloroform is used. In fact, the vibrational bands are broader, the relative intensities are not preserved and some peaks are missing. The chemical degradation of PFO deposited from a chloroform solution, which is a commonly used solvent in spin coating, is related to the presence of high reactive radicals containing Cl released during excimer laser irradiation [39]. Chemical reactions induced by chlorine were not detected by Bloisi et al. [63], who successfully used chloroform as diluent in their MAPLE depositions, most probably due to the very low fluence (136 mJ/cm^2) and relatively large wavelength (532 nm) of their laser. In contrast with the film deposited using chloroform as solvent, the FTIR spectrum of the MAPLE deposited PFO film using THF resulted very similar to the one of the spin coated film, since all the main peaks are present, preserving their relative intensities. It is important to notice that the peaks around 1,716 and 1,606 cm^{-1}, related to oxidized fluorene (fluorenone) [64], are not observed. This feature is particularly important as PFO is known to easily suffer photooxidation under light exposure, leading to yellow light emitting fluorenone defects. The increasing of the laser fluence has no effect on the peak presence or position, as it can be inferred from Fig. 12.7, where FTIR spectra of samples deposited at laser fluences ranging from 200 to 500 mJ/cm^2, using targets of PFO diluted in toluene, are shown. To investigate the role of the deposition conditions on the photoluminescence (PL) characteristics of the films, PL measurements were performed with a He–Cd laser ($\lambda = 325$ nm). The samples were excited at room

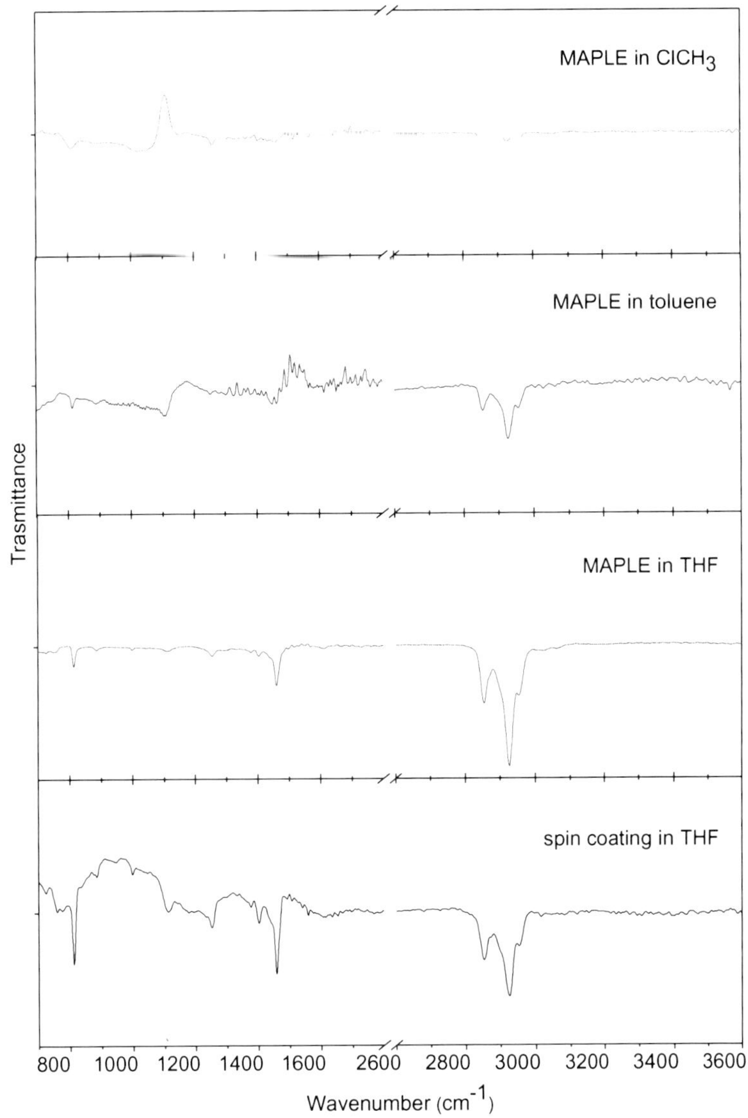

Fig. 12.6 FTIR spectra of Poly(9,9-dioctylfluorene)-(PFO) films deposited with a KrF laser at a fixed fluence (200 mJ/cm^2) using various solvents

temperature with a power density of about 40 W/cm^2. The PL spectrum of the MAPLE film deposited from toluene solution showed glassy phase PFO emission, while the reference spin-coated film showed β-phase emission only. As the β-phase is formed in spin-coated films from toluene due to interplay between aggregation in solution and solvent-induced chain planarization in the solid phase [65], the absence of β-phase emission in the MAPLE-deposited film suggests that

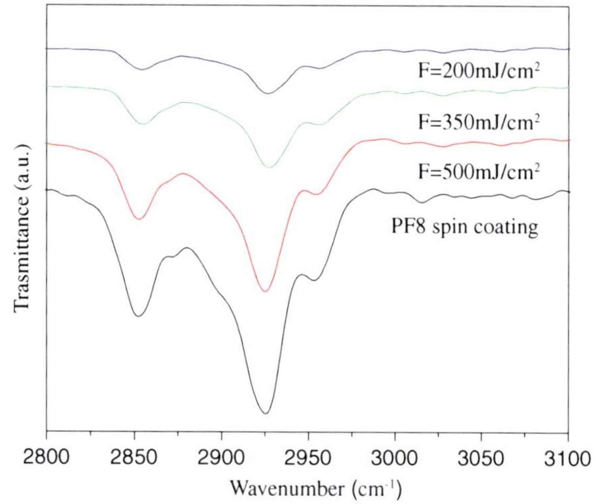

Fig. 12.7 FTIR spectra of Poly(9,9-dioctylfluorene)-(PFO) films deposited with a KrF laser at various fluences (200–500 mJ/cm^2) using toluene as solvent

molecular aggregation did not occur and that negligible interaction between PFO and toluene vapor took place during the deposition process. The PL spectrum of the MAPLE film from THF solution showed a superposition of the emission features of the glassy- and the β-phases. The absence of defect related emission in films deposited from toluene and THF solutions is strong evidence that no relevant chemical modification of the PFO emitting chromophors took place during MAPLE deposition.

A correct choice of the solvent is of great importance, not only to avoid chemical and photochemical reactions (e.g., chloroform) with the solute. Another important parameter is the solvent volatility. Solvents with low vapor pressure are not easily evacuated by the pumping system and are detected in the deposited films. In fact, when a small amount of trioctylphosphine was used in a toluene matrix, SEM inspection of the MAPLE-deposited film (Fig. 12.3b) showed that the film surface consisted of uniformly distributed elongated structures [66]. The FTIR spectrum showed different absorption bands, which are ascribed predominantly to trioctylphosphine, which was used to avoid solute precipitation. Trioctylphosphine has a vapor pressure of 120 Pa at 20 °C, which is much lower than that of toluene at the same temperature (2,900 Pa). So, it is not effectively pumped out during the MAPLE process and consequently it reaches the substrate, contributing to the composition of the deposited film. No trace of trioctylphosphine was present on the film after annealing at 400 °C.

To accurately test relationships between the solvent properties (solubility and other physical properties) and the morphology of the MAPLE-deposited films, Caricato et al. [67] deposited PFO films using different solvents. The PFO powder was diluted (0.5 wt% concentration) in THF, toluene and toluene-hexane (C_6H_{14}) mixtures. Two toluene-hexane mixtures were chosen (85 % toluene/15 % hexane and 61 % toluene/39 % hexane by volume) to gradually change solubility and other physical properties of the matrices. The physical properties, together with the

Table 12.1 Solubility parameters and boiling temperatures of the used solvents

Solvent 1	Solvent 2	Boiling temperature (°C)	Solubility parameters $(MPa)^{1/2}$			
			δ_d	δ_p	δ_h	δ
THF (C_4H_8O)		66	14.4	4.9	6.1	16.4
Toluene (C_7H_8)		110.6	17.3	1.03	0	17.4
Hexane (C_6H_{14})		69	14.6	0	0	14.6
85 % Toluene	15 % Hexane	103.8	16.9	0.88	0	16.92
61 % Toluene	39 % Hexane	94	16.37	0.63	0	16.39

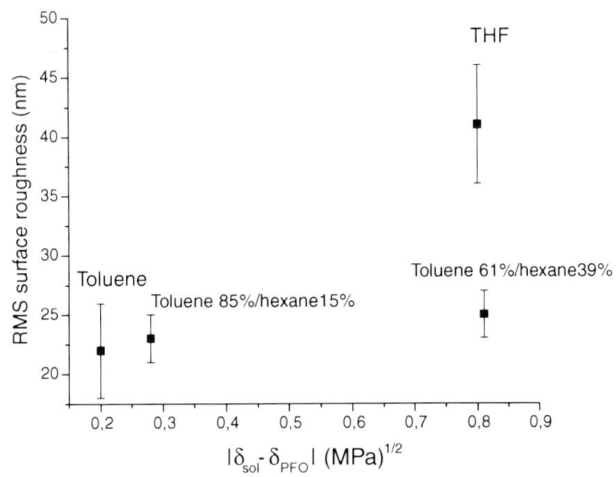

Fig. 12.8 Surface roughness of the deposited PFO films plotted as a function of the absolute value of the difference between the total solubility parameters

solubility parameters of the pure solvents and mixtures relevant for this study are reported in Table 12.1. PFO films were deposited on <100> Si substrates with 10,000 KrF laser pulses ($\tau = 20$ ns, 10 Hz, $F = 250$ mJ/cm^2). From the AFM scans, quite uniform substrate coverage was observed. However, some circular aggregates were present on the films' surface, whose dimensions decrease passing from the films deposited using pure toluene matrix to the films deposited using toluene-based mixtures with increasing amount of the hexane component. The film rms roughness values present an opposite behavior, passing from (22 ± 4) nm in the case of pure toluene matrix to (23 ± 2) nm for 85 % toluene/15 % hexane matrix, to (25 ± 2) nm for the 61 % toluene/39 % hexane matrix. The PFO film deposited using pure THF matrix was characterized by the presence of polymer filaments, randomly distributed on the surface. Consequently, the film surface presents a quite high rms roughness (41 ± 5) nm. As done by Bubb et al. [68], the rms roughness values were plotted against the solubility parameters (Fig. 12.8). The solubility parameters determine the effectiveness of the solvent in inducing dispersion forces (δ_d), its polarity (δ_p) and its tendency to hydrogen bond formation in solution (δ_h) [69]. The total solubility parameter is defined as

$$\delta = \sqrt{\delta_d^2 + \delta_p^2 + \delta_h^2}$$

The films prepared using the toluene-based matrices presented a very small increase, if any, of the rms roughness values, even if the absolute value of the difference between the total solubility parameters of the matrices and solute ($|\delta_{sol}-\delta_{PFO}|$) increases. Moreover, it results that the rms roughness of the PFO film deposited using the THF matrix is much higher (40 nm) than that of the sample prepared using the 61 % toluene/39 % hexane mixture (25 nm), although the $|\delta_{sol}-\delta_{PFO}|$ for the two solvents are equal ($\sim$0.8 $MPa^{1/2}$). It can be concluded that the solubility is not the only parameter to take into account. The difference is most probably determined by the different optical penetration depth of the laser energy into the targets and by the different boiling temperature of the solvents [67]. From molecular dynamic simulations, it results that the higher penetration depth is responsible for the ejection from the target surface of big clusters of polymer matrix [41]. At the same time the matrix-polymer clusters broke and reduce their dimension with time because of solvent evaporation. This process should be faster for solvents with lower boiling points, i.e. for THF with respect to toluene (66 vs. 110.6 °C, respectively). This feature should accelerate polymer entanglement and induce a higher surface roughness.

A further parameter affecting surface morphology is the substrate temperature. A proper choice of the substrate temperature can reduce the surface features. Indeed, the initial studies of the effect of the substrate temperature on the morphology of films deposited in MAPLE indicate that individual surface features become less distinct as the substrate temperature approaches or exceeds the polymer glass transition temperature, and disappear/merge at temperatures close to the polymer melting temperature [70–72]. The surface morphology of the MAPLE deposited films is also influenced by the number of laser pulses as reported in [72] in which, it was observed that the rms roughness of a PFO film prepared with 10,000 laser pulses was about half the value of the rms roughness of the film prepared with 16,000 laser pulses. This feature may be the consequence of significant structural, morphological, and compositional changes gradually accumulated in the surface region of a target irradiated by multiple laser pulses [41], leading to an increased roughness with increasing film thickness.

In general, it is possible to conclude that higher solubility of the solute in the solvent and higher substrate temperatures lead to lower surface roughness, while lower boiling point of the solvent and higher laser penetration in the target lead to higher surface roughness of the MAPLE-deposited films. However, each combination solute/solvent must be carefully investigated.

12.4.5 Deposition of Colloidal Nanoparticles/Nanorods

The MAPLE technique can be used to deposit nanoparticle and nanonorod films starting from solution of colloidal nanomaterials, which are relatively easy and cheap to fabricate. Colloidal nanoparticles of very different materials can be prepared by chemical routes with very small sizes and low size dispersion [73].

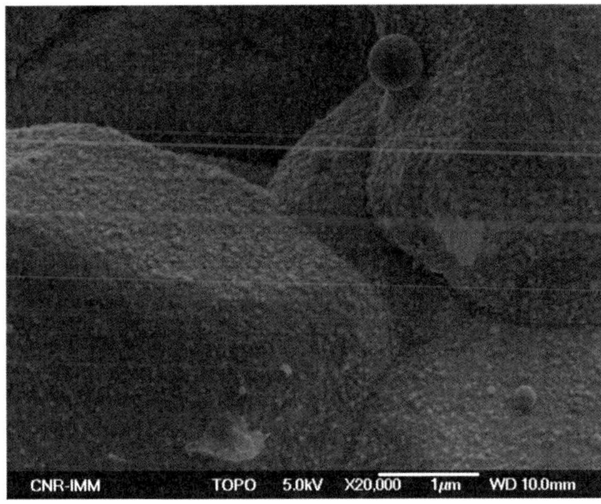

Fig. 12.9 SEM micrograph of TiO_2 nanoparticle thin film MAPLE-deposited onto a rough alumina substrate

MAPLE was seen as perspective method for fabricating nanoparticle/nanorod films. The nanoparticle/nanorod solution, once frozen to liquid nitrogen temperature, can be used as target to be laser-irradiated. With respect to the "traditional" use of MAPLE (polymer and biomaterial film deposition), some peculiarities emerged. First, the high thermal diffusivity of nanofluids (solid–liquid composite materials consisting of solid nanoparticles or nanofibers with sizes typically of 1–100 nm suspended in liquid), strong temperature dependence of thermal conductivity, greatly exceeding the prediction of well-established theories [74] that, if confirmed for solid solutions, could invalidate the assumption of thermal confinement following the laser pulse energy absorption. The second is the fact that nanomaterials are characterized by melting temperatures lower than that of the corresponding bulk materials [75]. This feature could lead to possible modification of the size and shape of the nanostructures to be transferred, as reported below.

First, films of carbon nanotubes were first MAPLE-deposited by Wu et al. [76]. After, TiO_2 and SnO_2 colloidal nanoparticle films were deposited by Caricato et al. [77]. Recently, TiO_2 nanorod films were also deposited by the same authors [78].

TiO_2 nanoparticle and nanorod films were intended for gas sensing applications. TiO_2 colloidal nanoparticles (diameter of 10 nm) in the anatase phase were prepared by using standard procedures [73], and then diluted in deionized water with a concentration of 0.2 wt%. The solution was then frozen at LN temperature and placed into a vacuum chamber on a rotating target holder, cooled with liquid nitrogen to guarantee a low and constant temperature (−160 °C). The frozen target was irradiated in vacuum (5×10^{-4} Pa) with an ArF ($\tau = 20$ ns, 10 Hz, $F = 550$ mJ/cm^2, 6,500 pulses/film) excimer laser. Films were deposited on silica, <100> Si and interdigitated alumina (Al_2O_3) slabs, for the different characterizations. High-resolution SEM images of the TiO_2 nanoparticle films showed that a quite uniform film of nanoparticles was deposited also onto rough Al_2O_3 substrates used for gas-sensing tests, following the morphology of the alumina grains (Fig. 12.9). The preservation of the anatase crystal phase was evidenced by X-ray

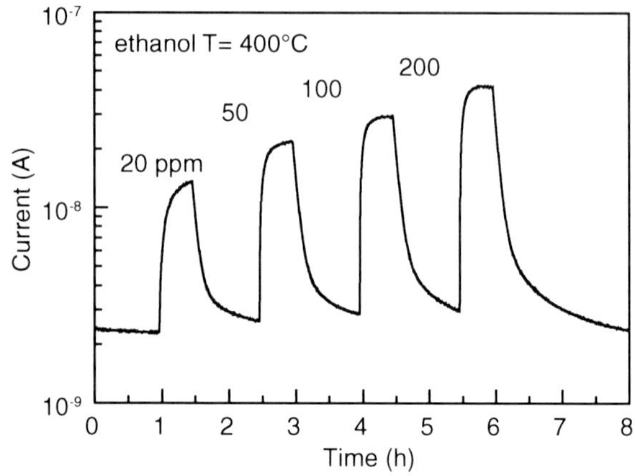

Fig. 12.10 Typical dynamic response of the MAPLE deposited TiO_2 sensing layer for different concentrations of ethanol vapors at the working temperature of 400 °C

diffraction (XRD) spectra, where the characteristic peaks of the anatase phase at $2\theta = 25^{\circ}$, corresponding to the reflection by the <101> crystallographic plane, were well evident.

Gas test measurements were carried out in the constant temperature mode by recording the dynamic changes of the electrical resistance caused by the exposition to different concentrations of ethanol and acetone vapors (20–200 ppm in dry air). The sensor responses were carried out by applying a constant voltage of 10 V between the sensor electrodes, and the resulting electrical current was monitored by means of an electrometer.The sensor working temperature was varied from 250 to 500 °C to find the best operating temperature. As an example, Fig. 12.10 shows the dynamic changes in electrical resistance of a MAPLE deposited TiO_2 nanoparticle film, at the working temperature of 400 °C, for ethanol vapors at different concentrations. Surface reactions with reducing chemical species as ethanol and acetone cause an increase in electrical conductance. The relative variation of signal in electrical current is very high (up to about 1 order of magnitude) even at very low concentrations of both the considered vapors. These very good gas-sensing properties towards ethanol and acetone may be attributed to the nanoscale dimensions of the TiO_2 particles.

Titanium dioxide nanorod thin films were also MAPLE deposited [78]. The starting materials were TiO_2 nanorods in the brookite phase, having a mean size of 3–4 × 20–50 nm (Fig. 12.11, left), prepared through a chemical route.

The nanorods were dissolved in pure toluene (0.016 wt%), then frozen at LN temperature and irradiated with a KrF excimer laser ($\tau = 20$ ns, 10 Hz, $F = 25$, 50, 150, 250 and 350 mJ/cm^2). <100> single-crystal Si wafers, silica slides, Cu carbon coated grids and alumina interdigital slabs were used as substrates to fully characterize the deposited layers. SEM and TEM analyses evidenced the occurrence of crystalline nanospheres mixed with individually distinguishable TiO_2 nanorods in the films deposited at $F = 350$ mJ/cm^2 (Fig. 12.11, right). The SAED pattern obtained from an area of the sample including both nanorods and

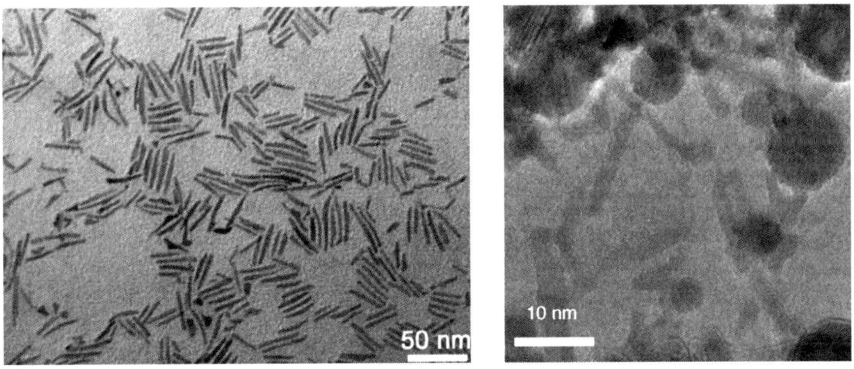

Fig. 12.11 Bright field TEM images of the as-prepared colloidal TiO_2 nanorods (*left*) and of the TiO_2 nanostructured film deposited by MAPLE ($F = 350$ mJ/cm^2) on a carbon coated Cu grid

nanospheres demonstrated that both types of shaped nanostructures are made of the brookite phase, like the starting nanorods. By reducing the laser fluence to 100 and 50 mJ/cm^2, a decrease of the fraction of spherical particles as well as of the deposition rate was observed. It seems that 50 mJ/cm^2 is close to the threshold fluence for the occurrence of the spherical NPs. In fact, the film deposited at $F = 25$ mJ/cm^2 resulted formed by nanorods only. This result confirms the reduction of the melting point of nanosized materials. In fact, the calculated temperature increase of the TiO_2 nanorods under laser irradiation at the highest used fluence (350 mJ/cm^2) is considerably lower than the melting temperature of bulk titanium oxide ($\sim$1,850 °C). Gas sensing tests were performed using MAPLE-deposited TiO_2 nanorods on Al_2O_3 interdigitated alumina slabs at the fluence of 350 mJ/cm^2. The gas sensing experiments were performed in dark condition and under illumination by using a UV xenon lamp delivering an optical power of 0.140 mW at the sample surface. Figure 12.12 shows the response of the sensing layer in the presence of an oxidising gas (NO_2) in a mixture with dry air at very low concentration (1 ppm) where beside a good sensor response, a significative increase in the signal was observed in the case of UV activation with respect to the dark condition at the same working temperature.

12.5 Discussion

MALDI has proven to be one of the most successful methods for mass spectrometric analysis and investigation of large molecules. The chemical matrix greatly facilitates the production of intact gas-phase ions from large, nonvolatile, and thermally labile compounds such as proteins, oligonucleotides, synthetic polymers and large inorganic compounds. The matrix plays a key role in this technique by absorbing the laser light energy and causing a small part of the target substrate to

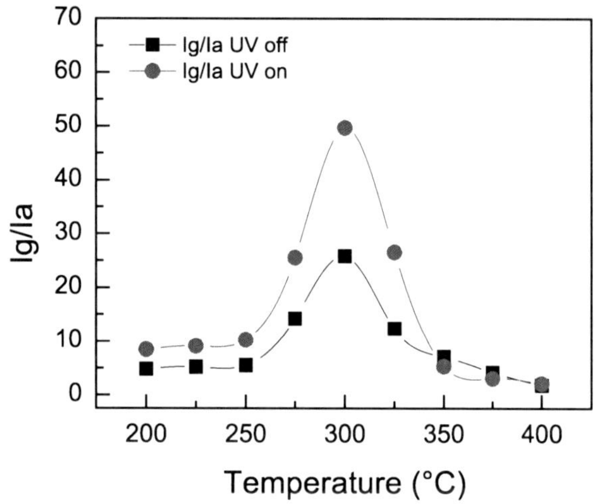

Fig. 12.12 Response of the TiO_2 nanorod sensor with and without UV illumination in the presence of 1 ppm NO_2 in mixture with dry-air at different working temperatures

vaporize. The MALDI matrix must meet a number of requirements simultaneously: to absorb at the laser wavelength, to be compatible with the analyte, to be vacuum stable and to cause co-desorption and ionization of the analyte upon laser irradiation, among others.

The sample preparation for MALDI is important for both sensitivity, reproducibility, and quantification of mass analysis. Important problems require much attention, like the extraction of inorganic salts, which are part of protein extracts, and could interfere with the ionization process. The salts can be removed by solid phase extraction or by washing the dried crystals with cold water. Attention must be paid to the fact that frequently matrix-protein mixtures are not homogenous because the polarity difference leads to separation of the two substances during co-crystallization. Most frequently, the diameter of the target is much larger than that of the laser spot. Consequently, many laser shots must be directed to different places of the target to get the statistical average of the substance concentration within the target. The matrix chemical composition, the laser wavelength, energy density and spot homogeneity and the impact angle of the laser on the target are among the most critical parameters for the quality and reproducibility of the MALDI method.

The MAPLE technique was introduced to deposit thin and ultrathin organic, bioorganic, and composite films with minimum chemical and structural modification of the target material, which occurs during traditional PLD. The goal can be obtained by dissolving the material of interest in a volatile solvent and freezing the solution to be used as the target for pulsed laser ablation. The main limit of the technique is the frequent appearance of clusters and droplets of the solute material in the deposited films. Computer simulations [41, 79] showed that the ejection of molecular clusters and droplets seems to be inherently connected to the basic mechanism of laser ablation-explosive decomposition of a surface region of the target overheated up to the limit of its thermodynamic stability. The aggregation of nanoparticles, observed in MAPLE deposited films, can be partially explained by this model although other phenomena must be considered when speaking of nanomaterials. These phenomena are particular evident when dealing with MAPLE deposition of nanorods where changes of shape, from nanorods to nanospheres, were observed (Fig. 12.11). The unexpected observation of TiO_2 nanospheres in films prepared starting from nanorods suggests occurrence of a laser-induced process, which induces a melting/coalescence of the nanorods, driving their transformation into the most thermodynamically stable spherical shape. Under laser pulses, the energy released into the frozen matrix induces solution vaporization. However, the laser light heats both the nanorods contained in the irradiated target and the nanorods leaving the target in the vapor phase. In any case, due to the low laser fluence (350–50 mJ/cm^2), the nanorod temperature should be far from the melting temperature of bulk TiO_2 ($\sim$1,850 °C). But, a decrease of the actual melting temperature for low-dimensional solids, compared to their corresponding bulk materials, is a well known phenomenon [75]. The preservation of the brookite phase is an important fingerprint. In fact, it was found that, by annealing the TiO_2 powders, the phase transformation brookite-to-rutile occurs at $\sim$650 °C, with the brookite phase completely transformed to the rutile phase at 750 °C [80]. This make us confident in assuming that at 350 mJ/cm^2 the temperature of 750 °C was not exceeded, even if melting of the nanorods occurred. So, the formation of the spherical particles can be explained on the basis of combined effects involving the absorption of laser energy by the nanorods, their low thermal-loss rate and their decreased melting temperature due to the nanometric dimension effects. It means that the laser fluence plays the most important role for the deposition of nanoparticles/nanorods with the same dimensions and morphology of the starting materials.

12.6 Conclusions

MALDI introduction determined a great progress in the analysis and structural determination of peptides. Resolution and accuracy of mass spectra in detecting down to femtomole levels is an almost routine task. Applications are increasing continuously and MALDI is established as a primary research tool due to its great sensitivity, precision, accuracy and throughput. Moreover, MALDI/TOF is becoming the standard method for species identification in medical microbiological laboratories.

MAPLE has proven to be a valuable technique to deposit thin films of polymers, organic and biological materials, with preservation of their complex structure and functionality. The thickness control and surface coverage obtained using

MAPLE cannot be achieved by other solvent-based coating methods, not to say of ability of depositing films on rough substrates.

The fabrication of very smooth films appears difficult to be obtained. High-resolution imaging and computer simulations clearly contest the original simple model of molecule-by-molecule deposition. Aggregation of polymers and presence of residual matrix molecules result in local corrugations of the deposits. The roughness of the growing films can be, at least partially, controlled by limiting the solute concentration. It was shown that the solvent properties influence the surface morphology of polymer films. Together with solubility parameters, the boiling temperature of the solvent influences the polymer surface morphology, which depends also on the laser light fluence and absorption depth. It was also shown that the moderate heating of the substrate reduces the polymer film roughness, which in contrast increases with increasing the laser pulse number. As regards the MAPLE-deposited nanoparticles and nanorods, formation of relatively large spherical particles can be explained on the basis of combined effects involving the absorption of laser energy by the nanorods, their low thermal-loss rates and their decreased melting temperature due to nanometric dimensions.

In conclusion, MAPLE opened very perspective roads for the deposition of performing films of complex molecules. The area of MAPLE applications is rapidly expanding from simple films to polymer blend deposition, multilayer structure deposition, nanoparticle and nanorod thin film deposition. As usual, for the deposition of each material the laser and solvent parameters require an accurate choice.

Many problems are still open, like the influence of the laser pulse length, laser spot dimensions, nature of the process not yet sufficiently studied.

Acknowledgments The author is very grateful to Prof. A. Luches for fruitful discussions and assistance.

References

1. T.H. Maiman, Nature **187**, 493 (1960)
2. P. Houldcroft, British Weld. J. 443 (1967)
3. W. Koechner, *Solid-State Laser Engineering* (Springer, New York, 1965), p. 2
4. R.F. Duhamel, C.M. Banas, *Lasers in Material Processing* (American Society of Metals, 1983)
5. A.L. Schawlow, Science **149**, 13 (1965)
6. H.M. Smith, A.F. Turner, Appl. Opt. **4**, 147 (1965)
7. J.P. Gavigan, D. Givord, A. Lienard, O.F.K. Mcgrath, J.P. Rebouillat, Y. Souche, MRS Proc. **191**, 49 (1990)
8. Y. Kawamura, K. Toyoda, S. Namba, Appl. Phys. Lett. **40**, 374 (1982)
9. R. Srinivasan, V. Mayne-Banton, Appl. Phys. Lett. **41**, 576 (1982)
10. J.G. Bednorz, K.A. Müller, Z. Physik B **64**, 189 (1986)

11. A.P. Caricato, G. Leggieri, A. Luches, F. Romano, G. Barucca, P. Mengucci, S.A. Mulenko, Appl. Surf. Sci. **254**, 1224 (2007)
12. J.M. Lackner, W. Waldhauser, T. Schöberl, Surf. Coat. Technol. **201**, 4037 (2006)
13. A.P. Caricato, M. Fernández, Z. Frait, S. Luby, A. Luches, E. Majkova, G. Majni, R. Malych, P. Mengucci, Appl. Phys. A **79**, 1251 (2004)
14. R. Tomita, H. Koga, T. Uchiyama, I. Iguchi, J. Phys. Soc. Jpn. **73**, 2639 (2004)
15. S. Acquaviva, G. Leggieri, A. Luches, A. Perrone, A. Zocco, N. Laidani, G. Speranza, M. Anderle, Appl. Phys. A **70**, 197 (2000)
16. E. D'Anna, M. Fernández, G. Leggieri, A. Luches, A. Zocco, G. Majni, Eur. Phys. J. Appl. Phys. **28**, 159 (2004)
17. A.P. Caricato, Y.V. Kudryavtsev, G. Leggieri, A. Luches, S.A. Mulenko, J. Phys, d. Appl. Phys. **40**, 4866 (2007)
18. M. Martino, A. Luches, M. Fernández, P. Anobile, V. Petruzzelli, J. Phys. D Appl. Phys. **34**, 2606 (2001)
19. G.B. Blanchet, S.I. Shah, Appl. Phys. Lett. **62**, 1026 (1993)
20. G.B. Blanchet, Macromolecules **28**, 4603 (1995)
21. M. Yudasaka, Y. Tasaka, M. Tanaka, H. Kamo, Y. Ohki, S. Usami, S. Yoshimura, Appl. Phys. Lett. **64**, 3237 (1994)
22. W.B. Lee, J.Y. Wu, Y.I. Lee, J. Sneddon, Appl. Spectrosc. Rev. **39**, 27 (2004)
23. K. Tanaka, Y. Ido, S. Akita, Y. Yoshida, T. Yoshida, in *Proceedings of 2nd Japan-China Joint Symposium. Mass Spectrometry*, (Osaka, 1987) p. 185
24. M. Karas, F. Hillenkamp, Anal. Chem. **60**, 1193 (1988)
25. A. Piqué, R.A. McGill, D.B. Chrisey, D. Leonhardt, T.E. Mslna, B.J. Spargo, J.H. Callahan, R.W. Vachet, R. Chung, M.A. Bucaro, Thin Solid Films. **355/356**, 536 (1999)
26. A. Piqué, P. Wu, B.R. Ringeisen, D.M. Bubb, J.S. Melinger, R.A. McGill, D.B. Chrisey, Appl. Surf. Sci. **186**, 408 (2002)
27. D.B. Chrisey, A. Piqué, R.A. McGill, J.S. Horwitz, B.R. Ringeisen, D.M. Bubb, P.K. Wu, Chem. Rev. **103**, 553 (2003)
28. J.Y. Oh, J.H. Moon, M.S. Kim, J. Am. Soc. Mass Spectrom. **15**, 1248 (2004)
29. R. Knochenmuss, L.V. Zhigilei, J. Mass Spectrom. **45**, 333 (2010)
30. F. Hillenkamp, M. Karas, Int. J. Mass Spectrom. **200**, 71 (2000)
31. C. Fenselau, P.A. Demirev, Mass Spectrom. Rev. **20**, 157 (2001)
32. K.J. Welham, M.A. Domin, D.E. Scannell, E. Cohen, D.S. Ashton, Rapid Commun. Mass Spectrom. **12**, 176 (1998)
33. Z.P. Wang, L. Russon, L. Li, D.C. Roser, S.R. Long, Rapid Commun. Mass Spectrom. **12**, 456 (1998)
34. B.J. Amiri-Eliasi, C. Fenselau, Anal. Chem. **73**, 5228 (2001)
35. K. Dreisewerd, Chem. Rev. **103**, 395 (2003)
36. M. Karas, R. Krüger, Chem. Rev. **103**, 427 (2003)
37. R. Knochenmuss, R. Zenobi, Chem. Rev. **103**, 441 (2003)
38. R.A. McGill, D.B. Chrisey, *Method of Producing a Film Coating by Matrix Assisted Pulsed Laser Deposition*, Patent No. 6,025,03 (2000)
39. D.M. Bubb, P.K. Wu, J.S. Horwitz, J.H. Callahan, M. Galicia, A. Vertes, R.A. McGill, E.J. Houser, B.R. Ringeisen, D.B. Chrisey, J. Appl. Phys. **91**, 2055 (2002)
40. A.L. Mercado, C.E. Allmond, J.G. Hoekstra, J.M. Fitz-Gerald, Appl. Phys. A **81**, 591 (2005)
41. E. Leveugle, L.V. Zhigilei, J. Appl. Phys. **102**, 074914 (2007)
42. B. Toftmann, M.R. Papantonakis, R.C.Y. Auyeung, W. Kim, S.M. O'Malley, D.M. Bubb, J.S. Horwitz, J. Schou, P.M. Johansen, R.F. Haglund Jr., Thin Solid Films **453/454**, 177 (2004)
43. A.P. Caricato, M. Lomascolo, A. Luches, F. Mandoj, M.G. Manera, M. Mastroianni, M. Martino, R. Paolesse, R. Rella, F. Romano, T. Tunno, D. Valerini, Appl. Phys. A **93**, 651 (2008)
44. G. Socol, I.N. Mihailescu, A.-M. Albu, S. Antohe, F. Stanculescu, A. Stanculescu, L. Mihut, N. Preda, M. Socol, O. Rasoga, Appl. Surf. Sci. **255**, 5611 (2008)

45. I.A. Paun, V. Ion, A. Moldovan, M. Dinescu, Appl. Phys. Lett. **96**, 243702 (2010)
46. A.P. Caricato, A. Luches, Appl. Phys. A **105**, 565 (2011)
47. Y. Fujii, H. Atarashi, M. Hino, T. Nagamura, K. Tanaka, Appl. Mat. Inter. **1**, 1856 (2009)
48. T.M. Patz, A. Doraiswamy, R.J. Narayan, N. Menegazzo, C. Kranz, B. Mizaikoff, Y. Zhong, R. Bellamkonda, J.D. Bumgardner, S.H. Elder, X.F. Walboomers, R. Modi, D.B. Chrisey, Mater. Sci. Engin. C **27**, 514 (2007)
49. V. Califano, F. Bloisi, L.R.M. Vicari, P. Colombi, E. Bontempi, L.E. Depero, Appl. Surf. Sci. **254**, 7143 (2008)
50. A. Stanculescu, M. Socol, G. Socol, I.N. Mihailescu, M. Girtan, F. Stanculescu, Appl. Phys. A **104**, 921 (2011)
51. A.P. Caricato, M. Cesaria, G. Gigli, A. Loiudice, A. Luches, M. Martino, V. Resta, A. Rizzo, A. Taurino, Appl. Phys. Lett. **100**, 073306 (2012)
52. A. Piqué, D.B. Chrisey, B.J. Spargo, M.A. Bucaro, R.W. Vachet, J.H. Callahan, R.A. McGill, D. Leonhardt, T.E. Mlsna, MRS Proc. **526**, 421 (1998)
53. R. Cristescu, I. Stamatin, D.E. Mihaiescu, C. Ghica, M. Albulescu, I.N.Mihailescu, D.B. Chrisey, Thin Solid Films **453/454**, 262 (2004)
54. B. Toftmann, K. Rodrigo, J. Schou, R. Pedrys, Appl. Surf. Sci. **247**, 211 (2005)
55. A. Purice, J. Schou, M. Dinescu, Chem. Phys. Lett. **427**, 251 (2006)
56. F. Bloisi, L. Vicari, R. Papa, V. Califano, R. Pedrazzani, E. Bontempi, L.E. Depero, Mater. Sci. Eng. C Biomim. Mater. Sens. Syst. **27**, 1185 (2007)
57. R. Cristescu, D. Mihaiescu, G. Socol, I. Stamatin, I.N. Mihailescu, D.B. Chrisey, Appl. Phys. A **79**, 1023 (2004)
58. B.R. Ringeisen, J. Callahan, P.K. Wu, A. Piqué, B. Spargo, R.A. McGill, M. Bucaro, H. Kim, D.M. Bubb, D.B. Chrisey, Langmuir **17**, 3472 (2001)
59. P.K. Wu, B.R. Ringeisen, D.B. Krizman, S.M. Hewitt, C.G. Frondoza, M. Brooks, D.M. Bubb, R.C.Y. Auyeung, H. Kim, A. Piqué, J.M. Fitz-Gerald, B. Spargo, R.A. McGill, D.B. Chrisey, Rev. Sci. Instrum. **74**, 2546 (2003)
60. M. Martino, A.P. Caricato, F. Romano, T. Tunno, D. Valerini, M. Anni, M.E. Caruso, A. Romano, T. Verri, J. Mater. Sci.: Mater. Electron. **20**, S435 (2009)
61. L. Stamatin, R. Cristescu, G. Socol, A. Moldovan, D. Mihailescu, I. Stamatin, I.N. Mihailescu, D.B. Chrisey, Appl. Surf. Sci. **248**, 422 (2005)
62. T. Tunno, A.P. Caricato, M.E. Caruso, A. Luches, M. Martino, F. Romano, D. Valerini, Appl. Surf. Sci. **253**, 6461 (2007)
63. F. Bloisi, A. Cassinese, R. Papa, L. Vicari, V. Califano, Thin Solid Films **516**, 1594 (2008)
64. M. Ariu, M. Sims, M.D. Rahn, J. Hill, A.M. Fox, D.G. Lidzey, M. Oda, J. Cabanillas-Gonzales, D.C. Bradley, Phys. Rev. B **67**, 195333 (2003)
65. M.E. Caruso, S. Lattante, R. Cingolani, M. Anni, Appl. Phys. Lett. **88**, 181906 (2006)
66. A.P. Caricato, M. Epifani, M. Martino, F. Romano, R. Rella, A. Taurino, T. Tunno, D. Valerini, J. Phys. D Appl. Phys. **42**, 095105 (2009)
67. A.P. Caricato, G. Leggieri, M. Martino, A. Vantaggiato, D. Valerini, A. Cretì, M. Lomascolo, M.G. Manera, R. Rella, M. Anni, Appl. Phys. A **101**, 759 (2010)
68. D.M. Bubb, M. Papantonakis, B. Collins, E. Brookes, J. Wood, U. Gurudas, Chem. Phys. Lett. **448**, 194 (2007)
69. C. Hansen, *Hansen Solubility Parameters: A User Handbook* (CRC Press, Boca Raton, 2000)
70. K. Rodrigo, P. Czuba, B. Toftmann, J. Schou, R. Pedrys, Appl. Surf. Sci. **252**, 4824 (2006)
71. A. Gutiérrez-Llorente, G. Horowitz, R. Pérez-Casero, J. Perrière, J.L. Fave, A. Yassar, C. Sant, Org. Electron. **5**, 29 (2004)
72. A.P. Caricato, M. Anni, M.G. Manera, M. Martino, R. Rella, F. Romano, T. Tunno, D. Valerini, Appl. Surf. Sci. **255**, 9659 (2009)
73. M. Epifani, J. Arbiol, R. Díaz, M.J. Perálvarez, P. Siciliano, R. Rella, Chem. Mater. **17**, 6468 (2005)
74. P. Keblinski, J.A. Eastman, D.G. Cahill, Mater. Today 36 (2005)
75. M.W. Cross, W.J. Varhue, Nanotechnology **19**, 435705 (2008)

76. P.K. Wu, J. Fitz-Gerald, A. Piqué, D.B. Chrisey, R.A. McGill, Mater. Res. Soc. Symp. Proc. **2–3**, 617 (2000)
77. A.P. Caricato, A. Luches, R. Rella, Sensors **9**, 2682 (2009)
78. A.P. Caricato, R. Buonsanti, M. Catalano, P.D. Cozzoli, A. Luches, M.G. Manera, M. Martino, R. Rella, A. Taurino, Appl. Phys. A **104**, 963 (2011)
79. E. Leveugle, L.V. Zhigilei, A. Sellinger, J.M. Fitz-Gerald, Appl. Surf. Sci. **253**, 6456 (2007)
80. M. Rezaee, S.M. Mousavi Khoie, J. Alloys Compd. **507**, 484 (2010)

Chapter 13
Laser Nanofabrication of Soft Matter

Marta Castillejo, Tiberio A. Ezquerra, Mohamed Oujja and Esther Rebollar

Abstract Laser based techniques constitute an advantageous versatile approach for the assembly and control at nanometer scale of polymers and biopolymers, fundamental components of soft matter. In this chapter, laser nanostructuring of thin films of these materials will be illustrated by studies on laser induced periodic surface structures (LIPSS) and on laser foaming and on their respective application for surface enhanced Raman spectroscopy based sensors and for scaffolds in tissue engineering.

13.1 Soft Matter

Soft Matter, a subfield of Condensed Matter, deals with the study of materials featuring basic length scales which are much larger than the constituent atoms and molecules. Mesoscopic sizes, below 100 nm, start revealing quantum mechanical properties due to quantum confinement effects and, in consequence, the electronic and optical properties of the nanostructured material differ from those of the bulk. The presence of many fluctuating degrees of freedom in soft materials causes the equilibrium structure to be very sensitive to external stresses and the physical states and properties to be dominated by thermal fluctuations. This sensitivity raises fascinating new problems in physics, chemistry and materials science, and therefore one of the fundamental focuses in soft matter research frontier is to find ways of tailoring, manipulating, and assembling building blocks into nanostructures to open up new possibilities for technological applications. Soft Matter

M. Castillejo (✉) · M. Oujja · E. Rebollar
Instituto de Química Física Rocasolano, CSIC, Madrid, Spain
e-mail: marta.castillejo@iqfr.csic.es

T. A. Ezquerra
Instituto de Estructura de la Materia, CSIC, Madrid, Spain

M. Castillejo et al. (eds.), *Lasers in Materials Science*,
Springer Series in Materials Science 191, DOI: 10.1007/978-3-319-02898-9_13,

involves the study of polymers, liquid crystals, colloids, foams, gels, biological structures such as DNA, proteins, and cell membranes and a wide variety of self-organizing materials. Soft Matter research boosted up with the contribution of P. G. de Gennes who was awarded the Nobel Prize in Physics in year 1991 for discovering that "methods developed for studying order phenomena in simple systems can be generalized to more complex forms of matter, in particular to liquid crystals and polymers" [1].

In particular, polymers constitute a fundamental component of Soft Matter and since the second half of the 20th century have become a primary component for the support of our everyday life [2]. Natural polymers, such as starch, cellulose, chitin and others [3], have been used by mankind from the very beginning of civilization. On the other hand, synthetic polymers and composites are an essential part of the technological revolution of the information age through their multiple applications ranging from lithographic masks to electronic connections and packaging. From the markets point of view, polymers can be divided into two broad groups: commodities and specialities. Commodity polymers are produced from a few simple starting compounds, they are very inexpensive and used in large amounts. In contrast, the more costly speciality polymers are made from many different compounds, can be tailored for nearly every use and are often at the base for the continuous progress in air, space, computer and medical technologies. In addition, the combination of polymer and non-polymer materials to give rise to new hybrids and composites is of great interest for conferring new properties or enhancing the ones of the starting substances [4].

Among commodity polymers, semicrystalline ones, polyethylene (PE), polypropylene (PP) and poly(ethylene terephthalate) (PET) represent approximately two-thirds of the annual production of synthetic polymers and their preparation has experienced a remarkable progress with the advent of new synthesis techniques [5]. The intrinsic structure of semicrystalline polymers consists in alternating regions of crystalline and amorphous phases with thicknesses in the range of nanometers [6] and depends heavily on how the material is processed. The end properties of the material can be consequently controlled by selection of the adequate processing protocols.

On the other hand, biopolymers hold promise to become the fundamental building blocks for organic photonics and electronics; they constitute important sources of novel functional materials and advantageous alternatives to synthetic polymers in biomedical applications, i.e. for the fabrication of implants and controlled drug delivery systems, and more recently to replace conventional polymers in packaging, coatings, disposable products, fibres and films.

Nanoconfinement of polymer materials in two dimensions by preparation of thin films (nanofilms) or in one dimension, by fabrication of nanotubes, nanogrooves, or nanocylinders, can induce large effects on both the structure and the dynamics of the material and it may even affect phase transitions or physical processes. The understanding of these processes is crucial, not only in the development of nanostructured materials for specific applications, but also of the physical basis of Nanotechnology [7].

13.2 Laser Nanofabrication

Nanofabrication allows the manipulation of nanoscale building blocks for a desired purpose by furnishing macroscopic materials with nanometer-scale structural motifs that confer modified physical and chemical properties and new functionalities. State-of-the-art techniques for generating polymer structures at the nanoscale are mainly based in soft lithography methods [8], like micro-contact printing or nanoimprint lithography (NIL), and templating-based techniques [9] such as those using inorganic nanoporous alumina membranes (Anodic Aluminium Oxide). These methods aim at simplicity, reproducibility and low-cost, and generally provide versatile processing strategies. Lithographic procedures suffer from limitations on spatial resolution and from several other drawbacks, as they are usually based on multiple-steps procedures, sometimes involve clean-room facilities, high vacuum or complex mask fabrication, and are difficult to scale to wafer-like size. The last few years have witnessed the emergence of a new family of nanofabrication techniques with specific capabilities based in the use of lasers. These techniques are good candidate tools in high-resolution patterning of soft materials as they afford the sought versatility and reliability.

Laser irradiation techniques, such as laser induced periodic surface structuring (LIPSS), laser foaming, and those based on laser ablation, as laser induced forward transfer (LIFT) [10], pulsed laser deposition (PLD) [11] and matrix-assisted pulsed laser evaporation (MAPLE) [12], are widely used for high precision, accuracy and control at the micro- and nanoscales. Further advantages are associated with the possibility of application to all sorts of materials under all types of environments by operating with non-contact, flexible set-ups. Using laser-based techniques, it is possible to select the most appropriate laser parameters, intensity, fluence, wavelength, total photon dose, and irradiation conditions that are most appropriate to the materials and processing protocols. In the case of femtosecond lasers, and because the pulse duration is shorter than the typical material relaxation times, it is the laser-material interaction, rather than the material thermal properties, what generally determines the outcome of the laser fabrication process. In addition, the possibility to temporally shape fs pulses with the corresponding time scale of processes involved offers new avenues for controlling and tailoring the features of the created structures [13, 14]. More recently, specific laser processing techniques, taking advantage of optical trapping or of plasmonic enhancement effects [15, 16], have been developed and applied to the nanopatterning of soft polymer materials.

This chapter will focus on studies carried out using lasers of nanosecond to femtosecond pulse lengths to develop superficial structures at the micro- and nanometer scales in thin films of polymers and biopolymers. In the first place, we will present recent studies of LIPSS imprinted on polymer thin films using a newly developed methodology which combines assessment of the structures in the real space by atomic force microscopy (AFM) and in the reciprocal space by grazing incidence X-ray scattering techniques. In the second place, we will report on recent research on the phenomenon of laser foaming and on the control that can be exerted on the superficial morphology by selecting the laser wavelength and pulse duration of the applied pulses. Application of these two laser nanofabrication approaches will be exemplified by respective studies of the use of rippled polymer films as sensing elements in surface enhanced Raman spectroscopy (SERS) and of cell growth on the laser foamed substrates.

13.3 Laser Induced Periodic Surface Structures of Thin Polymer Films

13.3.1 LIPSS Formation and Mechanisms

The phenomenon of laser induced periodic surface structures (LIPSS), in which a polarized laser source imprints a periodic structure on the surface of a substrate, has been studied in polymers using several laser wavelengths and various pulse durations [17–21]. Irradiation by a polarized laser beam induces self-organized ripple structure formation within a narrow fluence range below the ablation threshold [19, 22–25]. The period of the ripples L depends on the laser wavelength and on the angle of incidence of the radiation through the following relation [26]:

$$L = \frac{\lambda}{n - \sin(\theta)} \tag{13.1}$$

where λ is the laser wavelength, n the effective refractive index of the material and θ is the angle of incidence of the laser beam. LIPSS develop on the material surface by interference between the incoming and the surface scattered waves. The ensuing inhomogeneous intensity distribution, together with a feedback effect, results in the enhancement of the modulation depth [23, 27, 28]. However, the whole mechanism responsible for ripple formation is complex and still not well understood. For polymers irradiated with nanosecond laser pulses, different processes have been proposed to be involved in ripple formation, including thermal and non-thermal scissoring of polymer chains, spatially modulated melting and crystallization, amorphization of crystalline domains, local surface melting, ablation, photolytic shrinkage, photooxidation and material transport and rearrangement [22, 23, 27, 28, 29, 30]. The characteristic surface structures formed in this way, can be used to tailor a great variety of surface properties such as adhesion and friction [24, 31, 32], induced cell alignment [25], liquid crystal alignment [33, 34] and colour modulation due to superficial grating formation [35].

LIPSS with periods similar to the irradiation wavelength, and parallel to the laser polarization direction, were imprinted on thin films of several polymers aiming at control and tunability of the size and morphology of the periodic structures while ensuring photochemical integrity of the polymer material [17, 19, 36]. Thin films of PET, poly(trimethylene terephthalate) (PTT), poly(carbonate bisphenol A) (PC) and poly(vinylidene fluoride) (PVDF) were irradiated in air with linearly polarized

Table 13.1 Linear absorption coefficient and ranges of laser fluence and number of pulses for the generation of well-defined LIPSS in polymeric materials under various laser conditions of wavelength and pulse duration

Material and laser characteristics	α (cm^{-1}) × 10^3	Fluence (mJ/cm^2)	Number of pulses
PET/266 nm, 6 ns	18	4–10	300–3,000
PTT/266 nm, 6 ns	26	4–10	100–3,000
PC/266 nm, 6 ns	18	5–10	800–3,000
PVDF/266 nm, 6 ns	0.4	–	–
PET/265 nm, 260 fs	18	1.0–2.4	500–10,000
PTT/265 nm, 260 fs	26	1.0–2.4	500–10,000
PC/265 nm, 260 fs	18	1.0–2.4	500–10,000
PET/795 nm, 120 fs	0.8	35–40	5,000–50,000
PTT/795 nm, 120 fs	0.7	35–40	5,000–50,000
PC/795 nm, 120 fs	0.4	35–40	5,000–50,000

pulses from a Q-switched Nd:YAG laser (6 ns at 266 nm) and a Ti:Sapphire laser (120 fs at 795 nm and 260 fs at 265 nm). While PET, PTT and PC can be easily obtained in the fully amorphous state at room temperature, by fast quenching from the molten state, PVDF is typically semicrystalline. Polymer thin films of about 150 nm thickness were prepared by spin coating on silicon wafers (100) [19]. The linear absorption coefficients were measured with a UV-visible spectrophotometer at the wavelengths employed for irradiation and are reported in Table 13.1.

Analysis of the nanostructures was carried out by AFM (Nanoscope IIIA Multimode, Veeco) in tapping mode and images were analysed with the software Nanoscope Analysis 1.10. The original samples presented a flat surface, with mean roughness values $\leq$3 nm. Values of the period were derived from fast Fourier transform (FFT) analysis.

Grazing incidence X-Ray scattering techniques (grazing incidence small-angle X-ray scattering GISAXS, and grazing incidence wide-angle X-ray scattering GIWAXS) were applied for the study of LIPSS to mainly assess morphology order over large sample areas. The facilities of the BW4 beamline at HASYLAB (DESY, Hamburg) were used to that purpose.

Irradiated areas were analyzed by GISAXS, a technique which provides complementary structural information to that obtained by AFM [19, 37, 38]. A scheme of the experimental set-up is shown in Fig. 13.1. The information obtained from the GISAXS patterns can be interpreted on the basis of the two orthogonal scattering vectors $q_z = (2\pi/\lambda)$ $(\sin\alpha_i + \sin\alpha)$ and $q_y = (2\pi/\lambda)$ $\sin\omega \cos\alpha$, which provide information about structural correlations perpendicular and parallel to the film plane, respectively. Lateral correlation between scattering objects on the film surface induces the appearance of some scattered intensity out of the meridian (line *m–m* in Fig. 13.1) [39]. The analysis is carried out with an X-ray beam of $\lambda = 0.13808$ nm and size of 20 × 40 μm^2. The scattered intensity is recorded by a Mar CCD detector of 2,048 × 2,048 pixels with a resolution of 79.1 μm per pixel, and a distance sample-to-detector of 2.211 m. An incidence angle $\alpha_i = 0.4°$ is

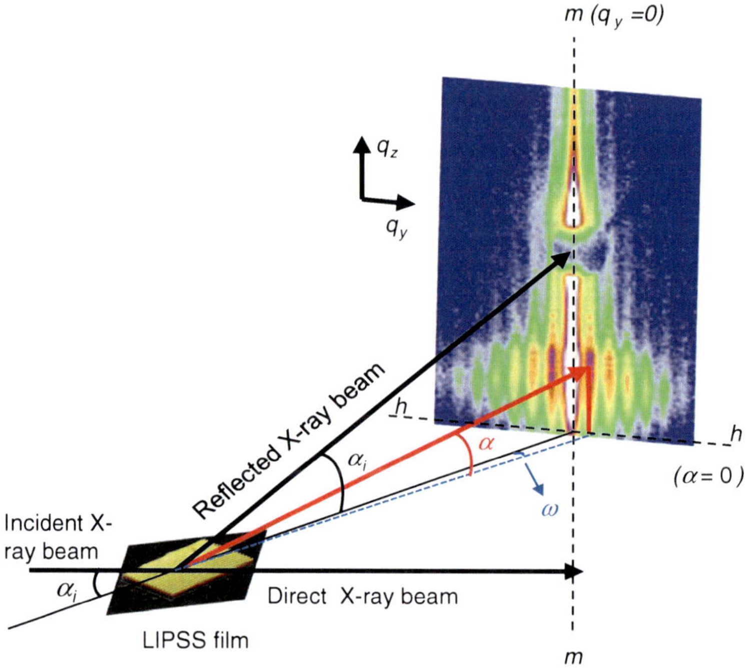

Fig. 13.1 Schematic view of a GISAXS experiment. The scattering plane, containing both the direct and the specular beams intersects the 2D detector along the meridian, *m–m* line, of the GISAXS pattern. The horizon, *h–h* line, is the intersection between the sample plane and the plane of the 2D detector that are perpendicular to each other. Each point on the GISAXS pattern can be characterized by the exit angle, α, and the out of scattering plane angle, ω. Reproduced with permission from [19]

chosen to ensure full penetration in the sample. Careful positioning of the samples is required to guarantee that the beam is parallel to the direction of the LIPSS and acquisition times between 40 and 600 s are typically used. Treatment of GISAXS images is performed with the software Fit2D [40]. GISAXS measurements provide an independent determination of the period, to be compared with the value obtained by AFM.

Figure 13.2 displays AFM topographic images of LIPSS generated on the polymer films under various laser irradiation conditions. As observed in all cases, the structures have a period close to the wavelength and are parallel to the polarization direction of the laser.

The range of laser fluences and number of pulses at which generation of well-defined LIPSS are observed are listed in Table 13.1. These ranges strongly depend on the material and on the laser wavelength and pulse duration. It has to be noted that using nanosecond pulses, absorption of laser light is essential to induce LIPSS formation. In fact, it was not possible to generate LIPSS on the studied polymer films by irradiation with nanosecond pulses of 532 nm, even if a wide range of

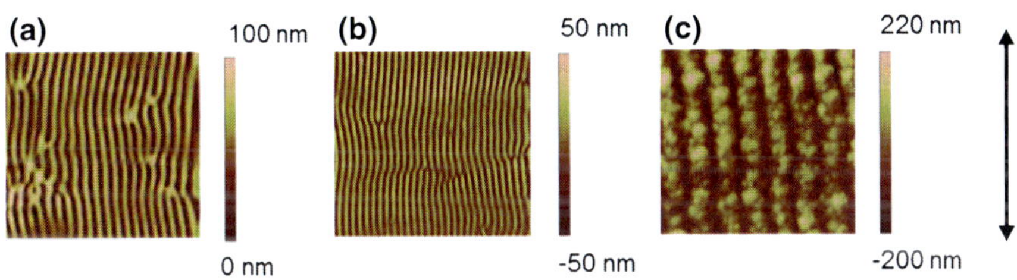

Fig. 13.2 AFM height images (5 × 5 μm² size) of LIPSS formed on polymeric substrates under the indicated laser conditions. The *double arrow* indicates the direction of the laser polarization. **a** PTT, 266 nm, 6 ns, 7 mJ/cm², 3 × 10² pulses. **b** PC, 265 nm, 260 fs, 1.2 mJ/cm², 5 × 10³ pulses. **c** PET, 795 nm, 120 fs, 37 mJ/cm², 2 × 10⁴ pulses

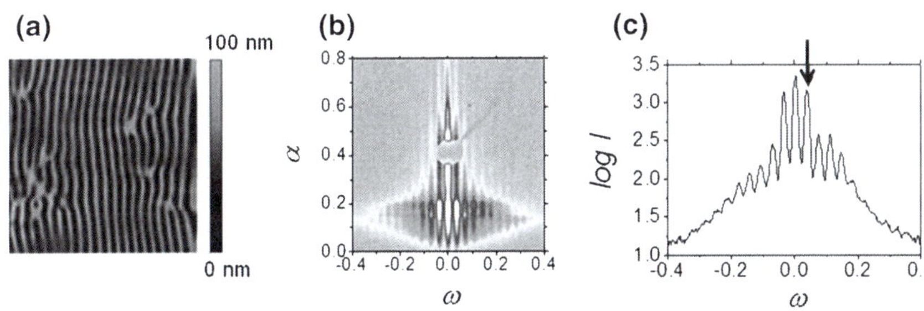

Fig. 13.3 **a** AFM height image (5 × 5 μm² size), **b** GISAXS pattern and **c** corresponding cut at $\alpha = 0.2^\circ$ of LIPSS generated in PTT with 300 pulses of 266 nm, 6 ns at 7 mJ/cm². LIPSS are parallel to the laser polarization direction. Each point on the GISAXS pattern is characterized by the exit angle, α, and the out of scattering plane angle, ω, both given in degrees. The *arrow* in **c** indicates the first intensity maximum

fluences was explored. The situation is different upon femtosecond irradiation, as even if absorption is negligible, i.e. 795 nm, good quality LIPSS emerge (Fig. 13.2c). This is ascribed to the large laser intensities involved which favour the coupling of laser light with the outer layer of the film through multiphoton absorption and ionization processes [17].

Figure 13.3 shows for a laser irradiated PTT sample, under conditions of LIPSS formation, the AFM height image (a) and the corresponding GISAXS pattern (b) characterized by the exit angle, α, and the out of scattering plane angle, ω. The figure also shows the corresponding intensity profile across the horizontal direction as a function of ω extracted from the GISAXS pattern (at a fixed $\alpha = 0.2^\circ$). Scattering maxima out of the meridian ($\omega \neq 0$) are clearly visible in the range of LIPSS formation. Spacing of the first maximum obtained from the GISAXS pattern can be determined through the expression $L = 2\pi/q_y^{Max}$, where q_y^{Max} is the q-value corresponding to the first intensity maximum, indicated by an arrow in Fig. 13.3c.

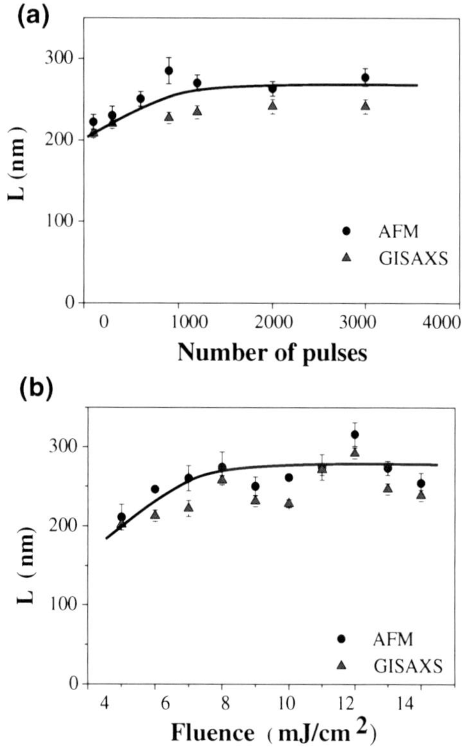

Fig. 13.4 Period of LIPSS on PTT films as a function of: **a** the number of pulses (6 ns at 266 nm) at 7 mJ/cm^2 and **b** the fluence for irradiation with 600 pulses. *Circles* and *triangles* indicate values derived from AFM and GISAXS analysis respectively. Reproduced with permission from [36]

The dependence of the period of LIPSS with the fluence of irradiation and the number of pulses is exemplified in Fig. 13.4. As observed, after an initial increase the period reaches a constant value around the laser wavelength employed. Spacing values derived from GISAXS for the irradiated polymers are in very good qualitative correlation with the period determination by AFM. This provides a validation of GISAXS as an appropriate technique for the analysis of this kind of nanostructures and paves the way for further studies involving the on-line monitoring of LIPSS by X-ray synchrotron techniques.

On the rippled polymer films GIWAXS measurements were also performed using the set up described (Fig. 13.1) and using a sample-detector distance of 0.17 m. Analysis of 2D-GIWAXS patterns provides information on possible changes of crystalline structure and crystal orientation induced by laser nanostructuring. Under the irradiation conditions explored, the absence of Bragg reflections in the GIWAXS patterns of PET, PTT and PC indicate that the nanostructured irradiated polymer films stay in the initial amorphous state and that LIPSS formation does not involve changes in polymer crystallinity.

Contrary to PET, PTT and PC films, which are amorphous at room temperature, PVDF is semicrystalline, and therefore the film surface presents the typical morphology of this kind of materials, constituted by spherulites of a few microns size

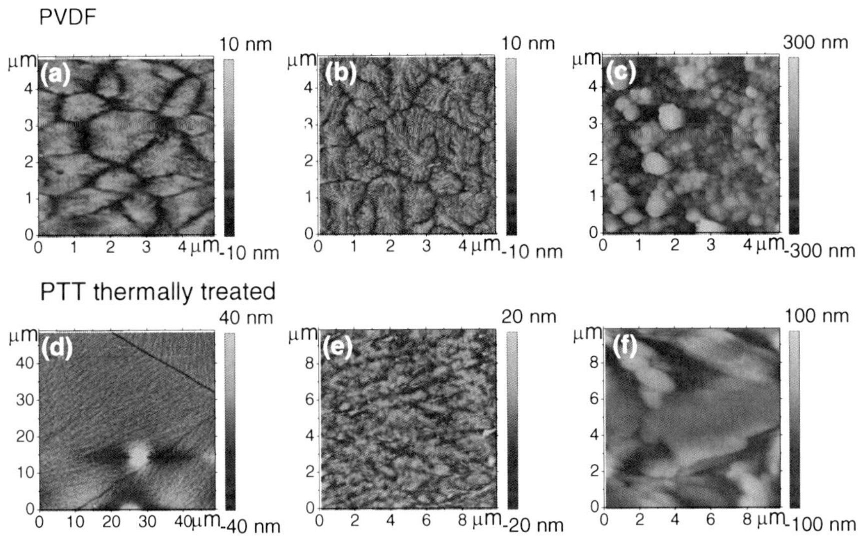

Fig. 13.5 AFM images of PVDF: **a** non-irradiated, **b** irradiated at 266 nm with 6,000 pulses of 6 ns at 85 mJ/cm^2 and **c** irradiated at 266 nm with 6,000 pulses at 140 mJ/cm^2; and of crystallized PTT: **d** non-irradiated, **e** irradiated at 266 nm with 6,000 pulses at 7 mJ/cm^2 and **f** irradiated at 266 nm with 6,000 pulses at 13 mJ/cm^2. Reproduced with permission from [19]

(Fig. 13.5a). It is worth noticing that irradiation of PVDF at 266 nm (pulses of 6 ns) does not result in formation of LIPSS. The linear absorption coefficient of this material at 266 nm (Table 13.1) is very low and the lack of ripples in this case is related either with the initial polymer crystalline structure or to its weak absorption [19].

In order to elucidate the effect of initial polymer crystallinity on LIPSS formation, amorphous PTT was thermally treated [19] to obtain a semicrystalline film. The characteristic spherulite-like morphology (Fig. 13.5d) does not become rippled (Fig. 13.5e) upon irradiation at the typical fluences which lead to LIPSS formation in the case of amorphous PTT, and rather signs of melting are evident on the film surface (Fig. 13.5f) at sufficiently high fluences. The thermal treatment does not modify the absorption coefficient of the material, as confirmed by UV–vis absorption spectroscopy. Therefore, one is led to conclude that the absence of LIPSS formation on semicrystalline PTT, in the fluence range where LIPSS appear for the amorphous specimens, is related with the restricted polymer dynamics associated with the polymer crystalline structure.

For nanosecond irradiation, estimation of the surface temperature increase by solving the one-dimensional heat conduction equation [19, 23] allows to understand the mechanism of LIPSS formation. The estimate of increase of temperature on the irradiated substrate follows the procedure described in detail in [19] and is shown in Fig. 13.6 for PET and PVDF.

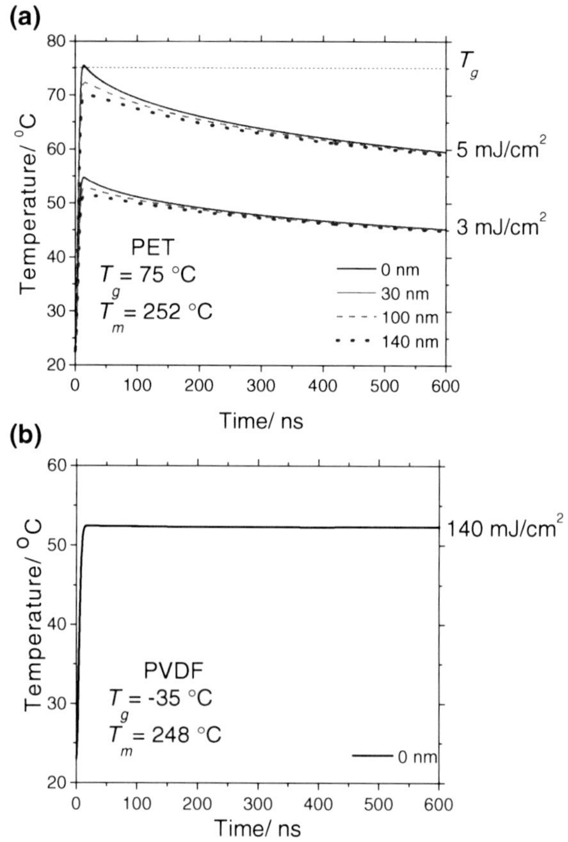

Fig. 13.6 Time dependence of the temperature reached at different depths for irradiation at 266 nm with pulses of 6 ns at the indicated fluences for: **a** PET (temperature curves at surface and at depth of 30 nm overlap) and **b** PVDF. The glass transition and melting points are indicated for each material. Adapted from [19]

For PET, when the fluence reaches 5 mJ/cm^2, a layer of around 30 nm heats up to a temperature above T_g (Fig. 13.6a). Overcoming T_g induces surface devitrification, a phenomenon understood as the transformation from a glassy to a soft state, and allows polymer segmental and chain dynamics. When the temperature overcomes T_g (75 °C for PET), the surface roughness rises as a consequence of capillary waves, which in turn enhance superficial inhomogeneities and facilitates the feedback mechanism involved in LIPSS formation. At higher fluences devitrification of the complete film develops under the effect of heating leading to film dewetting and distorsion. Therefore, the calculated temperatures, together with the GIWAXS results, suggest disregarding the role of melting and crystallization during LIPSS formation in amorphous spin-coated polymer films and rather LIPSS formation takes place by devitrification of the film surface at temperatures above their characteristic T_g.

Differently, the thermal properties of the semicrystalline polymer PVDF are governed by the melting point, T_m. Irradiation of films of this material causes heating of the film well below the melting point (Fig. 13.6b). The crystalline phase constitutes a robust scaffold that provides mechanical stability to the material impeding rearrangement into periodic structures.

In the femtosecond regime, the heating of the lattice is mediated by the energy transfer from the excited electrons and the temperature increase reached in the sample after irradiation cannot be estimated by the heat conduction equation and rather two-temperature models should be applied [41, 42]. However, the relative increase of temperature upon irradiation with a single laser pulse of the two wavelengths of 265 and 795 nm (Table 13.1) can be calculated by [43]

$$\Delta T = \varepsilon / \left(\rho_0 C_p\right) \tag{13.2}$$

where ρ_0 and C_p are the density and the heat capacity of the material respectively and ε is the plasma energy density at the end of the laser pulse

$$\varepsilon = \rho_{cr}(9/4)E_i \tag{13.3}$$

E_i is the molecular ionization potential (in eV) and ρ_{cr} the critical free electron density [44], needed to generate both a strongly reflective and absorbing plasma

$$\rho_{cr} = (2\pi/\lambda)^2 m_c \varepsilon_0 / e^2 \tag{13.4}$$

with ε_0 the vacuum dielectric permittivity, m_c the mass of the quasi-free electron in the conduction band and e the electron charge.

For a given polymer, ΔT should be an order of magnitude higher in the case of irradiation at 265 nm. This can explain why a larger fluence and number of pulses are needed for LIPSS formation upon fs irradiation at 795 nm (Table 13.1). The fact that multiple pulses are involved in LIPSS formation indicates the important role of a feedback mechanism in the process.

As described, the wide range of laser conditions explored results in a broad tunability of period dimensions. In particular by using femtosecond pulses, it is possible to control this parameter by changing the irradiation wavelength. Also the fact that, despite the pulse duration, the period of the nanostructures closely follows the irradiation wavelength upon normal incidence, strongly suggests a common formation mechanism based on the interference between the incident laser light wave and the surface-scattered waves created during irradiation [17, 19].

13.3.2 *Application of LIPSS Polymer Substrates for Surface Enhanced Raman Spectroscopy*

In order to develop the application of nanorippled thin polymer films as sensing elements, these films were coated with a uniform gold layer by PLD. A gold target was ablated in a stainless-steel vacuum chamber [45] with the 5th harmonic (213 nm) of a Q-switched Nd:YAG laser. The laser beam was focused by a 25 cm

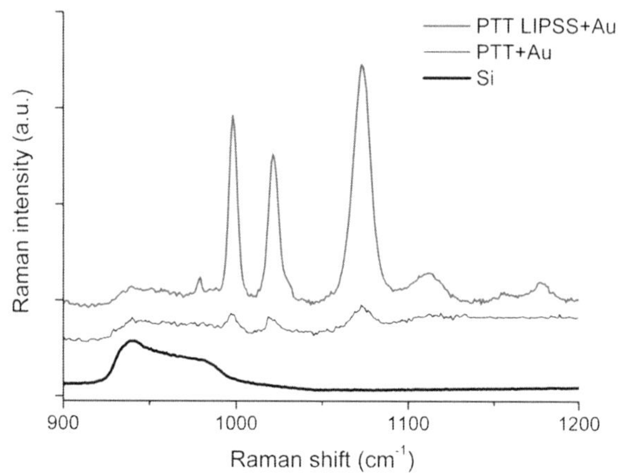

Fig. 13.7 Raman spectra of 10^{-3} M benzenethiol in aqueous solution on a silicon substrate and on the surfaces of gold coated PTT (PTT + Au) and gold coated nanostructured PTT (PTT + LIPSS + Au). LIPSS were fabricated at 266 nm with 600 pulses of 6 ns with a fluence of 7 mJ/cm^2. The spectra were recorded at the excitation wavelength of 785 nm at a power level of ca. 2 mW with spectral resolution of 2 cm^{-1}, and a × 100 magnification objective. Acquisition time and number of accumulations were 10 s and 10 respectively. Thickness of the gold layer is ca. 11 nm

focal length lens to yield fluences up to 2 J/cm^2 and the nanostructured polymer samples were placed at 4 cm in front of the target. Deposits grown at room temperature by delivering up to 36,000 pulses to the target resulted in gold layer thickness of up to 45 nm. The morphology of the gold coated polymer samples, characterized by AFM and GISAXS, retained the initial periodic relief after PLD coating. Raman spectra were acquired on drops of aqueous solutions of the model analyte benzenethiol (BT), in different concentrations, poured onto gold coated unrippled and rippled polymer substrates and dried in air. The Raman spectra were recorded in back scattering geometry with an InVia Raman microscope (Renishaw), equipped with a high sensitivity ultra-low noise CCD and a diode laser with emission at 785 nm as the excitation source. Additionally, for further comparison, control experiments were performed on an uncoated silicon substrate. Figure 13.7 shows the Raman spectra of benzenethiol (at a concentration of 98 %) on silicon and of the same analyte at a concentration of 10^{-3} M on gold coated unstructured and LIPSS structured PTT samples. While the Raman signal from benzenethiol is absent in the spectrum obtained from the silicon substrate, bands at 997, 1,020 and 1,071 cm^{-1}, assigned to the molecular ring, are clearly visible on the gold coated substrates. The enhancement factor of Raman signal for the gold-coated substrates has been estimated to be eight orders of magnitude, since BT at a concentration of ca. 9 M is not detected on the surface of an uncoated substrate, while the spectra of BT can be recorded at a concentration as low as 10^{-6} M. An additional enhancement of around 10 is observed, mediated by the presence of the periodic structures. Additional experiments have shown that even if signal enhancement is observed for all the thicknesses inspected (up to 45 nm) the effect seems to decrease for larger thickness. The relevance of this result lies in the fact that a very low amount of gold is enough to observe enhancement of Raman signal.

13.4 Laser Foaming of Biopolymer Films

13.4.1 The Role of Wavelength and Pulse Duration in Laser Foaming

Biopolymer foams and microstructures, using collagen, gelatine, chitosan, starch and other biomaterials, are currently investigated as 2D and 3D scaffolds for cell culture. Collagen is a major structural protein, a natural material of the extracellular matrix, and an essential structural component of all connective tissues. Gelatine is the product of denaturation of collagen. Chitosan is an aminopolysaccharide, the deacetylated derivative of chitin, which is the second most naturally occurring biopolymer after cellulose. This biopolymer offers a combination of properties uniquely suited for biofabrication, as its backbone provides sites that can be employed for the assembly of proteins, nucleic acids, and virus particles [46, 47]. Starch is a polysaccharide, which in its native state from plants, contains about 30 % amylose, 70 % amylopectine and less than 1 % lipids and proteins [48]. The tensile strength and the flexibility of starch films can be improved by incorporation of chitosan [49, 50], and blends of these two biopolymers have been studied due to their antibacterial activity [51].

A variety of processing techniques, including stamping, stereo lithography, two-photon polymerization, electrospinning, polymer demixing and stencil [3, 52–56], have been applied in order to control pore size and distribution, interconnectivity, adequate surface chemistry, etc. 2D scaffolds for cell culture are currently prepared by structuring the surface of polymers and biopolymers by some of these methods, based on the fact that the topography, together with chemical composition, have influence on cell adhesion and proliferation. In the literature, different types of nano- and microstructures have been reported, such as grooves, ripples, islands, pillars, particles, dots and fibrillar networks [3, 43, 53–55, 57–62, 63].

Superficial laser foaming on biopolymers films has recently been reported, a phenomenon that is induced by applying single pulses in the nanosecond to femtosecond domains [3, 43, 57–63]. The fast temperature rise causes a transient acoustic wave with both compressive and tensile components, followed by fast nucleation and bubble growth. It has been referred that these foam structures are good runners for fabrication of cell culture substrates, since, given the open interconnected pore structure, they exhibit increased availability of adhesion sites combined with permeability to fluids.

Table 13.2 Modification fluence thresholds (in mJ/cm^2) of biopolymer films at three wavelengths of irradiation with 90 fs pulses

Material	266 nm	400 nm	800 nm
Gelatine	330	900	1,200
Collagen	260	420	2,400

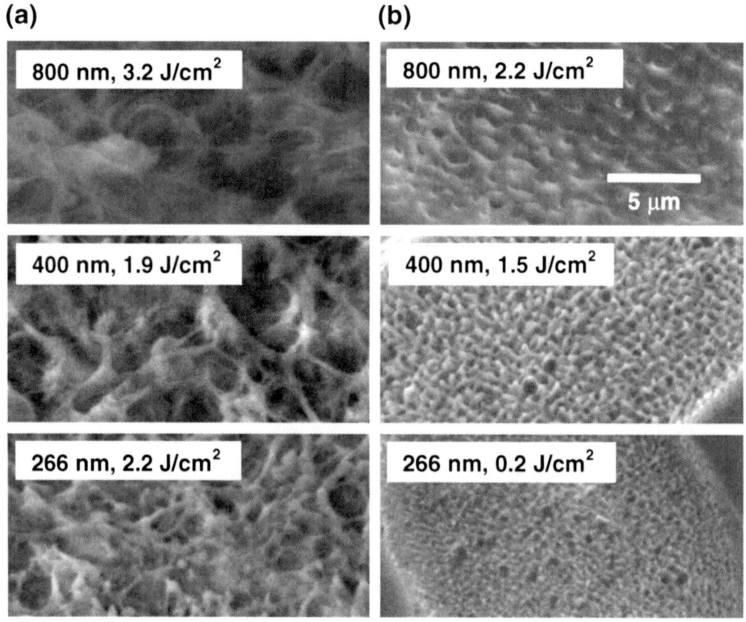

Fig. 13.8 Environmental scanning electron microscopy (ESEM, Philips XL30) graphs of irradiated areas of **a** collagen and **b** gelatine films after irradiation with a single laser pulse of 90 fs at the indicated wavelengths and fluences. Reproduced with permission from [43]

Laser foaming of biopolymers with fs pulses at 266, 400 and 800 nm was examined in self-standing films of 20–40 μm thickness of gelatine and collagen. Experiments made use of a Ti:Sapphire amplified laser system (Spectra Physics) delivering linearly polarized pulses of 90 fs at a maximum repetition rate of 1 kHz. Together with the fundamental output of the laser at 800 nm, frequency doubling/tripling provided shorter radiation wavelength at 400 and 266 nm. Single pulse irradiation using fluences exceeding a certain threshold (Table 13.2) lead to the formation of a nanofoamy and nanofibrous layer on films of collagen and gelatine. The superficial morphology of the affected layer (with a thickness of around 5 μm) is shown in Fig. 13.8.

The size and uniformity of the observed features are strongly dependent on irradiation wavelength and on the characteristics of the biopolymer (water content and mechanical strength). In collagen, the emerging fibrilar structure of interconnected voids and fibres is probably reminiscent of the pre-existing molecular organisation of the material. On gelatine, the irradiated region reveals a foamy layer with pores and bubbles of uniform size. The characteristic dimensions of the porous structures created on the films were found to scale linearly with wavelength; i.e. for gelatine the average bubble diameters are 0.3, 0.5 and 0.8 μm for irradiation at 266, 400 and 800 nm respectively. The typical feature size in collagen is about a factor of two higher than in gelatine. The laser fluence, in the range of foam formation, exerted negligible influence on their size.

As shown in the described studies, selection of irradiation wavelength served to control the characteristic size of the pores or bubbles generated by femtosecond pulses on the surface of biopolymer films. In order to assess the influence of the pulse duration, strongly absorbed UV laser pulses with durations ranging from nanoseconds to femtoseconds were used in [64] for the generation of superficial nanostructures on thin films of chitosan, starch and their blend. Laser irradiation with 500 fs pulses of 248 nm resulted in the formation of a layer in the form of interconnected pores and bubbles. However, when longer picosecond and nanosecond pulses were applied, this effect developed only in chitosan. The fact that a superficial foamy layer is not generated on the surface of the films of starch and of chitosan/starch blend is related with the different thermal properties of the materials, in particular their characteristic thermal transitions (i.e. glass transition and melting temperatures). The glass transition temperature of chitosan is around 140 °C [65] and this material is in amorphous state at room temperature. However, starch is a semicrystalline polymer and its glass transition takes place at around room temperature, or even at lower temperatures when the moisture content is over 20 % [66]. Similarly, as discussed above to explain the absence of LIPSS in semicrystalline PVDF, the absence of laser induced microstructures in starch is related to the restricted polymer dynamics within the mechanically stable scaffold constituted by the crystalline phase below the melting point.

The generation of nanostructures upon femtosecond laser irradiation of transparent biopolymer materials is discussed in relation to the generation of a plasma of quasi-free electrons by multiphoton and avalanche ionization and the subsequent transfer of the plasma energy to the lattice [43, 44]. The contribution of multiphoton ionization to the generation of quasi-free electrons is strongly wavelength dependent. In the studied biopolymers, absorption of two, three and five photons of 266, 400 and 800 nm respectively are required for ionization, assuming an ionization potential of ca. 6.5 eV. Also the avalanche ionization rate decreases strongly with decreasing wavelength [43]. The modification thresholds reported in Table 13.2 could be identified with the breakdown thresholds required to produce a critical free electron density [43], given by (13.4). The temperature increase reached in the irradiated volume after a single laser pulse estimated by (13.2) should scale inversely with wavelength through its dependence on ρ_{cr}. Thus, when irradiating the biopolymer films at the threshold fluence, the highest temperature increase should be expected in the case of irradiation at 266 nm. The fact that the size of the features (bubbles and pores) in this case are the smallest observed seems to support this trend, since smaller bubbles or pores can result from higher nucleation density. With regard to the pulse length effect, ultrashort

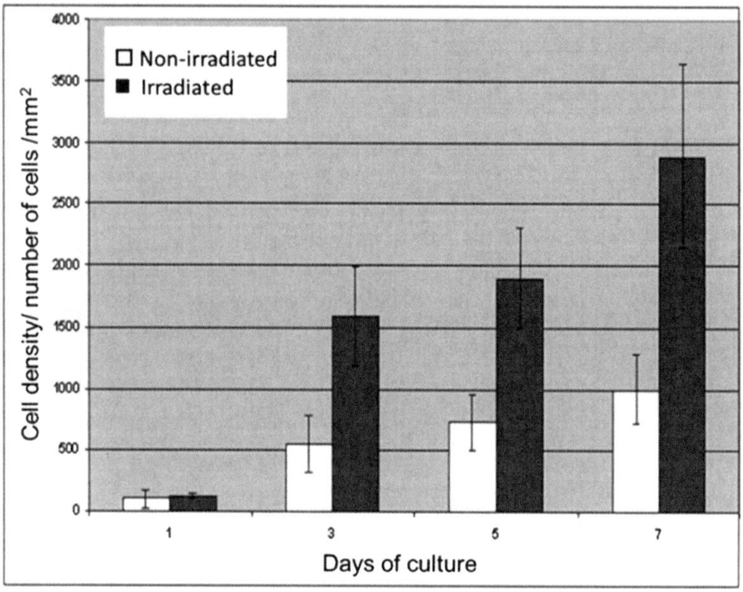

Fig. 13.9 Cell density on chitosan surface irradiated with a single 248 nm laser pulse of 20 ns, fluence of 2,800 mJ/cm^2 after 1, 3, 5 and 7 days of culture, compared with the corresponding to the non laser-treated one. Reproduced with permission from [64]

pulses of femtosecond duration ensure the strong spatial confinement within the thin layer below the film surface of temperature and pressure increase which drive the nucleation and growth of bubbles, finally leading to the observed foam in all of the studied materials [64].

13.4.2 Cell Culture on Laser Foamed Biopolymer Films

The viability of laser induced superficial foam structure for cell culture was investigated in [64]. Results of Mouse NIH/3T3 fibroblasts cell adhesion and proliferation on laser microfoamed chitosan substrates prepared by single nanosecond laser pulses provide an illustrative example of the possibility of the broader use of laser induced biopolymers foaming structures in biology.

It was observed that the cells do preferentially proliferate on the laser-irradiated areas of a chitosan film surface. The preferred adhesion of fibroblasts on modified chitosan spots was confirmed by Live-Dead Staining Kit assays by acquiring fluorescence microscopy images of live cells cultivated on the biopolymer film surface. Although cell culture has already been tested on porous chitosan [50], the fluorescence microscopy images indicated the viability and non-toxicity of laser-irradiated films of this material for cell culture. Figure 13.9 shows the evolution of

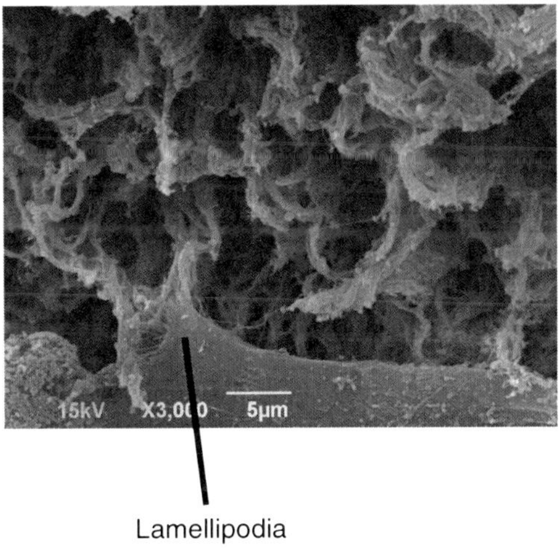

Fig. 13.10 Scanning electron microscopy image (acquired on a JEOL 7,000 field system) of the laser induced foaming on a chitosan film following single pulse irradiation at 248 nm, 20 ns, 2,800 mJ/cm^2 after cell culture of 3 days. Reproduced with permission from [64]

cell density with 1, 3, 5 and 7 days of culture and proves the preferential fibroblast growth on the foam structures on chitosan after irradiation.

Figure 13.10 illustrates the laser-foamed area of a chitosan film following single pulse irradiation at 248 nm with pulses of 20 ns after three days of cell culture. In general fibroblasts detect their substrate using the filopodia and their movement is facilitated by lamellipodium formation. As observed, the presence of lamellipodia, indicate the strong cell adhesion on the foam structured surfaces, ascribed to the increased substrate roughness. Higher superficial roughness supplies cells with extra attachment areas and configures a more favorable landscape for cell filopodia and adhesion proteins to migrate [67].

13.5 Conclusions

In conclusion, laser irradiation of thin films of synthetic polymers results in the formation of LIPSS using linearly polarized light at fluences well below the ablation threshold with laser pulse durations ranging from nanosecond to femtosecond. In all cases studied, the structures are parallel to the laser polarization direction and their period is of the order of the irradiation wavelength. While production of LIPSS with nanosecond pulses is only possible in materials which highly absorb at UV wavelengths, irradiation with femtosecond pulses results in ripple formation both in the UV and IR wavelengths, due to the efficient coupling of laser photons to the substrate mediated by multiphoton absorption. The presented strategies illustrate the possibility of control of the nanostructures created in polymer films in order to match the requirements derived from specific applications. As an example, it has been demonstrated that the rippled polymer films, once coated by pulsed laser deposition with tens of nanometer thick gold layer, serve as efficient SERS substrates. This has been illustrated using the model analyte benzenethiol, where the enhancement factor of the Raman signal has been estimated to be eight orders of magnitude.

On the other hand, laser irradiation of biopolymer films causes a superficial foaming effect. While it appears that the characteristics of the biopolymer material, including water content, mechanical strength and glass transition and melting temperatures, are crucial to determine the final morphology of the irradiated area it is observed that control of the size and uniformity of the observed foaming features can be readily controlled by selecting the irradiation wavelength (UV–IR) and the pulse duration (nanosecond to femtosecond). Successful mouse NIH/3T3 fibroblasts adhesion and proliferation on laser microfoamed biopolymer chitosan substrates prepared by single nanosecond laser pulses provide an illustrative example of the possibility of the broader use of laser induced biopolymers foaming structures in biology.

Acknowledgments Reported research has been funded by MICINN, Spain, Projects CTQ2010-15680 and MAT2009-07789. E. R. and M. O thank MICINN Ramón y Cajal and Consolider CSD2007-00058 Programs respectively for contracts. Collaborations with the groups of C. Domingo (Instituto de Estructura de la Materia, CSIC, Madrid, Spain), of P. Moreno (Universidad de Salamanca, Spain), S. Lazare (CNRS, Bordeaux) and C. Fotakis (IESL FORTH, Crete, Greece) are thankfully acknowledged.

References

1. G. Ekspong, in *Nobel Lectures in Physics 1991–1995*. (World Scientific Publishing Company, Singapore, 1997)
2. M. Raab, J. Kotek, Polimeri **26**, 172 (2005)
3. S. Lazare, Microfoams of biopolymers by laser-induced stretching: mechanisms and applications. in *Biopolymers*, ed. by M. Elnashar (Sciyo, 2010)
4. M. Moniruzzaman, K.I. Winey, Macromolecules **39**, 5194 (2006)
5. G.R. Strobl, *The Physics of Polymers* (Springer, Berlin, 1996)
6. R.H. Somani, B.S. Hsiao, A. Nogales, S. Srinivas, A.H. Tsou, I. Sics, F.J. Balta-Calleja, T.A. Ezquerra, Macromolecules **33**, 9385 (2000)
7. Special issue: Nanotechnology with soft matter. Small **7**(10) (2011)
8. H. Schift, J. Vac. Sci. Technol. B: Microelectron. Nanometer Struct. **26**, 458 (2008)
9. M. Steinhart, P. Göring, H. Dernaika, M. Prabhukaran, U. Gösele, E. Hempel, T. Thurn-Albrecht, Phys. Rev. Lett. **97**, 027801 (2006)
10. A. Palla-Papavlu, V. Dinca, C. Luculescu, J. Shaw-Stewart, M. Nagel, T. Lippert, M. Dinescu, J. Opt. **12**, 124014 (2010)
11. R. Eason, *Pulsed laser deposition of thin films: Applications-Led Growth of Functional Materials* (Wiley- Blackwell, New York, 2006)
12. A. Piqué, R.A. McGill, D.B. Chrisey, D. Leonhardt, T.E. Mslna, B.J. Spargo, J.H. Callahan, R.W. Vachet, R. Chung, M.A. Bucaro, Thin Solid Films **355**, 536 (1999)
13. L. Englert, M. Wollenhaupt, L. Haag, C. Sarpe-Tudoran, B. Rethfeld, T. Baumert, Appl. Phys. A Mater. Sci. Process. **92**, 749 (2008)
14. A. Klini, P.A. Loukakos, D. Gray, A. Manousaki, C. Fotakis, Opt. Express **16**, 11300 (2008)
15. Y.C. Tsai, K.H. Leitz, R. Fardel, A. Otto, M. Schmidt, C.B. Arnold, Nanotechnology **23**, 165304 (2012)

16. O. Vazquez-Mena, T. Sannomiya, M. Tosun, L.G. Villanueva, V. Savu, J. Voros, J. Brugger, ACS Nano **6**, 5474 (2012)
17. E. Rebollar, J.R. Vázquez de Aldana, J.A. Perez-Hernandez, T.A. Ezquerra, P. Moreno, M. Castillejo, Appl. Phys. Lett. **100**, 041106 (2012)
18. M. Csete, J. Kokavecz, Z. Bor, O. Marti, Mater. Sci. Eng. C **23**, 939 (2003)
19. E. Rebollar, S. Pérez, J.J. Hernández, I. Martín-Fabiani, D.R. Rueda, T.A. Ezquerra, M. Castillejo, Langmuir **27**, 5596 (2011)
20. M. Forster, W. Kautek, N. Faure, E. Audouard, R. Stoian, Phys. Chem. Chem. Phys. **13**, 4155 (2011)
21. S. Baudach, J. Krüger, W. Kautek, Rev. Laser Eng. **29**, 705 (2001)
22. M. Bolle, S. Lazare, M. Le Blanc, A. Wilmes, Appl. Phys. Lett. **60**, 674 (1992)
23. M. Csete, Z. Bor, Appl. Surf. Sci. **133**, 5 (1998)
24. P. Slepicka, E. Rebollar, J. Heitz, V. Svorcík, Appl. Surf. Sci. **254**, 3585 (2008)
25. E. Rebollar, I. Frischauf, M. Olbrich, T. Peterbauer, S. Hering, J. Preiner, P. Hinterdorfer, C. Romanin, J. Heitz, Biomaterials **29**, 1796 (2008)
26. D. Bäuerle, *Laser processing and chemistry* (Springer-Verlag, Berlin, 2000)
27. M. Bolle, S. Lazare, J. Appl. Phys. **73**, 3516 (1993)
28. U. Prendergast, S. Kudzma, R. Sherlock, C. O'Connell, T. Glynn, Photon processing in microelectronics and photonics VI. Proc. SPIE **6458**, 64581V (2007)
29. M. Csete, R. Eberle, M. Pietralla, O. Marti, Z. Bor, Appl. Surf. Sci. **208–209**, 474 (2003)
30. M. Csete, O. Marti, Z. Bor, Appl. Phys. A Mater. Sci. Process. **73**, 521 (2001)
31. M. Bolle, S. Lazare, Appl. Surf. Sci. **65–66**, 349 (1993)
32. M. Bolle, S. Lazare, Appl. Surf. Sci. **69**, 31 (1993)
33. H. Niino, Y. Kawabata, A. Yabe, Jpn. J. Appl. Phys. **28**, L2225 (1989)
34. H.W. Lu, Q.H. Lu, W.T. Chen, H.J. Xu, J. Yin, Mater. Lett. **58**, 29 (2004)
35. H. Lochbihler, Opt. Express **17**, 12189 (2009)
36. I. Martín-Fabiani, E. Rebollar, S. Pérez, D.R. Rueda, M.C. García-Gutiérrez, A. Szymczyk, Z. Roslaniec, M. Castillejo, T.A. Ezquerra, Langmuir **28**, 7938 (2012)
37. J.J. Hernández, D.R. Rueda, M.C. García-Gutiérrez, A. Nogales, T.A. Ezquerra, M. Soccio, N. Lotti, A. Munari, Langmuir **26**, 10731 (2010)
38. P.A. Müller-Buschbaum, Basic introduction to Grazing Incidence Small-Angle X-Ray Scattering. in *Lectures Notes in Physics* (Springer-Verlag, 2009)
39. D.R. Rueda, A. Nogales, J.J. Hernández, M.-C. García-Gutiérrez, T.A. Ezquerra, S.V. Roth, M.G. Zolotukhin, R. Serna, Langmuir **23**, 12677 (2007)
40. R. Lazzari, http://www.esrf.eu/computing/scientific/FIT2D/
41. M.E. Povarnitsyn, T.E. Itina, K.V. Khishchenko, P.R. Levashov, Appl. Surf. Sci. **253**, 6343 (2007)
42. T.J. Derrien, T. Sarnet, M. Sentis, T.E. Itina, J. Optoelectron. Adv. Mater. **12**, 610 (2010)
43. S. Gaspard, M. Oujja, R. de Nalda, C. Abrusci, F. Catalina, L. Bañares, S. Lazare, M. Castillejo, Appl. Surf. Sci. **254**, 1179 (2007)
44. A. Vogel, J. Noack, G. Hüttman, G. Paltauf, Appl. Phys. B: Lasers Opt. **81**, 1015 (2005)
45. M. Walczak, M. Oujja, J. Marco, M. Sanz, M. Castillejo, Appl. Phys. A Mater. Sci. Process. **93**, 735 (2008)
46. S.C.M. Fernandes, C.S.R. Freire, A.J.D. Silvestre, C.P. Neto, A. Gandini, J. Desbriéres, S. Blanc, R.A.S. Ferreira, L.D. Carlos, Carbohydr. Polym. **78**, 760 (2009)
47. H. Yi, L.-Q. Wu, W.E. Bentley, R. Ghodssi, G.W. Rubloff, J.N. Culver, G.F. Payne, Biomacromolecules **6**, 2881 (2005)
48. A. Buléon, P. Colonna, V. Planchot, S. Ball, Int. J. Biol. Macromol. **23**, 85 (1998)
49. D. Yumin, X. Zuyong, L. Rong, Wuhan Univ. J. Nat. Sci. **2**, 220 (1997)
50. C. Tangsadthakun, S. Kanokpanont, N. Sanchavanakit, T. Banaprasert, S. Damrongsakkul, J. Metals Mater. Miner. **16**, 37 (2006)
51. M. Zhai, L. Zhao, F. Yoshii, T. Kume, Carbohydr. Polym. **57**, 83 (2004)
52. M. Oujja, S. Perez, E. Fadeeva, J. Koch, B.N. Chichkov, M. Castillejo, Appl. Phys. Lett. **95**, 263703 (2009)

53. H. Jiankang, L. Dichen, L. Yaxiong, Y. Bo, L. Bingheng, L. Qin, Polymer **48**, 4578 (2007)
54. J.G. Fernandez, C.A. Mills, J. Samitier, Small **5**, 614 (2009)
55. W.-C. Hsieh, C.-P. Chang, S.-M. Lin, Colloids Surf. B **57**, 250 (2007)
56. S. Psycharakis, A. Tosca, V. Melissinaki, A. Giakoumaki, A. Ranella, Biomed. Mater. **6**, 045008 (2011)
57. S. Lazare, V. Tokarev, A. Sionkowska, M. Wiśniewski, Appl. Phys. A Mater. Sci. Process. **81**, 465 (2005)
58. E. Rebollar, G. Bounos, M. Oujja, S. Georgiou, M. Castillejo, J. Phys. Chem. B **110**, 16452 (2006)
59. S. Gaspard, M. Oujja, R. de Nalda, M. Castillejo, L. Bañares, S. Lazare, R. Bonneau, Appl. Phys. A Mater. Sci. Process. **93**, 209 (2008)
60. S. Gaspard, M. Oujja, C. Abrusci, F. Catalina, S. Lazare, J.P. Desvergne, M. Castillejo, J. Photochem. Photobiol. A **193**, 187 (2008)
61. S. Lazare, R. Bonneau, S. Gaspard, M. Oujja, R. De Nalda, M. Castillejo, A. Sionkowska, Appl. Phys. A Mater. Sci. Process. **94**, 719 (2009)
62. S. Lazare, I. Elaboudi, M. Castillejo, A. Sionkowska, Appl. Phys. A Mater. Sci. Process. **101**, 215 (2010)
63. M. Oujja, E. Rebollar, S. Gaspard, C. Abrusci, F. Catalina, S. Lazare, M. Castillejo, High-Power laser ablation VI. SPIE **6261**, 62611L (2006)
64. M. Castillejo, E. Rebollar, M. Oujja, M. Sanz, A. Selimis, M. Sigletou, S. Psycharakis, A. Ranella, C. Fotakis, Appl. Surf. Sci. **258**, 8919 (2012)
65. Y. Dong, Y. Ruan, H. Wang, Y. Zhao, D. Bi, J. Appl. Polym. Sci. **93**, 1553 (2004)
66. R.L. Shogren, Carbohydr. Polym. **19**, 83 (1992)
67. N. Saranwong, K. Inthanon, W. Wongkham, P. Wanichapichart, D. Suwannakachorn, L.D. Yu, Nucl. Instrum. Methods Phys. Res. Sect. B **272**, 386 (2012)

Chapter 14
Industrial Applications of Laser-Material Interactions for Coating Formation

Peter Schaaf and Daniel Höche

Abstract Laser Synthesis of TiN coatings on top of Ti pieces is performed by means of a free electron laser and also conventional lasers in reactive atmospheres. The produced coatings were investigated by various techniques. The results and properties of the resulting coatings are presented and discussed in connection with the different laser specialties. For the free electron laser treatment it was found that its ability to tune the pulse timing can be used to tailor the coating structure and properties (phases, hardness, strain, grain-size, etc.). This is discussed in connection with results of modeling the temperature, the plasma evolution, the mass transport, and the solidification behavior during and after the laser irradiation.

14.1 Introduction

Functional coatings and smart surfaces play an decisive role for the applicability and performance of modern materials. From an industrial point of view, there is great interest with respect to friction, wear, corrosion, and further properties. Many methods have been developed for the improvement of the respective surface and materials properties. Traditionally, these treatments range from simple PVD and CVD processes to complicated plasma methods and hybrid treatments [1, 2]. Recently, it has been established that short laser pulses of high energy can induce a direct laser synthesis of functional coatings if the material's surface is irradiated in

P. Schaaf (✉)
Chair Materials for Electronics and Electrical Engineering, TU Ilmenau, Institute of Materials Engineering and Institute for Micro- und Nanotechnologies MacroNano®, Gustav-Kirchhoff-Strasse 5 98693 Ilmenau, Germany
e-mail: peter.schaaf@tu-ilmenau.de

D. Höche
Department of Corrosion and Surface Technology, Institut für Werkstoffforschung/Institute of Materials Research, Helmholtz-Zentrum Geesthacht, Zentrum für Material- und Küstenforschung GmbH, Max-Planck-Straße 1 21502 Geesthacht, Germany

M. Castillejo et al. (eds.), *Lasers in Materials Science*, Springer Series in Materials Science 191, DOI: 10.1007/978-3-319-02898-9_14,

a reactive atmosphere [3, 4]. The mentioned process is based on a complicated combination of laser plasma—gas—material surface interactions [5]. Tests for steel, aluminum, magnesium, titanium and silicon in nitrogen, methane and hydrogen atmospheres have been carried out successfully. For these materials, interesting coatings can be produced by direct laser synthesis, e.g. AlN and SiC. Various laser types can be used for this purpose: Fiber, Excimer, Nd:YAG, CO_2 Laser, and even the free electron laser (FEL). Despite the simplicity of the treatment itself, up to now the process has neither been completely understood nor established as an industrial application, possibly due to the lack of high repetition rate pulsed high power lasers enabling fast and easy treatment of large areas and pieces. Here, the FEL with its unique properties is just the right tool to drive the mentioned process into applicability. Its high power and the flexibility in its temporal shaping was the motivation to do experiments with the FEL towards the direct laser synthesis of functional coatings. Some detailed points to this topic will be presented in the following, which represents an update to [3].

14.1.1 The Free Electron Laser

Surface irradiation with light from a free electron laser (FEL) is a new topic in material processing research. The Jefferson Lab FEL is a unique tool to modify materials and to study basic processes during the treatments. It is a light source, based on an energy recovered linac [6, 7]. Figure 14.1 shows the scheme of the FEL facility at the Jefferson lab (from [8]). Electron bunches were created by photoemission from GaAs and injected into the superconducting linac with up to 10 MeV acceleration voltage. Then, they were accelerated up to 150 MeV and aligned by several beam optics. Before the electrons emit light, bunches were compressed in a magnetic chicane. Afterwards a broadband THz beam is extracted, which can be used to investigate the functionality of the FEL. The short bunches itself were directed through the optical cavity and as a result of the electron acceleration in the wiggler, they emit a tunable narrow-band light known as the laser beam. Additionally, the light source is coherent and has good polarization properties with a ratio about 6,000:1. All these flexibilities initiate new application fields and research possibilities, especially for materials processing.

Currently, the 4th generation of the FEL is in use. The laser is capable to emit CW—mode like trains of sub-picosecond pulses (micro pulses of 0.2–2 ps in FWHM) up to an average power of 10 kW in a wavelength range of 1–14 μm. The frequency of the micro pulses is tunable between 4.68 and 74.85 MHz in steps of 2n where n is an integer from one to eight [9]. Alternatively, the setup can be switched to pulse-mode. In this mode, packages of micro-pulses (i.e. macro pulses) were emitted with frequencies up to 60 Hz (see Fig. 14.1). The pulse duration therefore is some hundred microseconds. Moreover, it is possible to run the FEL in the UV-branch, but with decreasing power output. Table 14.1 shows the actual beam parameters beam parameters at Jefferson Lab.

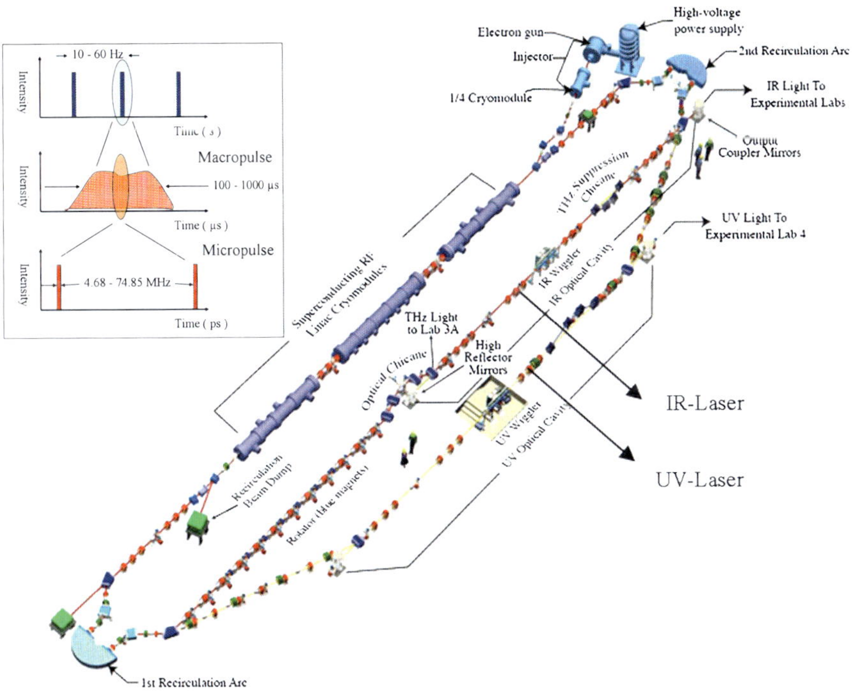

Fig. 14.1 Scheme of the Jefferson Lab free electron laser facility (from [8]). The inset shows the time structure in pulsed mode

Table 14.1 FEL beam parameters

	IR branch	UV branch
Wavelength range (μ)	1–14	0.25–1
Bunch length (FWHM ps)	0.2–2	0.2–2
Laser power/pulse (μJ)	100–300	25
Total laser power (kW)	>10	>1
Repetition rate (cw operation, MHz)	4.68–74.85	4.68–74.85

14.1.2 Direct Laser Synthesis

In relation to many surface cladding techniques, the direct laser synthesis is a simple and innovative process. The worksheets were placed in a chamber filled with a reactive environment, typically nitrogen or methane gas and then irradiated with the focused laser beam. As a result of the local heating, the induced melting and plasma formation, gas diffusion occurs into the sheets and a coating formation could be observed, whose thickness mainly will be determined by the diffusion coefficient and the melting depth. Many experiments and investigations have been performed successful for the nitriding of titanium with CO_2 lasers [10, 11] or

Nd:YAG lasers [12–16]. In the last case, the coatings have a thickness of about 2 μm. Other experiments show the successful synthesis of Fe_3N [17], AlN [18], TiC [19] and other compounds [20]. First synthesis of functional coatings by means of the FEL has been shown in [21–27]. The results show interesting coating properties and also dependencies on the scan parameter. This indicates the possibility of tailoring the treatments for industrial applications.

In contrary to the simple technical process, the governing physical mechanisms are very complex. Laser material interactions like absorption, induce melting and phase transformations takes place [5]. At high energy densities, the process will be assisted by plasma formation and its expansion into the ambient gas. Due to the shock wave and the laser light absorption in dense gases or plasmas, the gas molecules or atoms could be dissociated or ionized [28]. This rapidly increases the activity of the gas which results in an amplified adsorption at the liquid surface and lastly the coating formation.

At a timescale long enough (milliseconds), convectional flow in the melt pool occurs. This is determined by the Marangoni—and the recoil pressure induced forces. As a result, the surface quality decreases but contrarily the gas atom flux into deeper regions becomes higher. Concerning the tribological properties like wear resistance and hardness, the solidification process is the most important one. The nucleation and phase formation in the modified tracks mainly determine the solid properties like grain size and stress. The following enumeration shows the involved processes during laser nitriding [5]:

- Laser light absorption and heating
- Melting and evaporation (ablation, surface recession)
- Plasma expansion into the ambient gas (recoil pressure)
- Dissociation and ionization → increase of gas activity—plasma shielding
- Gas absorption (Sievert's law)
- Mass transport: melt (convective/diffusive) or in the solid (diffusive)
- Nucleation, solidification and phase formation → coating properties.

Due to the fact of complicated interactions between the different physical processes, controlling the process is a main problem. Therefore, it is necessary to get quantified data about the interactions by means of experiments and, in the case of experimental inaccessibility, by means of simulations. A recent review about this is given in [4].

14.1.3 Protective Coatings and TiN

Thin films, coating technologies and cladding techniques are of great interest in the current research. Thus, several methods like PVD, CVD or sputtering have been optimized and combined with other techniques (hybrid). New high-tech alloys have been developed with increasing mechanical, optical or electrical properties. They got extensive corrosion resistance, hardness or were used as thermal barriers.

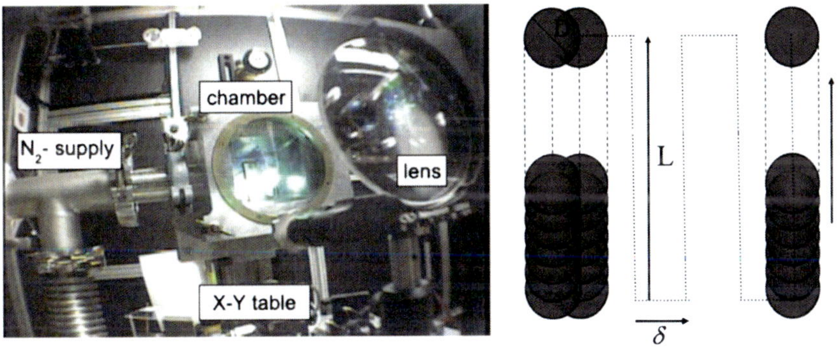

Fig. 14.2 Experimental setup at the FEL facility and meandering scan scheme

For industrial applications it is necessary to protect strained components and assemblies in an effective way. In the case of commercial metallic components which mostly based on iron, titanium or aluminum alloys, the simplest technique is nitriding or carburizing.

A typical example is the synthesis of titanium nitride. TiN can be used as a direct protection of titanium or as an additional coating on several alloys. The compound and its phase diagram have been studied as well. The nitrogen solubility in pure α-Ti was determined to be 23 at. %. In the case of stoichiometric δ-TiN, it has a cubic lattice (Fm3 m, 225) with a lattice constant of 4.24 Å. Pure TiN has a golden like color and has a hardness up to 25 GPa. Additionally, it has a strong chemical resistivity and a high melting point of 3,220 °C.

In order to carry out basic studies on the direct laser synthesis, the titanium—nitrogen system was chosen. It is a popular compound in material science and industries and allows comparisons with other coating methods.

14.2 Experiments

14.2.1 Sample Preparation and Setup

Sheets of commercial titanium (1 mm thickness, purity > 99.98 %) were cut into pieces of 15 × 15 mm^2 size. For the laser treatments, the samples were placed in a chamber, first evacuated and then filled with nitrogen (purity 99.999 %) to a pressure of 1–5 × 10^5 Pa. The focused beam reached the sample surface through a fused silica window. In order to treat the whole surface of the samples, the chamber was mounted onto a computer-controlled x–y table. The relative velocity v_{scan}, the lateral shift δ of the tracks, the spot size D and the pulse frequency f are the main scan parameters. Figure 14.2 shows the experiments at lab 2 at the FEL facility.

The treatments were performed in cw—and pulsed mode. Therefore, some blind tests have been executed, in order to get some information about the process

Table 14.2 Scan—and process parameters used during the treatments at the different modes

Parameter	cw	Pulsed
Wavelength range (μ)	1.6	3.1
Bunch length (FWHM psec)	0.2	0.5
Laser power/pulse (δJ)	125	20
Laser power (W)	650	160–750
Repetition rate (cw operation, MHz)	4.68	37.4
Scan velocity v_{scan} (mm/s)	24	0.5
Spot size D (μ)	600	440
Lateral shift δ (μ)	400–2,000	100–200
Macro pulse duration (μs)	–	250–1,000
Gas pressure (atm.)	1.15	1

parameters and their influences. Scan velocity, spot size, lateral shift, laser power or macro pulse duration have been varied in a parametric study. In pw-mode the sheets were treated in a meandering scheme. Table 14.2 presents the scan parameters at the different laser modes, where a successful coating formation was observed.

14.2.2 Analysis Methods

The microstructure of the synthesized coating was analyzed by X-ray diffraction in grazing incidence (GIXRD), Bragg–Brentano and Rocking curve geometry using a Bruker AXS diffractometer equipped with a Cu-K_α tube and a thin film attachment. Peak analyses yielded lattice constants, average nitrogen contents, stresses, textures and crystallite/grain size.

Nitrogen depth profiling was carried out by means of the resonant nuclear reaction analysis (RNRA) employing the reaction $^{15}N(p, \alpha\gamma)^{12}C$. The measurements were performed at the Göttingen IONAS accelerator. Details are given in references [29, 30]. The nitrogen depth profiles were limited to a depth of approximately 500 nm due to the limited proton energy of maximal 500 keV.

The microhardness depth profiles were measured with a Nanoindenter (Fischerscope HV100). It operates with a Vickers diamond tip and a maximum load of 1 N. Scanning electron microscopy (SEM) was performed for surface analyses attached with EDX measurements and for imaging cross section micrographs (FEI Nova 600).

14.3 Results

14.3.1 FEL Irradiation at CW-Mode

As demonstrated in Fig. 14.3, a melting track occurs during the irradiation. The nitrogen reacts with the melt and the synthesis of titanium nitride takes place. They got a golden like color and are quite inhomogeneous.

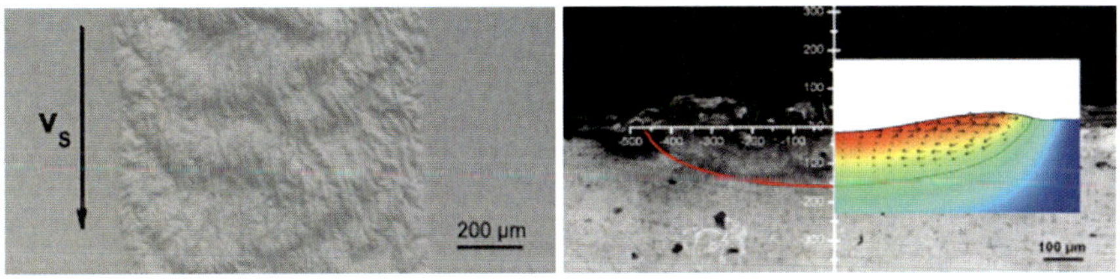

Fig. 14.3 Nitrated tracks in *top view* and as cross section

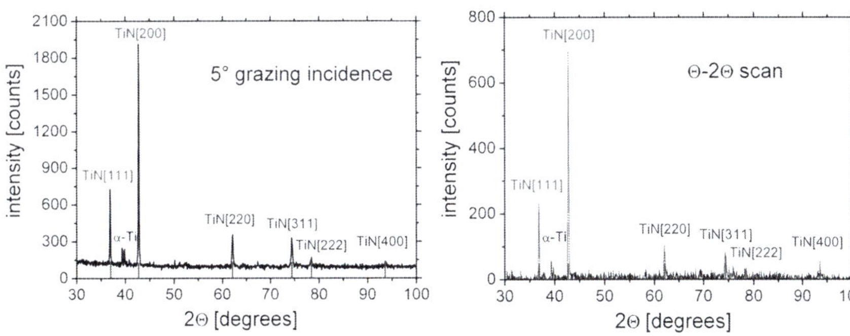

Fig. 14.4 Grazing incidence (*left*) and Bragg–Brentano (*right*) diffraction pattern of the selected sample

As expected, the track properties are mainly determined by the melt flow. Marangoni convection and pressure induced melt modifications resulted in a strong roughness. Humps and melt ejection were not being observed. Further a periodical structure is visible as a result of the equilibrium of the surface acting forces. Short wavelength structures could be observed, too. Due to the oscillations on the liquid titanium such modifications are developed. They are formally known as Rayleigh–Taylor instabilities.

Numerical studies have shown a strong influence of the convection on the track or respectively coating properties [31]. The Marangoni force induced flow velocities up to 1 m/s. Convective heat transfer becomes the main determining process. The describing number in fluid mechanics is the Peclet number (Pe). It reaches values about 60. That is the reason for the low aspect ratio of the melted tracks. Further the diffusive nitrogen transport can be assisted by the convection. Due to the mixing in the liquid pool the coating thickness will be determined by the melting depth. In the shown examples this depth was about 200 μm.

XRD measurement in grazing incidence geometry at 5° shows that only δ—TiN has been developed. The information depth is about 600 nm. Figure 14.4 presents the diffraction pattern of a selected sample with multiple tracks at a distance of 600 μm. The virgin α—Ti was observed too, but at a low content.

The Bragg–Brentano scan on the right-hand side of Fig. 14.4 shows a strong (200) peak which indicates the development of a weak (200) fiber texture. As a result, the titanium nitride lattice is directed perpendicular to the surface.

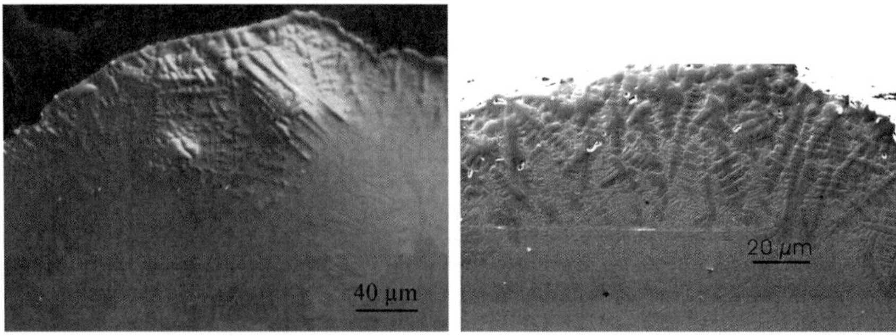

Fig. 14.5 Cross section micrographs of titanium nitride tracks at different scales

Cross section micrographs have been performed, too (Fig. 14.5), in order to take account the phase formation during solidification. They show the redirected TiN dendrites near the surface. Their distribution is quite inhomogeneous due to the varying temperature conditions during the treatments.

Dendritic solidification seems to be the main mechanism during CW irradiation titanium with the free electron laser. The top–down direction is a result of the difference in melting temperature of Ti and TiN of about 1,250 K. The directed lattice and the stoichiometric TiN phase are the reason for the improved tribological properties.

14.3.2 FEL Irradiation at Pulsed Mode

With pulsed mode treatments, coatings with much varying properties and thicknesses up to 20 µm have been generated. They exhibited interesting properties correlating to the scan and beam parameters. The most representable samples have been investigated and are presented here. Their scan parameters are shown in Table 14.3.

The surface properties are different and show a strong dependence on the scan parameters. Figure 14.6 presents SEM surface micrographs of the coatings. Sample 1 and 2 show solidified melting droplets and some cracks as a result of the remelted titanium and the resulting induced intrinsic stress. For longer macropulse durations, melting droplets could be avoided as a result of exceeding the evaporation point. The energy load for sample 3 was 4 times higher than for sample 4. As a consequence sample 3 shows many cracks and looks fragile. From a technical point of view, sample 4 has the best properties, where the coating is smooth and without any fractures.

In order to understand the melting behavior, numerical simulations have been performed by means of the finite element method (FEM). Heat transfer and phase transitions were studied. Detailed information is available in the literature [5, 22X].

Table 14.3 Macropulse duration τ_{ma}, macropulse repetition rate f_{ma} and lateral shift δ (in y-direction) used for the FEL treatments

Sample	τ_{ma} (μs)	f_{ma} (Hz)	δ (μm)
1	250	60	100
2	750	30	100
3	1,000	30	100
4	1,000	10	200

The scan velocity in x-direction is 0.5 mm/s

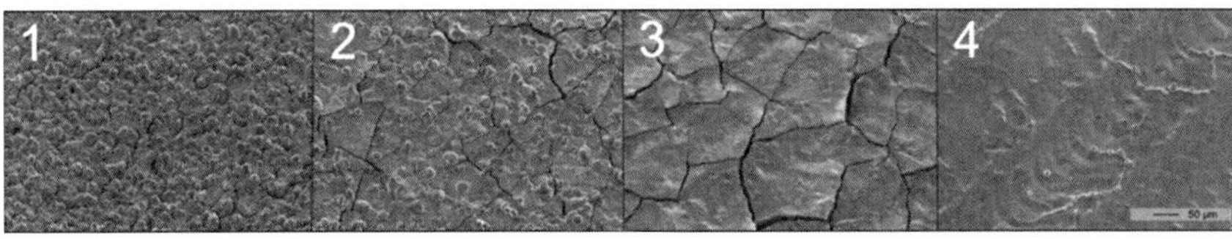

Fig. 14.6 Surface properties of selected samples

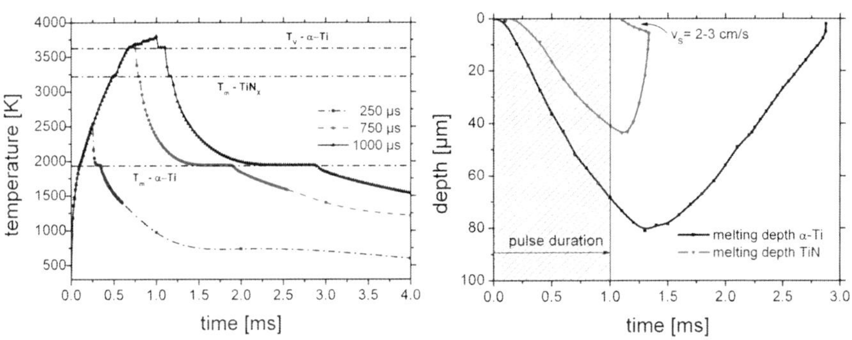

Fig. 14.7 Surface temperature development during a macropulse

The results show that the surface temperature during the treatment determines the properties. For example, Fig. 14.7 shows the temperature distribution during one macropulse. For longer pulse durations, the surface temperature is high enough to evaporate titanium and to remove the droplets. The increase in coating quality is a main result of these observations. Surface profiles of the samples show a decreasing roughness R_a from 3 to 1.2 μm.

On the right hand side of Fig. 14.7, the melting depth is shown for the pure titanium and the sample with a formed TiN coating. The main difference observed here is the solidification behavior of the TiN melt front. Due to the large difference in the thermal properties of titanium and its nitride, the solidification direction was achieved to be top–down for the TiN case. By means of the solidification velocity R (2–3 cm/s), the temperature gradient G ($\sim 10^4$ K/mm) and the cooling rate GR ($\sim 20 \times 10^4$ K/s) it is possible to determine the solidification mechanism.

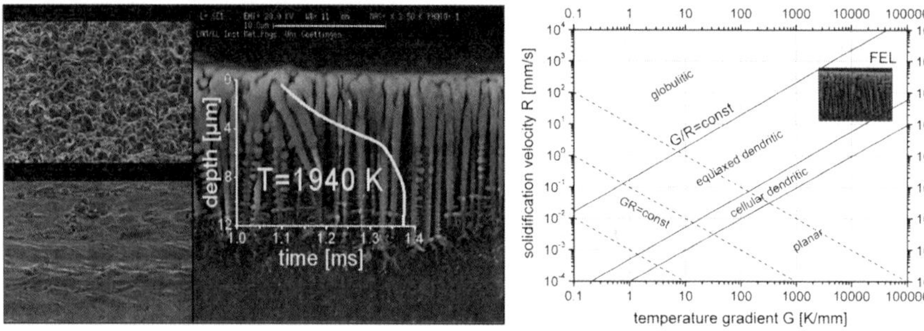

Fig. 14.8 Solidification behavior, melting temperature isotherm (TiN) of sample 4 (*left*) and cooling rate for FEL processing (*right*)

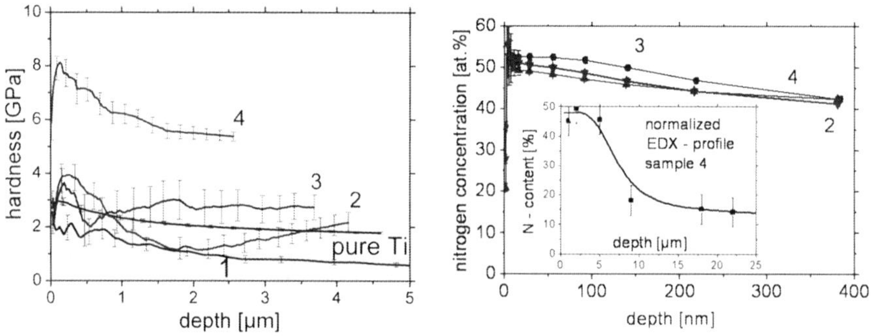

Fig. 14.9 Hardness depth profiles as measured by nanoindentation technique (*left*) and nitrogen depth profiles (*right*)

Dendritic growth was observed close to the surface. Figure 14.8 shows the solidification behavior and its reasons. The dendrites are equiaxed and got side branching and arm spacing, respectively, in deeper regions. Due the knowledge of G and R it is possible to control the solid structure and at least the tribological properties.

In relation to mechanical loading the coating hardness is the most important parameter. Wear resistance and friction coefficient can be improved in order to optimize several components or assemblies in technical applications. Figure 14.9 presents the results obtained for the selected samples. Due to the bad surface quality (rifts, droplets) samples 1–3 are very inhomogeneous. Sample 4 shows a strong improvement of the overall hardness to 8 GPa (film hardness 12 GPa). Those properties are mainly determined by the phase transitions which are strong related to the nitrogen content. For optimized coatings, stoichiometric TiN (50 at. %) has to be synthesized. The RNRA measurements show results in agreement to that. In Fig. 14.9, the stable titanium nitride was observed over the whole measured range. Additional EDX investigations show the diffusion like profiles at deeper regions.

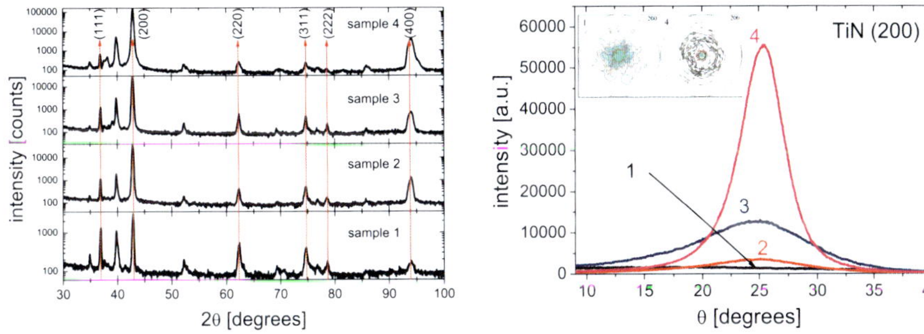

Fig. 14.10 Θ–2Θ scan, rocking curves and pole figures of the four selected samples. All indicated *peaks* are cubic TiN, the others belong pure titanium

X-ray diffraction measurements resulted in correlating lattice properties. The development of a strong (200) fiber texture was observed by means of Bragg–Brentano and Rocking curve scans. Figure 14.10 shows this orientation behavior assisted by pole figures.

These diffraction patterns verify the development of dendrites perpendicular to the surface. The directed lattice seems to be the reason for the improved mechanical properties and its strength.

14.4 Conclusions

This report reviews the synthesis of functional coatings by irradiating surfaces with a free electron laser and other Lasers in a reactive atmosphere. Various investigations and experiments have been performed by means of the model-system titanium and nitrogen. Thick and hard TiN could be achieved. The underlying physical processes and mechanisms have been identified and were correlated to the resulting synthesized titanium nitride. Coatings of 5–200 μm thickness have been produced. Their quality is mainly determined by the scan parameters. In the case of cw irradiation the coatings or tracks have a high thickness but as a result of the strong acting forces their roughness and homogeneity is still improvable. The processing time was in a range up to 20 s. Their tribological properties have been studied as well.

For the pulsed treatments very satisfying tests have been performed. The synthesis 20 μm thick coatings without fractures and melting droplets were achieved. They show a low roughness of about 1 μm and have a micro hardness up to 12 GPa [32]. Their properties are mainly determined by the dendritic solidification behavior and by the moderate heat entry, respectively. Nitrogen depth profiling confirms stoichiometric TiN at the near surface range. In deeper regions, under stoichiometric titanium nitride in a stable phase was observed. The lattice has a strong (200) fiber texture and possess an average induced strain at values up to 0.004.

Regarding that coating forming technique, the results offer conclusions which are universally valid. They can be used to design an effective synthesis of functional coatings in gas atmospheres by means of laser irradiation:

- Melting depth mainly determines the film thickness due to efficient diffusion in the liquid phase
- Exceeding the boiling point leads to plasma expansion and to activation of the reactive gas which ends in higher absorption rate and reduced roughness
- Convection can amplify the gas transport in the tracks and improve the mixing effect gradients
- Depth profiling is able to determine the solidification direction (top–down).

The synthesis process is an interaction of complicated physical and chemical mechanisms. For optimized processing, it is necessary to quantify them and to correlate them to the scan parameters. Heat load, respectively energy densities, have to be tailored in order to achieve the optimal conditions and to avoid fractures, stress, inhomogeneity and roughness. From an industrial point of view, this new method is an promising alternative to other techniques due to the reduced processing time of some seconds per square centimeter surface.

References

1. P. Schaaf, D. Höche, Free Electron Laser Synthesis of Functional Coatings, in *Laser-Surface Interactions for New Materials Production—Tailoring Structure and Properties*, vol. 130, ed. by Antonio Miotello, Paolo M. Ossi (Springer, Berlin-Heidelberg, 2010). (Springer Series in Materials Science)
2. D. Höche, P. Schaaf, Heat Mass Transf. **47**, 519–540 (2011)
3. C. Lange, M. Hopfeld, M. Wilke, J. Schawohl, Th Kups, M.W. Barsoum, P. Schaaf, Physica Status Solidi (A) **209**, 545–552 (2012)
4. E. Carpene, D. Höche, P. Schaaf, Fundamentals of Laser-Material Interactions, in *Laser Processing of Materials: Fundamentals, Applications and Developments*, vol. 139, ed. by P. Schaaf (Springer, Heidelberg, 2010), pp. 21–48. (Springer Series in Materials Science)
5. P. Schaaf (ed.), *Laser Processing of Materials: Fundamentals, Applications and Developments*, vol. 139 (Springer, Heidelberg, 2010). (Springer Series in Materials Science)
6. G. Williams, Rev. Sci. Instrum. **73**, 1461 (2002)
7. A. Thomas, G. Williams, Proc. IEEE **95**, 1679 (2007)
8. S. Benson, G. Biallas, J. Boyce, D. Bullard, J. Coleman, D. Douglas, F. Dylla, R. Evans, P. Evtushenko, A. Grippo, Nucl. Inst. Methods Phys. Res. A **582**, 14 (2007)
9. P. Evtushenko, J. Coleman, K. Jordan, J. Klopf, G. Neil, G. Williams, AIP Conf. Proc. **868**, 193 (2006)
10. I. Ursu, I.N. Mihailescu, A.M. Prokhorov, V.I. Konov, V.N. Tokarev, S.A. Uglov, J. Phys. D Appl. Phys. **18**, 2547 (1985)
11. M. Raaif, F. El-Hossary, N. Negm, S. Khalil, A. Kolitsch, D. Hoche, J. Kaspar, S. Mandl, P. Schaaf, J. Phys. D Appl. Phys. **41**, 085208 (2008)
12. D. Höche, H. Schikora, H. Zutz, A. Emmel, R. Queitsch, P. Schaaf, J. Coat. Tech. Res. **6** (2008)

13. D. Höche, H. Schikora, H. Zutz, R. Queitsch, A. Emmel, P. Schaaf, Appl. Phys. A Mater. Sci. Process. **91**, 305 (2008)
14. E. György, A. Perez del Pino, P. Serra, J.L. Morenza, Surf. Coat. Technol. **173**, 265 (2003)
15. H.C. Man, Z.D. Cui, T.M. Yue, F.T. Cheng, Mater. Sci. Eng. A **355**, 167 (2003)
16. J. Mori, P. Serra, E. Martianez, G. Sardin, J. Esteve, J. Morenza, Appl. Phys. A Mater. Sci. Process. **V69**, S699 (1999)
17. P. Schaaf, M. Han, K.-P. Lieb, E. Carpene, Appl. Phys. Lett. **80**, 1091 (2002)
18. E. Carpene, P. Schaaf, M. Han, K. Lieb, M. Shinn, Appl. Surf. Sci. **186**, 195 (2002)
19. B. Courant, J. Hantzpergue, L. Avril, S. Benayoun, J. Mater. Process. Tech. **160**, 374 (2005)
20. P. Schaaf, Prog. Mater Sci. **47**, 1 (2002)
21. E. Carpene, M. Shinn, P. Schaaf, Appl. Phys. A Mater. Sci. Process. **80**, 1707 (2005)
22. D. Höche, G. Rapin, J. Kaspar, M. Shinn, P. Schaaf, Appl. Surf. Sci. **253**, 8041 (2007)
23. D. Höche, M. Shinn, J. Kaspar, G. Rapin, P. Schaaf, J. Phys. D Appl. Phys. **40**, 818 (2007)
24. E. Carpene, J. Kaspar, M. Shinn, P. Schaaf, J. Laser Micro/Nanoeng. **1**, 129 (2006)
25. E. Carpene, M. Shinn, P. Schaaf, Appl. Phys. A **80**, 1707 (2005)
26. M. Han, K. Lieb, E. Carpene, P. Schaaf, M. Shinn, Appl. Surf. Sci. **186**, 195 (2002)
27. E. Carpene, J. Kaspar, P. Schaaf, M. Shinn, in International Conference on Laser Precision Microfabrication, LPM 2005, 3–8 April 2005, Williamsburg, VA, USA (Japan Laser Processing Society, Osaka, 2005)
28. D. Höche, G. Rapin, P. Schaaf, Appl. Surf. Sci. **254**, 888 (2007)
29. F. Landry, P. Schaaf, Nucl. Instr. Methods Phys. Res. Sect. B: Beam Interact. Mater. At. **179**, 262 (2001)
30. M. Uhrmacher, K. Pampus, F.J. Bergmeister, D. Purschke, K.P. Lieb, Nucl. Instr. Methods Phys. Res. Sect. B: Beam Interact. Mater. At. **9**, 234 (1985)
31. D. Höche, G. Rapin, S. Müller, M. Shinn, P. Schaaf, Metall. Mater. Trans. B **40**, 497–507 (2008)
32. E. Carpene, M. Shinn, P. Schaaf, Appl. Surf. Sci. **247**, 307 (2005)

Chapter 15
Ultrafast Laser Micro- and Nano-Processing of Glasses

Koji Sugioka

Abstract Ultrafast lasers can perform high-quality, high-precision surface micromachining of glasses through multiphoton absorption. When an ultrafast laser beam with a moderate pulse energy is focused into glass, multiphoton absorption is confined to a region near the focal point inside the glass. Ultrafast lasers can thus perform internal modification of glass as well as surface processing. Internal modification is widely used to write 3D optical waveguides and to fabricate micro-optical components and microfluidic channels buried inside glass, enabling functional microdevices such as 3D photonic, microfluidic, and optofluidic devices to be fabricated. Glass bonding based on internal melting is another interesting application of ultrafast lasers. Tailoring the temporal profiles of ultrafast laser pulses can improve the quality and efficiency of ultrafast laser processing and enhance the fabrication resolution. This chapter comprehensively reviews several applications of surface and volume processing of glass, including surface micromachining and the fabrication of photonic, microfluidic, and optofluidic devices. It also discusses pulse-shaping techniques for achieving high-quality, high-efficiency, and high-resolution processing.

15.1 Introduction

Ultrafast lasers emit light pulses shorter than a few tens of picoseconds. They thus include femtosecond and picosecond lasers. Since the first experiments in 1987 [1, 2], ultrafast laser processing has opened new avenues for materials processing because it offers great advantages over conventional laser processing using nanosecond or longer pulsed lasers. These advantages include nonthermal processing, reduction of the heat-affected zone (HAZ), the absence of plasma

K. Sugioka (✉)
Laser Technology Laboratory, RIKEN, Wako, Saitama 351-0198, Japan
e-mail: ksugioka@riken.jp

M. Castillejo et al. (eds.), *Lasers in Materials Science*,
Springer Series in Materials Science 191, DOI: 10.1007/978-3-319-02898-9_15,

shielding, the ability to process transparent materials through multiphoton absorption, the ability to modify the interior of transparent materials, and the ability to perform nanofabrication [3]. Ultrafast lasers are currently commonly used for both fundamental research and practical applications.

Ultrafast lasers can perform high-quality, high-precision surface micromachining of various materials including metals, ceramics, soft materials (e.g., polymers and biotissues), and even brittle materials (e.g., glasses) [4–6]. Of these materials, this chapter comprehensively reviews micro- and nano-processing of glasses. Ultrafast laser irradiation induces the following electron excitation and relaxation processes in glass [7]. Electrons are initially excited from the valence band to the conduction band either by multiphoton absorption of ultrafast laser light (multiphoton ionization) or by tunneling ionization when the Keldysh parameter γ (which is determined by the laser electric field strength, the wavelength, and the ionization potential of the material) is much smaller than 1 ($\gamma \ll 1$). The excited electrons can be further excited by laser photons to higher energy states where efficient free carrier absorption occurs (electron heating). Otherwise, at sufficiently high laser intensities, the excited electrons are accelerated by the intense electric field of the ultrafast laser beam and collide with surrounding atoms, generating secondary electrons (impact ionization). This process is repeated and finally avalanche ionization occurs. Electron heating and avalanche ionization can contribute to non-thermal processes such as bond breaking and ablation, which are critical processes in ultrafast laser processing. Meanwhile, some of the generated free electrons relax and localize their energy in electron–hole pairs, which form self-trapped excitons (STEs). This relaxation often commences a few picoseconds after laser irradiation stops [8]. Some STEs relax to form permanent defects after a few hundred picoseconds [9]. Glass heating also occurs a few tens of picoseconds after laser irradiation and the irradiated area returns to room temperature after several tens of microseconds, also causing modification or processing [10, 11].

When an ultrafast laser beam with a moderate pulse energy is focused by a lens into glass, multiphoton absorption is confined to a region near the focal point inside the glass. Ultrafast laser can thus perform internal modification of glass as well as surface processing [12, 13]. Internal modification is widely used to write 3D optical waveguides and to fabricate micro-optical components and microfluidic channels buried inside glass, enabling functional microdevices such as 3D photonic, microfluidic, optofluidic devices to be fabricated.

Tailoring the temporal profiles of ultrafast laser pulses is very attractive for micro- and nanoprocessing of glass and it is one of the hottest topics in this field. Since temporal manipulation of ultrafast laser pulses (i.e., pulse shaping) offers the possibility of controlling the transient free-electron density, it has been used to realize high-quality, high-efficiency, and high-resolution processing of glass [14, 15].

This chapter comprehensively reviews several applications of surface and volume processing of glass, including surface machining and the fabrication of photonic, microfluidic and optofluidic devices. It also discusses beam-shaping techniques for achieving high-quality, high-efficiency, and high-resolution ultrafast laser processing.

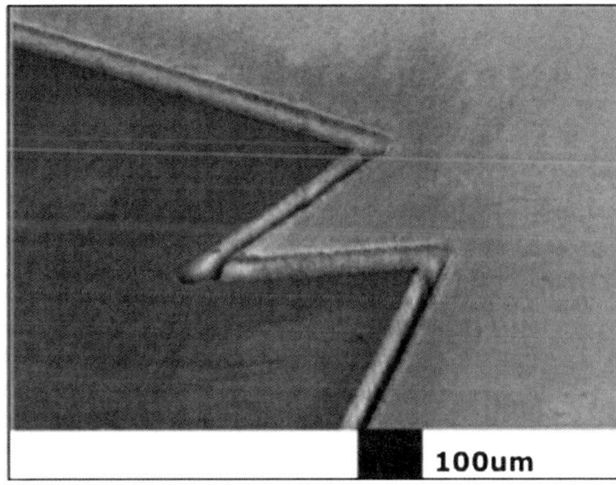

Fig. 15.1 SEM image of glass material cut by femtosecond laser ablation. (courtesy of M. Gower)

15.2 Surface Micromachining

Surface micromachining and patterning by laser ablation have been attracting growing interest due to their potential use in fabricating devices such as photovoltaic solar panels, flat panel displays, and light emitting devices. In contrast to ablation by nanosecond and longer pulses in which thermal processes dominate, ablation by ultrafast laser pulses involves nonthermal processes that enable high-precision material processing to be realized. This is due to rapid energy deposition in the material: it takes a few hundred femtoseconds to a few picoseconds for the electron distribution to reach thermal equilibrium after ultrafast laser irradiation [16, 17]. In contrast, the energy transfer time from the electron subsystem to the lattice, which induces thermalization, is of the order of 1–100 ps (depending on the electron–phonon coupling strength of the material); this time is much longer than the time for the electrons to reach thermal equilibrium [18, 19]. Thus, an ultrafast laser can efficiently cause electron heating or avalanche ionization and generate a hot electron gas, which is far from equilibrium with the lattice. Consequently, only a very small fraction of the laser pulse energy is converted into heat, resulting in non-thermal processing that enables high-precision microprocessing to be realized. Even though ultrafast laser pulses mainly induce nonthermal processes, they may still generate heat. However, ultrafast laser pulses do not produce a large HAZ because they have extremely short pulse widths of several tens of femtoseconds to a few picoseconds. This permits high-quality microfabrication, even for metals with high thermal conductivities.

Additionally, ultrafast lasers can induce strong absorption even in materials that are transparent to the laser wavelength [20]. When an extremely high density of photons (i.e., extremely high intensity light) is incident on a material, an electron can be excited by multiple photons even when the photon energy is smaller than the band gap; this phenomenon is known as multiphoton absorption. Ultrafast lasers can easily induce multiphoton absorption since their ultrashort pulses have extremely high peak powers. Consequently, ultrafast lasers enable high-quality microprocessing of even transparent materials such as glass. Figure 15.1 shows an scanning electron microscope (SEM) image of cutting of glass materials by femtosecond laser ablation; it reveals that clean ablation with sharp edges and with no cracks was achieved. There is currently a strong need to develop a high-quality glass-cutting technique for use in the manufacture of flat panel displays.

15.3 Internal Modification and 3D Micro/Nanofabrication

Multiphoton absorption is a nonlinear process that can be efficiently induced only at intensities exceeding a critical value, which depends on both the material and the pulse width. When an ultrafast laser beam with a moderate pulse energy is focused by a lens into glass, absorption is confined to a region near the focal point inside the material where the laser intensity exceeds this critical value. Thus, ultrafast laser pulses can modify the interior of transparent materials [12, 13]. Arbitrary 3D patterns can be generated by scanning a focused laser beam inside a fixed sample in the X–Y–Z directions (or equivalently by moving the sample while keeping the beam stationary). This enables 3D micro/nano-fabrication by an ultrafast laser to be realized inside glass (the wavelengths of $\sim$800 nm and $\sim$1 μm that are typically used for ultrafast laser processing are not linearly absorbed by glass). Internal modification and micro/nanofabrication is widely used to write 3D optical waveguides and to fabricate micro-optical components and microfluidic channels buried inside glass, as described in the following sections.

15.3.1 Photonic Device Fabrication

In 1996, Davis et al. used an ultrafast laser to permanently change the refractive index and they used this phenomenon to write an optical waveguide inside glass [12]. Many researchers are currently investigating writing of optical waveguides embedded in various glasses, including fused silica [10, 21], borosilicate glass [10, 22], and chalcogenide glass [23, 24]. Optical waveguides are typically written in glass using a transverse writing scheme in which the sample (or the laser beam) is translated perpendicular to the beam axis as shown in Fig. 15.2. This enables waveguides of arbitrary lengths and geometries to be written, making it particularly suitable for photonic device fabrication. Another scheme is the longitudinal geometry in which the sample is translated along the beam axis. This scheme can write waveguides with circularly symmetric cross-sectional shapes with dimensions of the order of the focal spot size. However, the total waveguide length is limited by the working distance of the objective lens employed. Optical waveguides written by the transverse writing scheme have core diameters (whose refractive index is increased by 10^{-4}–10^{-2}) ranging from 2 to 25 μm depending on

Fig. 15.2 Femtosecond laser writing of buried waveguides in glass materials using transverse writing scheme

the writing parameters used. An optical waveguide written in fused silica was estimated to have a propagation loss as low as ~0.2 dB/cm at a wavelength of 633 nm [25].

Refractive index modification was used to fabricate 3D photonic microdevices such as optical couplers and splitters [26], volume Bragg gratings [27], diffractive lenses [28], and compact and efficient single-longitudinal-mode lasers with a wavelength of 1.5 μm and maximum output powers of up to 55 mW [29].

As an example of a photonic device fabricated by ultrafast lasers, we consider multimode interference (MMI) devices based on the self-imaging phenomenon that occurs in multimode waveguides. An MMI device consists of input and output waveguides with a planar MMI waveguide in which several higher-order modes can propagate. MMI devices have several advantages over other power splitting devices based on directional couplers and Y branches: they have a simpler structure and they permit wide bandwidth light to propagate. Therefore, MMI devices reduce the size of photonic devices, which is beneficial for realizing high-density integration of photonic circuits. An MMI device was fabricated by femtosecond laser direct writing in silicate glass using a filamentation process [30]. Waveguide writing was performed by the longitudinal writing scheme since filamentation was employed. In the formation of the MMI waveguide, a multi-scan technique was employed to precisely control both the width and length of the multimode region (see Fig. 15.3a). The fabricated device was characterized by coupling light with different wavelengths (500, 550, 600, and 650 nm) and observing the output mode profiles obtained (see Fig. 15.3b). The results are in good agreement with a numerical simulation based on the beam propagation method. Such MMI devices can be used as compact power splitters with large fan-outs.

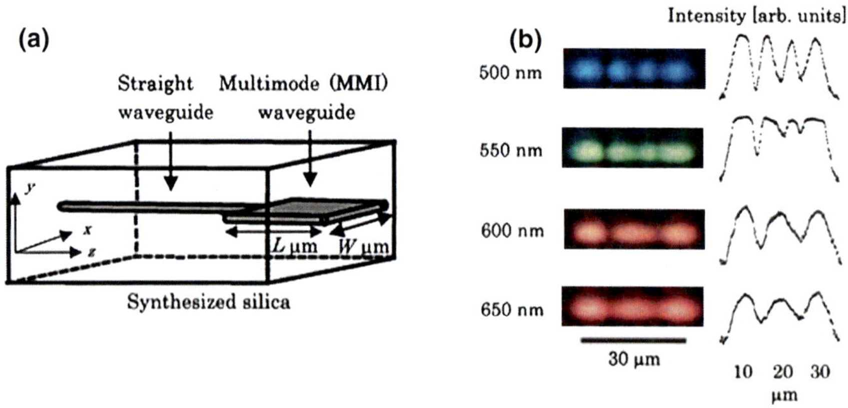

Fig. 15.3 **a** Schematic diagram of an MMI device in silica glass fabricated by the ultrafast laser. **b** Near-field patterns at MMI waveguide output for various wavelengths. The central wavelengths of the interference filters are indicated. ([30]—Reproduced with permission of The Optical Society of America)

15.3.2 Microfluidic Device Fabrication

In addition to refractive index modification, ultrafast lasers can also locally alter the chemical properties of photosensitive glass and fused silica [31–34]. Photosensitive glass consists of lithium aluminosilicate glass doped with trace amounts of silver, cerium, sodium, and antimony [35]. Ultrafast laser irradiation generates free electrons through multiphoton absorption processes in the photosensitive glass [36]. Silver ions doped in the photosensitive glass are then reduced by some of these free electrons and become silver atoms. Subsequent thermal treatment causes these silver atoms to agglomerate and form nanoclusters, which act as nuclei for growth of a crystalline phase of lithium metasilicate around the nanoclusters in the amorphous glass matrix. If the nanocluster density is sufficiently high, the crystalline regions will connect with each other and form a network. The etch rate of this crystalline phase of lithium metasilicate in dilute hydrofluoric (HF) acid is ~50 times higher than that of the unmodified glass matrix [32]. Figure 15.4 illustrates the three-step procedure for fabricating 3D microfluidic structures in photosensitive glass: (1) formation of a latent image in photosensitive glass by scanning a tightly focused ultrafast laser beam (laser direct writing); (2) transformation of this latent image to an etchable phase of lithium metasilicate by thermal treatment; and (3) removal of the modified material in the laser-exposed areas by wet chemical etching in a 5–10 % aqueous solution of HF acid in an ultrasonic bath. This technique was successfully applied to fabricate a complex microfluidic structure with an X-shaped channel embedded 200 µm below the surface of the photosensitive glass (Fig. 15.5) [37]. Notably, the width of ~45 µm is almost constant over the entire length (2.8 mm) of the fabricated channel.

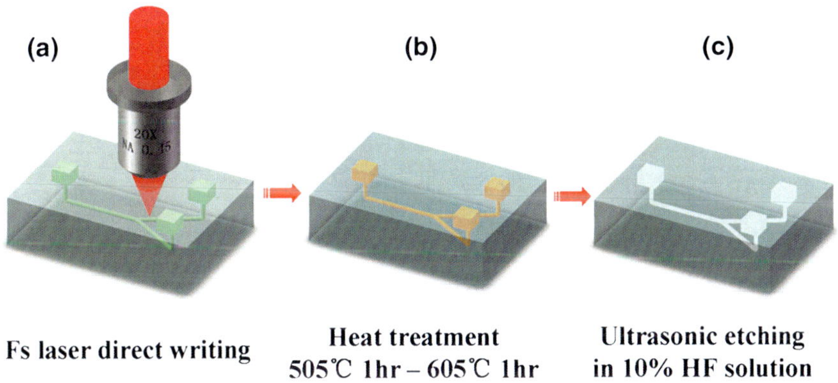

Fig. 15.4 Schematic diagrams of procedure for fabricating 3D microfluidic structures inside photosensitive glass. **a** 3D latent images are written inside photosensitive glass by femtosecond laser direct writing. **b** Samples are subjected to programmed heat treatment to develop modified regions. **c** Samples are soaked in an aqueous solution of hydrofluoric (*HF*) acid in an ultrasonic bath to selectively etch the laser-irradiated regions. Finally, hollow microstructures are formed inside the glass

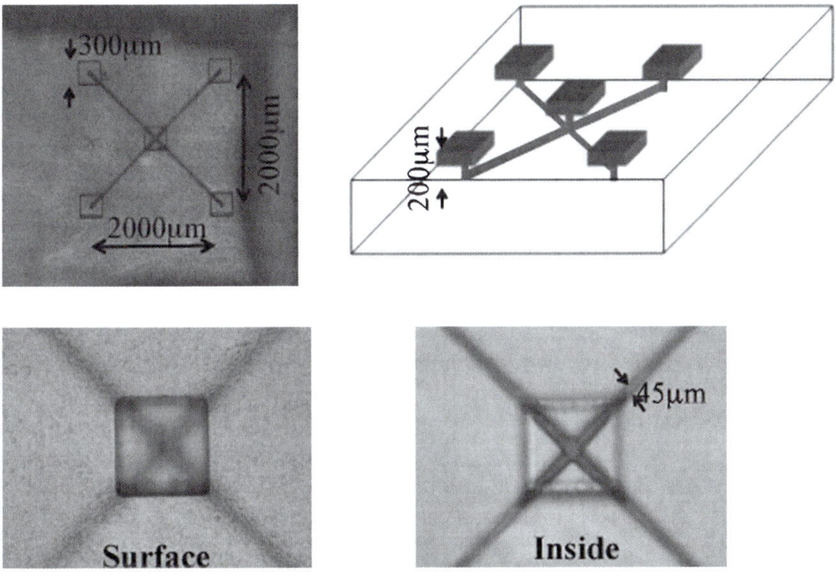

Fig. 15.5 Fabrication of a complex microfluidic structure with an X-shaped channel embedded 200 μm below the photosensitive glass surface by the procedure shown in Fig. 15.4

Unlike photosensitive glass, an enhanced etch rate in HF acid can be realized directly in fused silica by ultrafast laser irradiation without employing subsequent thermal treatment [31]. Although the mechanism of this phenomenon is currently not fully understood, current research suggests that it may be attributable to

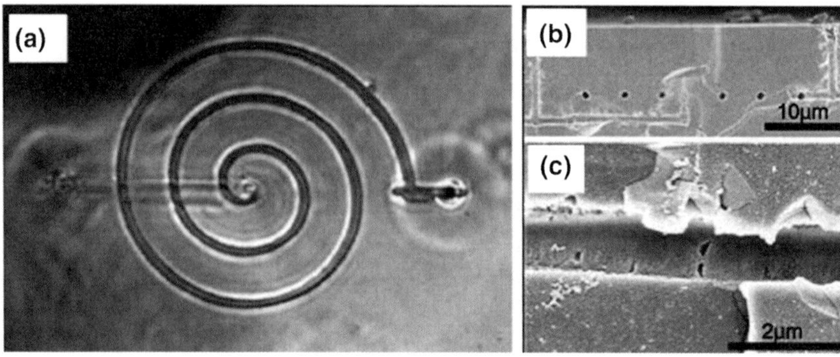

Fig. 15.6 Fabrication of ~700 nm-diameter subsurface nanochannels with arbitrary geometries by liquid-assisted femtosecond laser drilling. ([43]—Reproduced with permission of The American Chemical Society)

structural changes in the irradiated regions, such as a reduction in the Si–O–Si bond angle [34]. Additionally, it has recently been demonstrated that centimeter-long microfluidic channels with high aspect ratios (~1,000 or higher) can be fabricated by using a potassium hydroxide (KOH) solution as the etchant, which provides a higher etching selectivity between areas modified and unmodified by ultrafast laser irradiation [38].

Three-dimensional microfluidic structures can also be fabricated inside glass chips by ultrafast laser 3D drilling in distilled water from the rear surface of the glass; this process is frequently referred to as liquid-assisted femtosecond laser drilling [39]. The water introduced into the microchannel can help remove ablation debris, greatly enhancing the drilling length relative to that achievable by drilling in ambient air. Three-dimensional microdrilling is realized by translating the focal spot in the sample in the transverse direction. Since this technique does not rely on generating etch selectivity in materials by ultrafast laser irradiation, it can be applied to any material that is transparent to the writing pulses [40]. However, even with the assistance of water, debris generated by ultrafast laser ablation still clogs the microchannel when the depth/length of the channel reaches several hundreds of micrometers, restricting the size of fabricated microstructures [41]. Employing an ultrasonic bath can enhance the debris removal rate, allowing the channel length to be extended to nearly 1 mm [42], but this length scale is still too short for many microfluidic applications.

On the other hand, liquid-assisted femtosecond laser drilling can be used to fabricate microchannels with nanoscale diameters since the resolution of this process is determined by the focused spot size. In fact, subsurface nanochannels with diameters of ~700 nm and arbitrary geometries have been fabricated using low-energy femtosecond laser pulses tightly focused by a high-numerical-aperture objective lens as shown in Fig. 15.6 [43]. The channel lengths are again limited to several hundred micrometers because debris cannot be expelled from longer channels due to bubbles created in water by laser irradiation.

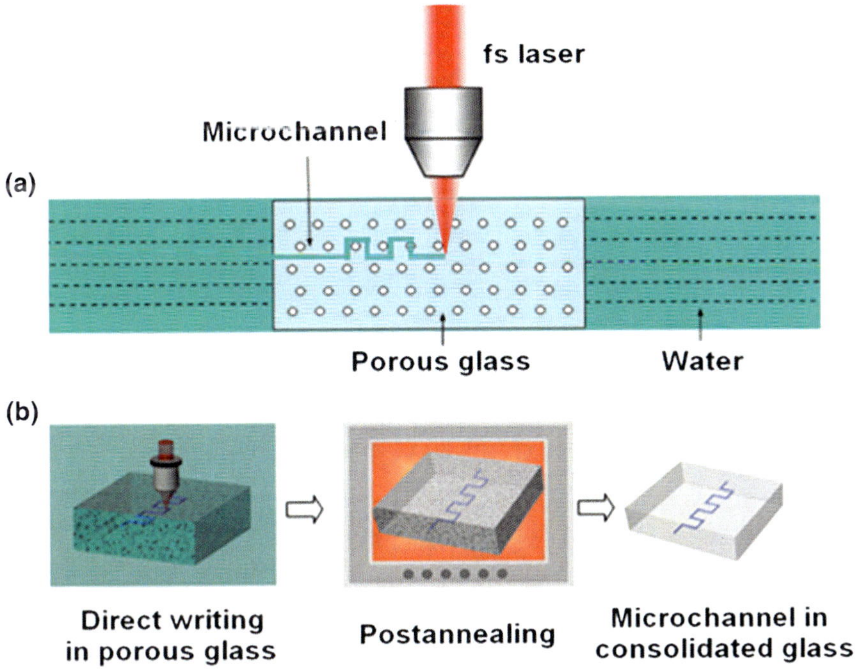

Fig. 15.7 **a** Schematic diagram of experimental setup and **b** flow diagram of fabrication process for liquid-assisted femtosecond laser drilling in porous glass. The main fabrication procedure involves two steps: (1) direct formation of hollow microchannels in a porous glass substrate immersed in water by femtosecond laser ablation and (2) post-annealing at $\sim$1,150 °C to consolidate porous glass substrate

The fundamental limit on the length of microfluidic channels fabricated by liquid-assisted femtosecond laser drilling has recently been exceeded by using a porous glass [44]. Figure 15.7a shows a schematic diagram of the experimental setup and Fig. 15.7b shows a flow diagram of the fabrication process. The main fabrication procedure involves two steps: (1) direct formation of hollow microchannels in a porous glass substrate immersed in water by femtosecond laser ablation and (2) post-annealing at $\sim$1,150 °C to consolidate the porous glass substrate. This technique permits 3D microfluidic channels to be fabricated inside glass with arbitrary lengths and geometries. This technique was used to fabricate a passive micromixer of fluids based on the baker transformation, which provides ideal mixing [45]. Such mixers generally require complex 3D geometries and are thus difficult to fabricate by conventional planar fabrication processes. Figure 15.8a and b show schematic diagrams of the designed 3D mixer [46]. It consists of a Y-shaped microchannel embedded 400 μm below the surface of the chip and a series of mixing units that are connected to two opening inlets and to an outlet, which are located on the backside of the chip. Three circular slots surrounding the inlets and outlet were also fabricated by ultrafast laser ablation. They were used as connection

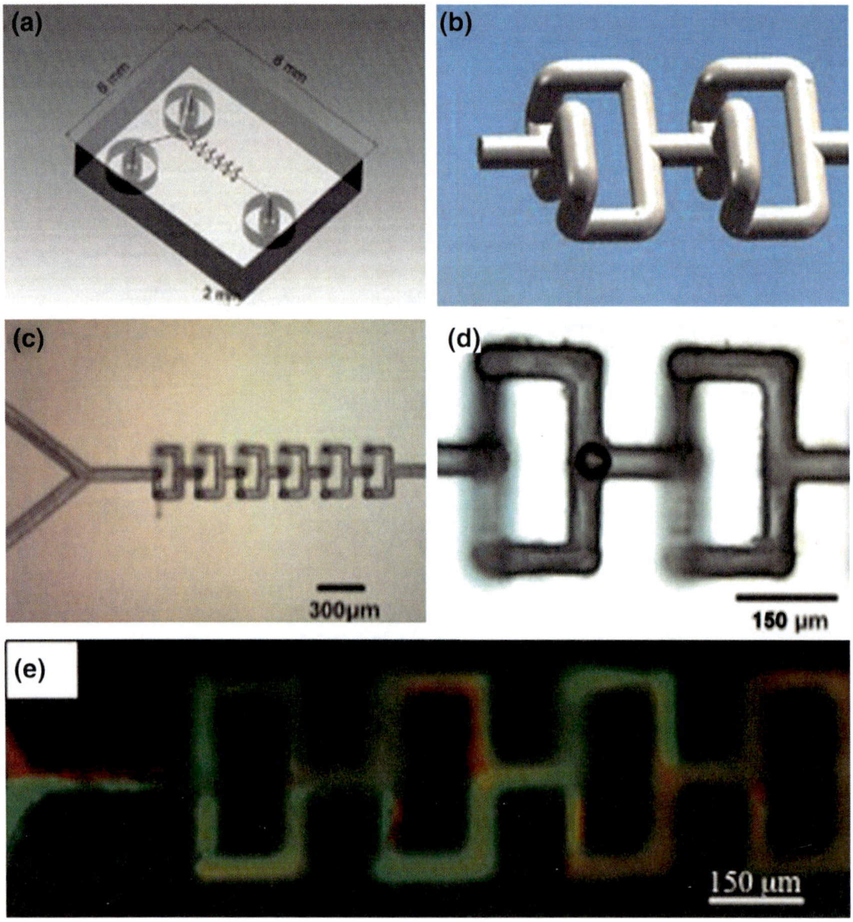

Fig. 15.8 Schematic diagrams of 3D passive microfluidic mixer: **a** overview and **b** enlargement. Optical micrographs of fabricated 3D microfluidic mixer: **c** overview and **d** high-magnification micrographs. **e** Demonstration of mixing of two fluids

interfaces for 0.9-mm-diameter plastic tubes. Figure 15.8c shows an overview of a fabricated 3D micromixer consisting of six mixing units. All the horizontal and vertical channels in the mixing units have lengths and widths of respectively 150 and 50 μm (Fig. 15.8d). Figure 15.8e shows the experimental results for mixing two fluorescent dye solutions (fluorescein sodium and rhodamine B) in the fabricated 3D microfluidic mixers. The two fluids are well mixed after passing through three mixing units (i.e., a length of 0.9 mm), which corresponds to a mixing time of about 10 ms.

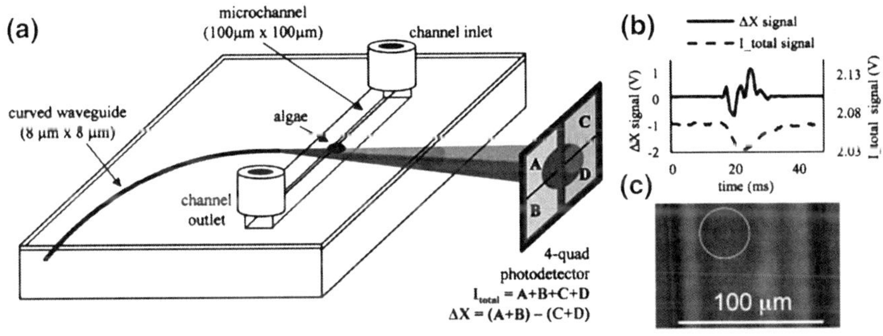

Fig. 15.9 **a** Schematic diagram of a system with a curved waveguide to direct laser light across a microchannel onto a photodetector; **b** typical signal obtained; **c** a photograph of a *Cyanothece* specimen (indicated by *circle*) in microchannel. ([50]—Reproduced with permission of The Royal Society of Chemistry)

15.3.3 Optofluidic Device Fabrication

Ultrafast lasers can directly fabricate 3D microfluidic structures and optical waveguides in glass. Each component can be integrated in a single glass chip with relatively easy alignment. Thus, ultrafast laser processing has several advantages over conventional methods for fabricating optofluidic devices. Optofluidic devices can reduce the volumes of samples and reagents used in chemical reactions, biological analysis, and medical tests. Moreover, they can be used to perform high-efficiency, high-accuracy, and high-sensitivity analysis [47, 48].

One of the simplest schemes for fabricating optofluidic devices by an ultrafast laser is in-plane integration of a straight microfluidic channel and a curved optical waveguide, as depicted in Fig. 15.9a. This scheme was used to fabricate a system for optically classifying algae species [49, 50]. It is very important to detect and identify submillimeter-sized phytoplankton to monitor environmental and climate change and to evaluate water for health reasons. A probe light (1,550 nm) delivered by the curved optical waveguide passed through algae-laden water flowing in a microchannel. The distance between the end of the optical waveguide and the channel is designed to be sufficiently long that light emitted from the end facet of the waveguide expands to illuminate the entire channel height. The curved waveguide is essential for preventing uncoupled light from interacting with the photodiode, which improves the signal-to-noise ratio. The probe light emitted from the end of the waveguide was introduced to a four-quadrant detector after passing through the microchannel. The photodetector generated two signals that varied depending on the size and shape of the algae species: the intensity of the total detected light ($I_{\text{total}} = \text{A} + \text{B} + \text{C} + \text{D}$) and the difference between the two upstream detectors and the two downstream detectors ($\Delta\text{X} = (\text{A} + \text{B}) - (\text{C} + \text{D})$) (see Fig. 15.9b for an example). A microscope and a camera were externally installed above the channel to enable manual identification of algae to verify the classification accuracy of the optofluidic system. Figure 15.9c shows an image of a *Cyanothece* specimen captured by this external observation system. This device categorized five algae species with an average positive identification rate of 78 %. In addition, it was applied in a field-deployable device that distinguished the toxin-producing cyanobacterium *Cyanothece* from detritus in field-collected water with a success rate of over 90 %.

Another optofluidic system in which two optical waveguides intersect the microfluidic channel was used to determine the functions of microorganisms. Such optofluidic systems have been termed nanoaquariums [51, 52]. Nanoaquariums were applied to elucidate the gliding mechanism of *Phormidium. Phormidium* is a soil-dwelling, unicellular, colonial cyanobacterium that forms motile filaments. These filaments migrate from the main biomass to nearby seedling roots where they form new colonies. *Phormidium* has an ectosymbiotic relationship with seedlings: seedling growth accelerates after forming an association with *Phormidium.* Therefore, understanding the factors that induce *Phormidium* gliding is very important for developing methods that accelerate seedling growth to improve crop yields, which will reduce food supply shortages. Preliminary experiments using the microfluidic devices suggested that CO_2 secreted by respiration of the root is a possible attractant that induces *Phormidium* gliding. To further investigate the gliding mechanism, nanoaquariums in which two optical waveguides were intersected by a microfluidic channel were fabricated to quantitatively analyze the CO_2 concentration in water containing a seedling root. In this case, the microfluidic channel was filled with water containing a pH indicator [bromothymol blue (BTB)] and white light from a halogen lamp was coupled to the entrance facet of waveguide I by an objective lens. The white light transmitted by waveguide I passed through the microfluidic channel, which was filled with a liquid sample, and was then coupled into optical waveguide II. The light transmitted by waveguide II was coupled into a spectrometer by another objective lens to obtain the spectrum of white light absorbed by the sample (Fig. 15.10a). The absorbance was calculated by subtracting the spectrum with the sample in the microfluidic channel from that observed without the sample. The green line in Fig. 15.10b indicates the absorption spectrum of the BTB solution. It has a large absorption peak at a wavelength of about 620 nm. The intensity of this peak decreases with increasing CO_2 concentration in the water due to the change in the pH. The spectrum of water containing the seedling root (yellow line) is comparable to that of 50 ml water mixed with 15 ml CO_2 (black line). This implies that the CO_2 concentration generated by the root's respiration is comparable to that of the carbonic water (15 ml CO_2:50 ml H_2O) used in this experiment. Interestingly, this CO_2 concentration is equal to the critical concentration at which *Phormidium* did not glide toward either the seedling root or the carbonic water in the T-shaped microfluidic channel [*Phormidium* glides toward the carbonic water (the seedling root) when the carbonic water concentration is higher (lower) than the critical concentration], indicating that CO_2 is the sole attractant for *Phormidium.* Similar optofluidic systems have also been used to detect [53], trap, and stretch [54] single cells.

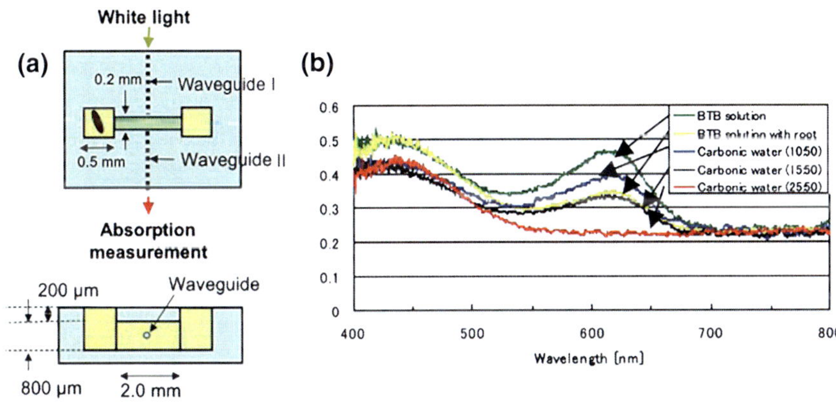

Fig. 15.10 **a** Nanoaquarium design and optical absorption measurement for identifying attractant that induces *Phormidium* gliding. Schematic illustration of microchip integrated with optical waveguides. Waveguides were written by modifying the refractive index using femtosecond laser irradiation after fabricating microfluidic channel. **b** Optical absorption spectra of water containing BTB solution (*green line*), with a seedling root (*yellow line*), and with carbonic water with different CO_2 concentrations [*blue line* (10 ml CO_2:50 ml H_2O); *black line* (15 ml CO_2:50 ml H_2O); and *red line* (25 ml CO_2:50 ml H_2O)]

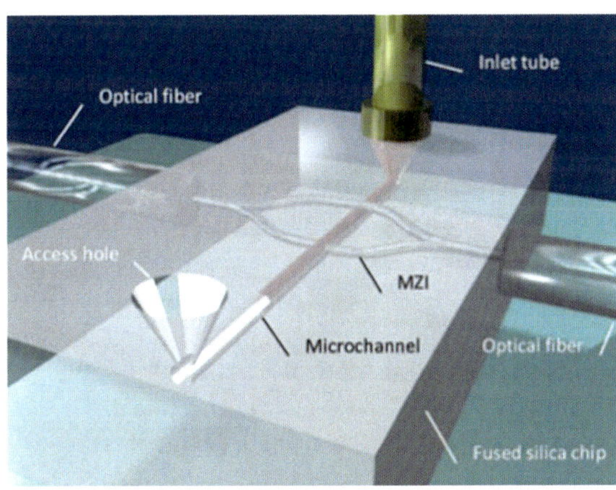

Fig. 15.11 Schematic diagram of a refractive index sensor fabricated by femtosecond laser direct writing. The *two arms* of the MZI lie in a tilted plane such that the reference arm lies above the microchannel and only the sensing arm intersects the microchannel. ([55]—Reproduced with permission of The Royal Society of Chemistry)

More attractive utilization of the unique 3D capability of ultrafast laser processing is out-of-plane integration of a waveguide-based Mach–Zehnder interferometer (MZI) and a microfluidic channel, as depicted in Fig. 15.11 [55]. In this specific case, the two arms of the interferometer lie in a plane tilted at 7° relative to the substrate plane. Thus, the reference arm is constructed above the microchannel and the sensing arm and the microchannel intersect at right angles. This optofluidic system is capable of measuring the refractive index with a spatial resolution of the order of the waveguide mode diameter (11 µm). The unbalanced MZI can detect fringes in the wavelength-dependent transmission when a sufficiently wide spectral

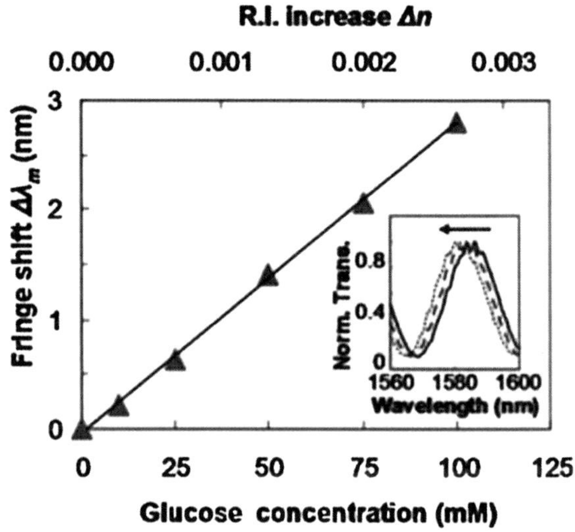

Fig. 15.12 Fringe shift for water with different glucose-D concentrations (in inset, *solid line* 0 mM; *dashed line* 50 mM; *dotted line* 100 mM) measured by optofluidic system with an unbalanced MZI. The *top axis* indicates the corresponding refractive index change. ([55]—Reproduced with permission of The Royal Society of Chemistry)

region is scanned by a tunable laser. The refractive index varies slightly with analyte concentration; this variation can be detected by a shift in the fringes, as shown in the inset of Fig. 15.12. The microchannel in the optofluidic system shown in Fig. 15.11 was filled with glucose-D solutions with various concentrations as test samples and its sensitivity was determined to be 10^{-4} refractive index units, which corresponds to a detection limit of 4 mM (see Fig. 15.12).

Although waveguide-based photonic components are very powerful, some functions (e.g., beam focusing and image formation) still have to be accomplished using free-space optics. Micro-optic components such as micromirrors and microlenses can be fabricated by the same technique as that used for fabricating 3D optofluidic structures, namely ultrafast laser direct writing followed by wet chemical etching [56–59]. However, in such cases, an additional postannealing step must be employed to smooth the etched surface to ensure high-performance of the fabricated micro-optics [56]. Figure 15.13 shows an optofluidic device in which microfluidic components, an optical waveguide, and microlenses are integrated. It is capable of performing both absorption and fluorescence spectroscopic detection of liquid samples [59]. In this device, a 6-mm-long optical waveguide is connected to a microfluidic reservoir that has dimensions of $1.0 \times 1.0 \times 1.0\ mm^3$. The waveguide is used to transfer either the fluorescence excitation light from a frequency-doubled YAG laser or a broadband beam from a white lamp for absorbance measurements to a liquid sample confined in the microreservoir. Since both the fluorescence emitted from the excited sample and the beam exiting the waveguide propagate in space with highly divergent angles, the signal collection efficiency will be relatively low. To counter this problem, two microspherical lenses (also with in-plane geometry and with a radius of curvature of 0.75 mm) were fabricated to the side of and behind the microreservoir to improve the fluorescence and transmitted light collection efficiencies, respectively.

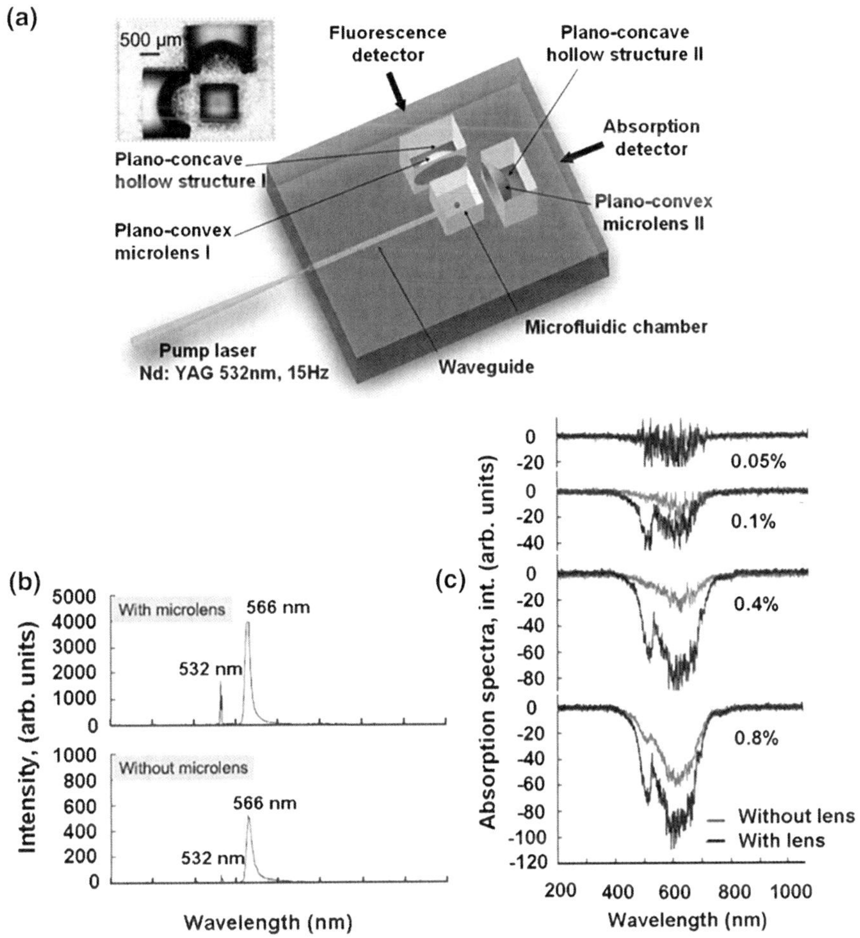

Fig. 15.13 **a** Schematic diagram of optofluidic microchip in which microoptic components (e.g., micro planoconvex lenses and an optical waveguide) are integrated with a microfluidic chamber in a single glass chip. Inset (*upper left corner*): *top view* of microchip fabricated by femtosecond laser direct writing. **b** Emission spectra of laser dye Rh6G pumped by a frequency-doubled Nd:YAG laser obtained using the microchip with (*top*) and without (*bottom*) microlens. **c** Optical absorption spectra of black ink diluted in water at different concentrations in the microchamber of a microchip integrated with (*black lines*) and without (*gray lines*) a microlens. The absorption spectra were obtained by subtracting the reference spectrum (i.e., the transmission spectrum without black ink) from the measured transmission spectra

To experimentally characterize the performance of this device, an Rh6G dye solution and water with various black ink concentrations were respectively used as fluorescence and absorbing samples. As shown in Fig. 15.13b and c, the two microlenses enhanced the detection efficiencies of fluorescence and absorption spectroscopic measurements by factors of 8 and 3, respectively.

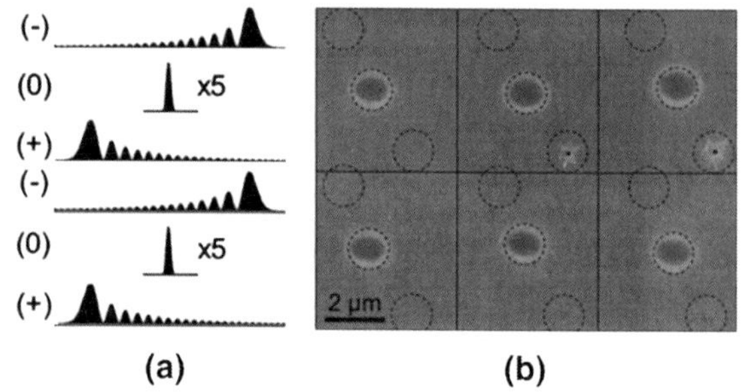

Fig. 15.14 Schematic diagram of transient laser intensities of shaped femtosecond laser pulse and corresponding SEM micrographs of a subset of laser-generated structures in fused silica. *Top row* surface in focal plane; *lower row* surface 1 μm below focal plane. ([62]—Reproduced with permission of The Optical Society of America)

15.4 Processing by Pulse Shaping Technique

As mentioned in Sect. 15.1, electron excitation processes in glass induced by ultrafast laser irradiation are associated with multiphoton ionization or tunneling ionization and subsequent electron heating or avalanche ionization. Tailored ultrafast laser pulses obtained by pulse-shaping an ultrafast laser beam have the potential to control the transient free-electron density, which can improve the fabrication quality, spatial resolution, and/or processing efficiency of ultrafast laser processing of glass.

The use of tailored ultrafast laser pulses was demonstrated in experimental ablation of fused silica and CaF_2. The experimental results reveal that irradiation of a conventional single-pulse train does not produce the best machined surface [60]. On the other hand, irradiation of a double- or triple-pulse train with a pulse separation of nearly 2 ps produced better ablation holes than a single-pulse train. This is probably because shaped pulses are more effective for brittle materials that have strong electron–phonon coupling because shaped pulses permit heating to be controlled, enabling relaxation of the induced stresses.

A series of ultrafast laser pulses (e.g., ~400 identical 1 ps pulses with a pulse separation of 7.5 ns) gave higher quality ablation with less microcracking and shock-induced effects than a single high-fluence laser pulse. Very smooth holes with diameters of 7–10 μm and depths of ~30 μm were formed in fused silica without the formation of fractures or cracks or surface swelling [61].

A shaped pulse that resembles an asymmetric series of pulses whose amplitudes decrease or increase amplitude over time (see Fig. 15.14a) was employed for nanoablation of fused silica [62, 63]. Control of multiphoton ionization and avalanche ionization by controlling the spatial free-electron density due to a Gaussian beam profile of the laser beam was combined with the different process dependencies on intensity. This resulted in the fabrication of ~100-nm-diameter holes by a 790 nm wavelength ultrafast laser beam, which was focused to a spot size of 1.4 μm (see Fig. 15.14b).

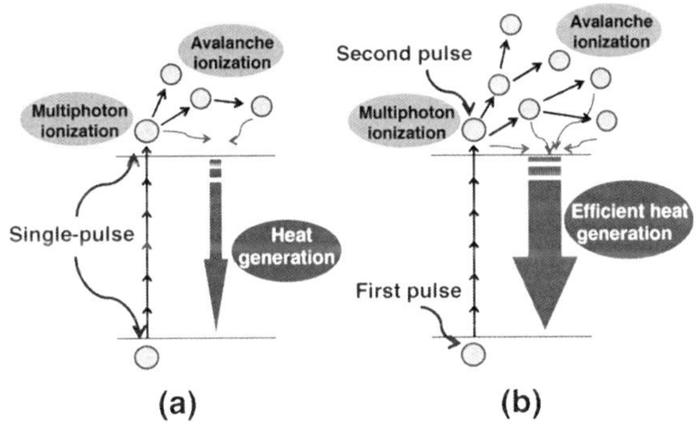

Fig. 15.15 Possible mechanism of free electron generation in glass by (**a**) single-pulse and (**b**) double-pulse irradiation of ultrafast laser beams for welding

Glass microwelding is another active research area in ultrafast laser processing [64–66]. Glass welding by ultrafast laser irradiation is considered to occur by melting induced at the interface between two glass substrates by irradiation of a focused laser beam. Melting occurs due to heating of the glass on relaxation of free electrons generated by two-step excitation; i.e., multiphoton ionization or tunneling ionization followed by electron heating or avalanche ionization, as described in Sect. 15.1. In a conventional laser irradiation scheme (i.e., single-pulse train irradiation), two-step electron excitation occurs during single pulse irradiation. This is shown in Fig. 15.15a, which illustrates the case of multiphoton ionization followed by avalanche ionization. To realize efficient heating that produces a large melted pool for efficient and high-quality welding, it is important to control electron excitation. Tailored ultrafast laser pulses in a double-pulse train can be used to control individual electron excitation processes (e.g., the first pulse induces multiphoton ionization, while the second pulse induces avalanche ionization) resulting in more efficient generation of free electrons and hence more efficient heat generation for glass welding (see Fig. 15.15b) [67]. Figure 15.16 shows optical microscopy images of laser-irradiated regions inside photosensitive glass. Laser beams consisting of double-pulse train (first pulse: p-polarization; second pulse: s-polarization) were focused at the same spot for 0.1 s for delay times of 10 ps (Fig. 15.16a). Figure 15.16b shows a sample irradiated by a single-pulse train of the p-polarized beam. All the images obtained show concentric circular patterns consisting of a dark inner region and a bright outer region. Both the dark inner and bright outer regions are considered to be HAZs [68]. The double-pulse train produces a significantly larger HAZ than the single-pulse train. In fact, the diameter of the HAZ produced by the double-pulse train at a delay time of 10 ps is estimated to be 36.7 μm, which is 27.4 % greater than that (28.8 μm) produced by a single-pulse train. For welding, the laser beam was focused on the interface between the two tightly stacked glass substrates and then scanned over an area of 1 × 1 mm at a scanning speed of 300 μm/s. The bonding strength of the sample prepared by the double-pulse train irradiation with a delay time of 10 ps was evaluated to be 22.9 MPa, which is approximately 22 % greater than that (18.7 MPa) obtained by irradiation of a single-pulse train.

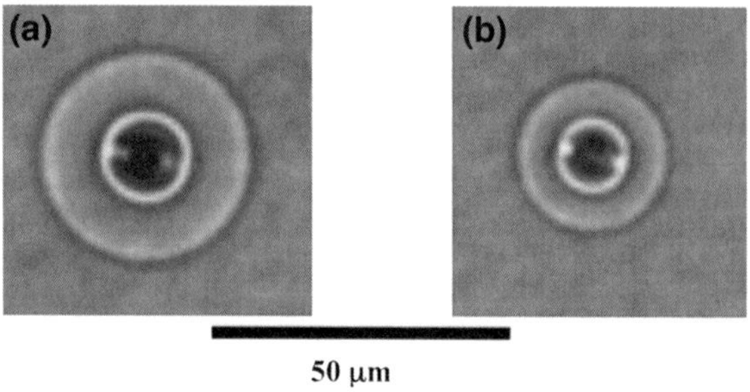

Fig. 15.16 Optical microscopy images of laser-irradiated regions; samples were prepared by (**a**) a double-pulse train (first pulse: p-polarization; second pulse: s-polarization) with a delay time of 10 ps and (**b**) single-pulse train of a p-polarized beam

15.5 Summary

Ultrafast laser is a promising tool for both surface and volume processing of glass materials. Strong absorption induced by nonlinear multiphoton absorption and a small HAZ due to the ultrashort pulse width allow high-quality surface micro-structuring and dicing of glass to be realized, which have a wide range of practical applications.

Additionally, ultrafast laser can modify the interior of glass in a spatially selective manner through multiphoton absorption. This enables the optical and chemical properties of glass to be simultaneously altered. Modification of the optical properties includes refractive index increases in laser-irradiated regions. This can be applied to fabricate 3D photonic microdevices, such as optical waveguides, optical couplers and splitters, volume Bragg gratings, MZI, and diffractive lenses.

The modified chemical properties of laser-irradiated regions in glass allow these regions to be selectively etched by subsequent wet etching using aqueous solutions of etchants such as hydrofluoric (HF) acid. This technique can be used to directly form 3D microfluidic systems with complex structures. The two-step process can also be used to fabricate free-space optical components such as micromirrors and microlenses inside glass. The unique ability of ultrafast laser processing to fabricate both microfluidic and optical components in glass opens up new avenues for fabricating various optofluidic microchips for biological analysis.

Additionally, controlling and manipulating the temporal properties of ultrafast laser pulses enhances the performance of ultrafast laser processing enabling a higher quality, higher efficiency, and higher fabrication resolution to be realized.

Ultrafast laser processing has benefited from the rapid development of ultrafast laser systems and ultrafast laser techniques. For example, compact, high-power, high-repetition-rate, ultrafast laser systems have been recently demonstrated, although they have much broader pulses than those generated by Ti:sapphire systems. Development of laser systems with high-reliability and high-stability pulses is expected to result in further advances in ultrafast laser processing. It is interesting to note that temporal pulse shaping was initially developed to realize coherent control of the dynamics of quantum systems (e.g., the outcomes of photochemical reactions) on ultrafast time scales. Hence, the exchange of ideas with other fields of ultrafast science is expected to increase the performance of ultrafast laser processing.

References

1. R. Srinivasan, E. Sutcliffe, B. Braren, Ablation and etching of polymethylmethacrylate by very short (160 fs) ultraviolet (308 nm) laser-pulses. Appl. Phys. Lett. **51**, 1285–1287 (1987)
2. S. Küper, M. Stuke, Femtosecond UV excimer laser ablation. Appl. Phys. **B44**, 199–204 (1987)
3. K. Sugioka, in *Laser-Surface Interactions for New Materials Production*, ed. by A. Miotello, P.M. Ossi (Springer, Berlin, 2010), pp. 279–293
4. C. Momma, B.N. Chichkov, S. Nolte, F. Alvensleben, A. Tünnermann, H. Welling, B. Wellegehausen, Short-pulse laser ablation of solid targets. Opt. Commun. **129**, 134–142 (1996)
5. M.F. Yanik, H. Cinar, H.N. Cinar, A.D. Chisholm, Y.I. Jin, A. Ben-Yakar, Neurosurgery: functional regeneration after laser axotomy. Nature **432**, 822 (2004)
6. N. Barsch, K. Korber, A. Ostendorf, K.H. Tönshoff, Ablation and cutting of planar silicon devices using femtosecond laser pulses. Appl. Phys **A77**, 237–242 (2003)
7. S.S. Mao, F. Quere, S. Guizard, X. Mao, R.E. Russo, G. Petite, P. Martin, Dynamics of femtosecond laser interactions with dielectrics. Appl. Phys **A79**, 1695–1709 (2004)
8. B.C. Stuart, M.D. Feit, S. Herman, A.M. Rubenchik, B.W. Shore, M.D. Perry, Nanosecond-to-femtosecond laser-induced breakdown in dielectrics. Phys. Rev. **B53**, 1749–1761 (1996)
9. K.S. Song, R. T. Williams, *Self-Trapped Excitons* (Springer, Berlin, 1993), pp. 269–394
10. S.M. Eaton, H. Zhang, P.R. Herman, F. Yoshino, L. Shah, kJ Bovatse, A.Y. Arai, Heat accumulation effects in femtosecond laser-written waveguides with variable repetition rate. Opt. Exp. **13**, 4708–4716 (2005)
11. S.M. Eaton, H. Zhang, M.L. Ng, J. Li, W.J. Chen, S. Ho, P.R. Herman, Transition from thermal diffusion to heat accumulation in high repetition rate femtosecond laser writing of buried optical waveguides. Opt. Exp. **16**, 9443–9458 (2008)
12. K.M. Davis, K. Miura, N. Sugimoto, K. Hirao, Writing waveguides in glass with a femtosecond laser. Opt. Lett. **21**, 1729–1731 (1996)
13. E.N. Glezer, M. Milosavljevic, L. Huang, R.J. Finlay, T.H. Her, J.P. Callan, E. Mazur, Three-dimensional optical storage inside transparent materials. Opt. Lett. **21**, 2023–2025 (1996)
14. L. Englert, B. Rethfeld, L. Haag, M. Wollenhaupt, C. Sarpe-Tudoran, T. Baumert, Control of ionization processes in high band gap materials via tailored femtosecond pulses. Opt. Exp. **15**, 17855–17862 (2007)
15. R. Stoian, M. Boyle, A. Thoss, A. Rosenfeld, G. Korn, I.V. Hertel, E.E.B. Campbell, Laser ablation of dielectrics with temporally shaped femtosecond pulses. Appl. Phys. Lett. **80**, 353–355 (2002)
16. C.K. Sun, F. Vallée, L.H. Acioli, E.P. Ippen, J.G. Fujimoto, Femtosecond-tunable measurement of electron thermalization in gold. Phys. Rev. **B50**, 15337–15348 (1994)
17. S.S. Wellershoff, J. Hohlfeld, J. Güdde, E. Matthias, The role of electron-phonon coupling in femtosecond laser damage of metals. Appl. Phys. **A69**, S99–S107 (1999)
18. J. Hohlfeld, S.S. Wellershoff, J. Güdde, U. Conrad, V. Jähnke, E. Matthias, Electron and lattice dynamics following optical excitation of metals. Chem. Phys. **251**, 237–258 (2000)
19. S.I. Anisimov, B. Rethfeld, Theory of ultrashort laser pulse interaction with a metal. Proc. SPIE **3093**, 192–203 (1997)

20. S. Küper, M. Stuke, Ablation of polytetrafluoroethylene (Teflon) with femtosecond UV exicimer laser pulses. Appl. Phys. Lett. **54**, 4–6 (1989)
21. J.W. Chan, T.R. Huser, S.H. Risbud, D.M. Krol, Modification of the fused silica glass network associated with waveguide fabrication using femtosecond laser pulses. Appl. Phys. **A76**, 367–372 (2003)
22. J.A. Dharmadhikari, A.K. Dharmadhikari, A. Bhatnagar, A. Mallik, P.C. Singh, R.K. Dhaman, D. Mathur, Writing low-loss waveguides in borosilicate (BK7) glass with a low-repetition-rate femtosecond laser. Opt. Comm. **284**, 630–634 (2011)
23. O.M. Efimov, L.B. Glebov, K.A. Richardson, E. Van Stryland, T. Cardinal, S.H. Park, M. Couzi, J.L. Bruneel, Waveguide writing in chalcogenide glasses by a train of femtosecond laser pulses. Opt. Mater. **17**, 379–386 (2001)
24. D. Le Coq, P. Masselin, C. Przygodski, E. Bychkov, Morphology of waveguide written by femtosecond laser in As2S3 glass. J. Non-Crystal. Solids **355**, 37–42 (2009)
25. H. Zhang, S.M. Eaton, P.R. Herman, Low-loss Type II waveguide writing in fused silica with single picosecond laser pulses. Opt. Exp. **14**, 4826–4834 (2006)
26. W. Watanabe, T. Asano, K. Yamada, K. Itoh, J. Nishii, Wavelength division with three-dimensional couplers fabricated by filamentation of femtosecond laser pulses. Opt. Lett. **28**, 2491–2493 (2003)
27. C. Florea, K.A. Winick, Fabrication and characterization of photonic devices directly written in glass using femtosecond laser pulses. J. Lightwave Technol. **21**, 246–253 (2003)
28. E. Bricchi, J.D. Mills, P.G. Kazamsky, B.G. Klappauf, J.J. Baumberg, Birefringent Fresnel zone plates in silica fabricated by femtosecond laser machining. Opt. Lett. **27**, 2200–2202 (2002)
29. G.D. Vella, S. Taccheo, R. Osellame, A. Festa, G. Cerullo, P. Laporta, 1.5 μm single longitudinal mode waveguide laser fabricated by femtosecond laser writing. Opt. Exp. **15**, 3190–3194 (2007)
30. W. Watanabe, K. Itoh, Fabrication of multimode interference waveguides in glass by use of a femtosecond laser. Opt. Lett. **30**, 2888–2890 (2005)
31. A. Marcinkevicius, S. Juodkazis, M. Watanabe, M. Miwa, S. Matsuo, H. Misawa, J. Nishii, Femtosecond laser-assisted three-dimensional microfabrication in silica. Opt. Lett. **26**, 277–279 (2001)
32. M. Masuda, K. Sugioka, Y. Cheng, N. Aoki, M. Kawachi, K. Shihoyama, K. Toyoda, H. Helvajian, K. Midorikawa, 3-D microstructuring inside photosensitive glass by femtosecond laser excitation. Appl. Phys. **A76**, 857–860 (2003)
33. Y. Kondo, J.R. Qiu, T. Mitsuyu, K. Hirao, T. Yoko, Three-dimensional microdrilling of glass by multiphoton process and chemical etching. Jpn. J. Appl. Phys. **38**, L1146–L1148 (1999)
34. Y. Bellouard, A. Said, M. Dugan, P. Bado, Fabrication of high-aspect ratio, micro-fluidic channels and tunnels using femtosecond laser pulses and chemical etching. Opt. Exp. **12**, 2120–2129 (2004)
35. P. Fuqua, S.W. Janson, W.W. Hansen, H. Helvajian, Fabrication of true 3D microstructures in glass/ceramic materials by pulsed UV laser volumetric exposure techniques. Proc. SPIE **3618**, 213–220 (1999)
36. T. Hongo, K. Sugioka, H. Niino, Y. Cheng, M. Masuda, I. Miyamoto, H. Takai, K. Midorikawa, Investigation of photoreaction mechanism of photosensitive glass by femtosecond laser. J. Appl. Phys. **97**, 063517 (2005)
37. Y. Cheng, K. Sugioka, M. Masuda, K. Toyoda, M. Kawachi, K. Shihoyama, K. Midorikawa, 3D microstructuring inside Foturan glass by femtosecond laser. RIKEN Rev. **50**, 101–106 (2003)
38. S. Kiyama, S. Matsuo, S. Hashimoto, Y. Morihira, Examination of etching agent and etching mechanism on femotosecond laser microfabrication of channels inside vitreous silica substrates. J. Phys. Chem. **C113**, 11560–11566 (2009)
39. Y. Li, K. Itoh, W. Watanabe, K. Yamada, D. Kuroda, J. Nishii, Y. Jiang, Three-dimensional hole drilling of silica glass from the rear surface with femtosecond laser pulses. Opt. Lett. **26** pp. 1912–1914 (2001)

40. T.N. Kim, K. Campbell, A. Groisman, D. Kleinfeld, C.B. Schaffer, Femtosecond laser-drilled capillary integrated into a microfluidic device. Appl. Phys. Lett. **86**, 201106 (2005)
41. R. An, Y. Li, Y. Dou, H. Yang, Q. Gong, Simultaneous multimicrohole drilling of soda-lime glass by water-assisted ablation with femtosecond laser pulses. Opt. Exp. **13**, 1855–1859 (2005)
42. D.J. Hwang, T.Y. Choi, C.P. Grigoropoulos, Liquid-assisted femtosecond laser drilling of straight and three-dimensional microchannels in glass. Appl. Phys. A **79**, 605–612 (2004)
43. K. Ke, E.F. Hasselbrink Jr, A.J. Hunt, Rapidly prototyped three-dimensional nanofluidic channel networks in glass substrates. Anal. Chem. **77**, 5083–5088 (2005)
44. Y. Liao, Y. Ju, L. Zhang, F. He, Q. Zhang, Y. Shen, D. Chen, Y. Cheng, Z. Xu, K. Sugioka, K. Midorikawa, Three-dimensional microfluidic channel with arbitrary length and configuration fabricated inside glass by femtosecond laser direct writing. Opt. Lett. **35**, 3225–3227 (2010)
45. S. Wiggins, J.M. Ottino, Foundations of chaotic mixing. Philos. Trans. R Soc. London, Ser. **A362**, 937–970 (2004)
46. Y. Liao, J. Song, E. Li, Y. Luo, Shen Yi, D. Chen, Y. Cheng, Z. Xu, K. Sugioka, K. Midorikawa, Rapid prototyping of three-dimensional microfluidic mixers in glass by femtosecond laser direct writing. Lab Chip **12**, 746–749 (2012)
47. M.A. Burns, B.N. Johnson, A.N. Brahmasandra, K. Handique, J.R. Webster, M. Krishnan, T.S. Sammarco, P.M. Man, D. Jones, D. Heldsinger, C.H. Mastrangelo, D.T. Burke, An integrated nanoliter DNA analysis device. Science **282**, 484–487 (1998)
48. P.S. Dittrich, K. Tachikawa, A. Manz, Micro total analysis systems. Latest advancements and trends. Anal. Chem. **78**, 3887–3907 (2006)
49. A. Schaap, Y. Bellouard, T. Rohrlack, Optofluidic lab-on-a-chip for rapid algae population screening. Biomed. Opt. Exp. **2**, 658–664 (2011)
50. A. Schaap, T. Rohrlack, Y. Bellouard, Optical classification of algae species with a glass lab-on-a-chip. Lab Chip **12**, 1527–1532 (2012)
51. Y. Hanada, K. Sugioka, H. Kawano, I.S. Ishikawa, A. Miyawaki, K. Midorikawa, Nano-aquarium for dynamic observation of living cells fabricated by femtosecond laser direct writing of photostructurable glass. Biomed. Microdevices **10**, 403–410 (2008)
52. Y. Hanada, K. Sugioka, I.S. Ishikawa, H. Kawano, A. Miyawaki, K. Midorikawa, 3D microfluidic chips with integrated functional microelements fabricated by a femtosecond laser for studying the gliding mechanism of cyanobacteria. Lab Chip **11**, 2109–2115 (2011)
53. M. Kim, D.J. Hwang, H. Jeon, K. Hiromatsu, C.P. Grigoropoulos, Single cell detection using a glass-based optofluidic device fabricated by femtosecond laser pulses. Lab Chip **9**, 311–318 (2009)
54. F. Bragheri, L. Ferrara, N. Bellini, K.C. Vishnubhatla, P. Minzioni, R. Ramponi, R. Osellame, I. Cristiani, Optofluidic chip for single cell trapping and stretching fabricated by a femtosecond laser, J. Biophotonics **3**, 234–243 (2010)
55. A. Crespi, Y. Gu, B. Ngamsom, H.J.W.M. Hoekstra, C. Dongre, M. Pollnau, R. Ramponi, H.H. van den Vlekkert, P. Watts, G. Cerullo, R. Osellame, Three-dimensional Mach-Zehnder interferometer in a microfluidic chip for spatially-resolved label-free detection. Lab Chip **10**, 1167–1173 (2010)
56. Y. Cheng, K. Sugioka, K. Midorikawa, M. Masuda, K. Toyoda, M. Kawachi, K. Shihoyama, Three-dimensional micro-optical components embedded in photosensitive glass by a femtosecond laser. Opt. Lett. **28**, 1144–1146 (2003)
57. Y. Cheng, H.L. Tsai, K. Sugioka, K. Midorikawa, Fabrication of 3D microoptical lenses in photosensitive glass using femtosecond laser micromachining. Appl. Phys. **A85**, 11–14 (2006)
58. Z. Wang, K. Sugioka, K. Midorikawa, Three-dimensional integration of microoptical components buried inside photosensitive glass by femtosecond laser direct writing. Appl. Phys. **A89**, 951–955 (2007)
59. Z. Wang, K. Sugioka, K. Midorikawa, Fabrication of integrated microchip for optical sensing by femtosecond laser direct writing of Foturan glass. Appl. Phys. **A93**, 225–229 (2008)

60. R. Stoian, M. Boyle, A. Thoss, A. Rosenfeld, G. Korn, I.V. Hertel, Advanced ultrafast laser material processing using temporal pulse shaping. Proc. SPIE **4830**, 435–441 (2003)
61. P.R. Herman, A. Oettle, K.P. Chen, R.S. Marjoribanks, Laser micromachining of 'transparent' fused silica with 1-ps pulses and pulse trains. Proc. SPIE **3616**, 148–155 (1999)
62. L. Englert, B. Rethfeld, L. Haag, M. Wollenhaupt, C. Sarpe-Tudoran, T. Baumert, Control of ionization processes in high band gap materials via tailored femtosecond pulses. Opt. Exp. **15**, 17855 (2007)
63. L. Englert, M. Wollenhaupt, L. Haag, C. Sarpe-Tudoran, B. Rethfeld, T. Baumert, Material processing of dielectrics with temporally asymmetric shaped femtosecond laser pulses on the nanometer scale. Appl. Phys. A **92**, 749 (2008)
64. T. Tamaki, W. Watanabe, J. Nishii, K. Itoh, Welding of transparent materials using femtosecond laser pulses. Jpn. J. Appl. Phys. **22**, L687–L689 (2005)
65. I. Miyamoto, A. Horn, J. Gottmann, Local melting of glass material and its application to direct fusion welding by ps-laser pulses. J. Laser Micro/Nanoengin. **2**, 7–14 (2007)
66. A. Horn, I. Mingareev, A. Werth, Investigations on melting and welding of glass by ultra-short laser radiation. J. Laser Micro/Nanoengin. **3**, 114–118 (2008)
67. K. Sugioka, M. Iida, H. Takai, K. Midorikawa, Efficient microwelding of glass substrates by ultrafast laser irradiation using a double-pulse train. Opt. Lett. **36**, 2734–2736 (2011)
68. M. Shimizu, M. Sakakura, M. Ohnishi, Y. Shimotsuma, T. Nakaya, K. Miura, K. Hirao, Mechanism of heat-modification inside a glass after irradiation with high-repetition rate femtosecond laser pulses. J. Appl. Phys. **108**, 073533 (2010)

Index

M. Castillejo et al. (eds.), *Lasers in Materials Science*,
Springer Series in Materials Science 191, DOI: 10.1007/978-3-319-02898-9,

Printed by Publishers' Graphics LLC
ESRO140123.15.16.3